Nitrogen Fixation at the Millennium

Nitrogen Fixation at the Millennium

Edited by

G. Jeffery Leigh

School of Chemistry
Physics and Environmental Science
University of Sussex
Brighton, BN1 9QJ
UK

2002

Elsevier
Amsterdam • Boston • London • New York • Oxford • Paris
San Diego • San Francisco • Singapore • Sydney • Tokyo

ELSEVIER SCIENCE B.V.
Sara Burgerhartstraat 25
P.O. Box 211, 1000 AE Amsterdam, The Netherlands

First edition 2002

Library of Congress Cataloging in Publication Data
A catalog record from the Library of Congress has been applied for.

British Library Cataloguing in Publication Data
A catalogue record from the British Library has been applied for.

ISBN: 0-444-50965-8

∞ The paper used in this publication meets the requirements of ANSI/NISO Z39.48-1992 (Permanence of Paper).
Printed in The Netherlands.

Foreword

The purpose behind the genesis of this book was two-fold. The first was to try to present a comprehensive and informed account of the whole field of nitrogen fixation at a specific time. The change from the second to the third millennium seemed to be a suitable occasion for a retrospective consideration of the subject, and so this was selected as appropriate. Accordingly, I approached a series of people generally acknowledged to be amongst the foremost experts in their fields, and asked them to present summaries of the state of the art in their fields at about the end of 2000. In the actual outcome, the literature coverage goes well into 2001, and sometimes even into 2002. This book represents the collected contributions.

The senior authors were asked to circulate the authors of all the other Chapters with details of what they intended to cover. This was to ensure that there is not much overlap. However, I decided against wielding too severe an editorial knife in order that each Chapter should be comprehensive and self-contained. Consequently, there is some repetition, for example, the structures of the nitrogenase metal-sulfur clusters are presented more than once, but this has the advantage of avoiding the necessity for much cross-referencing. In addition, I have declined to impose a uniform set of formalisms on all the Chapters because it is clear that in many cases there is no consensus within the scientific community as to what the preferred formalisms might be. My criteria have been comprehensibility and consistency within any one Chapter. Even less have I wanted to impose British or US spellings. Authors have used that which they prefer. The book as a whole presents a picture of an enormous subject that has blossomed to engulf a wide range of disciplines in the last fifty years. Anyone who wants to learn how nitrogen fixation studies have expanded and how science and agriculture have been enriched by those studies will find the answer in these pages.

The second stimulus behind this book is as follows. The study of nitrogen fixation has benefited greatly from the explosion in molecular biology. Before about 1970, the genetics of nitrogen fixation were hardly a subject at all. The isolation of cell-free extracts of nitrogenase in 1960 unleashed an extraordinary advance in the understanding of how nitrogen is fixed within living cells. In the mid-1960s, inorganic chemists discovered that dinitrogen is not the inert substance it had been believed to be for more than two centuries, and a wide range of dinitrogen reactivity has now been uncovered. Whether any of it is of direct relevance to nitrogenase is a matter of some disagreement, though the new horizons appearing in the distance (cheap fertiliser probably not amongst them) cannot be disputed. By about 1990, many people, myself included, felt that we only had to know the structure of nitrogenase in order to explain how it actually worked. Of course, when the structure appeared, it presented features rather different from those we had discovered in the study of reactive dinitrogen complexes, and this is still true today. The functioning of nitrogenase is still rather mysterious.

Funding for research in nitrogen-fixation chemistry, biochemistry and genetics has declined in recent years, perhaps because the expectations arising from the discovery of inorganic compounds that react with dinitrogen, and of nitrogen-fixing bacteria associated with cereals have not yet resulted in new industrial processes or in bacterial strains able to transfer effectively fixed nitrogen into agriculturally important cereal crops. While progress in practical applications may be disappointing, the potential remains. What is needed is an influx of new people into the field, with new ideas that will provoke fresh breakthroughs.

This book will provide a guide to those who appreciate the challenge of a difficult yet potentially rewarding research area. It will give them a complete picture of their selected field of study up to about the end of 2001 and into 2002, and will enable them easily to identify the relevant older literature. It is much to be wished that such people will be stimulated by what they find here. The field is still advancing, so that such readers must also consult the latest literature to be completely up-to-date. For example, consider the mysterious molybdenum-iron-sulfur cluster (FeMoco) that is at the heart of the usual nitrogenase enzyme, and that appears to contain the site at which dinitrogen is reduced. The structures of nitrogenases are discussed in considerable detail in Chapter 2 of this book, but the cluster itself has presented a puzzle, both to experimentalists and to theoreticians. How can it be synthesised in the laboratory and how is it synthesised *in vivo*? How does it interact with dinitrogen? An improved structure with a resolution to 1.16 Å (O. Einsle, F. A. Tezcan, S. L. A. Andrade, B. Schmid, M. Yoshida, J. B. Howard and D. C. Rees, Science 297 (2002) 1696) was published in September, 2002, after Chapter 2 had been written and edited. This new paper suggests that the apparent void at the centre of the cluster is actually filled by a single atom, possibly carbon, or maybe nitrogen, or even oxygen. The challenge posed by the new structure impinges on all areas of the study of nitrogen fixation. Is this atom an artefact in the sample studied or is it an integral part of the system? Is this atom merely part of the structure, contributing little to reactivity? How did it get there? Does this swing the balance of probability back to molybdenum being the centre of dinitrogen reduction? Clearly the conundrum of biological nitrogen fixation is far from being solved.

Finally, but not least, I wish to recall that the edifice of nitrogen fixation research, like all scientific endeavours, was built by an enormous number of people in many different disciplines. This book is a tribute to them, even if most have not been directly cited here. However, I wish to make special reference to one of the authors, Werner Klipp, who has died very recently, and was a major figure in the study of the genetics and regulation of nitrogen fixation in diazotrophs, including, in recent years, the photosynthetic nitrogen-fixing species *Rhodobacter capsulatus*. Werner was a meticulous experimental scientist with great insight and foresight, a fine teacher, a constructive collaborator, and friend to many members of the nitrogen-fixation community. His has not been the only loss. Others who have also passed on within the last year or so include Barbara Burgess, who, though not a contributor to this book, was one of the most stimulating and innovative workers on nitrogen fixation, and Dick Miller. Although he had formally retired, Dick was a true amateur of nitrogen fixation in the original sense of the word, a lover of the science, and he carried on his research after retirement, both collaboratively and in his own private lab. Those who knew these people feel their departure very deeply. To adapt a dedication from a writer of about a century ago, their names may even be forgotten in time, but the work remains.

G. Jeffery Leigh
University of Sussex, UK
17 October, 2002

Contents

Nitrogen Fixation at the Millennium.
G. Jeffery Leigh, Editor.

CHAPTER 1

Nitrogen Fixation – A General Overview

Karl Fisher and William E. Newton
Department of Biochemistry, The Virginia Polytechnic Institute and State University, Blacksburg, VA 24061, USA

1. INTRODUCTION

Although nitrogen gas (N_2) accounts for about eighty percent of the Earth's atmosphere, plants and animals do not have an easy time obtaining all the nitrogen they need for growth. This situation arises because the N_2 molecule is very stable chemically and so is unusable by most biological organisms. It must be "fixed" before it can be assimilated. The most commonly encountered "fixed" forms of nitrogen are ammonium (NH_4^+) and nitrate (NO_3^-). Nitrogen fixation is the process, which, as a component of the biogeochemical Nitrogen Cycle (see Figure 1), drives the conversion of atmospheric N_2 into ammonium ions. Other components of the Nitrogen Cycle are responsible for converting ammonium back into N_2 gas, which then returns to the atmosphere. These other components are called nitrification, which converts ammonium into nitrate, and denitrification, which converts nitrate into gaseous nitrogen oxides and N_2 gas. This constant cycling means that the pool of fixed nitrogen within the biosphere must be constantly replenished because ammonium is necessary for the formation of biologically crucial, nitrogen-containing compounds, such as amino acids, proteins and nucleic acids.

In nature, there are two main ways of "fixing" nitrogen. One way is by lightning. The enormous energy of lightning ionizes the molecules in the atmosphere and enables them to combine and form nitrogen oxides. These nitrogen oxides dissolve in rain, forming nitrite (NO_2^-) and nitrate, which are carried to earth. However, despite the large number of lightning flashes around the world, lightning is not a particularly effective means of producing nitrogenous compounds that are useful to plants and animals. Atmospheric nitrogen fixation probably contributes at most about 10% of the total annual yield of fixed nitrogen. By far the most important source of fixed nitrogen derives from the activity of certain soil bacteria that absorb atmospheric N_2 gas and convert it into ammonium. Some of these bacteria are widespread in soils, where they feed on dead organic materials. Other N_2-fixing bacteria are found growing in association with the roots of higher plants. The plant supplies the bacteria with the energy for growth, while the bacteria supply the plant with fixed nitrogen. Because both the plant and the bacteria benefit from this relationship, it is known as a symbiosis.

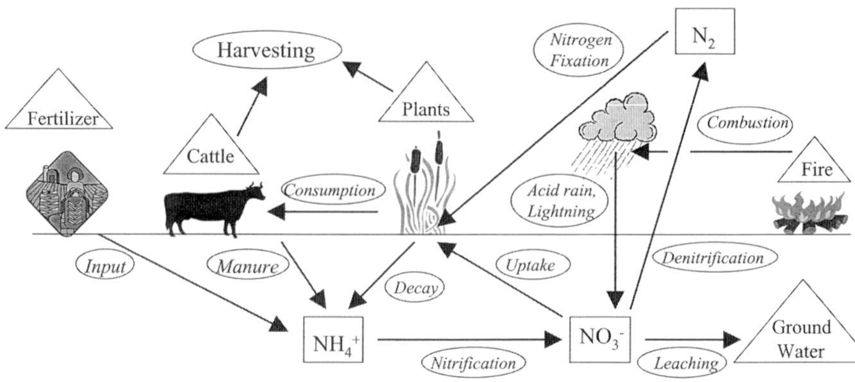

Figure 1. Biological pathways and processes involved in the Nitrogen Cycle. The rectangles (□) surrounding formulae designate chemical species; the triangles (Δ) designate sinks and sources; and the ellipses (o) define interconversion processes.

The biological process contributes about 65% of the total annual yield of fixed nitrogen. The quantification of nitrogen fixation is discussed in detail in Chapter 13.

Commercial ammonia synthesis is a rather different nitrogen-fixation process (see Chapter 11). Here, atmospheric N_2 gas (freed from contaminating O_2) and hydrogen (H_2) gas (usually derived either from natural gas or petroleum) are combined under a high pressure and high temperature to form ammonia (NH_3). The industrial process uses an inorganic catalyst and was developed originally by Fritz Haber and Carl Bosch at the time of the First World War. At about the same time, a catalyzed process for the oxidation of ammonia to nitrate was also developed. The impetus for this work was the wartime need for nitrates for explosives. The Haber-Bosch process uses N_2 and H_2 to produce an equilibrium mixture that contains *ca.* 20% ammonia. Nowadays, the Haber-Bosch process contributes about 25% of the total annual fixed-nitrogen production from all sources [1,2].

Other processes have also been exploited commercially in the past. The Birkeland-Eyde process oxidizes N_2 by passing air through an electric arc at temperatures above 3000°C. The nitric oxide (NO) generated is further oxidized on cooling to nitrogen dioxide (NO_2), which on absorption into water gives a mixture of nitric (HNO_3) and nitrous (HNO_2) acids [3].

$$N_2 + O_2 \rightarrow 2NO + O_2 \rightarrow 2NO_2 + H_2O \rightarrow HNO_3 + HNO_2$$

Its economics were only favorable in countries such as Norway, where it was developed and where substantial amounts of cheap hydroelectric power are available. However, even Norway could not sustain its operation in the face of the Haber-Bosch process. The low yield (about 2%) of NO, the tendency of NO to revert to reactants, the *ca.* 60,000 kWh required per metric ton of N_2 fixed, and the expense of sustaining the arc, all contributed to its demise.

At about the same time, a more energy-efficient process, which requires only about 20-25% of the energy used by the arc process, was commercialized [4]. The Frank-Caro cyanamide process starts with limestone ($CaCO_3$) to produce lime, which reacts with carbon to give calcium carbide (CaC_2) and then with N_2 to give calcium cyanamide. The last step is hydrolysis to produce ammonia and reform calcium carbonate.

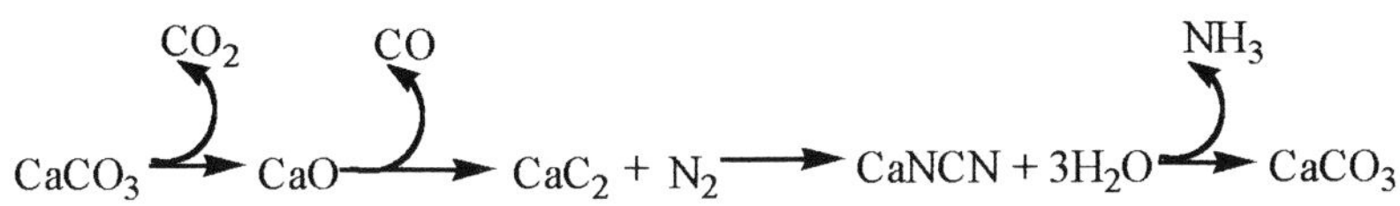

Again, the economics of the Haber-Bosch process proved to be superior to those of the cyanamide process and resulted in its downfall. Two other processes, the Wisconsin process, which was similar to the Birkeland-Eyde process, and the Serpak process for the catalytic nitriding of aluminum, were never commercially exploited to any significant degree.

As the foregoing discussion shows, biological nitrogen fixation is a central life-supporting process that provides most of the fixed nitrogen needed to sustain life. Animals, including humans, rely on plants to supply a great deal of the energy and nitrogenous compounds required for survival. Plants are likewise dependent upon the availability of nitrogenous compounds produced from atmospheric N_2 either commercially or biologically by microbes. In this way, nitrogen fixation assumes significant importance in agriculture because good crop yields depend on an adequate supply of fixed nitrogen. Useable nitrogen is likely to be in short supply for most non-legume crops and, therefore, to achieve the good yields required today, farmers must either supplement with legume crop residues or apply fertilizers, manure, or other high-nitrogen materials. Without nitrogen fertilizers, an estimated one-third of our current agricultural production would be lost (see Chapters 14 and 15).

Efforts to grow sufficient crops to feed the global population have led to increases in the application of commercial nitrogen fertilizers. Under most conditions, however, to achieve the best yield, farmers supply more than twice the fixed nitrogen required by a crop because much of it never reaches the crop. Unfortunately, soil nitrogen is in a continuous state of flux and much of the applied nitrogen is rapidly converted to nitrate, which now is mobile (the soil matrix is negatively charged) and so leaches away into groundwater, possibly contaminating drinking water supplies. There are other negative factors associated with applying nitrogen fertilizers, in particular, a concern with fossil fuel usage. For example, industrial fertilizer production: (a) consumes non-renewable fossil-fuel resources as both feedstock (as H_2) and energy source; (b) incurs significant storage costs for product manufactured year-round but applied seasonally; and (c) involves significant transportation costs due to the production site usually being remote from the site of use. Although nitrogen fertilizer is often viewed negatively as a potential pollutant, a health risk, and a significant user of fossil fuel, it is clear that there would be a major decline in crop productivity without it.

What is needed then is an alternative to the Haber-Bosch process for producing commercial quantities of nitrogen fertilizer. One longer-term possibility is through chemistry-based approaches, which might circumvent some of the problems associated with the Haber-Bosch process. To this end, chemists are developing ways to make the usually inert N_2 molecule react and these approaches are outlined in section 5 of this chapter and in Chapters 11 and 12. A complementary chemical approach is to build a molecule that approximates the structure, and therefore possibly the function, of the active site of the biological catalyst. In the shorter-term, a more likely possibility is to exploit the biological process itself and considerable effort is currently aimed at understanding the molecular mechanism of biological nitrogen fixation and developing targets for its beneficial use.

2. N_2-FIXING ORGANISMS

Only prokaryotes (Eubacteria, Archaea, and actinomycetes) can perform biological nitrogen fixation. The ability to fix N_2 is widely spread among bacterial genera [5,6] and, despite a number of claims, no eukaryote has been clearly established to fix N_2. Although crop rotations had been employed for many centuries in agricultural practice, it was not until the early 1800's that insight came into how rotations counteracted the depletion of soil nutrients resulting from continuous farming. The first studies [7,8] of rotations of clover with wheat and tuber crops showed that nitrogen was indeed made available to the following crop but these results were greeted skeptically. It took another 50 years before a convincing confirmation [9] also solved the perplexing question of the source of the fixed nitrogen that was available to the following crop. It was localized to the activity of the bacteria-filled nodules on the roots of the pea plants used [8,10].

2.1. Symbiotic associations

The most agriculturally important symbiotic relationship involves rhizobia bacteria with leguminous plants. This association maximizes the transfer of fixed nitrogen to the plant from the rhizobia residing in outgrowths, called nodules, on the plant's roots. The plant reciprocates by providing a protected environment and a supply of energy-rich compounds for the bacteria. The nodule also provides an environment with a low O_2 content, which is vital because nitrogenase, the enzyme that catalyzes biological nitrogen fixation, is denatured by O_2. Some O_2, however, must be provided so that the bacteria can respire and produce the energy required for both survival and to drive nitrogen fixation. A special O_2-transporting protein, called leghemoglobin supplies this carefully controlled amount of O_2.

A symbiotic relationship is usually a highly specific process between the plant host and the bacteria [11]. In legume symbiosis, the infection process starts when bacteria adhere to, and cause curling of, the tip of root hairs [12]. These curled structures then become an infection thread, which envelops the bacteria and penetrates the root cortex taking the invading bacteria into the cortical cells [13]. Simultaneously, cell division within the root cortex forms a nodule primordium into which the bacteria are released from the infection thread. Multiple divisions of the bacterial cells then fill the host cell with bacteria, which then differentiate into bacteroids that are specialized for fixing N_2.

Although root nodules are the usual result of N_2-fixing symbioses involving *Rhizobium*, some tropical legumes, like *Sesbania*, produce stem nodules when colonized by *Azorhizobium caulinodans* [14]. Moreover, some stem nodules, for example, on *Aeschynomene indica* [15], contain rhizobia capable of photosynthesis and this close relationship of photosynthesis to nitrogen fixation may alleviate some of the energy demand of nodules.

Other N_2-fixing, symbiotic relationships include those with lichens, bryophytes (mosses and liverworts), pteridophytes (ferns), cycads, and other non-leguminous angiosperms. Non-leguminous root nodules (called actinorhizae) are present on hundreds of species of smaller, woody, dicotyledonous trees and shrubs, like *Myrica* and alder, and also on larger trees, such as *Casuarina* [5], distributed throughout both tropical and temperate climates. The bacteria found in these nodules are members of the bacterial family, Streptomyces, and are classified under the generic name *Frankia*. These associations between non-legumes and *Frankia* are called actinorhizal associations (see Chapter 7).

2.2. Associative symbioses

Unfortunately, the nodule-based symbiotic relationship does not extend to the most important food crops, such as corn, wheat, rice, and root and tuber crops, none of which have N_2-fixing partners. For these crops to reach commercially acceptable productivity levels, extensive nitrogen fertilization is required. However, other less formal associations, called associative symbioses, are common among certain grasses and bacteria. The tropical grasses, *Paspalum* and *Digitaria*, associate with the bacteria, *Azotobacter paspali* and *Azospirillum brasilense*, respectively [16]. With *Paspalum*, the N_2-fixing bacteria live outside the plant tissue in a mucilaginous sheath around the root. The *Digitaria-Azospirillum* association involves root invasion but no nodule formation. How much the plants benefit from these associations is unclear. A temperate "cousin" of *Azospirillum brasilense*, called *A. lipoferum*, does associate with some corn and sorghum cultivars, but its effect may be minimal [17,18]. A more formal association of both *Acetobacter diazotrophicus* and *Herbaspirillum* spp. with sugar cane [19] completely satisfies the fixed-nitrogen requirement for the plant's growth. Similarly, the association of *Azoarcus* spp. with Kallar grass and possibly with rice [20] also indicates a significant agronomic and economic potential (see also Chapter 15).

2.3. Associations with animals

Animals as well as plants are colonized by N_2-fixing bacteria [21]. In many situations, like humans with *Klebsiella* and ruminant animals (cattle and sheep) with *Clostridium*, it is unlikely that the nitrogen status of the animal is much affected by the bacteria because the animal's diet likely contains sufficient fixed nitrogen to repress any N_2-fixing activity by the bacteria. However, with termites and shipworms, the associations are significant. Both *Citrobacter freundii*, which infects the intestinal tract of termites, and the cellulose-decomposing bacteria, which inhabits the Deshayes gland of wood-boring shipworms, can fix N_2 and provide benefit to these animals [22-24].

2.4. Free-Living N_2-Fixing Bacteria

Most N_2-fixing microbes, however, are not associated with either plants or animals. They live free in the soil and fix N_2 for their own benefit. Of these free-living bacteria, only the cyanobacteria are agriculturally important. The most extensively studied free-living bacterial species are *Azotobacter vinelandii* (an obligate aerobe), *Klebsiella pneumoniae* (a facultative anaerobe), *Clostridium pasteurianum* (an obligate anaerobe), *Rhodobacter capsulatus* (a photosynthetic bacterium) and various *Anabaena* and *Nostoc* species (heterocyst-forming cyanobacteria). The N_2 fixed by these organisms only becomes generally available to the surrounding ecosystem when the bacteria die.

3. NITROGENASES

A number of biochemically similar N_2-fixing enzymes are distributed among more than 100 species of N_2-fixing bacteria. The first nitrogenase extract to be prepared was from the anaerobe *Clostridium pasteurianum* in 1960 and, today, nitrogenase has been extracted from more than 30 different N_2-fixing microbes (generally called diazotrophs). With one exception [25], all known diazotrophs contain a nitrogenase (or nitrogenases) that is a complex of two distinct metalloproteins, neither active without the other [26-28].

3.1. Molybdenum nitrogenase (Mo-nitrogenase)

The individual nitrogenase component proteins of the most commonly occurring nitrogenase, usually referred to as Mo-nitrogenase, have been designated as the Fe protein (or component 2 or dinitrogenase reductase) and the MoFe protein (or component 1 or dinitrogenase). The trivial names for these proteins are derived from the composition of their associated metal centers. The dinitrogenase reductase nomenclature is often avoided because the Fe protein also has functions both in the biosynthesis of the iron-molybdenum cofactor, a prosthetic group in the MoFe protein, and in the maturation of the MoFe protein itself (see also Chapter 4).

The Fe protein is a homodimer (of about 64 kDa molecular mass), which contains two MgATP-binding sites and a single [4Fe-4S] cluster. The MoFe protein is a $\alpha_2\beta_2$ heterotetramer (of about 230 kDa molecular mass), which contains two pairs of two different metalloclusters, called P-clusters and iron-molybdenum cofactors (or FeMo-cofactors). Each individual $\alpha\beta$-dimer of the MoFe protein contains a FeMo-cofactor and P-cluster and is often considered as a functional unit. The three-dimensional structures of the Fe protein [29,30], the MoFe protein [31-35] and the Fe protein:MoFe protein complex stabilized by two different methods [36,37] have been solved and are described in detail in Chapter 2. This work has allowed reasonable structures of the biologically unique FeMo-cofactor and P-cluster to be modeled [33-35; see Figure 2].

For nitrogenase to catalyze substrate reduction (see below), it requires adenosine triphosphate (MgATP), a low-potential reductant, and an anaerobic environment (38,39). During catalysis, MgATP is hydrolyzed to give HPO_4^{2-} and MgADP, an inhibitor of nitrogenase action that must be either recycled or removed. The magnesium ion is essential for nitrogenase function. A fortunate discovery is that sodium dithionite is an effective reductant of the Fe protein and is most often used for *in vitro* enzyme studies [39]. In this case, the Fe protein is reduced by one electron and becomes a one-electron donor to the MoFe protein, which contains the site(s) of substrate reduction. Overall, the optimal stoichiometry for substrate reduction is four molecules of MgATP hydrolyzed for each pair of electrons transferred to substrate. The biological nitrogen-fixation reaction is, therefore, usually described by the following equation.

$$N_2 + 8H^+ + 8e^- + 16MgATP \rightarrow 2NH_3 + H_2 + 16MgADP + 16P_i$$

However, the Fe protein can be reduced by two electrons *in vitro* [40-42]. Moreover, this "super-reduced" Fe protein may transfer two electrons at a time to the MoFe protein with the hydrolysis of only two molecules of MgATP [43], an obvious saving in energy. Whether the Fe protein acts as either a one-electron or a two-electron donor during nitrogen fixation within the bacterial cell is not yet clear.

During reduction, the N_2 molecule is likely progressively reduced by either $1e^-/1H^+$ or $2e^-/2H^+$ steps to produce partially reduced intermediates, none of which is released from the enzyme, until finally NH_3 is formed and released. Hydrazine (N_2H_4) can be detected if, during turnover, the enzyme reaction is quenched with either acid or base [44]. H_2 evolution occurs concomitantly with N_2 reduction and may occur either by N_2 displacing H_2 from the active site [45] or as a separate event after N_2 binding [46]. In the absence of N_2, nitrogenase reduces H^+ in water to form H_2. Nitrogenase can also catalyze the reduction of many other small triply bonded molecules [47; see Table 1).

Substrate	Products	Substrate	Products
$2 H^+ + 2 e^-$	H_2	$N_2O + 2 H^+ + 2 e^-$	$N_2 + H_2O$
$N_2 + 6 H^+ + 6 e^-$	$2 NH_3$	$HCN + 6 H^+ + 6 e^-$	$CH_4 + NH_3$
$N_2H_4 + 2 H^+ + 2 e^-$	$2 NH_3$	$HCN + 4 H^+ + 4 e^-$	CH_3NH_2
$HN_3 + 6 H^+ + 6 e^-$	$NH_3 + N_2H_4$	$CH_3NC + 6 H^+ + 6 e^-$	$CH_3NH_2 + CH_4$
$N_3^- + 3 H^+ + 2 e^-$	$NH_3 + N_2$	$CH_3NC + 4 H^+ + 4 e^-$	CH_3NHCH_3
$N_3^- + 9 H^+ + 8 e^-$	$3 NH_3$	$C_2H_2 + 2 H^+ + 2 e^-$	C_2H_4

Table 1. The most commonly used substrates of Mo-nitrogenase, their electron and proton requirements, and their reduction products.

Among these so-called "alternative substrates", acetylene has become a particularly important substrate because its reduction product, ethylene, can be detected readily and quantified with great sensitivity by gas chromatography. The mechanism for acetylene reduction has been extensively studied [48-50]. This test for the presence of nitrogenase is quicker, cheaper and easier than that using $^{15}N_2$. Carbon monoxide (CO), although a small molecule with a triple bond like N_2, is not reduced. However, CO does interact with the enzyme as a potent non-competitive inhibitor of all nitrogenase-catalyzed substrate reductions with the exception of that of H^+ [51].

H_2 has a unique involvement with Mo-nitrogenase and with N_2 reduction in particular. It can play any of four well-documented roles, both *in vivo* and *in vitro*. First, H_2 is the sole product of nitrogenase turnover in the absence of other reducible substrates in a MgATP-dependent, CO-insensitive reaction [52]. Second, H_2 can be oxidized by the action of hydrogenase and so provide electrons for substrate reduction. Third, H_2 is a specific competitive inhibitor of N_2 reduction, affecting neither reduction of any other substrate nor its own evolution [53]. Fourth, HD is produced from D_2 by nitrogenase, but only in the presence of an N_2/D_2 atmosphere, in a CO-sensitive, MgATP-requiring reaction [54,55].

Mo-nitrogenases, from a variety of genera, display a high level of primary sequence identity. The sequence conservation is especially high in the region of the MgATP-binding and the metallocluster-binding sites [56]. The component proteins

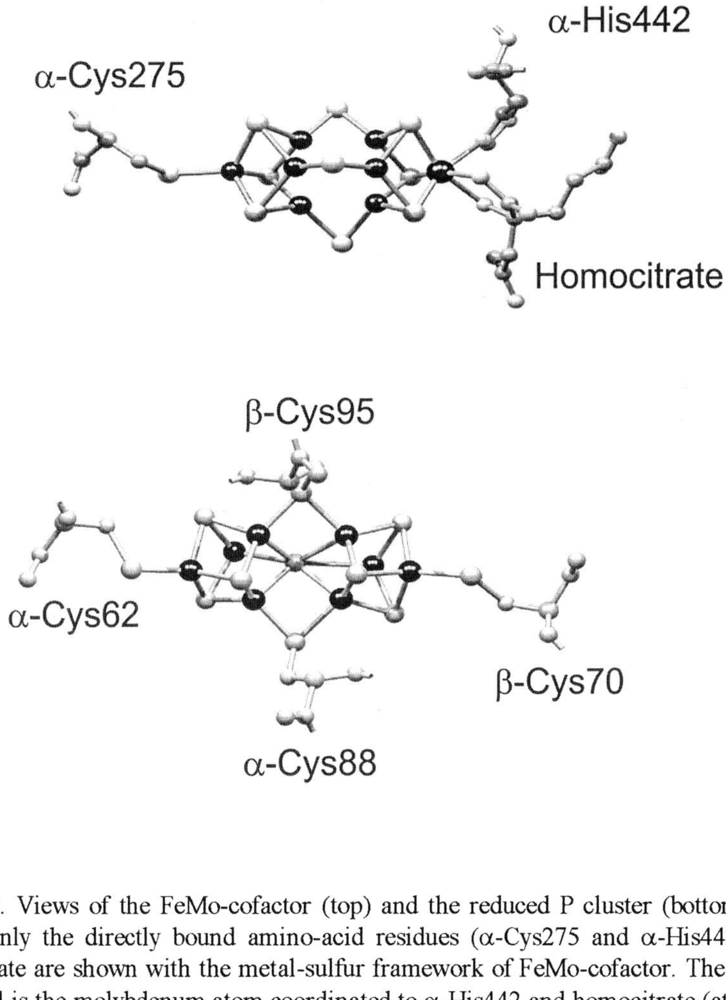

Figure 2. Views of the FeMo-cofactor (top) and the reduced P cluster (bottom). For clarity, only the directly bound amino-acid residues (α-Cys275 and α-His442) plus homocitrate are shown with the metal-sulfur framework of FeMo-cofactor. The largest black ball is the molybdenum atom coordinated to α-His442 and homocitrate (at right). The darker, intermediate-sized balls are the iron atoms, which are interspersed with the sulfur atoms (the lighter, smaller balls). For the P cluster, only two (α-Cys62 and β-Cys70) of the four terminally bound cysteine ligands, but both bridging cysteine residues (α-Cys88 and β-Cys95), plus the hexacoordinated central sulfur are shown. The larger, darker balls represent the iron atoms and the smaller, lighter balls represent the sulfur atoms.

from all Mo-based nitrogenases interact as heterologous crosses to form a fully active 'hybrid' nitrogenase enzyme [57], with the notable exception of the proteins from *C. pasteurianum*. The ability to form active heterologous complexes, although widespread, is not, therefore, universal.

3.2. Alternative nitrogenases

In the early 1980's, it became clear, if not widely accepted, that *A. vinelandii* forms a different kind of nitrogenase when starved of molybdenum [58]. The existence of alternative nitrogenase(s) was confirmed and accepted generally after it was shown that deleting the normal genes (called *nif* genes) for Mo-nitrogenase resulted in a mutant *A. vinelandii* strain that was still able to fix N_2 (see Chapter 8). Analysis of the alternative nitrogenase in *A. chroococcum* established that it contained vanadium (V) but no molybdenum [59,60]. To everyone's surprise, when the structural genes for the Mo-containing nitrogenase were similarly deleted from either *A. vinelandii* [61] or *Rhodobacter capsulatus* [62] and the strains then grown in the absence of Mo and V, yet another nitrogenase was discovered. This third nitrogenase system contains only Fe.

These alternative nitrogenases, like conventional Mo-nitrogenases, consist of two components, one large heteromultimeric protein and a smaller homodimeric Fe-protein component that is specific for its partner. Again like Mo-nitrogenase, both component proteins of both alternative nitrogenases are extremely sensitive to O_2. As expected from the metal analysis, the larger protein component of the V-containing nitrogenase contains normal P-clusters but has a V atom substituting for the usual Mo atom to produce a VFe protein that contains a FeV-cofactor. Similarly, the other alternative nitrogenase has the Mo atom substituted by Fe to produce a FeFe protein containing a FeFe-cofactor. Although these cofactors have yet to be structurally characterized, their structures are generally assumed to parallel that of the FeMo-cofactor.

The high level of primary sequence identity recognized among the Mo-nitrogenases from many microorganisms also extends to the alternative nitrogenases. This identity strongly suggests that all nitrogenases are likely to have common structural features and mechanistic similarities. One striking difference is that the VFe and FeFe proteins are hexameric rather than tetrameric as for the MoFe protein. The three types of subunit are organized as an $\alpha_2\beta_2\delta_2$ arrangement and are encoded by either *vnfDGK* [59,63] or *anfDGK* [61,62], respectively. The δ-subunit is a small ferredoxin-like peptide.

The distribution of these three nitrogenases among microorganisms defies logic [64]. For example, *K. pneumoniae* has only the Mo-nitrogenase, whereas *A. chroococcum* has both Mo- and V-nitrogenases and *R. capsulatus* has Mo- and Fe-nitrogenases. Moreover, all three nitrogenase types are found in *A. vinelandii*. The expression of each nitrogenase is under hierarchical control through the availability of either Mo or V in the growth medium [65,66]. Whenever Mo is available, expression of the Mo-dependent nitrogenase is stimulated and the expression of the others is repressed. Similarly, when V is available and Mo is absent, only expression of the V-nitrogenase occurs. If both metals are absent, then only the Fe-nitrogenase is expressed. Such control by metal availability is physiologically reasonable because Mo-nitrogenase is the most efficient catalyst for nitrogen reduction, followed by V-nitrogenase, with the Fe-nitrogenase being the least efficient and preferring to evolve H_2 rather than reduce N_2 [64,67]. Heterologous nitrogenases resulting from crosses among the component proteins from Mo-nitrogenase and V-nitrogenase are active, but crosses involving either component protein from the Fe-nitrogenase are ineffective [64].

3.3. *Streptomyces thermoautotrophicus* nitrogenase

S. thermoautotrophicus is a thermophilic autotroph that was isolated from the warm overlay of a charcoal-burning heap in the Black Forest in 1992 [25]. This bacterium contains a unique nitrogenase system. A detailed description of the structural genes, proteins and mechanism of nitrogenase action of this unusual system is given in Chapter 10. It consists of two component proteins, just like the more conventional nitrogenase, but here the larger component is an $\alpha\beta\gamma$ heterotrimer that contains Mo, Fe and sulfide. This protein has approximately half the molecular mass of the conventional Mo-containing protein and has a substantially different amino-acid sequence. The smaller component is a homodimeric manganese-containing superoxide oxidoreductase (of about 48 kDa molecular mass) rather than a Fe protein.

This unique nitrogenase oxidizes superoxide ($O_2^{\bullet-}$) to O_2, and then transfers the resulting electron to the MoFeS-containing protein at which N_2 is reduced by eight electrons to two ions of NH_4^+ and one H_2 molecule. This reaction appears to require hydrolysis of less MgATP per N_2 reduced than that required by the classical Mo-nitrogenase. Another difference is in the natural electron donor. In place of either flavodoxin or ferredoxin, which are the *in vivo* electron donors to the conventional Fe protein, a Mo-containing CO dehydrogenase operates by coupling the oxidation of carbon monoxide to carbon dioxide to the reduction of O_2 to give $O_2^{\bullet-}$ (superoxide).

$$CO + H_2O \rightarrow CO_2 + 2\,H^+ + 2\,e^-$$
$$2\,O_2 + 2\,e^- \rightarrow 2\,O_2^{\bullet-}$$

The uniqueness of this nitrogenase is also demonstrated by the observations that O_2, CO, H_2, and MgADP, all of which are potent inhibitors of N_2 reduction catalyzed by the classical nitrogenase systems, are completely without effect on its N_2-reduction activity. Finally, this nitrogenase does not catalyze the reduction of acetylene to ethylene.

3.4. Mechanism of nitrogenase action

As far as we know, the mechanism outlined below and in detail in Chapter 9, by which the Mo-, V- and Fe-nitrogenases catalyze the reduction of substrates, is the same for all organisms. During catalysis, one Fe protein with two bound molecules of MgATP combines transiently with an αβ-subunit of the MoFe protein. Each αβ-dimer is conceivably a functional unit that is serviced independently by Fe protein. While in this transient complex, the Fe protein donates an electron to the MoFe protein and the two molecules of MgATP are hydrolyzed to MgADP and HPO_4^{2-}. The complex then dissociates [45,68]. Both component proteins are required for productive MgATP hydrolysis because neither component protein alone will reduce substrate.

Based on a substantial amount of kinetic data, a numerical model has been developed to describe the process by which electrons are sequentially delivered to the MoFe protein and then to substrate [45]. This approach involves two interconnecting processes, which are called the Fe-protein cycle and the MoFe-protein cycle. The Fe-protein cycle describes how the Fe protein's [4Fe-4S] cluster cycles between its reduced 1^+ and its oxidized 2^+ redox states as it delivers an electron to the MoFe protein, coupled with MgATP hydrolysis, and is then re-reduced by another electron-transfer event. The MoFe-protein cycle is necessarily more complex because it involves the progressive reduction of the MoFe protein by up to 8 electrons for N_2 binding and reduction, which therefore requires eight turns of the Fe-protein cycle (see Figure 3).

In this model, N_2 binds to the active site only after three electrons accumulate within the MoFe protein. How and where these electrons are stored prior to the binding and reduction of substrate is unknown. However, electron transfer may occur within the MoFe-protein between the P-cluster and the FeMo-cofactor [69,70]. This concept finds support in the crystallographic model for the docked complex of two Fe proteins with one MoFe protein [35]. The model shows that docking occurs along the Fe protein's 2-fold symmetric axis, which bisects the single [4Fe-4S] cluster, and the pseudosymmetric αβ-interface of the MoFe protein, where the P-cluster is located. This arrangement places the Fe protein's [4Fe-4S] cluster in close proximity to the P-cluster of the MoFe protein and also situates the P-cluster between the Fe protein's [4Fe-4S] cluster and the FeMo-cofactor. This alignment suggests that electrons are transferred from the [4Fe-4S] cluster through the P-cluster to the FeMo-cofactor, where substrate reduction occurs.

Further support comes from structures of the MoFe protein from *A. vinelandii* nitrogenase in two different redox states [34]. The major difference between the two structures is at the P clusters. In both redox states, the P cluster is comprised of eight Fe and seven sulfur atoms. In the oxidized state, the eight Fe atoms are coordinated by six conserved cysteine residues (three from each subunit), serine-188 from the β-subunit (β-Ser188), and by the backbone amide-N of one of the coordinated cysteines (α-Cys88). In the as-isolated state, which is believed to represent its fully-reduced, all-

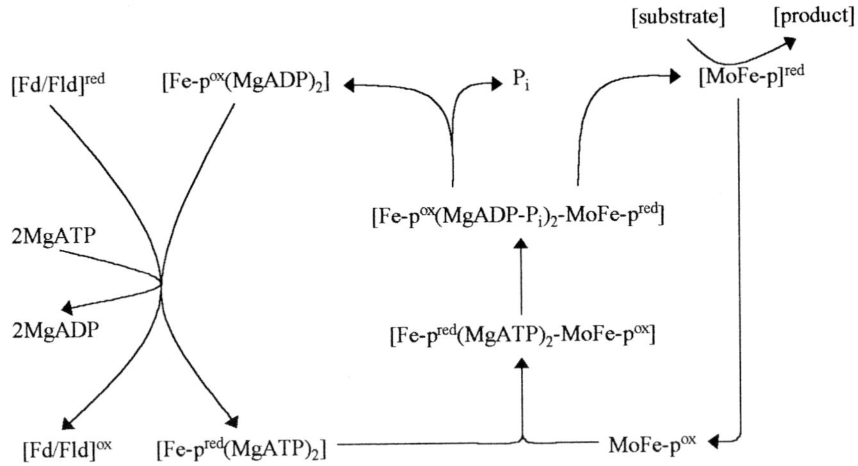

Figure 3. The interactions required for productive catalysis among the Fe protein (Fe-p) and MoFe protein (MoFe-p), including those with electron donors (either flavodoxin, Fld, or ferredoxin, Fd), with nucleotides (MgATP and MgADP), and with substrate. Because the catalyzed reduction of all known substrates requires at least two electrons, the Fe-protein cycle (the left half of the figure) must turnover at least twice for the MoFe protein to become competent for substrate reduction.

Fe^{2+} state, both the Oγ of β-Ser188 and backbone amide-N of α-Cys88 are no longer coordinated to the P-cluster Fe atoms. Instead, these ligands are exchanged for coordination of the Fe atoms to the central S atom of the P-cluster, creating a central hexacoordinated sulfur with distorted octahedral geometry. This latter coordinationgeometry was observed earlier in the MoFe-protein structures of *C. pasteurianum* [34] and *K. pneumoniae* [71]. This structural rearrangement results, upon reduction, in movement of two of the eight Fe atoms by 1.4 Å and 0.9 Å towards the central sulfur. This overall movement is consistent with X-ray absorption spectroscopy studies, which show a contraction of Fe-Fe distances in the reduced P-cluster [72]. This redox-dependent structural change of the P-cluster is likely to be mechanistically related to its role in accepting electrons from the Fe protein and coupling their delivery, along with the required protons, to the FeMo-cofactor and substrate [34].

Qualitative molecular modeling has identified three potential proton-transfer routes that run from the surface of the MoFe protein to the FeMo-cofactor, which lies about 10Å below [73]. These routes include a water-filled channel that terminates at the homocitrate ligand of the FeMo-cofactor and two hydrogen-bonded chains, which end at specific sulfides of the FeMo-cofactor. The water-filled channel may be used for multiple proton deliveries to the substrate as well as for diffusion of products into, and substrates out of, the active site. In contrast, the two hydrogen-bonded chains may each allow a single proton to be added to bound substrate, thus committing it to the reduction process.

Electron paramagnetic resonance (EPR) spectroscopy (see Chapter 3 for the use of EPR and other spectroscopic techniques in nitrogenase research) has been extensively used in the search for intermediates in the catalytic process. EPR utilizes the magnetic properties associated with unpaired electrons that are present in certain redox states of the nitrogenase component proteins, and is used to monitor electronic/structural/redox changes that occur at the active site during catalysis. The as-isolated dithionite-reduced states of both the MoFe and Fe proteins exhibit characteristic EPR spectra. When the Fe protein binds MgATP, the EPR spectrum undergoes a substantial change, which is accompanied by an increase in chemical reactivity. Effectively, it becomes a more powerful reducing agent and, as it transfers an electron to the MoFe protein and becomes oxidized, its EPR signal is completely bleached. The EPR spectrum of the MoFe protein changes too as the various reduced-intermediate forms are generated during catalysis [74]. Using specific conditions, EPR can both monitor the binding of certain substrates and inhibitors [75,76] and identify structural changes induced in the FeMo-cofactor environment by site-directed mutagenesis [77-80]. Recently, EPR spectroscopy has provided the first evidence for the binding of acetylene and cyanide to the resting state of a genetically altered MoFe protein [81; see Chapter 5].

If the MoFe protein is treated with acid and then re-buffered to neutral pH, the FeMo-cofactor can be extracted from the denatured protein by certain organic solvents [82]. This organic extract retains the characteristic EPR signature, indicating that the FeMo-cofactor remains essentially intact. Even though these extracts have been recalcitrant in attempts to demonstrate significant reactivity towards N_2 reduction, isolated FeMo-cofactor can be an effective catalyst for acetylene reduction in the presence of zinc amalgam and thiophenol as electron- and proton-donor, respectively. This reaction has many similarities to that observed with the holo-MoFe protein, including N_2 acting as a competitive reversible inhibitor of acetylene reduction [83]. Moreover, under reducing conditions, isolated FeMo-cofactor can bind CO and, when in acidified organic solvent, can catalyze the production of H_2 [84]. The exact composition and structure of extracted FeMo-cofactor remains uncertain, although it clearly retains many of the features of protein-bound FeMo-cofactor [85].

Within the protein, the FeMo-cofactor is composed of a [Mo-7Fe-9S] core plus one molecule of (*R*)-homocitrate (Figure 2). This basic framework consists of two partial clusters, [Mo-3Fe-3S] and [4Fe-3S], bridged by three sulfides. Homocitrate is coordinated to the Mo atom through its 2-hydroxyl and 2-carboxyl groups. Several lines of evidence support the hypothesis that the FeMo-cofactor is the substrate binding and reduction site. First, mutant strains that are unable to biosynthesize FeMo-cofactor are also unable to catalyze N_2 reduction [82]. However, when isolated FeMo-cofactor is added to crude extracts prepared from such mutant strains, the ability to reduce N_2 *in vitro* is restored. Second, citrate is found to replace homocitrate as the organic constituent of the FeMo-cofactor in certain mutant organisms and this substitution results in changes in catalytic properties [86-88]. Third, altered MoFe proteins, which have amino-acid substitutions within the FeMo-cofactor's polypeptide environment, also exhibit altered catalytic properties [55,77,89,90].

Although the way in which substrates interact with FeMo-cofactor during turnover is not currently known, either the six coordinately unsaturated Fe atoms [76] or the homocitrate-bound Mo atom [46,91], or both, may be involved in substrate and inhibitor binding. For example, multiple, possibly overlapping, binding sites have been proposed involving the Fe-S framework [78]; and the carboxylate group of homocitrate, which binds to the Mo atom, may serve as a leaving group to provide a substrate-binding site on Mo [46,91].

3.5. Role of MgATP in nitrogenase catalysis

Despite intensive research effort, the precise role(s) of MgATP hydrolysis are still unclear [92]. The consensus view is that MgATP is hydrolyzed to promote electron-transfer and provide the energy needed for completion of the catalytic cycle. The Fe protein is only capable of catalytic rates of nucleotide hydrolysis when complexed with the MoFe protein. At first sight, a role for MgATP as energy input appears unlikely because the overall reduction of N_2 to yield two molecules of NH_4^+ in aqueous solution is thermodynamically favorable. However, the formation of both the two-electron-reduced and the four-electron-reduced intermediates is not. Even so, because there is no direct evidence for formation (or release) of these possible intermediates, MgATP binding and hydrolysis during nitrogenase catalysis is unlikely to be simply a thermodynamic requirement.

A currently popular alternative role for MgATP (and its hydrolysis) is to ensure the unidirectionality of electron transfer, *i.e.*, preventing the electron delivered to the MoFe protein from returning to the Fe protein. This situation may be envisaged as a "gate" through which electrons flow and which is controlled by MgATP binding and hydrolysis. The "gate" is open and an electron is transferred when reduced Fe protein with bound MgATP complexes with the MoFe protein. When MgATP is hydrolyzed, the "gate" closes preventing back-flow of the electron. The binding of MgATP to the Fe protein is known to induce a change in its conformation and this reversible conformational change could be the basis for "electron gating" to ensure delivery to, and accumulation of, multiple electrons within the MoFe protein prior to their delivery to substrate. This concept is supported by primary amino-acid sequence and structural comparisons, which show that the Fe protein is a member of a large family of proteins, including, *e.g.*, H-ras p21, recA, and myosin, that have energy transduction mechanisms involving switching between conformational states upon nucleotide binding and hydrolysis.

Overall, the following sequence of events is thought to occur during each electron-transfer cycle [45]. First, the reduced Fe protein binds two molecules of MgATP, which causes the Fe protein to contract and induces a conformational change that makes it competent for interaction with the MoFe protein [30]. Second, the two proteins form a complex, which produces changes in the midpoint potentials of the metal-sulfur clusters such that movement of electrons towards the FeMo-cofactor is energetically favored. In the nitrogenase complex, the redox potentials are: -620 mV for the Fe protein's [4Fe-4S] cluster; -390 mV for the P cluster; and -40 mV for the FeMo-cofactor [93]. Third, in addition to electron-transfer, the docking of the component proteins triggers MgATP hydrolysis, although it is not clear whether hydrolysis occurs before, concomitantly with, or after electron-transfer or whether the timing varies and is dependant on other factors. Once the appropriate protein-protein complex has been formed, MgATP is hydrolyzed regardless of whether or not an electron is transferred. MgATP hydrolysis produces MgADP, H^+ and HPO_4^{2-}. Even though phosphate release is usually the work step (when energy transduction occurs), phosphate release from the nitrogenase complex does not drive the dissociation of this complex into its component proteins [94]. Fourth, somehow this hydrolytic conversion of the Fe protein from its MgATP-bound state to its MgADP-bound state causes complex dissociation. Dissociation of the complex is the rate-limiting step in nitrogenase catalysis when dithionite is used as reductant [45]. Lastly, the released Fe protein is re-reduced, exchanges its MgADP for MgATP, and is ready to re-enter the catalytic cycle.

This complex inter-communication between the two proteins, which is synchronized by sequential conformational changes induced by MgATP binding, component protein interaction, and MgATP hydrolysis, allows electrons to be accumulated within the MoFe protein for substrate reduction. However, because crystals of an altered Fe protein-MoFe protein complex are capable of exchanging MgADP for MgATP [37], dissociation of the complex is not an absolute requirement for nucleotide binding and exchange.

3.6. Electron transport to nitrogenase

A source of reducing equivalents of sufficiently low potential is required for regeneration of reduced Fe protein during the nitrogenase catalytic cycle. The ultimate donor of low-potential electrons within the cell has not been identified for most nitrogenases. However, both flavodoxins and ferredoxins (with molecular weights ranging from 6 to 24 kDa) are capable of serving this function *in vitro*. In some bacteria, *e.g.*, *A. vinelandii*, redundancy is apparent, with several flavodoxins and ferredoxins being capable of supporting nitrogenase catalysis. The ferredoxins that are involved in nitrogen fixation all contain at least one [4Fe-4S] cluster similar to that found in the nitrogenase Fe protein. In contrast, flavodoxins do not contain iron atoms but instead rely on a flavin prosthetic group for redox activity.

So, how do these ferredoxins and flavodoxins obtain their electrons? In *C. pasteurianum*, the ferredoxin that supplies the electrons to nitrogenase for the conversion of N_2 to ammonia is reduced by the pyruvic phosphoroclastic system. This last enzyme complex is responsible for pyruvate metabolism and, therefore, ties the ability to reduce N_2 directly to cell metabolism. In the heterocystous cyanobacteria, the primary electron-donor to nitrogenase is also a ferredoxin, but it receives electrons produced by the action of light on the photosynthetic apparatus. In *K. pneumoniae*, a flavodoxin is the ultimate electron-donor to the oxidized Fe protein. In this process, the fully reduced (hydroquinone) form of its flavin is oxidized by one electron to a semiquinone form (blue color). The re-reduction of flavodoxin is accomplished through the catalytic activity of a pyruvate-flavodoxin oxidoreductase, which couples the oxidation of pyruvate, yielding acetyl-CoA and CO_2, to the reduction of the bound flavin [95,96]. This last system is not found in other nitrogen-fixing genera and so these bacteria must have other means of providing the required reducing equivalents to nitrogenase.

3.7. Site-directed mutagenesis applied to nitrogenase

This technology allows a specific mutation to be introduced into a gene. The effect of the mutation is to substitute one amino-acid residue for another in the protein made from that gene. This substitution, if targeted to an important residue, will affect the structure and/or function of the protein. This method has been used to construct many mutant strains of *A. vinelandii* and *K. pneumoniae* that carry known amino-acid substitutions within the nitrogenase proteins [97,98]. The early goal of these studies, which were initiated before any X-ray-derived crystallographic structures were available, was to elucidate which amino-acid residues were located near each of the MoFe protein's prosthetic groups and to test whether or not these residues play a functional role in the catalytic process. The structural data now available have provided additional targets for substitution. These now include the residues that: (a) provide both covalent and non-covalent interactions with the FeMo-cofactor; (b) provide ligands to the P-cluster; (c) are situated between the two metalloclusters and so may provide an electron-transfer pathway; and (d) are involved in protein-protein docking. Even so, the characterization of the resulting altered MoFe proteins has not yet yielded the details of exactly how and where substrates bind and are reduced (see Chapter 5 for more discussion).

4. NITROGEN-FIXATION (*nif*) GENES

Early genetic research utilized the facultative species *K. pneumoniae* because of its similarities to *Eschericia coli*, which had proven to be a valuable tool for various molecular biological manipulations. This choice of organism was a very fortunate one because it later turned out to contain only the Mo-nitrogenase system. Because *K. pneumoniae* was the first organism to have its nitrogen-fixation genes cloned and sequenced, it is the paradigm for all sequence homology comparisons among N_2-fixing organisms. Its nitrogen-fixation genes are closely clustered on a 24 kb-long stretch of DNA. The individual genes are designated by the three-letter acronym, *nif*, followed by a single capital letter. Correspondingly, each gene product is designated by the acronym, Nif, followed by the same single capital letter.

4.1. Nitrogen-fixation gene cluster

Twenty *nif* genes exist in *K. pneumoniae* and they are organized into eight transcriptional units [56]. The specific designations for individual *K. pneumoniae nif* genes are also used to denote genes whose products have homologous functions in other N_2-fixing bacteria. For example, the *nif* structural genes from all N_2-fixing organisms are designated as *nifH, nifD,* and *nifK* and they encode the Fe protein subunit and the MoFe protein α- and β-subunits, respectively. In *Bradyrhizobium japonicum*, unlike in most other N_2 fixers, *nifHDK* are not clustered together. The *nifH* gene is separate from the *nifD* and *nifK* genes. Because the *nif* designation is reserved for only those genes that have functional counterparts in *K. pneumoniae*, nitrogen-fixation-related genes, which are present in other organisms but are without functional counterparts in *K. pneumoniae*, are given alternative designations, *e.g.*, the "*fix*" designation is used for such genes of the rhizobia.

A number of nitrogen-fixation genes associated with both of the genetically distinct, Mo-independent, alternative enzymes have also been identified. These genes include the structural genes for the V-nitrogenase (*vnfHDGK*), where *vnfH* is close to *vnfDGK* but not contiguous with them [63,99], and for the Fe-nitrogenase (*anfHDGK*) [100], where these structural genes are contiguous. The *vnf* and *anf* systems have their own regulatory genes, *vnfA* and *anfA*, respectively, which are quite similar to the *nifA* of the Mo-nitrogenase [101]. However, these genetic systems do not replicate all twenty *nif* genes. The products of at least the *nifM, B, V, U* and *S* genes are required for full activation of all three systems [64]. The *anf* and *vnf* systems also share the *vnfEN* genes [102].

4.2. Proposed functions for the *nif*-gene products

The nitrogenase protein structural genes produce immature inactive apo-forms of both the Fe protein (from *nifH*) and MoFe protein (from *nifDK*), which are acted upon by the products of other *nif*-specific maturation genes (see Figure 4). In fact, a majority of the genes in the nif-specific gene cluster have a role in metal-cluster biosynthesis, insertion, or maturation. For the Fe protein, only the *nifM* gene product is specifically required and it is involved in inserting the [4Fe-4S] center into the apo-Fe protein [103]. Although the NifM protein has not been isolated in an active form, sequence comparisons show it to be highly homologous to a family of peptidyl-prolyl isomerases, which are enzymes that assist protein folding by catalyzing the cis/trans isomerization of certain peptidyl-prolyl bonds.

Maturation of the MoFe protein is a much more complicated process and involves, at the least, the products of the *nifH*, *nifE*, *nifN*, *nifB*, *nifQ* and *nifV* genes. Biochemical complementation indicates that the FeMo-cofactor is synthesized elsewhere in the cell and is then inserted into an apo-MoFe protein that contains intact P clusters but no FeMo-cofactors [104]. The *nifH* product, the Fe protein, is required both for this independent synthesis of the FeMo-cofactor and for its insertion into the MoFe protein [105,106]. Exactly how the Fe protein functions in this procedure is unclear but neither its MgATP-binding/hydrolysis properties nor its electron-transfer capacity are required

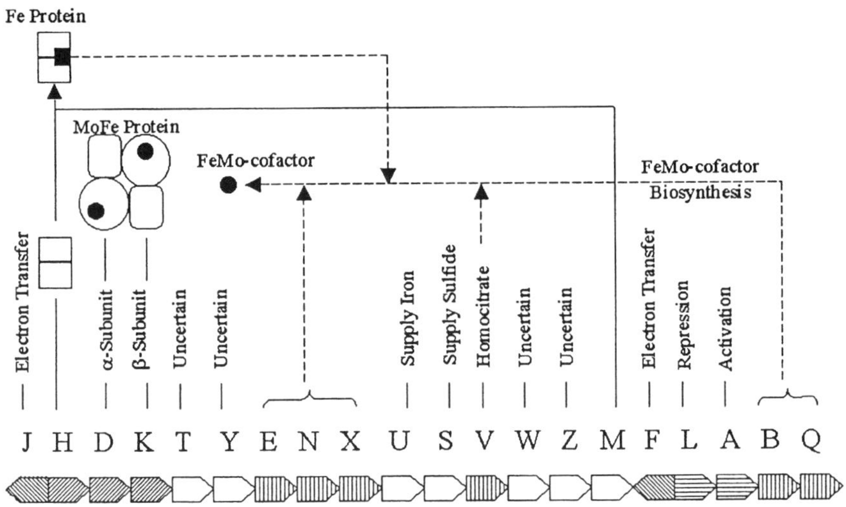

Figure 4. Organization of the *nif* genes in *Klebsiella pneumoniae* with the corresponding known or putative functions of the gene products. Individual genes with the same shading have related functions.

[107,108]. The *nifE* and *nifN* gene products express the macromolecular scaffold on which FeMo-cofactor is biosynthesized [109,110] and the *nifB* and *nifQ* gene products help in the early stages of building the FeMo-cofactor within that scaffold. NifB catalyzes the formation of a FeMo-cofactor precursor, called NifB-cofactor, which provides the basic Fe-S framework necessary for FeMo-cofactor construction within the NifEN complex [111,112]. The precise role of NifQ is unknown but it may be responsible for molybdenum assimilation into FeMo-cofactor especially under Mo-

deficient conditions [113]. The *nifV* gene product is homocitrate synthase, which catalyzes the condensation of acetyl-CoA and α-ketoglutarate to form homocitrate, a key organic constituent of the FeMo-cofactor whose function is still controversial [114,115].

The products of some of the other *nif* genes also have a role in either FeMo-cofactor biosynthesis or its insertion into the immature MoFe protein. The *nifX* gene product appears essential for FeMo-cofactor biosynthesis [116], but its role is unknown, The products of the *nifW* and *nifZ* genes might also be involved in a minor way. The products of both the latter genes affect MoFe protein activity but neither is required for the organism to grow on N_2. It is still unclear as to whether the *nifZ* product affects FeMo-cofactor synthesis or its insertion into the apo-MoFe protein [117], whereas the *nifW* product is somehow involved with the homocitrate moiety of the FeMo-cofactor [118,119]. Although not absolutely required for FeMo-cofactor synthesis, the gene products of *nifS* and *nifU* provide sulfide and iron, respectively, for nitrogenase iron-sulfur cluster synthesis. The *nifS* gene codes for a pyridoxal phosphate-containing cysteine desulfurase [120], whereas the mechanism of function for the *nifU* product is less apparent [121]. Because either the *nifU* or *nifS* gene can be deleted individually without complete loss of component-protein activity, homologues of *nifU* and *nifS* must exist that perform similar functions but whose expression is not under *nif*-specific control.

Other gene products are more intimately involved with the apo-MoFe protein during maturation. A low-molecular-weight protein, encoded by *nifY* remains associated with the *K. pneumoniae* apo-MoFe protein that is produced in strains lacking either *nifB* or *nifEN* activity [122]. NifY may stabilize the apo-MoFe protein in a conformation that is amenable to FeMo-cofactor insertion. Another small molecular-weight protein called gamma, which is not encoded by *nifY*, appears to serve the same function in *A. vinelandii* [123].

Quite remarkably, the products of the *nifMBVUS* genes are also essential for functional activity of the V-nitrogenase and Fe-nitrogenase as well as for the Mo-nitrogenase [64]. Thus, all three Fe proteins must be very similar because a single NifM protein can process them all. In addition, the common requirement of the *nifB* and *nifV* gene products suggests similar cofactors in all three systems. Further, the common *nifB* requirement suggests that the function of NifB cannot be Mo specific. The need for *nifB*, but not *nifEN*, by all three nitrogenase systems is surprising because *nifB* and *nifN* are fused in *C. pasteurianum* [124], which can express a Mo-independent nitrogenase [125]. Moreover, the *nifV* and *nifUS* requirements show that homocitrate, S^{2-} and Fe are constituents of all three cofactors.

The *nifF* and *nifJ* genes encode the ultimate electron-transfer components, which couple reduction of Fe protein to intermediary metabolism in *K. pneumoniae* [95,96]. The *nifF* gene encodes the flavodoxin and the *nifJ* gene encodes the pyruvate:flavodoxin oxidoreductase. The *nifJ* gene has not been found in other nitrogen-fixing genera, which must have other means of providing reducing equivalents to nitrogenase. Two other genes, *nifA* and *nifL*, are involved in the regulation of expression of all other *nif* operons by transcriptional control and are discussed separately below.

4.3. NifA-NifL regulatory system

Nitrogenase expression is under stringent transcriptional control because of the high ATP (energy) demand by nitrogen fixation and the extreme sensitivity to O_2 of the nitrogenase enzyme (see Chapter 6 for a more detailed description). NifA, a transcriptional activator protein, and NifL, a negative control element that operates by inactivating NifA, are responsible for regulating nitrogenase expression. The *nifLA* operon is controlled by the *ntr* (*ni*trogen *r*egulation) system.

The *ntrA* gene product is an alternative sigma factor (σ^{54}) that controls the expression of the *nif* and *ntr* genes [126]. The presence of NtrA allows RNA polymerase to recognize a consensus promoter sequence that is located -24 to -12 base pairs before the transcription initiation site. The normal RNA polymerase, which contains the abundant sigma factor called σ^{70}, recognizes a prokaryotic promoter sequence, at -35 to -10 base pairs before the transcription initiation site. NtrB and NtrC control regulation of *nifLA* expression. NtrB is a phosphatase/kinase sensor protein that acts on NtrC in response to changes in the α-ketoglutarate:ammonia ratio in the cell [127]. Under conditions of limiting fixed-nitrogen, NtrC is phosphorylated, then recognizes and binds to a consensus DNA sequence that is located upstream from the σ^{54}-RNA polymerase-binding site of *nifLA*. Bound, phosphorylated NtrC catalyzes an ATP-dependent conformational isomerization at the promoter site, which results in the initiation of transcription of the *nifLA* transcriptional unit [128]. Accumulation of the NifL and NifA proteins specifically controls the other *nif*-promoters.

NifA is structurally and functionally similar to NtrC [129,130]. It also binds to an upstream activator sequence, but one with a consensus sequence motif that is different to the NtrC-binding site. This sequence is located approximately 100 base pairs before each *nif* promoter except that for *nifLA* [131]. NifA activity is controlled by NifL, which acts as an anti-activator. Just how NifL interacts with NifA when either the O_2 level or the fixed-nitrogen level is high is not known. However, it is known that NifL is a flavoprotein that senses the redox status of the cell through conformational changes that are driven by the redox state of its FAD moiety [132,133].

5. CHEMICAL SYSTEMS

Chemists have approached the problem of how to reduce N_2 from two major directions. The first is to study the N_2 molecule and to devise chemical systems that might activate it. These systems are broadly classed as "reactivity models". The second approach is to take lessons from the enzyme and build active-site models that might simulate its activity. These systems are broadly classed as "structural models". No matter which direction one takes, the inertness of the N_2 molecule is the major concern (see Chapters 11 and 12). To reduce N_2 in aqueous solution adds an even greater hurdle at first glance because very potent reductants, such as lithium metal, appear to be required. However, as we know, lithium reacts very rapidly and vigorously with water to produce H_2 gas. Thus, discovering interactions energetic enough, especially ones likely to occur in the aqueous environment of the natural enzyme system, is very difficult. Even so, certain reductants can reduce N_2, even in aqueous solvents.

Binding to a metal center is the usual first step in activating N_2 toward reduction. Many metal-N_2 complexes have been synthesized [134,135] since the first compound of this type, $[Ru(NH_3)_5(N_2)]^{2+}$, was discovered [136]. Unfortunately, most metal-N_2 compounds are very stable and do not react further to produce nitrogen hydrides, like N_2H_4 and NH_3. Some, however, do contain a reactive N_2 molecule that can be reduced or cleaved.

5.1. Reactivity models - in non-aqueous systems

The first chemical system found to reduce N_2 was a non-aqueous mixture of titanium tetrachloride with a powerful reductant, such as ethylmagnesium bromide, in diethyl ether [137]. Although the reduction mechanism is unknown, this system likely cleaves N_2 to produce Ti-nitrides (Ti≡N), which decompose to produce NH_3 when a source of protons is added. Unfortunately, the protons also destroy the N_2-activating species.

Some insight into the possible intermediates formed came when organometallic derivatives of zirconium (Zr) were used in place of $TiCl_4$. Two major reaction products were isolated. One product had N_2 bound as a linear -N=N- bridge between two Zr^{II} ions, whereas the other product had a similar linear bridge plus one more N_2 ligand on each zirconium atom to give the moiety $-(N_2)Zr-N{=}N-Zr(N_2)-$. The complex with only the bound bridging N_2 produces neither hydrazine nor ammonia directly when H^+ is added but, in the presence of an excess of reductant, it does give NH_3. In contrast, the complex containing three N_2 molecules reacts directly with HCl to liberate two N_2 molecules and produce hydrazine from the third [138]. However, as a further complication, a related compound, with only one N_2 molecule bound between two Zr atoms, can incorporate H_2 to produce a bound nitrogen-hydride species [139].

Titanium can be truly catalytic as found when $TiCl_4$ is mixed with Al metal and $AlBr_3$ and heated to 50°C. A yield of 200 mol NH_3/g.atom Ti was obtained *via* the catalytic nitriding of aluminum followed by protolysis [140].

In a complementary approach to activating N_2, a number of metal-N_2 compounds have been isolated and well-characterized. Many of these have also been degraded to ammonia and/or hydrazine. In the early stages of these endeavors, there was no obvious connection between the nitriding systems described above, which used powerful reductants to produce either ammonia or hydrazine from N_2, and these well-defined, relatively stable metal-N_2 complexes. However, just as the highly reducing systems were found to yield metal-N_2 compounds, so a number of metal-N_2 compounds showed reactivity and produced ammonia and/or hydrazine from their coordinated N_2.

The mononuclear, tertiary phosphine complexes of Mo and W are the best studied examples of metal-N_2 complexes with activated N_2. Specifically, when $[M(N_2)_2(PR_3)_4]$ (M = Mo or W) is treated with acid, up to 0.7 mole per mole of complex is produced [141,142]. The process involves a stepwise protonation/reduction sequence to produce NH_3 from only one of the two N_2 molecules bound to the same metal atom. The other is lost as N_2 gas. An unfortunate consequence of acidification is the complete destruction of the complex. However, when a stabilizing tridentate phosphine is used together with a labile simple phosphine, the resulting system still produces NH_3 but now the metal-containing product of acidification can be recycled through the system [143]. This basic system has been developed into a cyclic process for the electrochemical synthesis of NH_3 [144].

These Ti/Zr and Mo/W systems share common features in the production of NH_3 from bound N_2. In both cases, although only one bound N_2 is reduced on acidification, both systems start with at least two N_2 molecules bound to the metal atom. Further, as protonation of N_2 proceeds, the ligands on the metal atom that stabilize these complexes are successively replaced by acid counter-ions that do not do so. As this ligand-exchange process proceeds, it encourages more and more electron-transfer from the metal atom onto N_2 and, thus, it promotes additional protonation. In the binuclear Zr^{II} system, only four electrons are available (two from each Zr) and so only hydrazine is formed. In the Mo/W systems, the ease with which the stabilizing ligands are replaced correlates with the extent to which the N_2 molecule is reduced. With chelating phosphines, which are not easily replaced, on the metal atom, ammonia is not readily produced, whereas with the more easily displaced, simple phosphines, all six electrons residing on the metal ion (in its zero redox state) are used to produce ammonia. In the electrochemical cycling system with the mixture of simple and chelating phosphines, ammonia is formed but only three electrons per atom of molybdenum are used.

Although these ammonia- and hydrazine-forming reactions appear to be conducted under mild conditions, the reducing power of, *e.g.*, magnesium metal, is already built into them during the preparation of the metal-N_2 complexes. The simple phosphine complexes of Mo and W that are completely degraded during ammonia formation are very unlikely candidates as catalysts. In contrast, the electrochemical system, the mixed-phosphine system, and the Zr^{II} system, where the product of acid degradation is the starting material for preparation of the metal-N_2 complex, all have possibilities for catalysis. To be effective, these systems will have to use significantly milder reductants than heretofore and to be stabilized to ensure prolonged cyclic or catalytic operation.

5.2. Reactivity models - in aqueous solution

Especially when metal atoms are present, many strong reducing agents may appear to produce minute amounts of ammonia from N_2 in aqueous solution. However, the results may be spurious for many reasons. For example, the color test used to detect ammonia is not specific; the system may scavenge traces of ammonia (or other compounds that which are subsequently reduced to NH_3) from the N_2 gas; and any nitrogen-containing substance present in the reaction mixture may be degraded to ammonia. Even so, there are aqueous systems that do reduce N_2 either to hydrazine or to ammonia and which have been substantiated by using $^{15}N_2$.

Many examples of this type of system also use the early transition metals, such as Ti, V, or Mo, either as catalyst or as reductant or as both. An early Mo example used a very basic (pH 10-14) aqueous-alcoholic solution of sodium molybdate, with $TiCl_3$ as reductant in the presence of Mg^{2+}, but it produced only a small amount of N_2H_4 at 25°C and 101 kPa (1 atm) N_2. However, when the temperature and pressure were increased, as much as 100 mole/mole Mo and some NH_3 were produced. Either V^{II} or Cr^{II} are equally effective as reductant. This system's efficiency is about 1% of the natural system. V^{II} can replace both Mo and Ti in this system and, when it is used alone, quantitative reduction to hydrazine, based on each of four V^{II} ions providing one electron each, occurs within minutes. By varying the conditions, NH_3 and H_2 can be produced through further reduction of the hydrazine by other V^{II} ions [145].

$$(V^{2+})_4 + N_2 + 4H_2O \rightarrow 4V^{3+} + N_2H_4 + 4OH^-$$

A related homogenous, aqueous-alcoholic system composed of V^{II}-catechol complexes reduces N_2 directly to ammonia and H_2. Only catecholates are active and the system is very sensitive to pH. This system may use a sequence of two four-electron reductions that results in the evolution of one H_2 for every N_2 reduced [146]. This H_2/N_2 ratio is similar to that found when Mo-nitrogenase is reducing N_2.

A third type of aqueous N_2-reducing system was developed using the knowledge that nitrogenase contains iron and molybdenum ions, sulfide, and thiols. A mixture of this kind produced only a trace of NH_3 under a high N_2 pressure [147]. These "molybdothiol" systems are suggested to produce diazene (N_2H_2), which disproportionates into N_2H_4 and N_2, and the N_2H_4 is then further reduced to NH_3. Later versions using either $[MoO(CN)_4(H_2O)]^{2-}$ and $NaBH_4$, which gives up to 0.3 mol NH_3 per mol complex [148], or a 6:1 MoO_4^{2-}:insulin mixture with $NaBH_4$, which produces up to 65 mol ammonia/g atom Mo in 30 min, were rather more successful [149].

The N_2-reducing systems discussed in this section require further considerable development if they are to become important N_2-reduction methods. They are successful to the extent that they provide important insight into both the binding and the activation of N_2, and the induction of internal redox reactions. The first step in N_2 reduction on nitrogenase undoubtedly involves binding N_2 to a metal ion and might involve either a change in the metal coordination sphere or protonation of bound N_2 (or both) as initiators of the redox process. It is clear, however, that none of these chemical systems for N_2 reduction in their present form will provide the basis for a commercially successful process for ammonia production in the near future.

5.3. Structural models

The second approach that was mentioned at the beginning of this section is to use the structure of the enzyme's active site as the target and to construct chemical compounds that duplicate as many of its features as possible. In this way, by carefully duplicating structure, function might also be duplicated! Inferential and circumstantial evidence has always pointed toward the FeMo-cofactor as the nitrogenase substrate-binding site.

The initial impetus for this work was the Mo X-ray absorption spectroscopic (XAS) studies of the FeMo-cofactor both within, and after extraction [150,151] from, the MoFe protein. These studies gave the first clues to the structural arrangement of its composite atoms. Analysis of the Mo-based spectrum indicated that the Mo atom was surrounded by 2-3 bound O or N atoms, 3-4 bound S atoms, and about 3 Fe atoms. These data were interpreted as indicating a hexacoordinated Mo-based cuboidal arrangement of $(O/N)_3$-[-Mo-S_3-Fe_3-]- [151] with the XAS-undetectable remainder of the molecule joined to the three Fe atoms in some unknown manner. This interpretation was substantiated by the X-ray structure of the MoFe protein and provides a set of structural features with which any realistic structural model must be compatible.

The chemical synthesis of mixed molybdenum-iron-sulfur cluster compounds is the goal of these efforts. The early prototypes consisted of two $[MoS_4Fe_3]$ cuboidal structures. Unfortunately, they were far from complete FeMo-cofactor models. They had two Mo atoms instead of one and both Mo atoms were located centrally rather than terminally. These complexes reproduced the Mo-Fe and Mo-S distances found in the MoFe protein, but they could not duplicate either the atomic ratio of molybdenum-to-iron (3:1 rather than 7:1) or the spectroscopy of the FeMo-cofactor. Since then, many other variations on the Mo/Fe/S-cluster theme have been synthesized (see, for example, [152,153]). One worthy variation is the single MoS_4Fe_3 cuboidal cluster, which is prevented from forming a Mo-(S_3)-Mo bridge by having a bulky ligand bound to the Mo atom. Compounds of this type bind CN^-, a Mo-nitrogenase inhibitor, among others. Other compounds with two of these ($MoFe_3S_4$) cores bridged to one another can bind N_2H_4 as a second bridge between the Mo atoms. None of these Mo/Fe/S cluster compounds binds N_2, although some of them act as catalyst precursors and produce small amounts of NH_3 from N_2 [154].

This lack of success in synthesizing a close structural model of FeMo-cofactor is because many of its structural features are unique in chemistry as well as in biology. Even so, attempts at synthesis continue and new creative techniques may well produce a synthetic FeMo-cofactor in the near future. In fact, such a synthesis might well be the only way to produce a Mo/Fe/S cluster capable of binding and reducing N_2. As fossil fuel supplies are again being recognized as a valuable, but scarce, commodity, the significance of these and related ventures increases.

6. SUMMARY AND OUTLOOK

Although the nitrogen-fixation phenomenon is becoming better understood, there is still no detailed mechanism that describes biological nitrogen reduction. Even so, synthetic reactivity models are available that bind N_2 and, in some cases, activate it toward reduction sufficiently so that ammonia and/or hydrazine, are produced on protonation. Although they are unlikely to either compete directly with or even impact the technology of the Haber-Bosch process, they could be developed as alternative low-temperature and low-pressure technologies. These alternative systems could be relatively small, might service only a limited locale, and be driven by electricity derived from renewable resources, such as solar, wind or water power. In fact, N_2 oxidation in an electric arc has been revisited under exactly these circumstances [155]. Such simple, inexpensive, small-scale systems may have a future in areas where fertilizer production capacity of a few tons per year can make a significant impact on agricultural production.

An obvious way to exploit agriculturally the benefits of biological N_2 fixation in the near future is to use more crop plants that have symbiotic relationships with N_2-fixing microbes. Moreover, a significant benefit would accrue if the leguminous symbiotic associations could be induced to start fixation earlier, continue to do so for a longer period, and be made insensitive to the presence of fertilizer (or other sources of) fixed nitrogen that repress microbial N_2 fixation. In addition, cyanobacteria are remarkably self-sufficient and their contribution to the productivity of rice culture should be exploited. An increasing number of new associative symbioses are also being evaluated as potential conduits for delivery of fixed nitrogen to crops. These less formalized associations may well hold the clues to how new associations might be engineered with the principal food crops that presently do have a N_2-fixing partner.

In the longer term, engineering either new or enhanced symbiotic associations may be the ultimate solution for increasing the fixed-nitrogen production from biological N_2 fixation [156]. Many plants have most, if not all, of the genes necessary to produce a nodule-like structure and the success in transferring the *nif* genes from one bacterial genus to another has opened up the possibility of their transfer to a crop plant [157]. However, transferring the genes is not sufficient. They must also be expressed in an appropriate environment within the plant. One possible location is within the chloroplasts of leaves where nitrogenase could take advantage of directly available reducing equivalents produced from sunlight. If the classical nitrogenase were used, it would have to be protected from O_2 evolved by photosynthesis. However, if the *S. thermoautotrophicus* nitrogenase (or an engineered derivative of this unusual nitrogenase) were relocated to chloroplasts, O_2 would cease to be a problem.

Finally, although many known associations involving animals exist, and more are sure to be discovered, it appears very unlikely that humans will develop their own symbiotic system for fixing N_2 and so escape dependence on other species for the vital supply of fixed nitrogen. Therefore, just as plants will continue to depend on microbes for a significant portion of their fixed-nitrogen intake, so too will animals, including mankind, continue to obtain fixed nitrogen by eating plants (and other animals). It is another example of the ironic situation in which we live, when survival of the most advanced organism on Earth, *i.e.*, mankind, depends ultimately on the simplest of organisms, in this case, the N_2-fixing microbes.

7. REFERENCES

1. W. E. Newton, in Kirk-Othmer Encyclopedia of Chemical Technology, 4th edition, vol. 17, John Wiley & Sons, New York, 1996, p. 172.
2. O. C. Bøckman, Plant and Soil 194 (1997) 11.
3. F. A. Ernst, Fixation of Atmospheric Nitrogen, van Nostrand Co., New York, 1928.
4. C. A. Vancini, Synthesis of Ammonia, CRC Press, Cleveland, Ohio, 1971.
5. J. I. Sprent and P. Sprent, Nitrogen Fixing Organisms: Pure and Applied Aspects, Chapman & Hall, New York, 1990.
6. J. P. W. Young in G. Stacey, R. H. Burris and H. J. Evans (eds.), Biological Nitrogen Fixation. Chapman and Hall, New York, 1992, p. 43.
7. J. B. Boussingault, Ann. Chim. et Phys., 2nd ser. 67 (1838) 5.
8. P. W. Wilson, The Biochemistry of Nitrogen Fixation, University of Wisconsin Press, Madison, 1940.

9. H. Hellriegel and H. Wilfarth, Beilageheft zu der Zeitschrift des Vereins f. d. Rübensucker-industrie d. Deutschen Reiches, 1888.
10. A Quispel, in H. Bothe, F. J. de Bruijn and W. E. Newton (eds), Nitrogen Fixation: Hundred Years After, Gustav Fischer, New York, 1988, p. 3.
11. J. Dénarié, F. Debellé and C. Rosenberg, Ann. Rev. Microbiol. 46 (1992) 497.
12. F. B. Dazzo and A. Gardiol, in D. P. S. Verma and Th. Hohn (eds.), Genes Involved in Microbe Plant Interactions, Springer, New York, 1984, 3.
13. D. Verma and S. Long, in K. W. Jeon (ed.), Intracellular Symbiosis, Academic Press, New York, 1983, p. 211.
14. B. Dreyfus and Y. R. Dommergues, Appl. Env. Microbiol. 41 (1981) 97.
15. A. R. J. Eaglesham, J. M. Ellis, W. R. Evans, D. E. Fleischman, M. Hungria and R. W. F. Hardy in P. M. Gresshoff, L. E. Roth, G. Stacey and W. E. Newton (eds.), Nitrogen Fixation: Achievements and Objectives, Chapman & Hall, New York, 1990, p. 805.
16. J. Döbereiner, J. M. Day and P. J. Dart, J. Gen. Microbiol. 71 (1972) 103.
17. J. F. W. von Bülow and J. Döbereiner, Proc. Natl. Acad. Sci. USA 72 (1974) 2389.
18. J. Döbereiner, V. M. Reis and A. C. Lazarini, in H. Bothe, F. J. de Bruijn and W. E. Newton (eds.), Nitrogen Fixation: Hundred Years After, Gustav Fischer, New York, 1988, p. 717.
19. J. Döbereiner, V. M. Reis, M. A. Paula and F. Olivares, in R. Palacios, J. Mora and W. E. Newton (eds.), New Horizons in Nitrogen Fixation, Kluwer Academic, Dordrecht, 1993, p. 671.
20. B. Reinhold-Hurek and T. Hurek, in R. Palacios, J. Mora and W. E. Newton (eds.), New Horizons in Nitrogen Fixation, Kluwer Academic, Dordrecht, 1993, p. 691.
21. G. J. Waughman, J. R. J. French and K. Jones in W. J. Broughton (ed.), Nitrogen Fixation, Vol. 1, Clarendon Press, Oxford, 1981, p. 135.
22. J. R. Benemann, Science 181 (1973) 164.
23. G. D. Prestwich and B. L. Bentley, Oecologia 49 (1981) 249.
24. J. B. Waterbury, C. B. Calloway and R. D. Turner, Science 221 (1983) 1401.
25. M. Ribbe, D. Gadkari and O. Meyer, J. Biol. Chem. 272 (1997) 26627.
26. J. B. Howard and D. C. Rees, Chem. Rev. 96 (1996) 2965.
27. B. K. Burgess and D. J. Lowe, Chem. Rev. 96 (1996) 2983.
28. J. W. Peters, K. Fisher and D. R. Dean, Annu. Rev. Microbiol. 49 (1995) 335.
29. M. M. Georgiadis, H. Komiya, P. Chakrabarti, D. Woo, J. J. Kornuc and D. C. Rees, Science 257 (1992) 1653.
30. S. B. Jang, L. C. Seefeldt and J. W. Peters, Biochemistry 39 (2000) 14745.
31. J. Kim and D. C. Rees, Science 257 (1992) 1677.

32. J. T. Bolin, N. Campobasso, S. W. Muchmore, T. V. Morgan and L. E. Mortenson, in E. I. Stiefel, D Coucouvanis and W. E. Newton (eds.), Molybdenum Enzymes, Cofactors and Model Systems, Symposium Series 535, American Chemical Society, Washington, D.C., 1993, p. 186.
33. M. K. Chan, J. Kim and D. C. Rees, Science 260 (1993) 792.
34. J. W. Peters, M. H. B. Stowell, S. M. Soltis, M. G. Finnegan, M. K. Johnson and D. C. Rees, Biochemistry 36 (1997) 1181.
35. S. M. Mayer, D. M. Lawson, C. A. Gormal, S. M. Roe and B. E. Smith, J. Mol. Biol. 292 (1999) 871.
36. H. Schindelin, C. Kisker, J. L. Schlessman, J. B. Howard and D. C. Rees, Nature 387 (1997) 370.
37. H.-J. Chui, J. W. Peters, W. N. Lanzilotta, M. J. Ryle, L. C. Seefeldt, J. B. Howard and D. C. Rees, Biochemistry 40 (2001) 641.
38. J. E. Carnahan and J. E. Castle, Ann. Rev. Plant. Physiol. 14 (1963) 125.
39. W. A. Bulen, R. C. Burns and J. R. LeComte, Proc. Natl. Acad. Sci. USA 53 (1965) 532.
40. G. D. Watt and K. R. N. Reddy, J. Inorg. Biochem. 53 (1994) 281.
41. H. C. Angove, S. J. Yoo, B. K. Burgess and E. Münck, J. Am. Chem. Soc. 119 (1997) 8730.
42. H. C. Angove, S. J. Yoo, E. Münck and B. K. Burgess, J. Biol. Chem. 272 (1998) 26330.
43 J. A. Erickson, A. C. Nyburg, J. L. Johnson, S. M. Truscott, A. Gunn, F. R. Nordmeyer and G. D. Watt, Biochemistry 38 (1999) 14279.
44. R. N. F. Thorneley, R. R. Eady and D. J. Lowe, Nature 272 (1978) 557.
45. D. J. Lowe and R. N. F. Thorneley, Biochem. J. 224 (1984) 877.
46. D. L. Hughes, S. K. Ibrahim, C. J. Pickett, G. Querne, A. Laouenen, J. Talarmin, A. Queiros and A. Fonseca, Polyhedron 13 (1994) 3341.
47. B. E. Smith, Adv. Inorg. Chem. 47 (1999) 159.
48. D. J. Lowe, K. Fisher and R. N. F. Thorneley, Biochem. J. 272 (1990) 621.
49. K. Fisher, M. J. Dilworth, C.-H. Kim and W. E. Newton, Biochemistry 39 (2000) 2970.
50. P. M. C. Benton, J. Christiansen, D. R. Dean and L. C. Seefeldt, J. Am. Chem. Soc. 123 (2001) 1822.
51. R. W. F. Hardy, E. Knight, Jr. and A. J. D'Eustachio, Biochem. Biophys. Res. Commun. 20 (1965) 539.
52. A. Bulen, J. R. LeComte, R. C. Burns and J. Hinkson, in A. San Pietro (ed.), Non-Heme Iron Proteins: Role in Energy Conversion, Antioch Press, Yellow Springs, OH, 1965, p. 261.
53. J. L. Hwang, C. H. Chen and R. H. Burris, Biochim. Biophys. Acta 292 (1973) 256.
54. G. E. Hoch, K. C. Schneider and R. H. Burris, Biochim. Biophys. Acta 37 (1960) 273.

55. K. Fisher, M. J. Dilworth and W. E. Newton, Biochemistry 39 (2000) 15570.
56. D. R. Dean and M. R. Jacobson, in G. Stacey, R. H. Burris and H. J. Evans (eds.), Biological Nitrogen Fixation, Chapman & Hall, 1992, p. 863.
57. D. W. Emerich and R. H. Burris, J. Bacteriol. 134 (1978) 936.
58. P. E. Bishop, D. M. L. Jarlenski and D. R. Hetherington, Proc. Natl. Acad.
59. R. L. Robson, R. R. Eady, T. H. Richardson, R. W. Miller, M. Hawkins and J. R. Postgate, Nature 322 (1986) 388.
60. B. J. Hales, E. E. Case, J. E. Morningstar, M. F. Dzeda and L. A. Mauterer, Biochemistry 25 (1986) 7251.
61. J. R. Chisnell, R. Premakumar and P. E. Bishop, J. Bacteriol. 170 (1988) 27.
62. K. Schneider, A. Müller, U. Schramm and W. Klipp, Eur. J. Biochem. 195 (1991) 653.
63. R. L. Robson, P. R. Woodley, R. N. Pau and R. R. Eady, EMBO J 8 (1989) 1217.
64. W. E. Newton, in R. Palacios, J. Mora and W. E. Newton (eds.), New Horizons in Nitrogen Fixation, Kluwer Academic, Dordrecht, 1993, p. 5.
65. M. R. Jacobson, R. Premakumar and P. E. Bishop, J. Bacteriol. 167 (1986) 480.
66. R. N. Pau, in G. L. Eichhorn and L. G. Marzilli (eds.), Advances in Inorganic Biochemistry, Vol. 10, PTR Prentice Hall, New Jersey, 1994, p. 49.
67. R. R. Eady, Chem. Rev. 96 (1996) 3013.
68. R. V. Hageman and R. H. Burris, Proc. Natl. Acad. Sci. USA 75 (1978) 2699.
69. H. D. May, D. R. Dean and W. E. Newton, Biochem. J. 277 (1991) 457.
70. J. W. Peters, K. Fisher, W. E. Newton and D. R. Dean, J. Biol. Chem. 270 (1995) 27007.
71. B. E. Smith, F. Yousafzai, R. R. Eady, C. A. Gormal, S. Mayer, S. M. Roe, D. M. Lawson, S. S. Hasnain and J. G. Grossmann, in C. Elmerich, A. Kondorosi and W. E. Newton (eds.), Biological Nitrogen Fixation for the 21st Century, Kluwer Academic, Dordrecht, 1998, p. 43.
72. J. Christiansen, R. C. Tittsworth, B. J. Hales and S. P. Cramer, J. Am. Chem. Soc. 117 (1995) 10017.
73. M. C. Durrant, Biochem. J. 355 (2001) 569.
74. K. Fisher, W. E. Newton and D. J. Lowe, Biochemistry 40 (2001) 3333.
75. D. J. Lowe, R. R. Eady and R. N. F. Thorneley, Biochem. J. 173 (1978) 277.
76. H.-I. Lee, L. M. Cameron, B. J. Hales and B. M. Hoffman, J. Am. Chem. Soc. 119 (1997) 10121.
77. D. J. Scott, H. D. May, W. E. Newton, K. E. Brigle and D. R. Dean, Nature 343 (1990) 188.
78. J. Shen, D. R. Dean and W. E. Newton, Biochemistry 36 (1997) 4884.
79. H.-I. Lee, K. S. Thrasher, D. R. Dean, W. E. Newton and B. M. Hoffman, Biochemistry 37 (1998) 13370.
80. M. Sørlie, J. Christiansen, B. J. Lemon, J. W. Peters, D. R. Dean and B. J. Hales, Biochemistry 40 (2001) 1540.

81. P. M. C. Benton, S. M. Mayer, J. Shao, B. M. Hoffman, D. R. Dean and L. C. Seefeldt, Biochemistry 40 (2001) 13816.
82. V. K. Shah and W. J. Brill, Proc. Natl. Acad. Sci. USA 74 (1977) 3249.
83. T. A. Bazhenova, M. A. Bazhenova, S. A. Mironova, G. N. Petrova, A. K. Shilova, N. I. Shuvalova and A. E. Shilov, Inorg. Chim. Acta 270 (1998) 221; and personal communication.
84. T. Le Gall, S. K. Ibrahim, C. A. Gormal, B. E. Smith and C. J. Pickett, Chem. Commun. (1999) 773; S. K. Ibrahim, K. Vincent, C. A. Gormal, B. E. Smith, S. P. Best and C. J. Pickett, Chem. Commun. (1999) 1019.
85. B. E. Smith, M. C. Durrant, S. A. Fairhurst, C. A. Gormal, K. L. C. Grönberg, R. A. Henderson, S. K. Ibrahim, T. E. Le Gall and C. J. Pickett, Coord. Chem. Rev. 185-186 (1999) 669.
86. T. R. Hawkes, P. A. McLean and B. E. Smith, Biochem. J. 217 (1984) 317.
87. J. Liang, M. Madden, V. K. Shah and R. H. Burris, Biochemistry 29 (1990) 8577.
88. W. E. Newton, K. Vichitphan and K. Fisher, in T. Finan, M. R. O'Brian, D. B. Layzell, J. K. Vessey and W. E. Newton (eds.), Nitrogen Fixation: Global Perspectives, CABI Publishing, Wallingford, 2002, in press.
89. M. J. Dilworth, K. Fisher, C.-H. Kim and W. E. Newton, Biochemistry 37 (1998) 17495.
90. J. Christiansen, V. L. Cash, L. C. Seefeldt and D. R. Dean, J. Biol. Chem. 275 (2000) 11459.
91. K. L. C. Grönberg, C. A. Gormal, M. C. Durrant, B. E. Smith and R. A. Henderson, J. Am. Chem. Soc. 120 (1998) 10613.
92. L. C. Seefeldt and D. R. Dean, Acc. Chem. Res. 30 (1997) 260.
93. W. N. Lanzilotta and L. C. Seefeldt, Biochemistry 36 (1997) 12976.
94. D. J. Lowe, G. A. Ashby, M. Brune, H. Knights, M. R. Webb and R. N. F. Thorneley, in I. A. Tikhonovich, N. A. Provorov, V. I. Romanov and W. E. Newton (eds.), Nitrogen Fixation: Fundamentals and Applications, Kluwer Academic, Dordrecht, 1995, p. 103.
95. D. Nieva-Gomez, G. P. Roberts, S. Klevickis and W. J. Brill, Proc. Natl. Acad. Sci. USA 77 (1980) 2555.
96. S. Hill and E. P. Kavanagh, J. Bacteriol. 141 (1980) 470.
97. K. E. Brigle, R. A. Setterquist, D. R. Dean, J. S. Cantwell, M. C. Weiss and W. E. Newton, Proc. Natl. Acad. Sci. USA 84 (1987) 7066.
98. H. M. Kent, I. Ioannidis, C. Gormal, B. E. Smith and M. Buck, Biochem. J. 264 (1989) 257.
99. R. D. Joerger, T. M. Loveless, R. N. Pau, L. A. Mitchenall, B. H. Simon and P. E. Bishop, J. Bacteriol. 172 (1990) 3400.
100. R. D. Joerger, M. R. Jacobson, R. Premakumar, E. D. Wolfinger and P. E. Bishop, J. Bacteriol. 171 (1989) 1075.
101. R. D. Joerger, M. R. Jacobson and P. E. Bishop, J. Bacteriol. 171 (1989) 3258.

102. E. D. Wolfinger and P. E. Bishop, J. Bacteriol. 173 (1991) 7565.
103. K. S. Howard, P. A. McLean, F. B. Hansen, P. V. Lemley, K. S. Koblan and W. H. Orme-Johnson, J. Biol. Chem. 261 (1986) 772.
104. R. A. Ugalde, J. Imperial, V. K. Shah and W. J. Brill, J. Bacteriol. 159 (1984) 888.
105. W. A. Filler, R. M. Kemp, J. C. Ng, T. R. Hawkes, R. A. Dixon and B. E. Smith, Eur. J. Biochem. 160 (1986) 371.
106. A. C. Robinson, D. R. Dean and B. K. Burgess, J. Biol. Chem. 262 (1987) 14327.
107. N. Gavini and B. K. Burgess, J. Biol. Chem. 267 (1992) 21179.
108. D. Wolle, D. R. Dean and J. B. Howard, Science 258 (1992) 992.
109. K. E. Brigle, M. C. Weiss, W. E. Newton and D. R. Dean, J. Bacteriol. 169 (1987) 1547.
110. T. D. Paustian, V. K. Shah and G. P. Roberts, Proc. Natl. Acad. Sci. USA 86 (1989) 6082.
111. V. K. Shah, R. M. Allen, N. J. Spangler and P. W. Ludden, J. Biol. Chem. 269 (1994) 1154.
112. J. T. Roll, V. K. Shah, D. R. Dean and G. P. Roberts, J. Biol. Chem. 270 (1995) 4432.
113. J. Imperial, R. A. Ugalde, V. K. Shah and W. J. Brill, J. Bacteriol. 158 (1984) 187.
114. T. R. Hoover, A. D. Robertson, R. L. Cerny, R. N. Hayes, J. Imperial, V. K. Shah and P. W. Ludden, Nature 329 (1987) 855.
115. L. Zheng, R. H. White and D. R. Dean, J. Bacteriol. 179 (1997) 5963.
116. V. K. Shah, P. Rangaraj, R. Chatterjee, R. M. Allen, J. T. Roll, G. P. Roberts and P. W. Ludden, in C. Elmerich, A. Kondorosi and W. E. Newton (eds.), Biological Nitrogen Fixation for the 21st Century, Kluwer Academic, Dordrecht, 1998, p. 51.
117. M. R. Jacobson, V. L. Cash, M. C. Weiss, N. F. Laird, W. E. Newton and D. R. Dean, Mol. Gen. Genet. 219 (1989) 49.
118. B. Masepohl, S. Angermüller, S. Hennecke, P. Hübner, C. Moreno-Vivian and W. Klipp, Mol. Gen. Genet. 238 (1993) 369.
119. S. Kim and B. K. Burgess, J. Biol. Chem. 269 (1994) 4215.
120. L. Zheng and D. R. Dean, J. Biol. Chem. 269 (1994) 18723.
121. W. Fu, R. F. Jack, T. V. Morgan, D. R. Dean and M. K. Johnson, Biochemistry 33 (1994) 13455.
122. M. J. Homer, T. D. Paustian, V. K. Shah and G. P. Roberts, J. Bacteriol. 175 (1993) 4907.
123. M. J. Homer, D. R. Dean and G. P. Roberts, J. Biol. Chem. 270 (1995) 24745.
124. J.-S. Chen, S.-Z. Wang and J. L. Johnson, in P. M. Gresshoff, L. E. Roth, G. Stacey and W. E. Newton (eds.), Nitrogen Fixation: Achievements and Objectives, Chapman & Hall, New York, 1990, p. 483.

125. M. J. Dilworth, R. R. Eady, R. L. Robson and R. W. Miller, Nature 327 (1987) 167.
126. S. Kustu, E. Santero, D. Popham and J. Keener, Microbiol. Rev. 53 (1989) 367.
127. B. T. Nixon, C. W. Ronson and F. M. Ausubel, Proc. Natl. Acad. Sci. USA 83 (1986) 7850.
128. D. L. Popham, D. Szeto, J. Keener and S. Kustu, Science 243 (1989) 629.
129. M. Drummond, P. Whitty and J Wooton, EMBO J. 5 :441-447, 1986.
130. W. J. Buikema, W. W. Szeto, P. V. Lemley, W. H. Orme-Johnson and F. M. Ausubel, Nucleic Acids Res. 13 (1985) 4539.
131. J. Beynon, M. Cannon, V. Buchanan-Wollaston and F. Cannon, Cell 34 (1983) 665.
132. S. Hill, S. Austin, T. Eydmann, T. Jones and R. Dixon, Proc. Natl. Acad. Sci. USA 93 (1996) 2143.
133. R. Dixon, Arch. Microbiol. 169 (1998) 371.
134. R. L. Richards, Coord. Chem. Rev. 154 (1996) 83.
135. D. J. Evans, R. A. Henderson and B. E. Smith, in J. Reedijk and E. Bouwman (eds.), Bioinorganic Catalysis, 2nd edit., Marcel Dekker, New York, 1999, p. 153.
136. A. D. Allen and C. V. Senoff, Chem. Commun. (1965) 621.
137. M. E. Vol'pin and V. B. Shur, Nature 209 (1966) 1236.
138. R. D. Sanner, D. M. Duggan, T. C. McKenzie, R. E. Marsh and J. E. Bercaw, J. Am. Chem. Soc. 98 (1976) 8358.
139. M. D. Fryzuk, J. B. Love, S. J. Rettig and V. G. Young, Science 275 (1997) 1445.
140. M. E. Vol'pin, M. A. Ilatovskaya, L. V. Kosyakova and V. B. Shur, J. Chem. Soc., Chem. Comm. (1968) 1074.
141. J. Chatt, G. A. Heath and R. L. Richards, J. Chem. Soc., Dalton Trans. (1974) 2074.
142. J. Chatt, A. J. Pearman and R. L. Richards, J. Chem. Soc., Dalton Trans. (1977) 1852.
143. J. A. Baumann and T. A. George, J. Am. Chem. Soc. 102 (1980) 6153.
144. C. J. Pickett and J. Talarmin, Nature 317 (1985) 652.
145. A. Shilov, N. Denisov, O. Efimov, N. Shuvalov, N. Shuvalova and A. Shilova, Nature 231 (1971) 460.
146. L. A. Nikonova, S. A. Isaeva, N. I. Pershikova and A. E. Shilov, J. Mol. Catal. 1 (1975-1976) 367.
147. G. N. Schrauzer, Angew. Chem., Int. Ed.Engl. 14 (1975) 514.
148. E. L. Moorehead, P. R. Robinson, T. M. Vickrey and G. N. Schrauzer, J. Am. Chem. Soc. 98 (1976) 6555.
149. B. J. Weathers, J. H. Grate, N. A. Strampach and G. N. Schrauzer, J. Am. Chem. Soc. 101 (1979) 925.

150. S. P. Cramer, W. O. Gillum, K. O. Hodgson, L. E. Mortenson, E. I. Stiefel, J. R. Chisnell, W. J. Brill and V. K. Shah, J. Am. Chem. Soc. 100 (1978) 3814.
151. S. D. Conradson, B. K. Burgess, W. E. Newton, L. E. Mortenson and K. O. Hodgson, J. Am. Chem. Soc. 109 (1987) 7507.
152. S. M. Malinak and D. Coucouvanis, Prog. Inorg. Chem. 49 (2001) 599.
153. F. Osterloh, C. Achim and R. H. Holm, Inorg. Chem. 40 (2001) 224.
154. K. Tanaka, Y. Hozumi and T. Tanaka, Chemistry Letts. (Japan) (1982) 1203.
155. R. W. Treharne, D. R. Moles, M. R. Bruce and C. K. McKibben, in Proc. Second Review Meeting of I.N.P.U.T.S. Project, East-West Center, Honolulu, Hawaii, 1978, p. 279.
156. W. E. Newton, in F. O. Pedrosa, M. Hungria, M. G. Yates and W. E. Newton (eds.), Nitrogen Fixation: From Molecules to Crop Productivity, Kluwer Academic, Dordrecht, 2000, p. 3.
157. M. Merrick and R. A. Dixon, Trends in Biotechnology 2 (1984) 162.

Nitrogen Fixation at the Millennium.
G. Jeffery Leigh, Editor.

CHAPTER 2

Nitrogenase Structure

Paul M. C. Benton, Sanchayita Sen, and John W. Peters
Department of Chemistry and Biochemistry, Utah State University, Logan
UT 84322, USA

1. INTRODUCTION

The structural characterization of the nitrogenase components provides a very appropriate introduction for the description of the biochemical and biophysical characterization of nitrogenase. The structural characterization of the individual components of the Mo-nitrogenase system are a significant contribution to understanding the catalytic process, but they do not provide all the answers to some of the key questions. However, the structural glimpses offered in these studies do provide a basis to refine our proposed hypotheses for different aspects of nitrogenase catalysts and allow us to design better experiments to test these hypotheses. These experiments can involve a variety of biochemical and biophysical methods, including the structural characterization of defined captured states of the enzyme that are relevant to catalysis. Recently, several structures of nitrogenase captured during nucleotide binding and complex formation offer significant advances toward our understanding of the role of nucleotide binding and hydrolysis in nitrogenase catalysis. The focus of this chapter will be to provide a description of the culmination of results of the structural studies (Table I) on nitrogenase in the context of the biochemical and biophysical properties of the components described in subsequent Chapters.

Nitrogenase exists as two separable components in which an exclusive electron-transfer component transfers electrons to the substrate-reduction component in a macromolecular complex. The iron (Fe) protein serves as the exclusive electron donor to the substrate-reduction component, which exists in three varieties containing molybdenum and iron, vanadium and iron, or iron alone. The majority of biochemical and biophysical studies on nitrogenase, including structural studies, have focused on the characterization of the components of the molybdenum-containing nitrogenase system. Mo-nitrogenase is comprised of two discrete metalloprotein components, the iron (Fe) protein and the molybdenum iron (MoFe) protein. The Fe protein is a homodimer with a molecular mass of ~ 60 kDa that is composed of two chemically identical subunits that are bridged by a [4Fe-4S] cluster. The MoFe protein is a more complex $\alpha_2\beta_2$ heterodimer of molecular mass ~ 250 kDa that contains two distinct metalloclusters, a [8Fe-7S] cluster (P cluster) and a [7Fe-9S-Mo] cluster coordinated by homocitrate (FeMo-cofactor). During the catalytic process the two component proteins of nitrogenase combine and electron-transfer occurs from the Fe protein to the MoFe protein where substrates are then bound and reduced. Integral to this electron-

Structure	PDB#	Resolution Å	Ref
Azotobacter vinelandii Fe protein	1NIP	2.90	1
A. vinelandii MoFe protein	NA	2.70	2,3
Clostridium pasteurianum MoFe protein	1MIO	3.00	4-6
A. vinelandii MoFe protein	1MIN	2.20	2,3,7
A. vinelandii MoFe protein (oxidized)	2MIN	2.03	5
A. vinelandii MoFe protein (reduced)	3MIN	2.03	5
A. vinelandii Fe protein - MoFe protein complex stabilized with MgADP-AlF_4	1N2C	3.00	8
A. vinelandii Fe protein	2NIP	2.20	9
C. pasteurianum Fe protein	1CP2	1.93	9
Klebsiella pneumonia MoFe protein (oxidized)	1QH1	1.60	6
K. pneumonia MoFe protein (reduced)	1QGU	1.60	6
K. pneumonia MoFe protein (mixed oxidation state)	1QH8	1.60	6
A. vinelandii Fe protein ($Phe^{135}\rightarrow Trp^{135}$)	1DEO	2.40	10
A. vinelandii Fe protein - MgADP	1FP6	2.15	11
A. vinelandii Fe protein $[4Fe\text{-}4S]^0$	1G1M	2.25	12
A. vinelandii Fe protein $\Delta^{Leu}127$-MoFe protein Complex	1G20	2.20	13
A. vinelandii Fe protein $\Delta^{Leu}127$-MoFe protein Complex with MgATP	1G21	3.00	13
A. vinelandii MoFe protein ($\alpha\text{-}His^{195}\rightarrow\alpha\text{-}Gln^{195}$)	1FP4	2.50	14

Table 1. Chronological listing of nitrogenase structures determined to date (end 2001).

transfer event is the concomitant hydrolysis of two molecules of MgATP. Once the one electron transfer process and MgATP hydrolysis are complete the two component proteins dissociate. This process is repeated until the MoFe protein has accumulated sufficient electrons to reduce the substrate present.

2. THE MoFe PROTEIN

The structures of the nitrogenase MoFe protein from three different organisms (*A. vinelandii, C. pasteurianum,* and *K. pneumoniae*) have been crystallographically determined (Table 1). The first MoFe protein to be structurally characterized was that isolated from *A. vinelandii.* The highest resolution structure currently available is the structure of the MoFe protein from *K. pneumoniae*. The salient structural features of the MoFe proteins are the same (Figure 1).

2.1. Overall fold

The MoFe proteins from the microorganisms *A. vinelandii*, *C. pasteurianum*, and *K. pneumoniae* exist as $\alpha_2\beta_2$ tetramers. The subunits share amino acid sequence homology, are of comparable size (with 491 and 522 amino acids, respectively), and exhibit

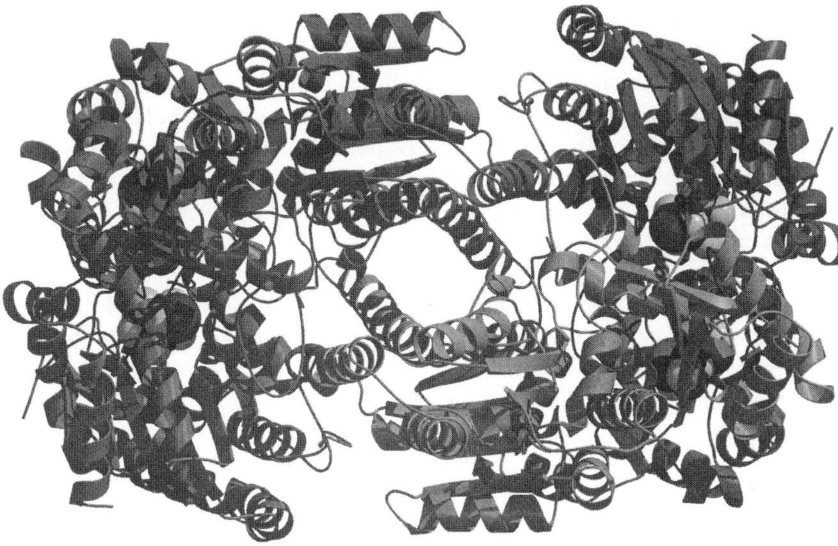

Figure 1. The *A. vinelandii* nitrogenase MoFe protein.

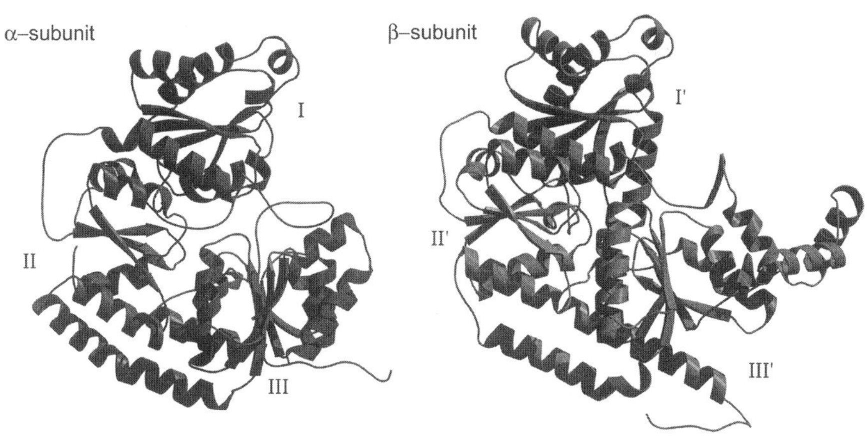

Figure 2. Comparison of the topology of the MoFe protein α and β subunits.

similar polypeptide folds comprised of three domains of the parallel β-sheet/α-helical type (Figure 2). These domains are designated I, II, III, and I′, II′, III′ in the α and β subunits, respectively [3]. The MoFe protein contains two types of metallocluster, the

FeMo-cofactor [Mo-7Fe-9S-homocitrate] and a set of two P-clusters [8Fe-7S]. FeMo-cofactor is located within the α-subunit, at the bottom of a wide, shallow cleft that interfaces the three domains. The corresponding site of the β-subunit is occupied by the side-chains of four amino acid residues, β-His^{193}, β-Gln^{294}, β-His^{297}, and β-Asp^{372} *(A. vinelandii)*. The P-cluster resides at the interface between domain I of the α subunit and domain I′ of the β subunit. The α- and β-subunits are related by a pseudo-2-fold rotation axis that passes between the two halves of the P-cluster. Shallow clefts around the P-cluster pair provide sites for the interaction of the dimeric Fe protein during nitrogenase complex formation.

The $\alpha_2\beta_2$ tetramer is formed essentially by the association of a pair of αβ dimers that are related by the molecular 2-fold rotation axis. At their positions of closest approach, the αβ 2-fold axis and the tetramer 2-fold axis are separated by ~ 12 Å. The two axes are not arranged exactly perpendicular to one another as their orientations differ by ~97°. The tetramer interface is dominated by extensive interactions between helices from domains II' and III' of the two β-subunits, along with some interface contributions from helices in domain III of each α-subunit. The stabilization rendered by these helical interactions appears to provide the predominant incentive for the formation of the tetramer. Additional stabilization of the tetramer interface is probably contributed by a cation-binding site that is created by residues from both β subunits (Figure 3). The coordination of the cation, modeled as calcium in the *Azotobacter vinelandii* MoFe protein structure, consists of carboxyl of Asp^{353} and Asp^{357} of one β-subunit and the

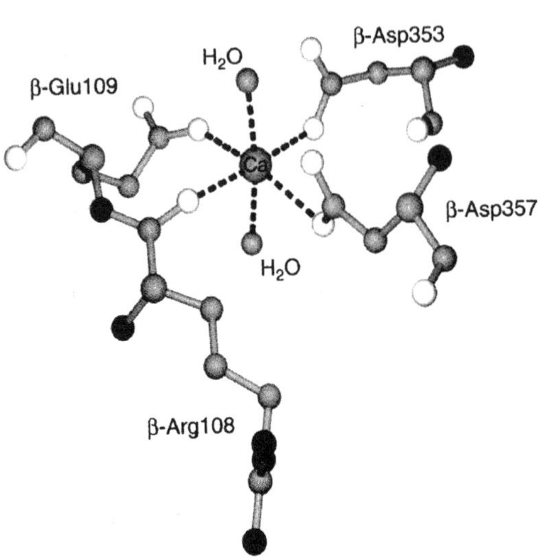

Figure 3. One of two equivalent divalent metal binding sites in the *A. vinelandii* MoFe protein. The metal is provisionally presumed to be a Ca^{2+} ion.

carboxyl oxygen of Glu[109] and the carbonyl oxygen of Arg[108] of the other β subunit. Additional ligands are likely provided by two water molecules such that each cation exists in an octahedral coordination environment. The cation-binding site is greater than 20 Å from either the P-clusters or FeMo cofactor. The functional significance of this site is not yet known. However, its location at the β-subunit interface may suggest a role in structural stabilization of the tetramer. Interestingly, instead of being occupied by residue side chains, an open channel ~8 Å in diameter exists between the two pairs of αβ dimers with the tetramer 2-fold axis extending through the center.

2.2. FeMo-cofactor

The FeMo-cofactor is comprised of two partial cubanes of composition [4Fe-3S] and [Mo-3Fe-3S] that are bridged by three non-protein S atoms (Figure 4). The FeMo-cofactor is located ~10 Å beneath the protein surface in a polypeptide environment predominantly consisting of amino acids of the α-subunit. The FeMo-cofactor of the *A. vinelandii* MoFe protein is coordinated by side chains from only two residues, α-Cys[275] (to Fe1) and α-His[442] (to Mo). The Mo atom of the FeMo-cofactor is also bound by hydroxyl and carboxyl oxygens from homocitrate which completes the octahedral coordination sphere of the Mo atom. With the lack of any additional protein or ligand interactions with the FeMo-cofactor, the six Fe atoms (Fe2-Fe7) bridged by the three non-protein S atoms remain three coordinate and possess approximate trigonal planar geometry. While three coordinate Fe atoms bearing S ligands is not unprecedented, it remains a possibility that their trigonal coordination geometry may somehow be related to the catalytic function of the FeMo-cofactor. In fact, a recent study demonstrated that three coordinate Fe centers are capable of reducing both the bond strength and bond order of the N_2 triple bond of bound dinitrogen [15].

One of the first models proposed for the binding of dinitrogen to the FeMo-cofactor was based on the trigonal coordination geometry of the six Fe atoms contained within the midsection of the FeMo-cofactor (Fe2-Fe7) [7]. It was suggested that if the substrate N_2 molecule bound within the internal cavity of the FeMo-cofactor, then it would be possible for dinitrogen to coordinate simultaneously to all six of the coordinately unsaturated trigonal Fe atoms. The formation of six *sp*3 hybridized Fe-N bonds would therefore afford all six of the Fe atoms approximately ideal tetrahedral geometry and would also weaken the *sp* hybridized N-N triple bond. Therefore, the overall result of this binding would be the increased stabilization of the six trigonal Fe atoms and the lowering of the activation energy for the nitrogenase-catalyzed reduction of dinitrogen. It was noted that the N_2 molecule was too large to fit inside the cavity of the FeMo-cofactor in such a manner, although it was speculated that, upon reduction, the bridging Fe-Fe distances of the FeMo-cofactor would increase, thereby allowing the accommodation of dinitrogen. However, the solution of the crystal structure of the MoFe protein from *K. pnemoniae* to 1.6 Å resolution indicates that there is no expansion of the FeMo-cofactor dimensions upon reduction, suggesting that even a reduced FeMo-cofactor would be incapable of accommodating the N_2 molecule internally. Furthermore, other alternative nitrogenase substrates, such as acetylene, are

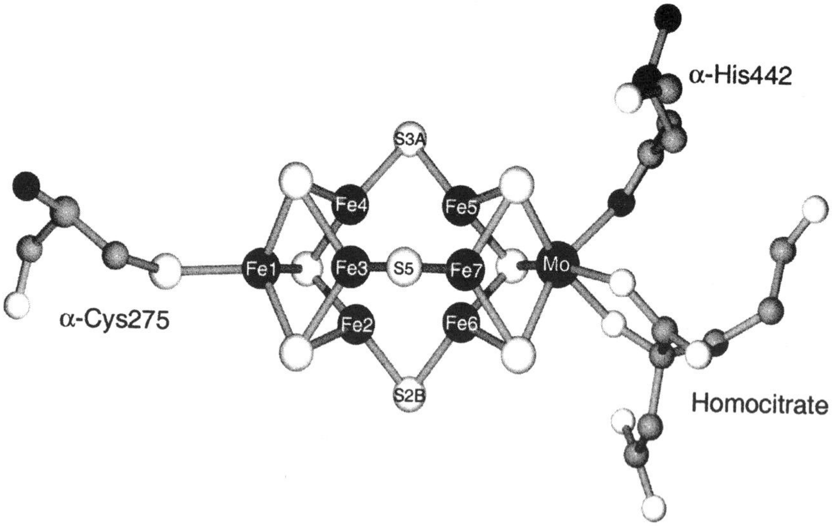

Figure 4. The *A.vinelandii* Nitrogenase FeMo-cofactor.

undoubtedly too big to bind to the FeMo-cofactor in this fashion. Although different substrates may adopt different binding modes, it seems that this potential binding model may never be used.

Numerous other binding mechanisms have been proposed in an attempt to describe how and where substrates interact with the FeMo-cofactor. Most hypotheses offer variations on the binding of substrates to one or more members of the trigonal prism of Fe atoms located within the midsection of the FeMo-cofactor, although the notion of substrates binding to the Mo atom has also been asserted. A number of possible N_2 substrate binding modes have been suggested as a result of theoretical studies [16-19]. Three of these binding models involve the replacement of one of the three bridging sulfides by N_2, three models describe N_2 binding to one of the three geometrically identical [4Fe-4S] faces, one model represents N_2 terminally bound to one of the trigonally coordinated Fe atoms, one model illustrates N_2 binding within the internal cavity of the FeMo-cofactor, and one model features N_2 binding to the Mo atom with the associated loss of one of its terminal ligands.

Although the theoretical study of these potential binding modes leads to no firm conclusion as to which is the most feasible, a simple and direct assessment can be made based on the known chemistry of the FeMo-cofactor. Models involving displacement of the central bridging sulfides seem unlikely since they require the breaking and making of Fe-S bonds in order to complete the catalytic cycle making these models energetically unfavorable. Binding in the internal cavity as discussed above seems unlikely because there appears to be insufficient space within

Metal-Metal	Reduced (Å) Kp	Reduced (Å) Av	Oxidized (Å) Kp	Oxidized (Å) Av	Bond Distances	Reduced (Å) Kp	Reduced (Å) Av	Oxidized (Å) Kp	Oxidized (Å) Av
Mo1-Fe1	6.97	6.90	6.97	6.99	Mo1-S1B	2.31	2.22	2.32	2.35
Mo1-Fe2	5.04	4.96	5.05	4.99	Mo1-S3B	2.36	2.23	2.37	2.12
Mo1-Fe3	5.05	4.95	5.06	4.98	Mo1-S4B	2.33	2.28	2.34	2.29
Mo1-Fe4	5.08	5.01	5.09	5.02	Fe1-S1A	2.31	2.31	2.30	2.27
Mo1-Fe5	2.72	2.69	2.72	2.73	Fe1-S2A	2.22	2.61	2.21	2.22
Mo1-Fe6	2.68	2.64	2.67	2.64	Fe1-S4A	2.26	2.26	2.26	2.30
Mo1-Fe7	2.67	2.63	2.68	2.54	Fe2-S1A	2.26	2.16	2.26	2.22
Fe1-Fe2	2.66	2.62	2.66	2.69	Fe2-S2A	2.23	2.26	2.18	2.27
Fe1-Fe3	2.65	2.58	2.64	2.73	Fe2-S2B	2.20	2.21	2.20	2.16
Fe1-Fe4	2.67	2.65	2.64	2.74	Fe3-S2A	2.26	2.24	2.21	2.28
Fe1-Fe5	5.02	4.93	5.00	5.03	Fe3-S4A	2.24	2.24	2.23	2.29
Fe1-Fe6	5.00	4.89	4.99	4.97	Fe3-S5	2.23	2.12	2.25	2.09
Fe1-Fe7	5.00	4.89	5.00	5.07	Fe4-S1A	2.30	2.21	2.26	2.28
Fe2-Fe3	2.70	2.57	2.69	2.66	Fe4-S4A	2.27	2.22	2.28	2.24
Fe2-Fe4	2.70	2.63	2.69	2.66	Fe4-S3A	2.22	2.20	2.21	2.11
Fe2-Fe5	3.74	3.62	3.74	3.68	Fe5-S1B	2.29	2.23	2.27	2.25
Fe2-Fe6	2.59	2.51	2.60	2.53	Fe5-S4B	2.23	2.21	2.23	2.17
Fe2-Fe7	3.74	3.58	3.75	3.64	Fe5-S3A	2.25	2.14	2.25	2.18
Fe3-Fe4	2.68	2.56	2.68	2.65	Fe6-S1B	2.26	2.14	2.27	2.14
Fe3-Fe5	3.73	3.58	3.73	3.65	Fe6-S3B	2.20	2.17	2.20	2.26
Fe3-Fe6	3.71	3.54	3.70	3.62	Fe6-S2B	2.19	2.19	2.19	2.17
Fe3-Fe7	2.61	2.53	2.62	2.62	Fe7-S3B	2.24	2.18	2.22	2.19
Fe4-Fe5	2.62	2.55	2.62	2.54	Fe7-S4B	2.20	2.13	2.23	2.18
Fe4-Fe6	3.74	3.63	3.74	3.61	Fe7-S5	2.20	2.13	2.18	2.16
Fe4-Fe7	3.74	3.63	3.74	3.67	Mo1-α-His-Nδ	2.48	2.15	2.47	2.16
Fe5-Fe6	2.68	2.57	2.66	2.58	Mo1-O5-homocit.	2.29	2.05	2.29	2.09
Fe5-Fe7	2.65	2.56	2.65	2.54	Mo1-O7-homocit.	2.35	2.02	2.35	2.09
Fe6-Fe7	2.63	2.46	2.62	2.47	Fe1-α-Cys-Sγ	2.32	2.34	2.33	2.25

Table 2. FeMo-cofactor distances in the (Kp) *K. pneumoniae* (1.6 Å resolution) and (Av) *A. vinelandii* (2.0 Å resolution) MoFe protein structures [5,6]. All distances represent the average between pairs of the FeMocos related by noncrystallographic symmetry.

FeMo- cofactor to accommodate the N_2 molecule. Binding at Mo is a possibility but the binding without dissociation of a ligand would disrupt the Mo atom's normal octahedral coordination sphere. Since the FeMo-cofactor contains six coordinately unsaturated Fe atoms, the involvement of these Fe atoms in substrate binding seems to be most likely.

The bond distances determined between the constituent metal atoms of the [4Fe-3S] and [Mo-3Fe-3S] cubane fragments are reasonably typical for such interactions (Table 2) [5,6]. The Mo to Fe2, Fe3, and Fe4 distances, and the

Fe1 to Fe5, Fe6, and Fe7 distances, are ~5.0 Å. The Mo to Fe1 separation of ~7.0Å is the largest metal-metal distance within the FeMo-cofactor. The distances separating pairs of three coordinate Fe atoms between the two partial cubane fragments is either 2.5-2.7 Å (for Fe2-Fe6, Fe3-Fe7, & Fe4-Fe5) or 3.6-3.7 Å (for Fe2-Fe5, Fe2-Fe7, Fe3-Fe5, Fe3-Fe6, Fe4-Fe6, and Fe4-Fe7).

The structure of the MoFe protein from *A. vinelandii* has been refined at 2.0 Å resolution in two oxidation states (spectroscopically assigned oxidized and dithionite-reduced) [5]. It was noted in this study that the Fe-Mo and Fe-Fe distances in the FeMo-cofactor, particularly those of the [4Fe-3S] partial cubane, decrease slightly on going from the oxidized to the reduced state. This result is consistent with those obtained from EXAFS studies in which shorter metal-metal distances are observed in the FeMo-cofactor upon reduction of the MoFe protein [20]. However, these observed differences in the *A. vinelandii* MoFe structure are on the order of the expected individual coordinate errors for a refined structure at this resolution of X-ray diffraction data (2.0 Å). In addition, a similar trend was not observed in the comparison of the analogous (1.6 Å resolution) oxidized and reduced structures of the MoFe protein from *K. pneumoniae* [6] and thus the trend observed in the *A. vinelandii* structures may be fortuitous.

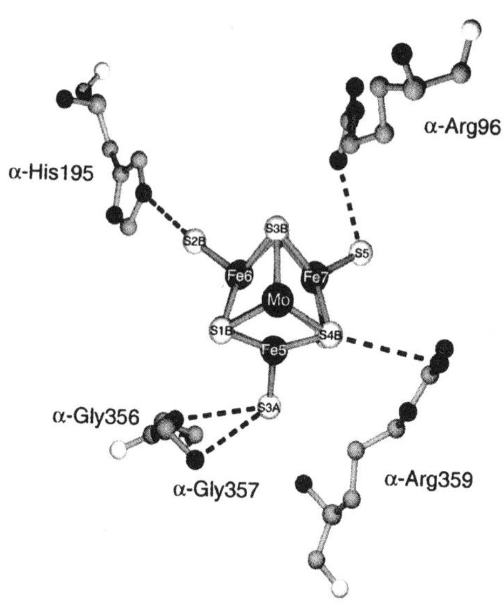

Figure 5. The *A vinelandii* nitrogenase FeMo-cofactor as it appears looking down the metal cluster three-fold symmetric axis. In this view, the [4Fe-3S] subcluster is eclipsed by the [Mo-3Fe-3S] sub-cluster. Also shown are the protein atoms that interact with the sub-cluster-bridging S atoms through H-bonding.

It is generally accepted that the FeMo-cofactor provides the site for substrate binding and reduction [21-23]. Furthermore, numerous mutagenesis experiments have established that the immediate polypeptide environment of the FeMo-cofactor affects FeMo-cofactor's ability to perform its catalytic function [24-29]. For the *A. vinelandii* MoFe protein, the protein domain surrounding the FeMo-cofactor is comprised predominantly of hydrophilic residues, with the exception of α-Val^{70}, α-Tyr^{229}, α-Ile^{231}, α-Ile^{355}, α-Leu^{358}, and α-Phe^{381}, which are hydrophobic. Aside from residues α-Cys^{275} and α-His^{442}, which directly coordinate the FeMo-cofactor, several other residues interact with the metallocluster. The side chains of the residues α-Arg^{96}, α-His^{195}, α-Arg^{359} are all capable of forming hydrogen-bonds to sulfur atoms of FeMo-cofactor, as are the NH groups of α-Gly^{356} and α-Gly^{357} (Figure 5). The exact function of these interactions has not yet been established unequivocally, but it seems likely that they contribute toward maintaining the FeMo-cofactor's stability within the polypeptide domain, as well as possibly facilitating the catalytic process *via* some, as yet undetermined, mechanism.

Furthermore, these residues, along with α-Val^{70} and α-Phe^{381}, all reside around the central midsection of the FeMo-cofactor that is comprised of three-coordinate Fe atoms and the bridging S atoms. Recent studies involving altered MoFe-proteins with substitutions at the α-Val^{70} position (which is located over one of three geometrically identical [4Fe-4S] faces of the FeMo-cofactor) reveal that when Val is changed to a residue with a smaller side chain, such as Ala, the resulting MoFe protein variant possesses an expanded substrate range and is capable of reducing propyne, propargyl alcohol and other alkyne substrates larger than acetylene, effectively [29] (see Chapter 5). Since alkynes larger than acetylene are poor substrates for the wild-type MoFe protein, it would appear that changing Val to a residue that possesses a side chain of less steric bulk has opened up the binding pocket at the active site thereby permitting the binding and reduction of larger acetylenic compounds. The α-Arg^{96} side chain is also located over the same [4Fe-4S] face of the FeMo-cofactor as is the α-Val^{70} (Figure 6). It was recently reported that when this residue is changed to Leu, Gln, Ala, or His, both cyanide and acetylene are able to interact with the FeMo-cofactor of the resulting altered MoFe proteins under resting state, non-turnover conditions [28]. These results certainly imply that the [4Fe-4S] face that is adjacent to these two residues accommodates the site for substrate binding and reduction. If small molecule substrates interact with the FeMo-cofactor in this region, then it is certainly feasible for one or more of the residues located within the immediate vicinity to contribute to this interaction.

Other residues that are located within close proximity of the FeMo-cofactor are α-His^{195}, α-Gln^{191}, α-Glu^{440}, and α-Glu^{427}. It has been postulated that α-His^{195} may participate in proton-transfer reactions while the other residues are all situated near homocitrate. The α-His^{195} residue is able to form an NH-S hydrogen-bond to one of the bridging sulfides of the FeMo-cofactor and it has been suggested that this interaction

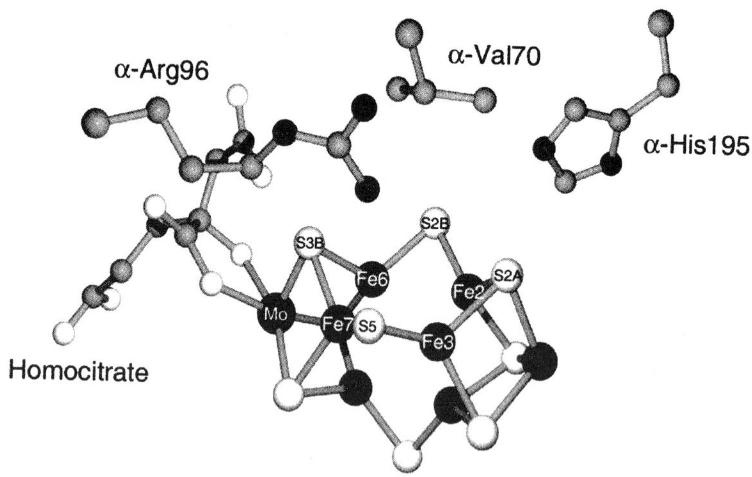

Figure 6. The *A. vinelandii* nitrogenase FeMo-cofactor and residues in the vicinity of the proposed 4Fe substrate-binding region.

may help to properly position the FeMo-cofactor within its immediate protein environment.

Two altered MoFe proteins with substitutions at the α-His^{195} position, α-Gln^{195} and α-Asn^{195}, show altered catalytic activity compared to that of wild-type MoFe protein [24-26,30]. The α-Gln^{195} MoFe protein is incapable of reducing N_2, but N_2 does act as an effective competitive inhibitor of acetylene and proton reduction by this MoFe protein. In contrast, the reduction of acetylene and proton reduction by the α-Asn^{195} MoFe protein is not inhibited by N_2.

The crystal structure of the altered α-Gln^{195} MoFe protein has been solved at 2.5 Å resolution [14]. The structure shows that the α-Gln^{195} substituted residue is also capable of forming a hydrogen bond to the same bridging sulfide of the FeMo-cofactor as the α-His^{195} residue of the wild-type MoFe protein. This result suggests that the functional and spectroscopic changes exhibited by the altered α-Gln^{195} MoFe protein are induced as a direct result of the loss of the histidine imidazole group of the α-His^{195} residue as the source of the hydrogen-bond to the bridging sulfide of the FeMo-cofactor. Since substitution of the α-His^{195} residue of wild-type MoFe protein exerts a profound effect on the spectroscopic properties and catalytic activity of the resulting altered MoFe proteins, this residue undoubtedly performs an important role in the proper catalytic functioning of the nitrogenase MoFe protein.

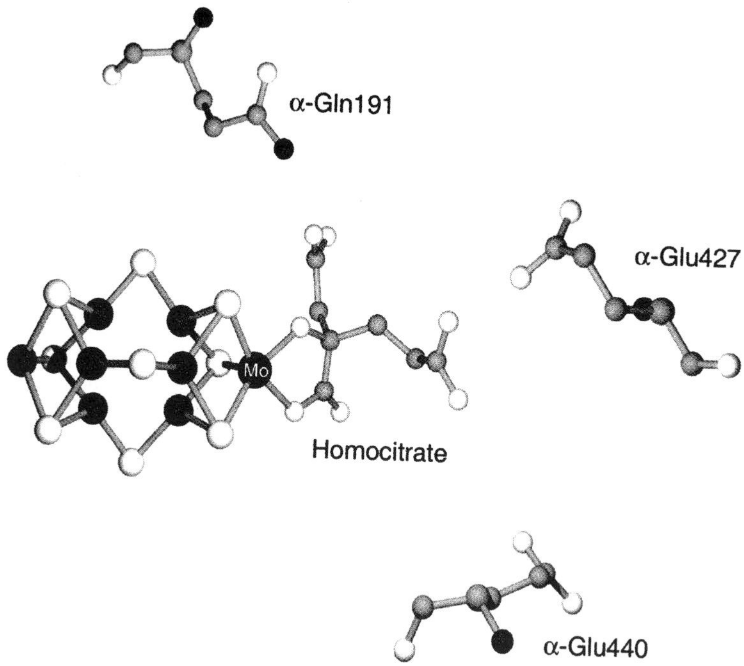

Figure 7. The *A. vinelandii* nitrogenase FeMo-cofactor highlighting the location of amino acid residues that interact with homocitrate.

The homocitrate is stabilized by hydrogen bonding interactions of α-Gln^{191}, and α-Glu^{440} (via a water molecule) at the terminal carboxylates and a large number of water molecules are located at the cluster distal side of the homocitrate (Figure 7).

Since the presence of homocitrate is important for catalysis, any residue that is able to interact with homocitrate may therefore influence or assist the correct catalytic functioning of the FeMo-cofactor. Substitution of α-Gln^{191} with α-Lys^{191} produces an altered MoFe protein that is unable to reduce N_2 but is still capable of reducing acetylene and protons, although at a reduced capacity [24,25]. Additionally, in contrast to the wild-type MoFe protein, the altered α-Lys^{191} MoFe protein exhibits proton reduction activity that is sensitive to carbon monoxide.

The FeMo-cofactor is situated approximately 10 Å beneath the protein surface of the MoFe protein. In order for nitrogenase substrates to bind and be reduced they have to be able to interact with the FeMo-cofactor, but they must first penetrate the MoFe protein surface and negotiate the FeMo-cofactor's surrounding polypeptide environment. However, as yet, no distinct channels for substrate access or product release have been identified in the region of the MoFe protein that accommodates the FeMo-cofactor. Despite this, a theoretical assessment of the available MoFe protein crystal structures, using density functional theory (DFT), has identified three possible proton-transfer routes from the MoFe protein surface to the FeMo-cofactor [31]. One is a water-filled channel that connects the protein exterior to the homocitrate of the FeMo-cofactor, with the other two being hydrogen-bonded chains to bridging sulfides of the FeMo-cofactor. It is suggested that the water filled channel is used to deliver protons to the partially reduced substrate as well as allowing products and substrates to diffuse between the FeMo-cofactor and the surface bulk solvent. It is proposed that the additional two proton channels allow a single proton to be transferred to, or from, the FeMo-cofactor during catalysis. Although these theoretical hypotheses have not yet been experimentally tested, they nevertheless provide insight into the potential substrate/product routes between the FeMo-cofactor and the MoFe protein surface.

2.3. P-cluster

The P-clusters exist as [8Fe-7S] clusters that are located ~10 Å beneath the protein surface, on the pseudo-two-fold symmetric axes that approximately relate each of the α and β subunits. The structural characterizations of nitrogenase MoFe protein from *A. vinelandii* and *K. pneumoniae* in multiple oxidation states have revealed a very intriguing mechanism of redox dependent conformational change [5,6]. Although the overall structure of the MoFe protein is essentially identical in each oxidation state, the structure of the P-cluster changes significantly between the states. Interconversion between the two oxidation states results in structural changes within the P-cluster, involving the movement of two Fe atoms and an exchange of protein coordination for ligands provided by a central S atom (Figure 8). In the oxidized state the P-cluster can be considered to be comprised of a [4Fe-4S] cluster and a [4Fe-3S] cluster that are predominately bound to the α-subunit and β-subunit, respectively. In the oxidized form, the central S atom of the P-cluster (S1) is within bonding distance of four Fe atoms, three from the [4Fe-4S] α-subunit portion of the cluster, and one from the [4Fe-3S] β-subunit part of the cluster. Upon conversion to the reduced form of the P-cluster an additional Fe atom, Fe5 from the [4Fe-3S] β-subunit part of the cluster, is shifted to within bonding distance of S1. Additionally, Fe6, also from the β-subunit partial cubane, moves to the central S atom (S1) forming a putative bond, although this distance is longer than the normal Fe-S distance in the *A. vinelandii* reduced P cluster structure in the *K. pneumoniae* is closer to a normal Fe-S bond distance (Fe6-S1 in Table 2). This observation may indicate that the *A. vinelandii* Fe protein is not completely reduced. As a result of these conformational changes, the central S1 sulfur of the P-cluster is surrounded by a distorted octahedral arrangement of six Fe atoms in the reduced form. Therefore, for the *A. vinelandii* MoFe protein structure, the oxidized P-cluster average Fe-Fe distance is ~2.88 Å, while for the reduced form the average Fe-Fe distance is ~2.59 Å (Table 3). A similar trend is observed for the P- cluster metal-metal distances of the *K. pneumoniae* MoFe protein. This observed decrease in the average Fe-Fe distance upon interconversion from the oxidized to the reduced form is

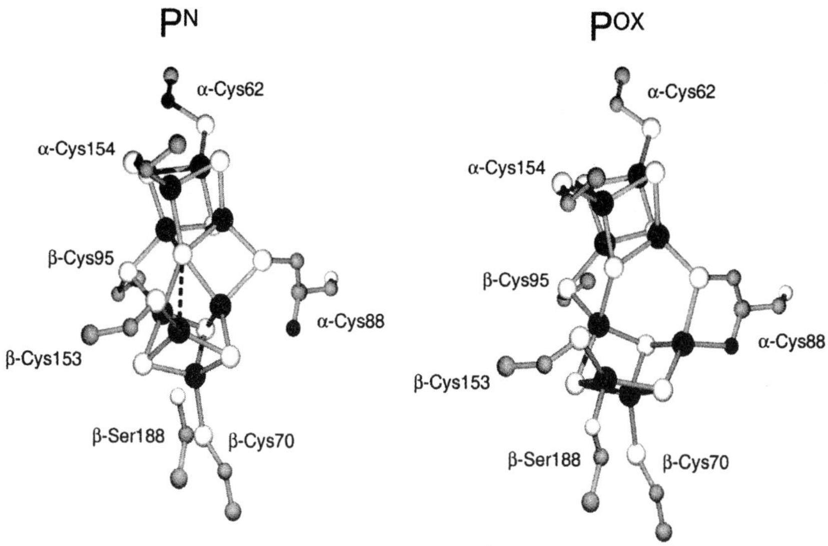

Figure 8. The structure of the *A. vinelandii* nitrogenase P-cluster in the reduced and oxidized state.

consistent with EXAFS studies that indicate a contraction of metal-metal distances upon reduction [20].

The P-cluster is covalently coordinated to the α- and β-subunits of the MoFe protein through six cysteinyl ligands, with three being provided by the α-subunit and three from the β-subunit. For the *A. vinelandii* MoFe protein P cluster, the α-subunit, α-Cys[62] and α-Cys[154] provide individual ligands to Fe atoms, while α-Cys[88] serves as a bridging cysteinyl ligand between two Fe atoms of the cluster. Similarly for the β-subunit, β-Cys[70] and β-Cys[153] coordinate Fe atoms individually, and β-Cys[95]acts as a bridging ligand coordinating two Fe atoms (Figure 8). As described earlier, the change of redox state from oxidized to reduced is accompanied by movement of two Fe atoms together with an exchange of covalent ligands. In the oxidized state, two additional protein ligands coordinate the P-cluster. For the *A vinelandii* structure α-Cys[88] coordinates an Fe atom by virtue of a backbone amide nitrogen atom (as well as maintaining the original cysteinyl ligand coordination), while the Oγ atom of β-Ser[188] coordinates the Fe atoms which is coordinated by β-Cys[153]. In the reduced state, these two non-cysteinyl ligands are replaced by interactions with the central S1 sulfur. Hence, in both the oxidized and reduced states, all Fe atoms in the P-cluster remain four-coordinate.

In contrast to the immediate protein domain of the FeMo-cofactor, the protein environment of the P-cluster is comprised mainly of hydrophobic residues such as

Metal-	Reduced (Å)		Oxidized (Å)		Bonds	Reduced (Å)		Oxidized (Å)	
Metal	Kp	Av	Kp	Av	distances	Kp	Av	Kp	Av
Fe1-Fe2	2.52	2.42	2.52	2.40	Fe1-S1	2.42	2.42	2.34	2.21
Fe1-Fe3	2.79	2.76	2.80	2.78	Fe1-S2A	2.42	2.22	2.37	2.24
Fe1-Fe4	2.53	2.50	2.58	2.61	Fe1-S3A	2.28	2.26	2.29	2.32
Fe1-Fe5	4.51	4.59	5.77	5.79	Fe2-S1	2.52	2.28	2.51	2.25
Fe1-Fe6	4.80	5.02	6.09	6.05	Fe2-S2A	2.31	2.18	2.29	2.22
Fe1-Fe7	6.70	6.64	6.80	6.78	Fe2-S4A	2.28	2.26	2.31	2.26
Fe1-Fe8	4.46	4.46	4.36	4.39	Fe3-S2A	2.32	2.22	2.32	2.22
Fe2-Fe3	2.75	2.70	2.74	2.71	Fe3-S3A	2.38	2.33	2.36	2.31
Fe2-Fe4	2.58	2.53	2.55	2.50	Fe3-S4A	2.28	2.26	2.25	2.17
Fe2-Fe5	3.75	3.77	4.86	4.83	Fe4-S1	2.40	2.32	2.37	2.23
Fe2-Fe6	4.59	4.76	5.82	5.69	Fe4-S3A	2.25	2.19	2.26	2.33
Fe2-Fe7	5.45	5.42	5.51	5.46	Fe4-S4A	2.29	2.23	2.30	2.21
Fe2-Fe8	2.94	3.05	2.97	2.99	Fe5-S1	2.41	2.39	3.86	3.88
Fe3-Fe4	2.67	2.60	2.69	2.68	Fe5-S2B	2.30	2.26	2.24	2.20
Fe3-Fe5	5.34	5.37	6.20	6.24	Fe5-S4B	2.28	2.19	2.35	2.31
Fe3-Fe6	6.67	6.83	7.93	7.86	Fe6-S1	2.47	2.76	3.83	3.96
Fe3-Fe7	7.69	7.61	7.72	7.71	Fe6-S2B	2.32	2.27	2.33	2.36
Fe3-Fe8	5.51	5.57	5.48	5.53	Fe6-S3B	2.36	2.27	2.35	2.30
Fe4-Fe5	2.92	3.03	3.74	3.79	Fe7-S2B	2.37	2.23	2.30	2.20
Fe4-Fe6	4.49	4.70	5.62	5.62	Fe7-S3B	2.31	2.27	2.32	2.31
Fe4-Fe7	5.59	5.57	5.58	5.60	Fe7-S4B	2.34	2.34	2.31	2.27
Fe4-Fe8	3.99	4.06	3.80	3.93	Fe8-S1	2.49	2.38	2.37	2.50
Fe5-Fe6	2.52	2.56	3.91	3.84	Fe8-S3B	2.35	2.28	2.28	2.30
Fe5-Fe7	2.82	2.66	2.81	2.77	Fe8-S4B	2.24	2.31	2.27	2.39
Fe6-Fe8	2.58	2.52	3.25	3.15	Fe1-Cys-α-Sγ	2.39	2.38	2.39	2.34
Fe7-fe8	2.72	2.62	2.75	2.71	Fe2- Cys-α-Sγ	2.40	2.26	2.36	2.30
					Fe3- Cys-α-Sγ	2.39	2.21	2.38	2.27
					Fe4- Cys-α-Sγ	2.44	2.24	2.39	2.18
					Fe5- Cys-α-Sγ	2.47	2.32	2.40	2.43
					Fe5- Cys-α-N	3.47	3.35	2.28	2.17
					Fe6- Cys-α-Sγ	2.33	2.32	2.31	2.27
					Fe6- Ser-α-Oγ	3.50	2.82	2.18	1.98
					Fe7- Cys-α-Sγ	2.36	3.34	2.36	2.42
					Fe8- Cys-α-Sγ	2.32	2.28	2.33	2.21

Table 3. P-cluster distances in the (Kp) *K. pneumoniae* (1.6 Å resolution) and (Av) *A. vinelandii* (Av-2.0 Å resolution) MoFe protein structures [5,6]. All distances represent the average between pairs of the FeMocos related by noncrystallographic symmetry. *Some of the distances involving Fe5 and Fe6 are not considered to imply bonding depending on oxidation state.

α-Tyr^{64}, β-Pro^{72}, α -Pro^{85}, α-Tyr91, β-Tyr^{98}, β-Phe^{99}, β-Met^{154}, α-Pro^{155}, α-Phe^{186}, and β-Phe^{189} (*A. vinelandii* MoFe protein). The location of the P-cluster near the N-terminal ends of six helices (α-63 – α-74, α-88 – α-92, α-155 – α-159, β-71 – β-81, β-93 – β106, and β-153 – β-158) may serve to provide an electrostatic contribution to cluster stability.

3. THE Fe PROTEIN

The Fe protein is the key component of the bacterial nitrogen fixation process. The reduced form of the Fe protein acts as a specific electron donor for the MoFe protein. This electron transfer is coupled to ATP hydrolysis. In addition to its role as the exclusive electron donor in the nitrogen reduction process, the Fe protein is also required for MoFe protein maturation and synthesis of FeMo-cofactors [32,33]. The function of the Fe protein in FeMo cofactor biosynthesis is independent of its role in nitrogen reduction reaction. This has been established *via* several mutagenesis studies in which variants of the Fe protein with altered capacity for nucleotide binding and hydrolysis are ineffective in nitrogen catalysis but are able to function in FeMo cofactor biosynthesis [34-38].

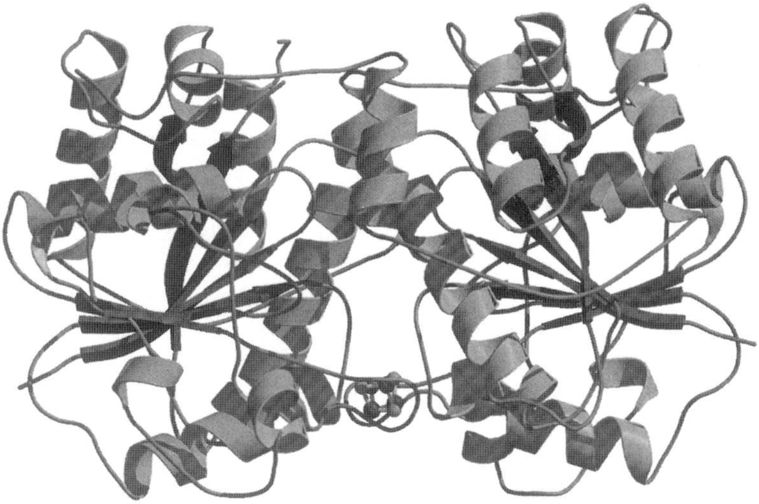

Figure 9. The *A. vinelandii* nitrogenase Fe protein.

3.1. Overall structure

The protein fold observed for the Fe protein is similar to that observed for the nucleotide binding proteins Ras [39,40], elongation factor Tu [41,42], transducin [43], and RecA [44,45]. These proteins are representative of a general class of proteins that utilize nucleotide binding and hydrolysis to drive conformational changes that occur within a macromolecular complex [46,47]. Characteristic structural features of these proteins include predominantly β structure and conserved nucleotide binding motifs situated at topology switch points at the termini of several β-strands of the β-sheet core. For the Fe protein these regions serve to link the sites for nucleotide binding both to the [4Fe-4S] cluster and to the region responsible for nitrogenase complex formation.

The Fe protein is clearly the most interesting member of this class due its dimeric arrangement and its covalent association with the [4Fe-4S] cluster. The first structure of the nitrogenase Fe protein reported was the Fe protein from *A. vinelandii* [1]. The Fe protein exists as a dimer of identical subunits covalently bridged by a single [4Fe-4S] cluster (Figure 9). Each subunit is folded into an α/β type domain consisting of an eight-stranded β-sheet flanked by nine α-helices. The twisted β-sheet core of each subunit consists of seven parallel β-strands and a single short antiparallel strand. A comparison of the available Fe protein structures indicates that the structure of the β-sheet core is highly conserved with conformationally variable regions concentrated among the flexible loop regions that connect the secondary structure elements. In comparing the nucleotide free structures of the Fe proteins from *A. vinelandii* and *C. pasteurianum*, r.m.s. deviations of ~0.5 Å of corresponding carbon-α atoms are observed in the β -sheet core while in the loop regions and carboxy termini the r.m.s. values exceed 1.5 Å in some positions [9]. This conformational variability most likely occurs as a result of deletions of certain amino acids and variation in the amino acid sequences in these regions.

3.2. The [4Fe-4S] cluster

The Fe protein's FeS cluster exists as a regular [4Fe-4S] cubane coordinated to the dimer through two cysteine residues from each subunit. The symmetric coordination environment involves covalent coordination by Cys^{97} and Cys^{132} from each subunit. This covalent association of identical subunits is rare, although a similar covalent attachment and pseudo-symmetric coordination of a [4Fe-4S] has recently been observed in the structures of carbon monoxide dehydrogenases from *Carboxydothermus hydrogenoformans* and *Rhodospirillum rubrum* [47,48]. The cluster is located near the surface of the protein and it was surmised from this structure that docking of the Fe protein with the MoFe protein would occur at the face of the Fe protein in which the cluster is most surface exposed. A docking model which was based primarily on the structures of the nitrogenase components but also on the results of covalent cross-linking and mutagenesis experiments was rationalized by pairing the two-fold symmetric axis of the Fe protein with the pseudo two-fold symmetric axis that relates each α- and β- subunit such that two equivalent sites for Fe protein interaction exist on the MoFe protein [23,50,51]. Docking in this manner minimizes the distance between the [4Fe-4S] cluster of the Fe protein and a P cluster, consistent with a number of studies that suggested that the P cluster serves as an intermediate in electron-transfer from the Fe protein to the FeMo-cofactor substrate-reduction site. Characterization of Fe protein and MoFe protein complexes [8,13] described herein have confirmed this assertion and have shown that through the reorientation of Fe protein subunits, the structural environment of the Fe protein [4Fe-4S] cluster undergoes conformational changes. The location of the cluster at the dimer interface facilitates this conformational flexibility and no doubt contributes to the range of midpoint potentials observed in the various nucleotide bound states of the Fe protein and during Fe protein-MoFe protein complex formation. In addition, its location results in a greater degree of solvent accessibility than observed in other [4Fe-4S] cluster-containing proteins. Most [4Fe-4S] clusters in proteins are completely buried, but the Fe protein the cluster is accessible within a surface area on the order of ~14 $Å^2$, potentially contributing to its oxygen sensitivity.

Detailed biochemical and biophysical analysis of the properties of the [4Fe-4S] cluster indicate that the binding of nucleotides results in changes in the cluster environment and accompanying changes in the midpoint potential of the cluster [52-60]. The prevailing view concerning the structural changes that dominate the control of midpoint potential of the cluster concerns the proximity of hydrogen bonding amide

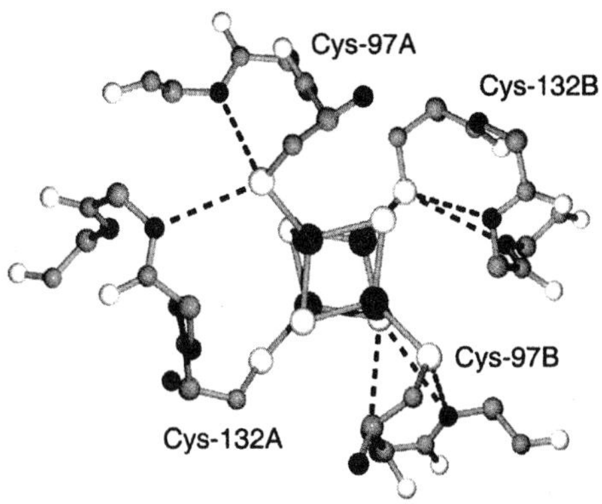

Figure 10. The coordination environment of the [4Fe-4S] of the *A. vinelandii* nitrogenase Fe protein with several amide hydrogen-bond interactions to the cluster and ligand S atoms indicate by dashed lines.

groups to the cluster and cysteine ligand sulfur atoms [10,12,61,62]. The premise is that the presence of these hydrogen bonding groups stabilizes the oxidized state of the cluster resulting in more positive midpoint potentials. It is reasonable to think that the binding of nucleotide and the associated conformational changes should result in changes in the overall polarity of the cluster environment including a reshuffling of various peptide bond amides in the cluster environment, providing for the stabilization of the different oxidation levels to varying degrees. In the most recent structure of the nucleotide-free form of the *A. vinelandii* Fe protein, six potential peptide amide-to-sulfur interactions were observed, three from each subunit, arising from amino acids Ala^{98}, Gly^{99} and Gly^{134} [9]. In addition to this, amino acids Ala^{98}, Val^{130}, and Phe^{135} from each subunit are found to be in Van der Waals contact with the [4Fe-4S] cluster. Along these lines, differences are observed in the proximity of amide hydrogen-bond donors in the structures of the free Fe protein, Fe protein with MgADP bound, and within the Fe protein – MoFe protein complexes [1,8,9,11,13]. In addition, the structural characterization of an Fe protein variant with Phe^{135} substituted by Trp, in which more positive midpoint potentials are observed, revealed that the amide groups of the Trp indoles are within hydrogen-bonding distance of two cluster S atoms [10].

It has been demonstrated through previous biochemical and spectroscopic characterization that this cluster can undergo a one-electron redox cycle from the $[4Fe-4S]^{2+}$ to the $[4Fe-4S]^{1+}$ oxidation level or a two-electron reduction from the $[4Fe-4S]^{2+}$ to the $[4Fe-4S]^{0}$ oxidation level [63-68]. The bulk of the biochemical results have been interpreted assuming that each electron-transfer event from the Fe protein to the MoFe protein occurs at a ratio of 2MgATP molecules hydrolyzed per electron transferred, with a transition from the $[4Fe-4S]^{1+}$ to the $[4Fe-4S]^{2+}$ levels. Recent studies have revealed that the Fe protein can exist stably in the $[4Fe-4S]^{0}$ redox level and can undergo an analogous set of nucleotide-dependent conformational changes. This all-ferrous form of the Fe protein can serve as an electron donor for nitrogenase-substrate reduction and does so at a 1:1 ratio of MgATP hydrolyzed per electron transferred [69]. The recent structural characterization of the all ferrous Fe protein indicates that although the formal covalent coordination of the cluster remains unchanged there are changes in the cluster geometry, including longer S-S distances and an overall increase in the volume of the [4Fe-4S] cluster in comparison to the previously presented Fe protein structures [12]. In addition, substantial differences in the number of amide interactions with the cluster and ligand sulfur atoms and an increase in overall cluster solvent accessibility were observed, suggesting that a number of factors may contribute to stabilization of the various [4Fe-4S] cluster oxidation levels and related midpoint potentials.

3.3. Nucleotide binding sites

The class of nucleotide binding proteins to which the Fe protein belongs all possess characteristic amino acid sequences called Walker motifs [46,47,50]. In the major class of nucleotide-binding G-proteins, the sequence Gly-X-X-X-X-Gly-Lys/Ser found between residues 9 to 16 of the *A. vinelandii* Fe protein, known as Walker's motif A, adopts a β-strand-loop-α-helix structure which is a characteristic phosphate binding site contacting the α- and β-phosphates of the nucleotide. A second conserved sequence Asp-X-X-Gly, also known as Walker's motif B, is found between residues 125-128 of the *A. vinelandii* Fe protein. In the G proteins this motif is responsible mainly for the binding of Mg^{2+} and γ-phosphate of GTP.

A comparison of the Ras protein and the Fe protein indicates a close structural similarity, with their Walker-motif regions almost superimposable. However, unlike Ras, the dimeric structure of the Fe protein perhaps allows for cross-subunit interactions that may have some consequences for the coupling of electron-transfer and ATP hydrolysis, and on the binding of different nucleotides (*e.g.*,. MgATP/MgADP) to the Fe protein. This situation is never observed in the monomeric nucleotide-binding proteins such as the Ras protein [39,40]. An important parallel between the nitrogenase and the G protein families is that most of the nucleotide-dependent proteins have a basal rate of hydrolysis which is greatly accelerated by association with a second protein, *e.g.*, GAP (GTPase activating proteins) for Ras, RGS (Regulators of G-protein signaling) for G-proteins. In contrast, the Fe protein alone has no intrinsic hydrolysis rate and nucleotide hydrolysis requires both the Fe protein and the MoFe protein.

The structure of the nitrogenase Fe protein in the presence of MgADP has been determined and refined to 2.15 Å resolution [11]. The two MgADP molecules bound to the Fe protein are located at the dimer interface, oriented parallel to the two-fold symmetry axis with the phosphates directed toward the [4Fe-4S] cluster. This binding mode is analogous to the GDP binding mode observed for Ras. The MgADP binding mode differs from the binding mode reported earlier for low-occupancy ADP bound to the Fe protein observed in the 1NIP Fe protein structure (see Table 1). The nucleotide-binding sites on the Fe protein are accessed from the face opposite the [4Fe-4S] cluster through a channel that terminates in a region of significant negative charge. The majority of the negative charges come from three aspartate residues (Asp^{39}, Asp^{43}, and Asp^{125}) that participate in the coordination of the Mg^{2+} ion, either directly or through bound water molecules (Figure 11). The extent of the channel is clearly large enough for MgADP molecules to access freely the site in the MgADP-bound state of the protein.

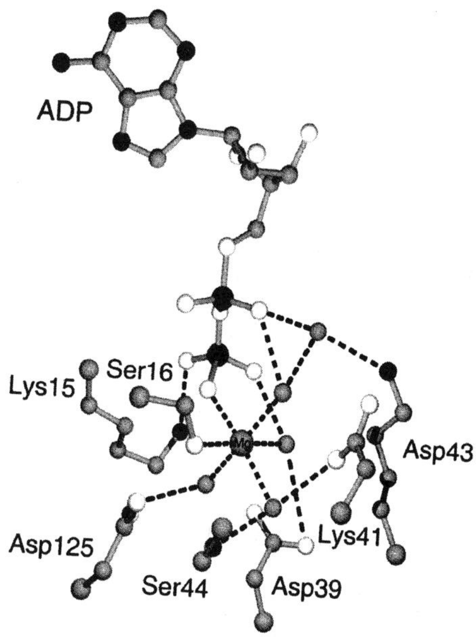

Figure 11. The MgADP environment in the structure of the *A. vinelandii* nitrogenase Fe protein in the presence of bound MgADP.

Several amino acids are involved in the binding of MgADP. The adenine base is bound to the Fe protein through interactions with the side chains of Asp^{185}, Gln^{218}, and Gln^{236}. In addition, N1 of the adenine ring approaches to within hydrogen-bond distance from the amide nitrogen and carbonyl oxygen of the peptide-bond of Asp^{214}.

The ribose moiety is hydrogen-bonded to the protein through the side chain oxygen of Glu^{221}. The phosphate-binding loop region (P-loop) of the Fe protein, which is a common amino acid sequence motif Gly-X-X-Gly-X-Gly found in other nucleotide-binding proteins, is found to contribute to all the direct interactions of the protein with the phosphates, through hydrogen-bond interactions provided by residues 10 through 17. Although the P-loop is involved in the direct interaction with the phosphates, only small differences in the P-loop conformation in comparison to the Fe protein structure obtained in the absence of nucleotides have been observed. Superimposition of the P-loop regions of the MgGDP "off" state of Ras and the MgADP-bound Fe protein indicates that the position and orientations of MgADP and MgGDP in the two structures are strikingly similar, indicating the interactions between the P-loop and the bound Mg^{2+} ion and phosphates of nucleotide diphosphate are similar for these two proteins.

The structure of the MgADP bound state of the Fe protein provides insights into how the binding of nucleotides is linked to protein conformational changes that modulate the properties of the [4Fe-4S] cluster, and control the association of the Fe protein with the MoFe protein. On comparing the MgADP nucleotide-bound state of the Fe protein with the structure of the Fe protein in the absence of nucleotide, two notable regions within the Fe protein are observed to be conformationally different. One region, termed switch II by analogy to a homologous region found in Ras, links Asp^{125}, which interacts with the bound Mg^{2+}, to the [4Fe-4S] cluster ligand Cys^{132}. A second region of the Fe protein, termed switch I, links Asp^{39}, which interacts with the bound Mg^{2+}, with amino acids in the 59-69 region at the MoFe protein docking interface. Switch II provides a mechanism for communication between the nucleotide-binding site and the [4Fe-4S] cluster. A comparison of the nucleotide-free Fe protein structure with that of the MgADP Fe protein indicates that the Cys^{132} ligands of the cluster experience a change in conformation upon MgADP binding, while the Cys^{97} ligands remain largely unchanged. Conformational changes are communicated through the binding of MgADP at the P-loop regions, and these, in turn, are communicated to the cluster through a hydrogen-bonding interaction of the main chain amide of Gly^{14} to the side chain of Asp^{129}. Switch I provides a mechanism for communication from the nucleotide-binding site to the Fe protein surface involved in interactions with the MoFe protein. It appears that the hydrolysis of MgATP to MgADP lessens the intensity of the association of the Fe protein with the MoFe protein following electron-transfer, thus allowing the two proteins to dissociate. Switch I provides a mechanism for propagation to the surface of the Fe protein of the change from the MgATP-bound state to the MgADP-state, to control protein-protein interaction.

4. NITROGENASE COMPLEXES

Recently, X-ray structures for two different stable nitrogenase Fe protein-MoFe protein complexes have been solved, providing important insights into several aspects of the nitrogenase mechanism (Figure 12). These structures are essential to understanding ATP hydrolysis and inter-component electron-transfer, since both events occur only within the Fe protein-MoFe protein complex. The barrier to obtaining a structure of the complex was of stabilizing the two proteins to obtain diffraction quality crystals. As mentioned earlier, the Fe protein (γ_2 dimer), with two ATP molecules bound, normally binds transiently to one $\alpha\beta$ dimer of the tetrameric MoFe protein ($\alpha_2\beta_2$). Thus, a catalytic unit may be described as an $\alpha\beta\gamma_2$ complex with one ATP molecule bound to each γ subunit. During this association, ATP hydrolysis on the Fe protein is activated and is coupled to the transfer of a single electron from the 4Fe-4S cluster of the Fe protein to the P-cluster of the MoFe protein and ultimately to the FeMo-cofactor, where substrates are bound and reduced. Following ATP hydrolysis and electron-transfer, the Fe protein dissociates from the MoFe protein in the slow step of the overall reaction.

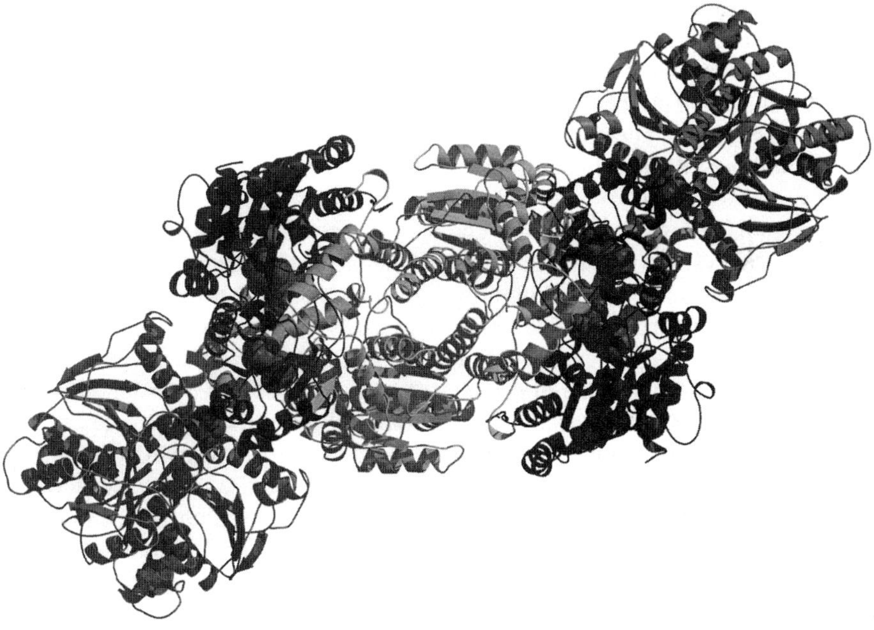

Figure 12. The *A. vinelandii* nitrogenase ΔLeu^{127}-Fe protein variant - MoFe protein complex.

Given that the Fe protein associates only transiently with the MoFe protein, it was essential for studies on the complex (including crystallography) to find ways to stabilize the Fe protein-MoFe protein bound state. To date, this has been accomplished by three different methods. Early on, it was recognized that when the Fe protein purified from *C. pasteurianum* was mixed with the MoFe protein purified from *A. vinelandii*, a stable 2:1 complex was formed [70]. Repeated attempts at crystallization of this complex have not been successful. A second method for stabilizing a nitrogenase complex employed MgADP and AlF_4^-. It had been found earlier with other nucleotide binding proteins that ADP-AlF_4^- binds to the nucleotide binding site, possibly mimicking the ADP + Pi transition state. When added to the nitrogenase components, ADP-$AlF_4^-$was found to stabilize a nitrogenase Fe protein-MoFe protein complex [71,72]. The ADP-AlF_4^--stabilized nitrogenase complex with component protein derived from *A. vinelandii* was crystallized and the structure has been determined and refined to 3 Å resolution [8]. A third successful strategy to stabilize a nitrogenase complex involved amino acid deletion of a Leu residue at position 127 in *A. vinelandii* Fe protein [73,74]. This residue is located in the switch II nucleotide-signal-transduction pathway within the Fe protein. Deleting this amino acid results in a protein that mimics the ATP-bound state. Importantly, addition of this amino acid-substituted variant of *A. vinelandii* Fe protein to the *A. vinelandii* MoFe protein results in a non-dissociating complex. The structure of this complex has been determined and refined to 2.2 Å resolution [13]. This last complex has been shown to undergo electron-transfer from the Fe protein to the MoFe protein P-cluster, suggesting that the electron-transfer functions are operational. Additionally, it was possible to obtain this structure with bound MgATP, providing insights into the nucleotide binding interactions.

An analysis of these two complex structures has provided a wealth of information relevant to understanding; (1) the nature of the interactions between the Fe and the MoFe proteins, (2) the conformationals changes that occur in each protein in the complex, (3) the possible electron-transfer paths between the proteins, and (4) the ATP interactions. The conclusions drawn from these two different complexes are generally consistent, but some important differences are observed, which have helped to refine further some of the details. In addition, analysis of the two complex structures reveals many details of component protein interactions. Consistent with earlier studies, there are a number of electrostatic interactions between specific amino acid side chains. Figure 13 shows the electrostatic surfaces for the Fe protein and the MoFe protein that are involved in association. There are complementary regions of positive charge and negative charge on the Fe protein and MoFe protein at the docking interface. The interaction between these two proteins is extensive, covering ~3500 $Å^2$ in each Fe protein – MoFe protein interface (Table 4). In addition to electrostatic interactions, hydrophobic interactions are apparent and a number of hydrogen bonds are implied.

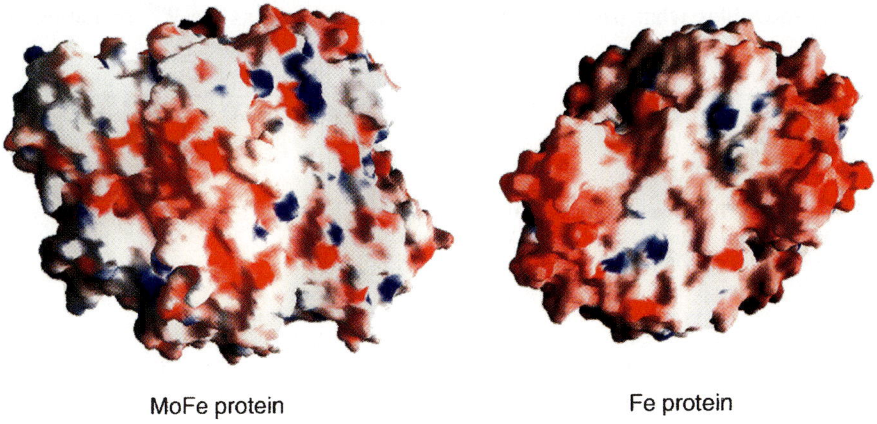

Figure 13. The complementary electrostatic interactions at the nitrogenase Fe protein-MoFe protein docking interface.

4.1. Fe Protein conformational change

Two striking features are immediately evident upon initial inspection of the Fe protein-MoFe protein MgADP-AlF4$^-$ (Av2MgADP-AlF$_4^-$:Av1) and the ΔLeu^{127}-Fe protein variant - MoFe protein (Av2ΔLeu^{127}:Av1) complex structures. First, structural changes observed in the complex when compared to the isolated native proteins are primarily manifested in the Fe protein, with few obvious structural changes in MoFe protein. Second, overall complex structures (Av2MgADP-AlF$_4^-$:Av1 and Av2ΔLeu^{127}:Av1) are very similar, including the larger conformational changes observed of the Fe protein in both complex structures (Figure 14). The structure of the MoFe protein in the complex structures is remarkably similar to that of free MoFe protein, evidenced by low r.m.s. deviations for corresponding Cα atoms of < 0.5 Å (Table 4). The only evident changes occur in the positions of the side chains involved in specific interactions with Fe protein. Most importantly, no obvious structural

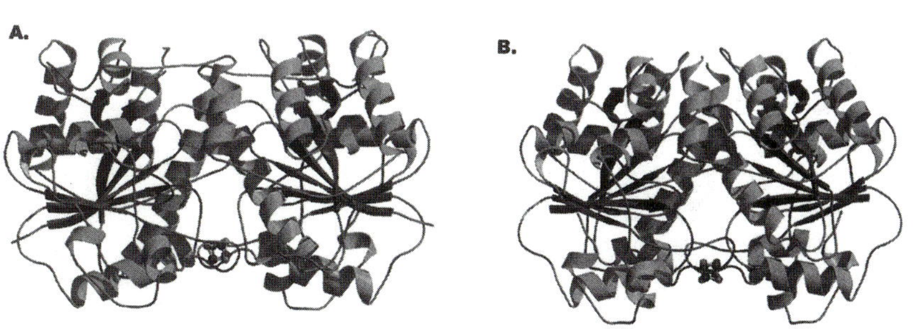

Figure 14. Comparison of (A) the native nitrogenase Fe protein and (B) the nitrogenase Fe protein within the nitrogenase complex.

changes have occurred around the FeMo-cofactor, while the [8Fe-7S] cluster (P-cluster) resembles what has been previously characterized as the P^{OX} oxidation state [5].

Superimposition	**r.m.s deviation (Å)**	
MoFe protein		
Av2ΔLeu^{127}-MgATP:Av1 on Av2MgADP-AlF$_4^-$:Av1	0.36	
Av2ΔLeu^{127}:Av1 on Av2MgADP-AlF$_4^-$:Av1	0.40	
Av2ΔLeu^{127}-MgATP:Av1 on Av2MgADP-AlF$_4^-$:Av1	0.47	
Av2ΔLeu^{127}:Av1 on Av1	0.28	
Av2ΔLeu^{127}-MgATP:Av1 on Av1	0.34	
Av2MgADP-AlF$_4^-$:Av1 on Av1	0.41	
Fe protein	monomers	dimers
Av2ΔLeu^{127}:Av1 on Av2ΔLeu^{127}:Av1	0.13	0.39
Av2ΔLeu^{127}-MgATP:Av1 on Av2ΔLeu^{127}-MgATP:Av1	0.12	0.28
Av2MgADP-AlF$_4^-$:Av1 on Av2MgADP-AlF$_4^-$:Av1	0.01	0.04
Av2 on Av2	0.63	-
Av2ΔLeu^{127}:Av1 on Av2ΔLeu^{127}-MgATP:Av1	0.71	0.90
Av2-ΔLeu127:Av1 on Av2MgADP-AlF$_4^-$:Av1	1.07	1.83
Av2ΔLeu^{127}-MgATP:Av1 on Av2MgADP-AlF$_4^-$:Av1	0.88	1.80
Av2ΔLeu^{127}:Av1 on Av2	1.54	3.53
Av2ΔLeu^{127}-MgATP:Av1 on Av2	1.50	3.20
Av2MgADP-AlF$_4^-$:Av1 on Av2	1.58	4.11
Buried surface area upon complex formation	**surface area/Av2 dimer (Å^2)**	
Av2ΔLeu^{127}:Av1	3909	
Av2ΔLeu^{127}-MgATP:Av1	3984	
Av2MgADP-AlF$_4^-$:Av1	3855	
Comparison of Fe protein surface area		
Av2	19433	
Av2ΔLeu^{127}:Av1	17431	
Av2ΔLeu^{127}-MgATP:Av1	17359	
Av2ΔLeu^{127}-MgATP:Av1 with MgATP removed	17882	
Av2MgADP-AlF$_4^-$:Av1	16299	
Av2MgADP-AlF$_4^-$:Av1with MgADP-AlF$_4^-$ removed	16444	
MgATP X 2	1280	
MgADP-AlF$_4^-$ X 2	1424	
Angular relationship of two-fold axes (°)		
Av2ΔLeu^{127}:Av1/ Av2ΔLeu^{127}-MgATP:Av1 and Av2		11.2
Av2MgADP-AlF$_4^-$:Av1and Av2		12.5
Av2ΔLeu^{127}:Av1/ Av2ΔLeu^{127}-MgATP:Av1 and Av2MgADP-AlF$_4^-$:Av1		4.4

Table 4. Structural features and surface area calculations of the nitrogenase complexes and of the free nitrogenase components.

In contrast, the Fe-protein components of the complexes have undergone major conformational changes, the most notable being the rigid-body re-orientation of the subunits, leading to a more compact and enclosed interface. Close approach of the subunits results in a decrease of the exposed surface area of the Fe protein within the complexes. The pivot points for closure of the dimer interface appears to be near the [4Fe-4S] cluster ligands Cys^{132}, resulting in the repositioning of the cluster toward the surface that is in contact with MoFe protein within ~14 Å of the MoFe protein P-cluster. Biochemical studies involving site-directed mutagenisis of amino acid residues near the Fe sulfur cluster have suggested that the [4Fe-4S]-cluster domain acts as a hinge involved in repositioning the nucleotide binding regions of Fe protein for cooperative MgATP binding [75]. The degree of rigid-body movement can be indicated by a comparison of the r.m.s. deviations of corresponding Cα atoms of the Fe protein dimers of the complex compared to the native Fe protein dimer (Table 4). This should be a valid comparison since the analogous comparison using individual monomers of Fe protein results in very similar r.m.s. deviations ranging from 1.50–1.54 Å (Table 4). This analysis indicates that the Fe protein monomers of the ADP-AlF_4^- -stabilized complex have the largest rigid-body movement towards one another relative to native Fe protein, with an overall r.m.s deviation of corresponding Cα atoms of 4.1 Å.

The similar comparison of the Av2ΔLeu^{127}-Av1 complex with and without MgATP also indicates a significant repositioning of Fe protein monomers relative to each other compared to the native Fe protein, with overall r.m.s. deviations of 3.5 Å and 3.2 Å, respectively. This is also illustrated by comparing the angles relating the two-fold symmetry axes of the Fe protein monomers and superimposing a single monomer of the Fe protein from each complex upon native Fe protein monomer. This indicates that there is an 11.2° angle relating the two-fold axis of the native Fe protein to the Fe protein within the Av2ΔLeu^{127}:Av1 complexes.

As predicted using the aforementioned comparison of Fe protein dimers within the complexes, the angle relating the two-fold axis of the native Fe protein and Fe protein within ADP•AlF_4^- -stabilized complex is slightly larger, 12.5° (Table 4). Comparison of the ADP•AlF_4^--stabilized complex to the Av2ΔLeu ^{127}:Av1 complex in this superimposition shows a 4.4° angle between two-fold axes, which is larger than would be expected from simply comparing the relationships of the Fe protein within the complexes with the native Fe protein. This suggests that the rigid body rotation of the Fe protein occurring within the complexes does not occur along the same plane.

4.2. Component-protein interactions

In addition to repositioning the [4Fe-4S] cluster, the closure of the Fe protein dimer allows the correct positioning of residues for component-protein interaction. The component-protein interface in the complex is significantly more extensive than expected from simple rigid-body docking of the individual native structures and amounts to a ~10% (~3900 Å^2) decrease in the exposed surface area of the individual components. The large contact surface at the interface is afforded by the conformational changes in the Fe protein compared to the native nucleotide-free Fe protein; without the changes, most of the hydrogen bonds and electrostatic interactions in the interface would not be possible (Figure 15). The interface between the two components includes the long symmetrical α-helixes that flank the [4Fe-4S] cluster of the Fe protein and the groove above and to either side of the MoFe protein P-cluster.

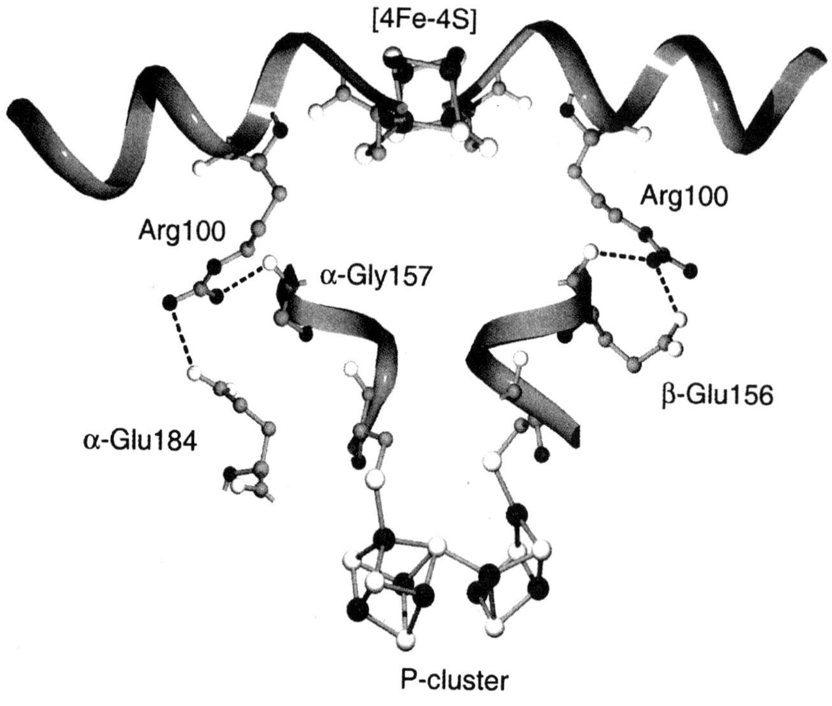

Figure 15. Electrostatic interactions at the docking interface of the nitrogenase Av2Leu^{127}:Av1 complex.

The residues of MoFe protein that form bonds with the Fe protein [4Fe-4S] cluster and its ligands undoubtedly contribute to changes in the properties of the[4Fe-4S] cluster within the complex. The four glycine residues (96 and 133, from each subunit) which are adjacent to the four cysteinyl ligands of the Fe protein in the complex (residues 97 and 132, from each subunit), form hydrogen bonds with the peptide backbone of the MoFe protein (residues α-Ile^{159} and β-Ile^{158}). This undoubtedly contributes to the correct orientation of the cluster to affect electron-transfer.

Two sets of contacts at the interface have been identified that suggest potential paths for through-bond electron-transfer from the Fe protein [4Fe-4S] cluster to the MoFe protein P-cluster. The peptide bond backbone carbonyls of α-Ile^{123} and α-Val^{124} and the corresponding pseudo-symmetrical β-Ala^{123} and β-Val^{124} form hydrogen bonds to the amide nitrogens of the two Fe protein Cys^{97} [4Fe-4S] cluster ligands. In addition, the peptide bond carbonyls of α-Leu^{158} and β-Val^{157} are hydrogen-bonded by an intervening water to the sulfides of the [4Fe-4S] cluster of the Fe protein. This path is completed by the intervening three peptide bonds to the P-cluster ligands, α-Cys^{154} and β-Cys^{153}. In *K. pneumoniae* the residue analogous to β-Phe^{125} (β-Phe^{124}) was shown to be susceptible to digestion by chymotrypsin [75]. Substitution of this residue resulted in altered proteins that exhibited kinetic properties consistent with its involvement in component protein interaction. In the Av2ΔLeu^{127}:Av1 complex β-Phe^{125} and its pseudo-symmetric mate α-Phe^{125} are in close proximity to the Fe protein Met^{58} residues and may be participating in hydrophobic interactions important for the stability of the complex.

The peptide-bond carbonyl oxygen atoms of adjacent residues α-Ile^{159} and β-Ile^{158} also interact with the Fe protein by hydrogen bonding to the peptide bond amide nitrogen atoms of the Gly^{133} residues and to the backbone carbonyl oxygen atoms of the Gly^{96} residues. Mutagenesis studies involving Arg^{140} of the Fe protein have been interpreted as indicating that this residue is important for component protein interaction during nitrogenase turnover [77]. The complex structure reveals that the Arg^{140} residues of the Fe protein monomers interact with the MoFe protein by hydrogen bonding at the peptide backbone carbonyls of α-Gly^{160} and β-Gly^{159}. In addition, the Fe protein Arg^{140} side chain of one monomer participates in an ionic interaction with α-Asp^{162} and α-Glu^{164} resulting in a significantly different structural conformation for residues 139 through 143 in the complexes.

Interestingly, an altered Av1 with Asn substituted for Asp at α^{162} has little effect on catalytic activity [78]. However, this may indicate that this interaction is not critical for productive complex formation or that Asn may be able to substitute productively for Asp in providing a similar interaction. Substitution of the adjacent Asp residue α-Asp^{161} [78] does result in altered catalytic activity, although the side chain is not participating in interaction at the Fe protein/MoFe protein interface. Perhaps substitution at this residue disrupts the interaction at α-Asp^{162}. Since the analogous substitution in the β-subunit (β-Asp^{160}) has little effect, this is consistent with the observed asymmetry of interactions at the MoFe protein and Fe protein interface in the complexes.

The Fe protein Arg^{100} residues in certain photosynthetic nitrogen-organisms have been shown to undergo reversible covalent ADP-ribosylation, which results in inhibition of nitrogenase activity [79]. Biochemical characterizations of altered Fe proteins with amino acid substitutions at Arg^{100} are consistent with its involvement in component protein interaction [80,81]. It is clear that Arg^{100} plays a significant role in the complex structures in stabilization at the interface, in both symmetric and asymmetric hydrogen-bonding and electrostatic interactions with amino-acid side chains of the MoFe protein. Each Arg^{100} has a unique binding pocket composed of residues from both α- and β-subunits of the MoFe protein. The electrostatic interactions involve the side chain of Arg^{100} of one subunit interacting with the side chains of β-Glu^{120} and β-Glu^{156}. The side chain of Arg^{100} of the other Av1 monomer is involved in a similar set of interactions with side chains of α-Glu^{120} and α-Glu^{184}. Further, these interactions are stabilized by hydrogen-bonding interactions of the Arg^{100} side chains to the peptide bond carbonyl of α-Gly^{157} and β-Glu^{156}, respectively. One turn further along the helix are Thr^{104} (one from each subunit) which also participate in hydrogen-bonding interactions with α-Glu^{120} and β-Glu^{120}.

4.3. Nucleotide binding

It is emphasized that in the following discussion five structures of the Fe protein are now available for structural comparison: a) native, nucleotide-free Fe protein; b) Av2MgADP-AlF_4^-:Av1; c) Av2ΔLeu^{127}:Av1; d) Av2ΔLeu^{127}:Av1 with MgATP bound; and e) the native Fe protein with bound MgADP. There are a number of differences at the dimer interfaces of the different Fe protein molecules, apparently resulting from the repositioning of the Fe protein monomers with respect to each other and/or the binding of, MgADP, MgATP or MgADP•AlF_4^- (Table 5). The structures of the native Fe protein dimer and monomer serve as convenient reference points for comparison.

The [4Fe-4S] cluster represents the site of covalent attachment of the Fe protein monomers, thus restricting the type and magnitude of conformational change observed. The loop regions, which support the cysteinyl ligands to the [4Fe-4S] and its adjacent helix (specifically residues 88 through 118) are sites of significant conformational change at the dimer interface. As mentioned previously, the pivot point for the near-rigid-body reorientation of the Fe protein monomers is apparently near the Cys^{132} ligands, which are those distal to the Fe protein/MoFe protein interface. Pivoting about these ligands results in a closure of the Fe protein dimer interface in regions proximal to these ligands and movement apart of regions that are close to Fe protein/MoFe protein interface. In all the observed complexes discussed here, the loop regions containing the Cys^{97} ligands, located close to the Fe protein/MoFe protein interface compensate for the conformational changes of the dimer by movement toward the dimer interface. This is required to accommodate movement of the distal regions of the Fe protein closer together while maintaining the distance required for coordination of the Cys^{97}. The repositioning of these loops and adjacent regions also results in greater accessibility of the Fe protein [4Fe-4S] cluster and is also likely to be important in allowing residues at the docking interface to interact productively in the complex with residues of MoFe protein.

	FeP-ADP	FeP-ADP.ALF$_4^-$	ΔLeu^{127}mon 1	ΔLeu^{127}mon 2
phosphate				
O1A	Thr17 (N, Oγ)	Thr17 (N, Oγ)	Thr17 (N, Oγ)	Thr17(N, Oγ)
O2A		Lys10 (Nζ)		
O3A	Gly14 (N)	Gly14 (N)		
O1B		Lys10 (Nζ)	Lys15 (N)	Lys15 (N)
			Ser16 (N)	Ser16 (N)
O2B	Ser16 (N, Oγ)	Ser16 (N, Oγ)	Gly12 (N)	G12 (N)
	Lys15 (N)	Lys15 (N)		
O3B	Lys15 (Nζ)	Gly12 (N)		
		Gly14 (N)		
		Lys15 (Nζ)		
O1G			Lys15 (Nζ)	Lys15 (Nζ)
O2G			Lys10 (Nζ)	
			Lys 41 (Nζ)	
F1		Gly128 (N)		
		Lys15 (Nζ)		
F2		Lys10 (Nζ)		
		Lys41 (Nζ)		
F3		Gly12 (N)		
F4		Lys10 (Nζ)		
		Ser16 (Oγ)		
sugar				
O2*				Glu221 (Oε1)
O3*	Glu221 (Oε2)	Glu221 (Oε1)	Glu221 (Oε1, Oε2)	
base				
N1	Asp214 (N)	Asp214 (N)		
N6	Pro212 (O)	Pro212 (O)	Pro212 (O)	Gln236 (Oε1)
N7	Asn185 (Nδ1)	Asn185 (Nδ2)	Asn185 (Oδ2)	

Table 5. The interactions between the Fe protein and in the presence of MgADP and in the nitrogenase complexes. Slight differences were observed in the Fe protein monomers of the Av2ΔLeu^{127}:Av1 complex. Monomer 1 indicates the Fe protein monomer that interacts with the MoFe protein α-subunit and monomer 2 that interacts with the β-subunit. Residues are indicated by the single-letter code for amino acids.

There are significant differences in the conformation of the loop region containing the other pair of cysteinyls (Cys^{132}). The region corresponds to a canonical Walker B nucleotide-binding motif (Asp^{125}-X-X-Gly^{128}) found in a number of proteins, including Ras and other members of the G-protein family, which harness the energy from nucleotide-binding and hydrolysis to drive conformational changes. Additionally, this region represents a direct linkage of the nucleotide-binding site to potential pathways of intermolecular electron-transfer *via* the Fe protein [4Fe-4S] cluster. In the structure of the nitrogenase complex stabilized by MgADP•AlF_4^-, the peptide bond carbonyl of Gly^{128} interacts directly with AlF_4^- and Asp^{129} is poised for potential interaction with AlF_4^- through a water molecule. Since this region seems to be at or near the point at which the monomers pivot with respect to each other, it is likely that these interactions are critical in driving the conformational changes that result in the repositioning of Fe protein monomers with respect to each other.

Interestingly, in the Av2ΔLeu^{127}:Av1 complex, in which a similar repositioning of the Fe protein monomers occurs, the aforementioned interactions with the terminal phosphate are not possible due to a shortening of this loop region. The inability of Fe protein in the Av2ΔLeu^{127}:Av1 to form these interactions with bound nucleotide probably contributes to the markedly reduced rate of MgATP hydrolysis within the complex [82]. In Fe protein in the Av2ΔLeu^{127}:Av1 complex the deletion of Leu^{127} results in an entirely different conformation of residues 128 and 129 compared to that of the complex stabilized by MgADP•AlF_4^-.However, the conformation of the residues that flank them, (Asp^{125} and Phe^{135}) and the orientation of the side chains of Val^{130} and Val^{131} residues, are nearly the same. Reversal of the orientation of Val^{130} and Val^{131} residues relative to native Fe protein results in the side chains of the Val^{130} residues pointing towards the Fe protein [4Fe-4S] cluster. This may contribute to the reduction in the redox potential of this center that is observed in the complex [55].

In general, other regions of that are involved with nucleotide binding in the complex structures are quite different in the nucleotide-bound structures compared to the nucleotide-free state. This is so even in the case of the nucleotide-free Av2Leu^{127}Δ:Av1 complex, in which the altered Fe protein Leu^{127} has a number of biochemical properties that resemble those of nucleotide-bound Fe proten in the absence of nucleotides [73]. Upon addition of MgATP to crystals of the Av2ΔLeu^{127}:Av1 complex, several regions that are interacting directly with MgATP adopt a conformation very similar to that observed in the Fe protein bound to MgADP•AlF_4^- and complexed with the MoFe protein. The P-loop, phosphate binding loop, or Walker A nucleotide-binding motif (residues 9 through 15) exists in different conformations in the five Fe protein structures. On comparing the native Fe protein with that of the Fe protein within the Av2ΔLeu^{127}:Av1 complex, it is apparent that the differences appear to result from an obligatory conformational change in the P-loop of the native Fe protein to allow close contact between the Fe protein monomers. Several residues of the P-loop of comparing the Fe protein within the Av2ΔLeu^{127}:Av1 complex are not well-ordered, suggesting flexibility and the possibility of multiple conformations (Figure 16). The P-loop regions of MgATP-bound Av2ΔLeu^{127}:Av1 complex and of the Fe protein–MoFe protein complex stabilized with MgADP•AlF_4^- are in nearly the same conformation. Most significantly, Lys^{10} interacts with the terminal phosphate of MgATP or with AlF_4^- across the subunit interface, so that Lys^{10} of one Fe protein subunit is interacting with the terminal phosphate of the other Fe protein subunit.

Another region where structural differences are evident is the region around residue Asp^{39} that is proposed to stabilize the attacking nucleophile for MgATP hydrolysis and that extends to Val^{75}. In general, it appears that the changes observed in the in this region of the Av2ΔLeu^{127}:Av1 complex relative to native Fe protein are linked to the changes observed in residues 88 through 118 and in the Fe protein–MoFe protein complex stabilized with MgADP•AlF_4^- because this region is similar in all the complex structures. However, there are two areas within this region of the complex structures that differ. The area of Asp[39] through Asp[43] in the Av2ΔLeu[127]:Av1 complex is very different from that observed in the Fe protein – MoFe protein complex stabilized with MgADP•AlF_4^-. However, upon MgATP addition these residues adopt a conformation resembling that observed in the Fe protein–MoFe protein complex stabilized with MgADP•AlF_4^-.

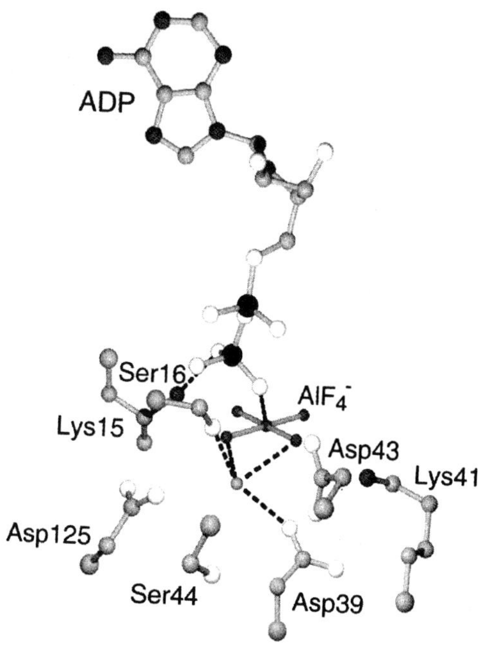

Figure 16. The environment of MgADP•AlF_4^- in the structure of the nitrogenase Fe protein in the Av2MgADP-AlF_4^-:Av1 complex.

Changes are also observed in the region surrounding Asp[154] of the Av2ΔLeu[127]:Av1 complexes relative to native Fe protein. These changes in Av2ΔLeu[127]:Av1, nucleotide-free or nucleotide-bound, approximate to what is observed in the Fe protein-MoFe protein complex stabilized with MgADP•AlF_4^-. However, the Asp[154]side-chain does not form an inter-subunit salt bridge with Arg[213], as observed for the Fe protein-MoFe protein complex stabilized with MgADP•AlF_4^-. This is likely to be a result of the differences observed in the degree of rigid-body repositioning of the Fe protein monomers with respect to one another. These changes occur to a lesser degree in the Fe protein of the Av2ΔLeu[127]:Av1 complexes relative to the Fe protein in the Fe protein–MoFe protein complex stabilized with MgADP•AlF_4^-. These differences are amplified since Asp[154] occurs in a region that is distal from the Fe protein-MoFe protein interface. The salt bridge between Asp[154] and Arg[213] in the Fe protein-MoFe protein complex stabilized with MgADP•AlF_4^- was suggested to be involved in stabilizing the interaction between Arg^{213} and the bound adenosine ring. The observation that this salt bridge is absent in the Av2ΔLeu^{127}:Av1 complexes as well as the absence of the interaction of Arg^{213} with the adenosine may contribute to its reduced rates of MgATP hydrolysis.

A closer look at the interactions between the Fe protein and the Mg^{2+} and phosphate portions of the bound MgADP reveals likely mechanisms for communication of the ATP hydrolytic event to changes in switches I and II. It was proposed earlier that the change of the Mg^{2+} from interaction with the β- and γ-phosphates to interaction with the α- and β-phosphates is a driving-step for nucleotide hydrolysis. Analysis of the structures of the Fe protein bound with MgADP and MgADP•AlF_4^- supports the hypothesis that protein interactions with the Mg^{2+}.are essential to the transduction of the nucleotide hydrolysis event. Changes in the coordination of the Mg^{2+} ion transduce the different signals to the docking interface through direct interactions with residues of switch I. Similarly, the interaction of the phosphates of the MgADP molecules results in a change in the conformation of the P-loop, which, in turn, is communicated to the bridging [4Fe-4S] cluster through the adjacent Asp^{129} side chain through a direct hydrogen-bonding interaction to the main chain of the P-loop.

4.4. Implications for ATP hydrolysis and electron transfer

The structure of the nucleotide-free Av2ΔLeu^{127}:Av1 complex shares some features with that of the nitrogenase complex stabilized by MgADP•AlF_4^-, namely the interactions at the docking interface and the gross conformational change observed in the Fe-protein. On the other hand, many of the interactions across the Fe protein dimer interface are different, especially those related to nucleotide binding and hydrolysis. Many of the interactions at the dimer interface adopt conformations which allow nucleotide binding in the Av2ΔLeu^{127}:Av1 complex but do not change the integrity or the overall structure of the complex. This raises the questions: why is this complex stable?; and how is nucleotide binding/hydrolysis utilized in the native, catalytically-active enzyme?

The hydrolysis of MgATP by the native Fe protein occurs only in the presence of the MoFe protein. The MoFe protein participates in the process of nucleotide hydrolysis by stabilizing a transition state that cannot be attained by the Fe protein alone. The Fe protein Leu^{127} is unable to achieve this state even in the presence of the MoFe protein and thus it is locked together with the MoFe protein in much the same manner as the nitrogenase complex stabilized by MgADP•AlF_4^-. Although the deletion of Leu^{127} of the Fe protein results in a conformational state of the Fe protein that will participate in productive, stable complex formation in the absence of any bound nucleotide, it occurs at the expense of the ability to bring about the conformational state of Fe protein that results from the hydrolysis of MgATP, that is, the state of the Fe protein that dissociates and participates in another cycle of complex formation and electron-transfer. This is reminiscent of the interaction of actin and myosin, in which the movement of myosin relative to actin has been suggested to occur by a reversal of the nucleotide-triphosphate-induced conformational changes [83].

Electron transfer has been shown to occur within the Av2ΔLeu^{127}:Av1 complex, albeit at a markedly reduced rate compared to the native enzyme [74]. This suggests that a pathway exists for electron transfer, but it is unclear whether this pathway is biochemically relevant to the functioning of the native enzyme. The pathway is essentially analogous to that conceived for the Fe protein–MoFe protein complex stabilized with MgADP•AlF_4^-. It is conceivable that even very slight changes in the structural conformations of these pathways could result in dramatic changes in rates of electron-transfer, for example, effecting through-space jumps. Another factor is the driving force generated by differences in the environment of the [4Fe-4S], which can be affected dramatically, for example, by the proximity of hydrophobic residues or of dipoles to the cluster. Upon complex formation, the redox potential of the [4Fe-4S] cluster shifts dramatically, from ~-400mV to ~-600mV [55]. Obviously factors such as the decrease in solvent accessibility and the location of hydrophobic residues (Val^{130}) contribute to the increased driving force and reduction potential, but there are many details that are not entirely understood.

5. FUTURE DIRECTIONS

Clearly there are still areas remaining in which structure-determination by X-ray diffraction methods can advance our understanding of nitrogenase function. The substrate-reduction site is well characterized, but rationalizing the details of dinitrogen binding and reduction at this site from this structural information is not obvious. The structural characterization of a substrate- or inhibitor-bound form of the MoFe protein could offer a great deal in terms of our understanding of the process of dinitrogen reduction at this site. Unfortunately, capturing the MoFe protein in a substrate- or inhibitor-bound form for structure determination has not yet been possible. This is, in part, due to the inability to attain the reduced states required for substrate and inhibitor binding. The ability to reach these states requires reduction of the MoFe protein by the Fe protein and this is not possible in crystals of the nitrogenase MoFe protein. Recently, results have been presented that indicate that certain MoFe proteins with defined amino acid substitutions in the polypeptide environment of the FeMoco can bind substrates and inhibitors under non-turnover conditions. These site-directed variants of the MoFe protein may provide the means to generate a crystal structure of an inhibitor- or substrate-bound form of the MoFe protein, which would be of profound importance and interest.

In the area of understanding the role of MgATP binding and hydrolysis in catalysis, the structural characterization of nucleotide-bound forms of the Fe protein and nitrogenase complexes has offered a great many insights. In this process, conformational changes in the Fe protein brought about by the binding of MgATP result in a high affinity of the Fe protein for the MoFe protein, promoting nitrogenase complex formation. It is generally accepted that, through complex formation, additional conformational changes are imposed upon the Fe protein by the MoFe protein, leading to MgATP hydrolysis. It is likely that the complexes represent some intermediate stage of the process and that at least some MoFe protein-dependent Fe protein conformational changes are being observed. A more definitive assessment of the conformational changes observed in the complexes requires a structural description of the MgATP-bound form of the Fe protein. With this structure in hand, conformational changes brought about by the interaction of MgATP can be differentiated from those resulting from complex formation. It has not been possible to capture a MgATP bound state of the Fe protein. This is partly due to the instability of MgATP in aqueous solution, in which it is likely to be hydrolyzed during crystallization. The expected structural changes of the Fe protein by MgATP-binding probably precludes addition of MgATP to crystals of the native Fe protein, since such changes would be likely to disrupt the crystal lattice. A resolution may lie in the use of nucleotide analogs. However, the interaction of these analogs with the Fe protein has been.shown only to approximate the conformational changes that result from MgATP binding. Another possibility could be to characterize structurally the Leu^{127} variant of the Fe protein, which has been shown to possess biochemical and biophysical properties of an Fe protein with MgATP bound but in the absence of nucleotides. The structure of the Leu^{127} Fe protein–MoFe protein complex will provide the basis for direct comparisons.

6. ACKNOWLEDGEMENTS

We thank Lance C. Seefeldt of the Department of Chemistry and Biochemistry at Utah State University for comments. The nitrogenase research in the group is supported by the United States Department of Agriculture Grant (2001-35318-10122). to J.W.P. John W. Peters is the recipient of a Camille Dreyfus Teacher/Scholar Award.

7. REFERENCES

1. M. M. Georgiadis, H. Komiya, P. Chakrabarti, D. Woo, J. J. Kornuc, and D. C. Rees, Science 257 (1992) 1653.
2. J. Kim, and D. C. Rees, Science 257 (1992) 1677.
3. J. Kim, and D. C. Rees, Nature 360 (1992) 553.
4. J. Kim, D. Woo and D. C. Rees, Biochemistry 32 (1993) 7104.
5. J. W. Peters M. H. B. Stowell, S. M. Soltis, M. G. Finnegan, M. K. Johnson, and D. C. Rees, Biochemistry 36 (1997) 1181.
6. S. M. Mayer, D. M. Lawson, C. A. Gormal, S. M. Roe and B. E. Smith J. Mol. Biol. 292 (1999) 871.
7. M. K. Chan, J. Kim, and D. C. Rees, Science 260 (1993) 792.

8. H. Schindelin, C. Kisker, J. L. Schlessman, J. B. Howard and D. C. Rees Nature 387 (1997) 370.
9. J. L. Schlessman, D. Woo, L. Joshua-Tor, J. B. Howard and D. C. Rees, J. Mol. Biol. 280 (1998) 669.
10. S. B. Jang, L.C. Seefeldt and J. W. Peters, Biochemistry 39 (2000) 641.
11. S. B.Jang, L. C. Seefeldt, and J. W. Peters, Biochemistry 39 (2000) 14745.
12. P. Strop, P. M. Takahara, H. J. Chiu, H. C. Angove, B. K. Burgess and D. C. Rees, Biochemistry 40 (2001) 651.
13. H. Chiu, J. W. Peters, W. N. Lanzilotta, M. J. Ryle, L. C. Seefeldt, J. B. Howard and D. C. Rees, Biochemistry 40 (2001) 641.
14. M. Sørlie, J. Christiansen, B. J. Lemon, J. W. Peters, D. R. Dean and B. J. Hales, Biochemistry 40 (2001) 1540.
15. J. M. Smith, R. J. Lachicotte, K. A. Pittard, T. R. Cundari, G. Lukat-Rodgers, K. R. Rodgers, P. L. Holland, J. Am. Chem. Soc. 123 (2001) 9222.
16. H. Deng, and R. Hoffmann, Angew. Chem., Int. Ed. Engl. 32 (1993) 1062.
17. I. G. Dance, Aust. J. Chem. 47 (1994) 979.
18. I. Dance, J. Biol. Inorg. Chem. 1, (1996) 581.
19. D. Sellmann, and J. Sutter, J. Biol. Inorg. Chem. 1 (1996) 587.
20. J. Christiansen, R. C. Tittsworth, B. J. Hales, and S. P. Cramer, J. Am. Chem. Soc. 117 (1995) 10017.
21. B. K. Burgess, and D. J. Lowe, Chem. Rev. 96 (1996) 2983.
22. R. R. Eady, Chem. Rev. 96 (1996) 3013.
23. J. B. Howard, and D. C. Rees, Chem. Rev. 96 (1996) 2965.
24. D. J. Scott, D. R. Dean and W. E. Newton, J. Biol. Chem. 267, (1992) 20002.
25. D. J. Scott, H. D. May, W. E. Newton, K. E. Brigle, and D. R. Dean, Nature 343 (1990) 188.
26. C. H. Kim, W. E. Newton and D. R. Dean, Biochemistry 34 (1995) 2798.
27. J. Shen, D. R. Dean and W. E. Newton, Biochemistry 36 (1997) 4884.
28. P. M. C. Benton, S. M.Mayer, J. Shao, B. M. Hoffman, D. R. Dean and L. C. Seefeldt, Biochemistry 40 (2001) 13816.
29. S. M. Mayer, W. G. Niehaus and D. R. Dean, J. Chem, Soc., Dalton Trans. (2002) 802.
30. K. Fisher, M. J. Dilworth, and W. E. Newton, Biochemistry 39, (2000) 15570.
31. M. C. Durrant, Biochem. J. 355 (2001) 569.
32. A. C. Robinson, D. R. Dean and B. K. Burgess, J. Biol. Chem. 262 (1987) 14327.
33. V. K. Shah, Methods Enzymol. 118 (1986) 511.
34. N. Gaviniand and B. K. Burgess, J. Biol. Chem. 267 (1992) 21179.
35. W. N. Lanzilotta, M. J. Ryle, and L. C. Seefeldt, Biochemistry 34 (1995) 10713.
36. L. C. Seefeldt, T. V. Morgan, D. R. Dean, and L. E. Mortenson, J. Biol. Chem. 267 (1992) 6680.
37. V. K. Shah and W. J. Brill, , Biochim. Biophys. Acta 292 (1973) 246.

38. D. Wolle, D. R. Dean and J. B. Howard Science 258, (1992) 992.
39. E. F. Pai, U. Krengel, G. A. Petsko, R. S. Goody, W. Kabsch and A. Wittinghofer, EMBO J. 9 (1990) 2351.
40. L. Tong, A. M. DeVos, M. V. Milburn and S. H Kim, J. Mol. Biol. 217 (1991) 503.
41. F. Jurnak, Science 230, (1985) 32.
42. M. Kjeldgaard and J. Nyborg, J. Mol. Biol. 223 (1992) 721.
43. J. Sondek, D. G. Lambright, J. P. Noel, H. E. Hamm and P. B. Sigler, Nature 372 (1994) 276.
44. R. M. Story, I. T. Weber, and T. A. Steitz, Nature 355 (1992) 318.
45. R. M. Story and T. A. Steitz, Nature 355 (1992) 374.
46. E. V. Koonin, J. Mol. Biol. 229 (1993)1165.
47. G. E. Shulz, *Curr. Opin. Struct. Biol. 2* (1992) 61.
48. C. L. Drennan, J. Heo, M. D. Sintchak, E. Schreiter and P. W. Ludden, Proc. Natl. Acad. Sci. USA, 98 (2001) 11973.
49. Dobbek, H., Svetlitchnyi, V., Gremer, L., Huber, R., and Meyer, O. (2001) *Science 293*, 1264-1265.
50. J. B. Howard and D. C. Rees, Ann. Rev. Biochem. 63 (1994) 235.
51. J. Kim and D. C. Rees, Biochemistry 33 (1994) 389.
52. J. H. Spee, A. F. Arendsen, H. Wassink, S. J. Marritt, W. R. Hagen and H. Haaker, FEBS Lett. 432 (1998) 55.
53. W. G. Zumft, and L. E. Mortenson, Biochim. Biophys. Acta 292 (1973) 413.
54. G. D. Watt, R. B. Frankel, G. C. Papaefthymiou, K. Spartalian and E. I.Stiefel, Biochemistry 25 (1986) 4330.
55. W. N. Lanzilotta and L. C. Seefeldt, Biochemistry 36 (1997) 12976.
56. P. J. Stephens, C. E. McKenna, B. E. Smith, H. T. Nguyen, M. C. McKenna, A. J. Thomson, F. Devlin and J. B. Jones, Proc. Natl. Acad. Sci. U.S.A. 76 (1979) 2585.
57. P. A. Lindahl, N. J. Gorelick, E. Münck, and W. H. Orme-Johnson, J. Biol. Chem. 262 (1987) 14945.
58. J. Meyer, J. Gaillard and J. M. Moulis, Biochemistry 27 (1988) 6150.
59. W. N. Lanzilotta, R. C. Holz and L. C. Seefeldt, Biochemistry 34 (1995) 15646.
60. M. J. Ryle W. N. Lanzilotta, L. C. Seefeldt, R. C. Scarrow, and G. M. Jensen, J. Biol. Chem. 271 (1996) 1551.
61. R. Langen, G. M Jensen, U. Jacob, P. J. Stephens, and A. Warshel, J. Biol. Chem. 267 (1992) 25625.
62. P. J. Stephens, D. R. Jollie and A. Warshel, Chem. Rev. 96 (1996) 2491.
63. G. D. Watt and K. R. N. Reddy, J. Inorg. Biochem. 53 (1994) 281.
64. H. C. Angove, S. J. Uyoo, B. K. Burgess, and E. Munck, J. Am. Chem. Soc. 119 (1997) 8730.
65. H. C. Angove, S. J. Yoo, E. Munck, and B. K. Burgess, J. Biol. Chem. 273 (1998) 26330.

66. K. B. Musgrave, H. C. Angove, B. K. Burgess, B. Hedman, and K. O. Hodgson, J. Am. Chem. Soc. 120 (1998) 5325.
67. S. J. Yoo, H. C. Angove, B. K. Burgess, E. Munck, and J. Peterson, J. Am. Chem. Soc.*120*, (1998) 9704.
68. S. J. Yoo, H. C. Angove, V. Papaefthymiou, B. K. Burgess, and E. Munck, J. Am. Chem. Soc. 122 (2000) 4926.
69. J. A. Erickson, A. C. Nyborg, J. L. Johnson, S. M. Truscott, A. Gunn, F. R. Nordmeyer, and G. D. Watt, Biochemistry 38 (1999) 14279.
70. D. W. Emerich and , R. H. Burris, Proc. Natl. Acad. Sci. U.S.A. 73 (1976) 4369.
71. M. G.Duyvis, H. Wassink, and H. Haaker, FEBS Lett. 380 (1996) 233.
72. K. A. Renner, and J. B. Howard, Biochemistry 35 (1996) 5353.
73. M. J. Ryle and L. C. Seefeldt, Biochemistry 35 (1996) 4766.
74. W. N., Lanzilotta, K. Fisher, and L. C. Seefeldt, Biochemistry 35 (1996) 7188.
75. M. J.Ryle, W. N. Lanzilotta and L. C. Seefeldt, Biochemistry 35 (1996) 9424.
76. K. Fisher, R. N. F. Thorneley, D. J. Lowe and R. N. Pau in New Horizons in Nitrogen Fixation, R., Palacios, J. Mora and W. E. Newton (eds.), Kluwer Academic Publishers, Norwell, Mass., 1992, pp 135-135.
77. L. C. Seefeldt, Protein Sci. 3 (1994) 2073.
78. C.-H. Kim, L. Zheng, W. E. Newton, and D. R. Dean in New horizons in nitrogen fixation, R. Palacios, J. Mora, and W. E. Newton (eds.), Kluwer Academic Publishers, Norwell, Mass., 1992, pp 105-110.
79. P. W. Ludden and G. P. Roberts, Curr. Top. Cell Regul. 30 (1989) 23.
80. R. G. Lowery, C. L. Chang, L. C. Davis, M. C.McKenna, P. J. Stephens, and P. W. Ludden, Biochemistry 28 (1989) 1206.
81. D. Wolle, C. Kim, D. Dean and J. B. Howard, J. Biol. Chem. 267 (1992) 3667.
82. W. N.Lanzilotta, V. D. Parker and L. C. Seefeldt, Biochim. Biophys. Acta 1429 (1999) 411.
83. A. J. Fisher, C. A. Smith, J. B. Thoden, R. Smith, K. Sutoh, H. M. Holden and I. Rayment, Biochemistry 34 (1995) 8960.

Nitrogen Fixation at the Millennium.
G. Jeffery Leigh, Editor.

CHAPTER 3

Spectroscopy of Nitrogenase

Brian J. Hales
Department of Chemistry, Louisiana State University,
Baton Rouge. LA, 70803-1804, USA

1. INTRODUCTION

The study of metals in biology is a very technique-oriented field. The presence of transition metals introduces the available use of many spectroscopic techniques, which would ordinarily be useless in the study of organic molecules. The low energy of d-d and metal-ligand charge-transfer transitions allows a study in the visible and infrared, regions normally void of spectra in basic hydrocarbon systems. The presence of metal atoms opens the way for the investigation with techniques such as X-ray Absorption Spectroscopy (XAS) and Mössbauer Spectroscopy. Finally, unpaired electrons in certain oxidation states of metals can be probed by a wealth of spectroscopic techniques such as Electron Paramagnetic Resonance (EPR), Electron Nuclear Double Resonance (ENDOR), Electron Spin Echo Envelope Modulation (ESEEM) and Magnetic Circular Dichroism (MCD), all specific to the study of paramagnetic systems. For all of these reasons, spectroscopy has been invaluable in the study of metalloenzymes, and the rich used of it is best illustrated in the history of the study of nitrogenase.

Many excellent reviews have appeared on the background of all the different spectroscopic techniques discussed here. The reader unfamiliar with one or more is encouraged to reads these, which will be referenced in the text. A brief overview of the strengths of a few, however, will be presented. In certain situations, a more detailed discussion will appear in the text when it is deemed necessary to explain a result.

As mentioned, paramagnetism can be investigated by a variety of techniques. Paramagnetism arises from a net electronic magnetic moment, almost always associated with unpaired electrons (i.e., $S > 0$). By far the simplest and most sensitive technique for the study of paramagnetic samples is EPR spectroscopy, which measures the energy of transition between different spin sublevels (*i.e.*, m_s states) split in an external magnetic field [1]. The sublevels of half-integer states (e.g., $S = 1/2, 3/2, 5/2$) are grouped into a series of $\pm\ m_s$ Kramers doublets, which are degenerate in zero magnetic field. As a general rule, half-integer states are almost always detectable by EPR spectroscopy, although not all sub-level transitions may be observed when $S > 1/2$. The predicted g-factors or magnetic field positions for the transitions of high-spin states are predicted by "Rhombograms", as outlined by Hagen [2].

Unlike half-integer states, the energy levels of ± m_s states for integer spin systems are split even in zero magnetic field. The magnitude of this "zero-field" splitting for metalloenzymes is often much larger than the energy of an X-band (9.5 GHz) EPR spectrometer. Because of this, the spectrometer does not have enough energy to induce transitions between these levels and the sample, although paramagnetic, is said to be "EPR silent". Exceptions to this occur in some sublevels of high spin systems ($S \geq 2$) where the separation between upper ± m_s states is smaller than 9.5 GHz and transitions can be detected, especially in parallel mode EPR cavities.

The second technique commonly used on paramagnetic samples is ENDOR spectroscopy [3]. One can think of ENDOR as performing a Nuclear Magnetic Resonance (NMR) measurement in an EPR spectrometer. While positioned at the magnetic field of an EPR absorption, radio frequency is varied about the sample, as in an NMR spectrometer, to induce nuclear transitions. Because electrons are magnetically coupled to nuclei bound to the paramagnet, the nuclear transitions will induce a change in the intensity of the EPR signal. Obviously, ENDOR is very helpful in determining both the type and number of magnetic nuclei bound to the paramagnet.

To detect magnetic nuclei neighboring but not directly bound to the paramagnet, ESEEM spectroscopy is used [3]. In this technique, the relaxation profile of the paramagnet is recorded. The magnetic moment of a neighboring nucleus will precess about the spectrometer's external magnetic field and induce a modulation of the relaxation profile. Since the precession frequency (and resultant modulation frequency) is a characteristic parameter of each nucleus, the identity of the coupled neighboring element can be determined. Additional information on the environment of neighboring nuclei can be obtained from the quadrupole splitting term for nuclei with spin greater than 1/2.

The last magnetic technique, MCD spectroscopy, detects the circular dichroism induced by an external magnetic field [4]. Although it can be used to study diamagnetic samples, it is most informative when paramagnetism is present. There are two reasons for this. First, the MCD component of a paramagnetic ground states will dominate at very low temperatures (~ 1.5 K) and high field (7 Tesla) and can be easily detected over background CD or diamagnetic MCD signals. Second, since the paramagnetism is associated solely with the metal center, MCD only probes the electronic nature of that site without being influenced by the spectra of the protein backbone or medium. Finally, unlike EPR, ENDOR and ESEEM, MCD spectroscopy also detects integer spin systems that are EPR silent. This technique has been very beneficial in predicting both cluster type and spin state, especially among FeS centers.

Mössbauer spectroscopy does not require paramagnetism but the interpretation of the spectrum is aided when it is present [5]. This technique measures the isomer shift (δ) and quadrupole splitting (ΔE_Q) of a nucleus relative to a standard. Information about the sample is gained because both of these terms are related to the electronic environment of the nucleus. While there are several Mössbauer active nuclei, ^{57}Fe yields the most intense spectrum and is the only one biologically relevant.

Finally, of all the techniques commonly used in the study of metalloenzymes, XAS is the only one that yields direct structural information [6]. To use XAS, high-energy photons (usually generated by a synchrotron source) are use to excite and eventually ionize K- or L-shell electrons. In theory any atom can be irradiated, but in the study of metalloenzymes, metal or sulfur atoms are the most common targets. An XAS spectrum is divided into two regions, XANES and EXAFS, acronyms for X-ray Absorption Near Edge Structure and Extended X-ray Absorption Fine Structure, respectively. The XANES spectrum is most useful in obtaining a metal's oxidation state and ligand field environment, while the EXAFS region is used to identify ligands and metal-ligand distances.

2. RESTING STATE STUDIES

2.1 Fe-Protein

By far, the majority of spectroscopic studies on nitrogenase have been performed on the component proteins in their as-isolated or resting state, usually in the presence of dithionite as a reductant [7]. When we are discussing the MoFe-protein and Fe-protein of Mo-nitrogenase, the accepted shorthand notation of Sg1 and Sg2 will be used where "S" and "g" refers to the bacterial source species and genus, respectively. For example, the MoFe-protein (also called component 1) and Fe-protein (component 2) from *Azotobacter vinelandii* will be noted as Av1 and Av2, respectively.

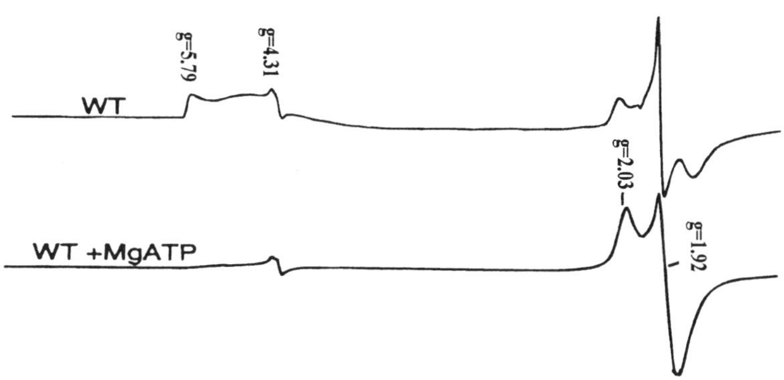

Figure 1. EPR spectra of Av2 without (top) and with (bottom) MgATP. Inflections at the right (g = 2) arise from the S = 1/2 signals while those at the left (g = 5.79) is associated with the S = 3/2 signal. Spectra were taken from ref. 16.

The analogous proteins from *Klebsiella pneumoniae* would be Kp1 and Kp2, and so on. Unless otherwise specified, our discussion will focus on the component proteins of Mo-nitrogenase.

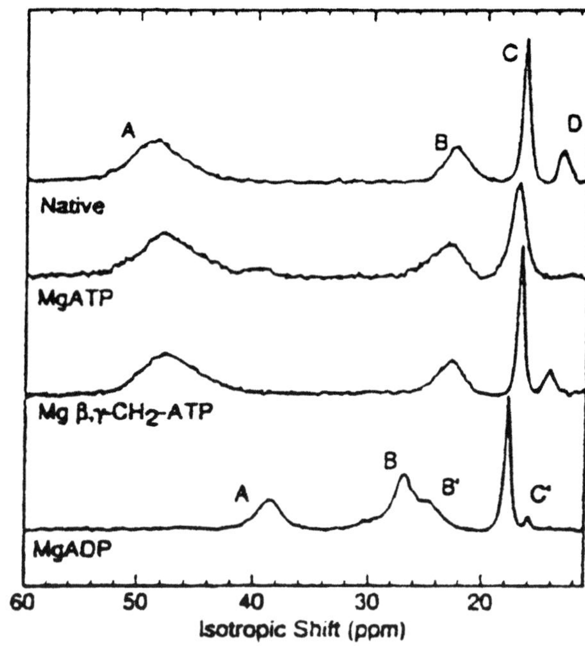

Figure 2. NMR spectra of Av2 as the native form and in the presence of MgATP, Mg β,γ-CH_2-ATP and MgADP. Figure was taken from ref. 16.

When discussing the proteins of alternative enzymes, a superscript will be used to designate the heterometal. For example, $Av1^V$ and $Av1^{Fe}$ refer to the component 1 proteins of V-nitrogenase and Fe-nitrogenase, respectively, from *A. vinelandii*. For simplicity, Av1 instead of $Av1^{Mo}$ will be used for the MoFe-protein from that species.

As with most metalloenzymes, the initial spectroscopic studies performed on the Fe-protein used EPR [8-15]. This is often the technique of choice because it is sensitive yet easy to use. The EPR spectrum of the Fe-protein contains a dominant rhombic S = 1/2, g 1.94 signal (Fig. 1), which obvious arises from the protein's FeS center. This signal has several interesting properties. The shape of the signal changes from rhombic to axial upon MgATP binding. It is important to remember that most spectroscopic techniques measure only the energy of electronic or magnetic transitions (an exception is an EXAFS where the spectrum is governed by metal-ligand distances).

Therefore, the change in shape of the Fe-protein spectrum in the presence of MgATP implies a change in the electronic environment of the cluster. However, because the EPR spectrum arises from a complex metalloenzyme, the magnitude or source of the change in the electronic environment (or corresponding structural change) cannot be identified. In other words, the shift in the spectrum from rhombic to axial means that something has changed in the electronic environment of the FeS cluster, but it doesn't tell us what has changed. That is the advantage and disadvantage of using spectroscopy to investigate complex systems. Spectroscopy is very sensitive at showing that something has changed while often poor at identifying the source of the change. This situation should improve in the future as more, well-defined model metalloenzyme systems are investigated by spectroscopy.

^{1}H-NMR spectroscopy has detected an MgATP-induced shift in the 60-100 ppm region (Figure 2) for the reduced paramagnetic protein [16]. In Figure 2 signals A (49 ppm) and B (23 ppm) arise from four β-CH_2 protons of cysteinyl ligands while signals C (17 ppm) and D (14 ppm) arise from two cysteinyl α-CH protons. The addition of MgATP to Av2 induces a small shift of each of the four resonances, with signal D shifting such that it is no longer observable. Quantitation of these data suggests that signal D has shifted downfield and is concealed beneath signal C.

The CD spectrum of the Fe-protein also shows significant shifts upon MgATP binding [11,17-21]. On the other hand, Fe-XAS detects no measurable difference in Fe-S or Fe-Fe distances in the Fe-protein with or without MgATP [22], suggesting that the observed EPR, NMR and CD changes are associated with the protein environment and not the cluster. This interpretation is consistent with the results of small-angle X-ray scattering studies [23]. Thus the large EPR spectral change observed upon MgATP binding does not appear to be related to major changes in the orientation of the cluster ligands.

The second interesting property of the $S = 1/2$ signal of the Fe-protein is its spin concentration. Whenever an EPR spectrum of a new metalloenzyme is recorded, the spin concentration is also measured. The spin concentration allows one to determine if the signal arises from an impurity (low concentration) or from one or more paramagnetic sites in the enzyme (1 or more spins per protein). When the concentration of the $g = 1.94$ signal was initially determined it was found to be much less than 1 per protein, typically in the range 0.2 to 0.3 spins [9,12]. A detailed investigation [9,10,14,15] of the entire EPR spectrum of the Fe-protein revealed the presence of a set of small inflections in the $g = 5$ region corresponding to an $S = 3/2$ state (Figure 1). Due to the use of magnetic field modulation to enhance the signal-to-noise ratio of a spectrum, EPR spectrometers record the first derivative rather than the straight absorption spectrum. As such, the spin concentration is related to the signal amplitude times the square of the spectral width. Therefore, broad signals, even at high concentration, will have small amplitudes and can be easily unobserved. Spin quantitation of both the $g = 1.94$ and $g = 5$ signals yielded a total of 1 spin per protein molecule, demonstrating that the Fe-protein exhibits a mixed $S = 1/2, 3/2$ spin state. The relative amounts of these states appear to be solvent dependent. In 50% ethylene glycol the cluster exists mainly in the $S = 1/2$ state, but this shifts to the $S = 3/2$ state when 0.4 M urea is present.

Fe-XAS was unable to detect any spectral differences when the cluster was in the $S = 1/2$ or $3/2$ states, suggesting no major structural changes about the Fe atoms in this conversion [22]. Mössbauer spectroscopy was also unable to resolve a difference, mainly due to the difficulty of deconvoluting overlapping signals [14,15]. On the other hand, MCD spectroscopy did exhibit spectral shifts associated with this spin-state conversion [24]. Similar shifts were observed in two synthetic analog clusters, $[Fe_4S_4(SEt)_4]^{3-}$ and $[Fe_4S_4(SC_6H_{11})_4]^{3-}$, which possess dominant $S = 1/2$ and $3/2$ states, respectively. The physiological relevance of the mixed spin state of the Fe-protein has been questioned by room temperature magnetic susceptibility and NMR results, which show spectral shifts associated only with an $S = 1/2$ cluster suggesting that the $S = 3/2$ state may be induced by freezing [25].

2.2 MoFe-Protein

The structural complexity of the two types of metal clusters in the MoFe-protein (Chapter 2) has made them well suited for a vast array of spectroscopic investigations. For decades prior to the determination of these structures, spectroscopy was the only means of obtaining insight into their complexity. The first definitive information on the metal clusters came from Mössbauer studies of Kp1, Av1 and Cp1 (the MoFe-protein from *Clostridium pasteurianum*) [13,26-33]. Mössbauer, being a highly quantitative spectroscopic technique, was able to identify 30 Fe atoms in two types of metal cluster. One of the clusters is termed the M-center, or is sometimes referred to as the FeMo cofactor (FeMoco). In the resting state (M^N), the M-center is paramagnetic, exhibiting a unique $S = 3/2$ signal with low rhombicity (Figure 3) [12,26,34-38]. As mentioned above, the m_s sublevel levels of half-integer spin states ($S > 1/2$) are split in zero magnetic field into $\pm\, m_s$ degenerate doublets. For $S = 3/2$, the separation (Δ) between the $\pm 1/2$ and $\pm 3/2$ doublets depends on the rhombicity, λ ($0 \leq \lambda \leq 1/3$) of the signal, and is given by Equation 1.

$$\Delta = 2D(1 + 3\lambda^2)^{1/2} \tag{1}$$

In Equation 1, D is the axial zero-field splitting term. When $D > 0$, the $m_s = \pm 1/2$ doublet is the lower level with predicted g-factors as shown in Equations 2, 3 and 4.

$$g_x = g_0[1 + (1 - 3\lambda)/(1 + 3\lambda^2)^{1/2}] \tag{2}$$

$$g_y = g_0[1 + (1 + 3\lambda)/(1 + 3\lambda^2)^{1/2}] \tag{3}$$

$$g_z = g_0[(2/(1 + 3\lambda^2)^{1/2} - 1] \quad (4)$$

For MoFe-proteins, where the rhombicity is small (λ = 0.045 - 0.055) and $g_0 \approx 2.00$, these expressions simplify as shown.

$$g_x \approx 4.00 - 6\lambda \quad (5)$$

$$g_y \approx 4.00 + 6\lambda \quad (6)$$

$$g_z \approx 2.00 \quad (7)$$

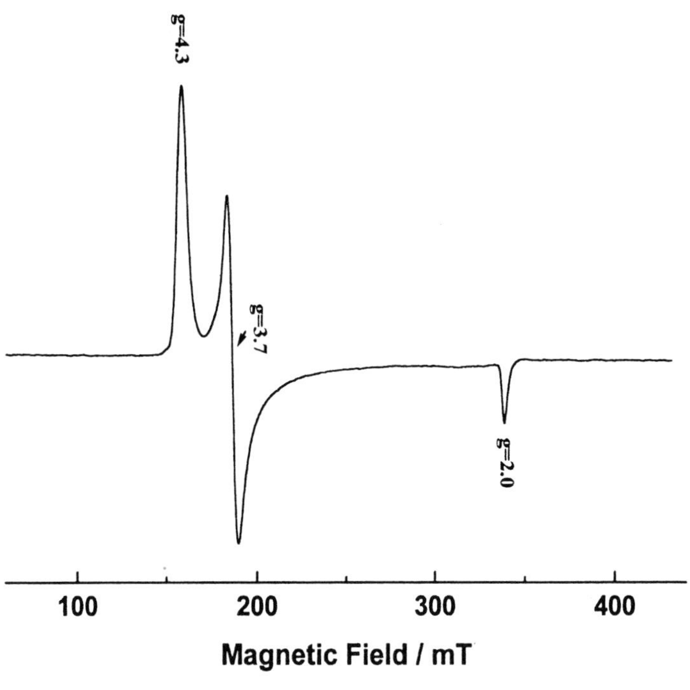

Figure 3. EPR spectrum of MoFe-protein from *Azotobactervinelandii*. The signal arises from the paramagnetic M-center in an $S = 3/2$ state.

Variations observed in the published EPR spectra of MoFe-proteins from different bacteria and mutant strains can be easily interpreted by these equations in terms of small fluctuations of the spectral rhombicity, λ. While $S = 3/2$ states have been observed with other metalloproteins, the low rhombicity of the M^N signal is highly unusual. Because of this low rhombicity, all three inflections of the ground state are sharp and easily detected. Of the three, the g_x inflection is the most interesting since it occurs in a spectral region (g = 3-4) typically void of EPR signals. This property has been used to detect the MoFe-protein both in cell extracts and in whole cells [39].

Early Mössbauer studies [13,26-33] of the M-center suggested six Fe sites grouped into two classes, according to the signs of the ^{57}Fe magnetic hyperfine coupling constants, A_{iso}. According to these studies, there are three sites, A^1-A^3, with negative A_{iso}, and three essentially equivalent B-sites, B^1 (2 equivalent) and B^2, having positive A_{iso}. The hyperfine parameters for these sites, along with those determined from ENDOR spectroscopy, are shown in Table 1. More recent Mössbauer studies [33] on isotopic hybrids of the MoFe-protein now reveal a seventh Fe site (A^4) with unusually small, highly anisotropic A-tensor components and a negative A_{iso} (~ -4 MHz). Because of its small A-tensor, the A^4-site eluded detection by the ENDOR spectroscopic studies described below [40-42]. The small hyperfine interaction of the A^4 site arises from the local spin of this site being oriented perpendicular to the cluster spin. To date, none of the sites identified by Mössbauer (or ENDOR) spectroscopy has been assigned to a particular crystallographic iron site. Finally, isomer shifts (δ) depend on the coordination environment and the oxidation states of the iron atoms. For a fixed coordination, δ is a reliable oxidation-state marker. Using model FeS clusters for reference, the metal ion valence of M^N has been predicted [33] from the isomer shifts to be [Mo^{4+}, $4Fe^{2+}$, $3Fe^{3+}$]. This assignment differs from that proposed from ENDOR studies [43], as described below.

Fe site	A^1	A^2	A^3	A^4	B^1	B^2
δ (mm/s)	0.39	0.48	0.39	0.41	0.33	0.50
ΔE_Q (mm/s)	-0.69	-0.94	-0.56	0.68	-0.66	-0.65
$A_{x'}$ (MHz)	-13.8 (-13)	-17.3 (-18)	-13.9 (-14)	-1.5	+11.7 (11)	+8.5(11)
$A_{y'}$ (MHz)	-21.1 (-21)	-15.1 (-14)	-11.6 (-12)	-9.5	+14.1 (14)	+11.0 (9)
$A_{z'}$ (MHz)	-19 (-10)	-19 (-19)	-10 (-10)	0	+9.2 (9)	+8.2 (19)
A_{iso} (MHz)	-18.0	-17.1	-11.8	-3.7	+11.7	+9.3

[a]Data from ref. 35. [b]Shown in parenthesis are the ENDOR values from ref. 43.

Table 1. 4.2 K Hyperfine Parameters of the S = 3/2 State of M^N [a,b].

Mössbauer spectroscopy of the resting-state of the MoFe-protein also detected a second class of cluster, the P-cluster ("P" for protein-bound), composed of 16 Fe atoms. Early spectral and redox titration data implied that these Fe atoms were grouped into four [4Fe-4S] centers, which may be spin-coupled [26,28,44]. Later EPR [45] and Mössbauer [46] data suggested that the P-clusters were actually 8Fe-clusters, as verified by subsequent X-ray diffraction studies [47-52]. An interesting aspect of these clusters is that they are diamagnetic in the resting state (P^N) and appear to contain all iron(II) atoms which, until recently, was unprecedented in a large FeS cluster.

ENDOR spectroscopy similarly has been successful in identifying the different iron sites in the cofactor of Av1, Kp1 and Cp1 [41]. This technique detected six Fe atoms in 5 nonequivalent sites (Av1 sites are listed in Table I). In addition to iron, spectra have been detected for ^{1}H, ^{33}S and $^{95/97}Mo$ [40,41]. While multiple ^{1}H resonances were detected, only 1-2 represent exchangeable protons resulting from direct H_2O or OH^- binding to the center. Even though the M-center was known to contain 9±1 acid-labile S atoms, the simple ^{33}S spectrum suggested basic bridging S^{2-} ions similar to those found in [4Fe-4S] clusters. The ^{95}Mo spectrum shows the presence of small hyperfine interaction between that atom and the unpaired electron and suggested Mo to be in a highly asymmetric coordination environment corresponding to a single Mo atom in a nonmagnetic Mo^{4+} state. The *A*-tensors for all of these nuclei are very similar in Av1 and Kp1, yet are distinctly different to those in Cp1. The source of these differences and the association of the different ENDOR and Mössbauer Fe hyperfine constants with specific structural sites will require future structural analysis.

As mentioned above, ESEEM spectroscopy detects the precession of the magnetic moment of interacting nuclei neighboring a paramagnetic center. An ESEEM study on Av1 identified N modulation [53], initially proposed to arise from a histidine residue coupled to the M^N state and tentatively assigned [54] to hydrogen bonding of α-His195. The use of variant MoFe proteins with selected amino acid substitutions in the α-His195 position later negated this suggestion and showed that the origin of the deep modulation must be another amino acid nitrogen [55]. Recent ESEEM work [56] suggests the source of the modulation to be α-Arg359 and shows the existence of a second smaller N modulation possibly associated with the amide nitrogen of α-Gly356 and/or α-Gly357. All three of these amino acid residues interact (possibly by hydrogen bonds) with sulfur in the waist region of the FeMo cofactor.

In early studies on the MoFe-protein, CD and MCD spectroscopy were used to investigate the electronic structure of the paramagnetic M-center [17,57-59]. The most significant contribution made by MCD was the demonstration that this center was much more complex than a simple [4Fe-4S] cluster. This finding was important because several early structural models portrayed the M-center as cubane clusters tethered together with a Mo atom substituted into one of the Fe sites. The MCD spectrum reveals that the cluster is more complex than this.

Structural information on the metal clusters in the MoFe-protein finally emerged from XAS studies. The first studies were performed on the Mo edge of lyophilized protein and suggested that Mo exists as Mo^{5+} in the resting state of the enzyme with no Mo-O bonds [60-63]. Later XANES and EXAFS studies revised these assignments to the 4+ state for Mo with 2-3 Mo-O ligands and a similar number of Mo-S bonds, in agreement with the six coordination (3S, 2O and 1N) shown in subsequent X-ray diffraction studies on the enzyme [64-71]. In addition to a short Mo-Fe distance at 2.70 Å and a Mo-S distance of 2.37Å, similar to those observed in cubane [4Fe-4S] clusters, these studies also demonstrated the existence of a long Mo-Fe distance at 5.06 Å illustrating long-range structure. Additional long Fe-Fe distances at 3.74 Å and 4.7 Å have been detected in Fe-EXAFS as well as a long Fe-S distance at 4.3 Å [69,71,72]. Analogous distances are observed in a model Fe_6S_6 prismane cluster [69]. Comparing these data with the X-ray diffraction structure shows that the 3.74 Å Fe-Fe distance is associated only with the cofactor. These measurements have allowed a refinement of the less-precise distances obtained from X-ray structural studies.

It was early shown that the M-center can be extracted intact from the MoFe-protein, at which point it is referred to as FeMoco [73]. Spectroscopic studies analogous to those outlined above on the MoFe-protein, have been undertaken on isolated FeMoco. Since FeMoco has not been crystallized, its structure is not known. Therefore, a prime focus of the spectroscopic studies has been the determination of the similarities and differences between the protein-bound FeMoco (*i.e.*, the M-center) and the extracted cofactor center in NMF (*N*-methylformamide). The isolated FeMoco exhibits an $S = 3/2$ EPR signal similar to, but much broader than, the protein-bound M-center, implying a distribution of structures in the extracted cofactor. A similar broadness is observed in the EPR spectra of other, simpler model FeS clusters. The EPR spectrum of FeMoco sharpens upon addition of thiophenol or imidazole, presumably due to the ligation by these molecules to the terminal tetrahedral Fe or Mo atoms, respectively. In other words, thiophenol and imidazole substitute for the α-Cys275 and α-His442 ligation in Av1 (see Chapter 2). Other than minor deviations, the CD, XAS and MCD spectra of the isolated FeMoco are essentially unchanged from those of M^N, implying that FeMoco has the same overall structure as that of the protein-bound cluster.

2.3 Alternative Nitrogenases

When first isolated in the mid-1980's, the VFe-proteins of V-nitrogenase from *Azotobacter chroococcum* [74,75] and *A. vinelandii* [76] were extensively investigated with different spectroscopic techniques. The common goal of these investigations was the determination of the similarities of the metal clusters in these proteins with those of the conventional MoFe-protein. The first studies, those involving EPR spectroscopy, showed spectra more complex than the single $S = 3/2$ M^N signal in the MoFe-protein. The spectrum of $Ac1^V$ [75] contains a mixture of $S = 3/2$ species along with an axial $S = 1/2$ signal. This latter signal appears in low integrated concentration (~0.2 spins per protein) and is similar to the de-molybdo species often observed in early purification steps of the MoFe-protein [77]. In $Ac1^V$ the intensity of this signal does not correlate with specific activity and can be greatly diminished with extensive chromatographic purification [78]. This is not the case in $Av1^V$, where this signal appears to follow activity. Other than the presence of the $S = 1/2$ species, the EPR spectrum of $Av1^V$ is simpler than that of $Ac1^V$ and contains only a single, highly rhombic ($\lambda = 0.29$ with $g = 5.68$, 5. 45 inflections) $S = 3/2$ species, presumably arising from the protein-bound VFe cofactor [79]. A second form of $Av1^V$ (termed $Av1^V{}_A$) has been isolated [80] and shown to contain an $\alpha\beta_2$ structure of the major subunits only, instead of the usual tetrahedral $\alpha_2\beta_2$. The EPR spectrum of $Av1^V{}_A$ exhibits an additional $S = 1/2$ signal, which has been proposed to arise from an incomplete P-cluster on the second, isolated β subunit. It is interesting to note that the spectrum of this additional species is identical to that recently published for the NifU protein where the signal has been assigned to an FeS cluster used in the construction of protein-bound FeS clusters [81].

MCD and Mössbauer spectroscopic techniques have been performed on only the $Av1^V$ form of the VFe-protein and clearly show the presence of M-centers and P-clusters [82,83]. These techniques, and subsequent EPR studies, also illustrated the present of a small amount of oxidized P-clusters in the resting state of $Av1^V$, even when dithionite and methyl viologen are present. This suggests that the P-cluster has a lower mid-point potential in $Av1^V$ than in Av1. Structural studies using Fe- and V-XAS on $Ac1^V$, $Av1^V$ and extracted FeVco have confirmed the presence of M-centers [69,72,84-87]. These spectra also show the presence of a long 3.7 Å Fe-Fe distance, similar to that observed in M^N and FeMoco and associated only with the cofactor and not the P-cluster. This observation suggests a similarity in the FeMoco and FeVco structures.

Cofactor	δ (mm/s)	ΔE_Q (mm/s)	Area (%)
FeMoco[b]	0.40	0.76	40
FeVco[b]	0.39	0.94	48
FeFeco[c]	0.37/0.42[d]	0.74/1.00[d]	46

[a] Data from ref. 92. [b] From *A. vinelandii.* [c] From *R. capsulatus*
[d] Spectrum of FeFeco is split into two subspectral components M_1 and M_2.

Table 2. Isomer shifts and quadrupole splitting of the three cofactors in their respective, dithionite-reduced, nitrogenase proteins[a].

Nitrogenase	Fe-N/O	Fe-S	Fe-Fe	Fe-Fe'	Fe-(Mo,V, Fe)	Fe-S'
$Av1^{Mo}$	1.842	2.317	2.641	3.78	4.00	4.28
$Av1^V$	1.864	2.322	2.674	3.76	3.99	4.30
$Rc1^{Fe}$	2.16	2.29	2.63	3.68	4.05	4.20

Table 3. R (Å) from Fe K-edge EXAFS. Data taken from refs. 74 and 92.

The second alternative nitrogenase, Fe-nitrogenase, has been less extensively studied than Mo-nitrogenase and V-nitrogenase. It has been isolated from *A. vinelandii* [88], *Rhodobacter capsulatus* [89,90] and *Rhodospirillum rubrum* [91,92]. Until recently, spectroscopic studies on component 1 of Fe-nitrogenase were limited to EPR, which showed minor components at *g* 4.3 and 2, both of which are probably due to contaminants [93].

The lack of an EPR signal suggests that the M^N center of Fe-nitrogenase is an integer-spin and EPR-silent state. This result was recently confirmed by Mössbauer studies on $Rc1^{Fe}$, which suggest that the cofactor of $Rc1^{Fe}$ contains 4 Fe^{2+} and 4 Fe^{3+} ions in the resting state of the protein with $S = 0$ [94]. This oxidation-state assignment is consistent with that of FeMoco from Mössbauer spectra and differs from the assignment based on ENDOR and theoretical studies [95] of only one Fe^{3+} in FeMoco. The Mössbauer spectra also show that the isomer shift and quadrupole splitting of the M-center (Table 2) and P-cluster are conserved in all three forms of the enzyme. Fe-XAS studies [94] similarly show a consistent spectral pattern in all three forms, including the unique long 3.7 Å Fe-Fe distance in the cofactor (Table 3). These data indicate that, except for the presence of the heterometal in the cofactor, the structures of the metal clusters in Mo-, V- and Fe- nitrogenases are essentially the same.

3. OXIDIZED AND REDUCED STATES

3.1 Fe-Protein

The [4Fe-4S] cluster of the Fe-protein, like other ferredoxins, shuttles between the 1+ and 2+ states. As described above, $[4Fe\text{-}4S]^{+}$ is a mixed paramagnetic state with $S = 1/2$, 3/2 signals exhibited by the isolated protein in the presence of dithionite. When the cluster is oxidized to $[4Fe\text{-}4S]^{2+}$, either by an external oxidant or following electron donation to the MoFe-protein, the cluster contains pairs of high-spin iron(III) and iron(II) ions antiferromagnetically coupled to generate a diamagnetic $S = 0$ states. Being diamagnetic, the oxidized cluster has been investigated only by Mössbauer, CD and XAS spectroscopies. Mössbauer spectra show a significant decrease in the quadrupole splitting of two doublets from 1.22 and 0.83 mm/s to 0.91 and 0.58 mm/s when MgATP binds to the oxidized protein [15]. Similarly, CD spectra show large, yet different, changes upon binding of MgATP and MgADP to the oxidized protein [18]. On the other hand, Fe-Fe and Fe-S distances obtained from Fe-EXAFS studies are typical of those measured for $[4Fe\text{-}4S]^{2+}$ clusters in other proteins [14].

Cluster state	ΔE_Q (mm/s)	δ (mm/s)[b]	no. of sites
$[Fe_4S_4]^{2+}$ $S = 0$[a]	1.22	0.45	3
	0.83	0.44	1
$[Fe_4S_4]^{1+}$ $S = 1/2$[a]	1.60	0.59	2
	0.98	0.53	2
$[Fe_4S_4]^{0}$ $S = 4$	3.08	0.68	1
	~1.5	0.68	3

[a] Data from ref. 94 . [b] Shift relative to Fe metal at 298 K.

Table 4. ΔE_Q and δ values for *Av*2 $[Fe_4S_4]^{2+,1+,0}$ cluster at 4.2 K.

Of greater interest is the recent demonstration that the metal cluster of the Fe-protein can also exist in the $[4Fe\text{-}4S]^0$ state, unprecedented for this class of cluster. This state was first predicted in 1994 when it was shown that both Ti^{III} citrate and flavodoxin are able to reduce the Fe-protein in the $[4Fe\text{-}4S]^{2+}$ state to an EPR-silent state [96]. Dithionite, the normal reductant used in this system, is only able to reduce the cluster to the $[4Fe\text{-}4S]^+$ state. A subsequent Mössbauer investigation [97,98] verified the existence of an $[4Fe\text{-}4S]^0$ state containing only iron(II) ions (Table 4).

Surprisingly, even though all of the sites are iron(II), the Mössbauer parameters can be broken into two sets, one representing 3 Fe atoms and one representing the remaining Fe atom [98]. This was unexpected since X-ray diffraction studies clearly show that the cluster is symmetrically bound to the two identical polypeptides of the protein. Coupling among the Fe atoms yields a cluster with a non-zero integer spin. Using parallel mode EPR spectroscopy, a signal has been detected at $g = 16.4$ and assigned to a transition of the ground $m_s = \pm 4$ double of an $S = 4$ state [97].

A dramatic difference is observed in the Fe-EXAFS Fourier transform [99] of the $[4Fe\text{-}4S]^0$ state compared to the other two more oxidized states (Figure 4). The transform is significantly shifted to longer distances with decreased amplitude. The decreased amplitude most likely represents significantly less regular cluster structure. Specifically, the EXAFS parameters could not be used to fix an [4Fe-4S] model with longer distances. A fit was achieved only when two different Fe-Fe and Fe-S distances were used. A proposed structural model consistent with this fit is one with 1:3 short-to-long Fe-S distribution and a 2:1 short-to-long Fe-Fe distribution. This suggests a compression of the cluster along the axis in which the Fe-Fe and Fe-S distances are shorter relative to the perpendicular direction where these and the Fe-S thiolate distances are longer.

MCD and CD spectroscopy have also been used to study the unique $[4Fe\text{-}4S]^0$ cluster [100]. Because this state has no iron(III) ions, there will be no intervalence or thioloate-to-iron(III) charge-transfer transition bands. The MCD spectrum revealed bands in 490-1250 nm region previously void of transitions in simple $Fe^{II}S_4$ sites. These transitions may arise from d-d transitions associated with the unusual Fe-Fe compression detected in the Fe-EXAFS Fourier transform. Similarly, the CD spectrum contains a band at 520 nm, unusual for a reduced rubredoxin FeS_4 site [97].

Some of the questions concerning the structure of the all-iron(II) form of the Fe-protein have now been answered with the recent X-ray diffraction study on Av2 [101]. While the Mössbauer studies described above suggest one unique Fe site in the all-iron(II) cluster, such a site could not be identified in the crystal structure. The distortion of the all-iron(II) [4Fe-4S] cluster cannot be easily classified; two S-S

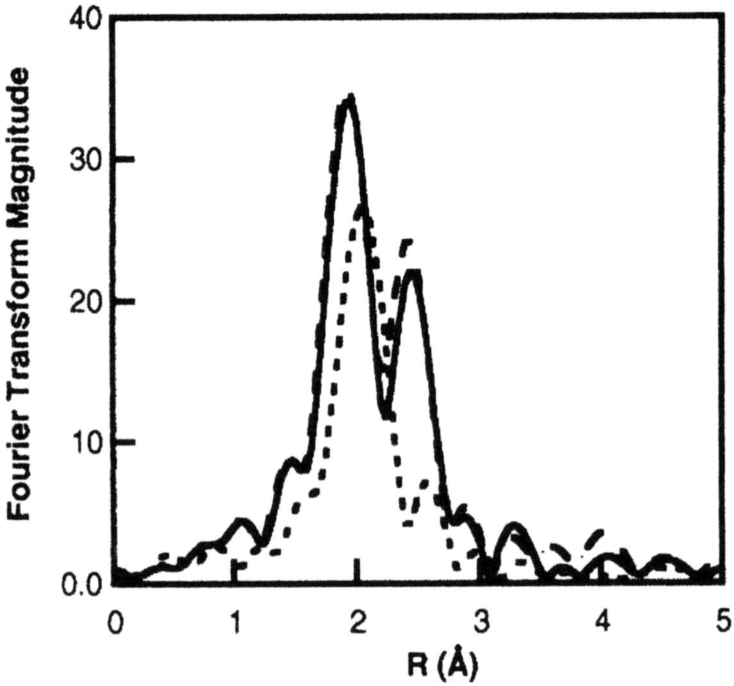

Figure 4. Fe-EXAFS Fourier transform of Av2 in the $[4Fe\text{-}4S]^{2+}$ state (dashed line), $[4Fe\text{-}4S]^{1+}$ state (solid line) and the $[4Fe\text{-}4S]^{0}$ state (dotted line). Figure was taken from Ref. 96.

distances become slightly longer, but the overall volume of the cluster remains nearly constant. Furthermore, the number of NH···S hydrogen bonds to the cluster and thiolgroups of cysteine ligands is higher in the all-iron(II) Av2 than in most other [4Fe-4S] proteins. An interesting property of the all-iron(II) cluster is its solvent accessibility. The Fe protein has the most surface accessible cluster of all known [4Fe-4S] proteins, suggesting that the solvent accessibility of the cluster might play a role in its ability to exist in all three oxidation states.

3.2 P-Cluster

According to Mössbauer studies, P-clusters in the P^N state of the as-isolated MoFe-protein have all of the Fe atoms in the ferrous valence [26]. To date, no further reduced state has been observed. Several oxidized states, however, have been detected.

When first investigated by Mössbauer spectroscopy, the P-clusters were proposed to be four [4Fe-4S] structures, possibly spin coupled in pairs. Oxidative titrations revealed that four oxidation equivalents per protein were needed to convert these clusters to the P^{OX} state, a predicted half-integer state [28]. The unusual aspect of this state is that, even though it is half-integer, it is EPR silent. The four [4Fe-4S] cluster structural assignment was accepted for many years but this came into question when it was shown [45] that oxidation of the P-clusters by 6 equivalents generated a mixed $S = 1/2$, 7/2 spectrum. From this study it was proposed that P-clusters were actually 8Fe structures, two per protein. In this situation, the P^{OX} state represents a two-equivalent oxidation and, therefore, an EPR-silent integer state. Subsequent X-ray diffraction studies demonstrated that this structural assignment is correct [47-52,102]. The earlier error in the spin-state assignment by Mössbauer spectroscopy (*i.e.*, half-integer rather than integer) arose from the near-degeneracy of the ground $\pm m_s$ state, which yielded spectral parameters similar to half-integer states [46].

Using a parallel-mode EPR cavity, the spectrum of P^{OX} has been detected at $g = 11.9$ for Av1, Cp1 and Kp1 and at $g = 15.6$ for the MoFe-protein of *Xanthobacter autotrophicus*, Xa1 [46]. While the latter signal arises from a ground state, the other three signals are excited states transitions. For a given spin system, the largest observable g-factor comes from the $\pm m_s$ transition where $m_s = S$ (or $\Delta m_s = 2S$) and is observed at $g = 4S$. For example, for an $S = 2$ system, it would be the $m_s \pm 2$ transitions at $g = 8$. This g-factor assignment occurs because we are treating the separate ± ms levels as isolated $S = 1/2$ states (D is much larger than the microwave energy and, therefore, we only observe inter-level $\pm m_s$ transitions). The Zeeman expression for the energy of an EPR transition is given below,

$$\Delta E = g\beta B\Delta m_s$$

where B is the magnetic field flux and β is the Bohr magneton. For an $S = 1/2$ system, $g = 2$ and $\Delta m_s = 1$, or $\Delta E = 2\beta B\Delta m_s = 2\beta B$. When $S > 1/2$, the energy change for the greatest Δm_s transition is the same even though Δm_s differs ($\Delta m_s = 2S$). Therefore, $g\beta B = 2\beta B(2S)$ and this reduces to $g = 4S$.

At first glance, it would appear that the transition observed for Xa1 corresponds to an $S = 4$ state while the signals of the other proteins are excited state $\Delta m_s = \pm 3$ transitions also of an $S = 4$ states. A possible error in this reasoning is that g-factors for integer states don't strictly obey the above rule and can be greater than $4S$ due to zero-field splitting of $\pm m_s$ levels. In other words, the $g = 11.9$ inflection could also arise from the $\Delta m_s = \pm 2$ transition of an $S = 3$ system. Because of this, the spin state of P^{OX} is unknown and usually listed as $S = 3$ or 4. Mössbauer and MCD [103] studies are consistent with this assignment although tend to favor the $S = 3$ state.

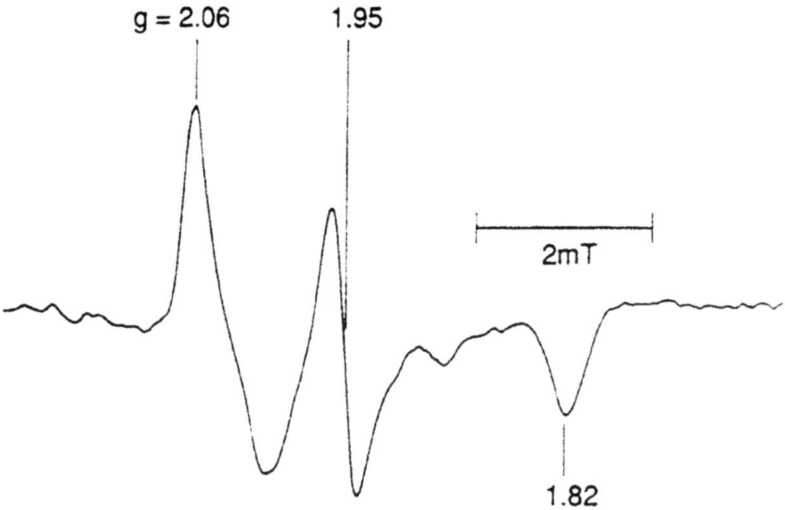

Figure 5. $S = 1/2$ EPR signal of P^+ state in Av1. Negative inflection between $g = 2.06$ and $g = 1.95$ peaks arises from $g = 2.01$ inflection of M^N signal.

Because P^{OX} (or P^{2+}, relative to $P^N = P^0$) represents a two equivalent oxidation of P^N, there must also be a one-equivalent oxidized $P^{semi\text{-}OX}$ (or P^+) state. EPR studies [104] during an oxidative titration of the MoFe-protein showed the formation of signals consistent with $S = 1/2$ and 5/2 spin states (Figure 5) prior to the formation of the integer spin signal of P^{OX}. The relative concentration of each spin state is pH- and protein-dependent. For example, $Av1^V$ exhibits only the $S = 5/2$ system [105]. It has been suggested that the simultaneous formation of these two spin states arises from two near-degenerate wave functions of P^+ associated with spin density predominantly on one or the other F_4S_3 cuboidal halves of the P-cluster [104]. If this is assignment is correct, then the spin state of P^{OX} should represent the vector sum of these two spin states or $S = 3$.

The P^{OX} state can be further oxidized to a mixed spin $S = 1/2$, 7/2 state of P^{3+} [45]. As mentioned above, it was the discovery of this state, which led to the prediction that P-clusters consist of 8Fe units. Other than P^{OX}, which has been studies by EPR,

MCD and Mössbauer spectroscopic techniques, the other oxidized states of the P-cluster have been detected only by EPR spectroscopy.

3.3 M-center

The Lowe/Thorneley scheme for nitrogen fixation contains a series of 8 one-electron reduction steps, $E_n \rightarrow E_{n+1}$, $0 \leq n \leq 7$ (Chapter 9). In this scheme E_0 represents the resting state of the MoFe-protein with the P-clusters in their diamagnetic P^N $S = 0$ state and the M-centers in the M^N $S = 3/2$ state. To date, only two other redox states of the M-center have been detected, the one-electron-oxidized (M^+ or M^{OX}) and the one-electron-reduced (M^- or M^R) state. The state M^+ can be achieved by chemical oxidation, but it is probably not germane to the mechanism of nitrogen fixation. Both Mössbauer and MCD spectroscopy show this state to be diamagnetic ($S = 0$) [28,82,103].

The M^R state can be achieved only by a one-electron enzymatic reduction. Other than Mössbauer spectroscopy [26], which shows that this state to be a paramagnetic integer spin state, XAS is the only other technique to have been used to investigate it. Fe- and Mo-EXAFS of Av1 and $Av1^V$ during enzymatic turnover characterize M^R to have a nearly identical, but slightly constricted, structure compared to M^N and M^+ [72]. These results have been questioned in subsequent studies, which do not show a contraction [106]. Mössbauer parameters suggest that the Fe valence of M^R is the same as M^N, implying that the added electron has reduced Mo from Mo^{4+} to Mo^{3+}.

Recently a second form of M^- (called M^I) has been detected by Mössbauer spectroscopy [33]. This form was generated by radiolytic reduction of frozen Av1 using X-rays from a synchrotron source. The Mössbauer parameters of M^I differ from those of M^R and suggest that there is reduction of one of the cluster Fe atoms instead of Mo, as in M^R.

4. INTERMEDIATE STATES

4.1 Turnover Under Argon

Data from decades of spectroscopic analysis of the different metal clusters of nitrogenase greatly aided the eventual X-ray crystallographic determination of their structure. Following this determination, the use of spectroscopy shifted from characterization of the metal clusters in their resting states to probing the clusters during enzymatic turnover. Traditionally, mechanism and kinetics of a reaction are most often monitored with spectroscopy. Past difficulties in using spectroscopy to investigate nitrogen fixation occurred because the substrates, being small molecules, are poor chromophores, which don't often exhibit optical signals in the visible or near ultraviolet region.

On the other hand, EPR spectroscopy depends on paramagnetism and not on optical absorption and, as such, has been useful in monitoring substrate interactions with the paramagnetic M-center. In fact, except for a few examples, EPR and ENDOR have been the only spectroscopic techniques used to study intermediate states during turnover. In general, substrates do not interact significantly with the FeMo cofactor in the M^N state of the wild-type enzyme. Interaction requires turnover conditions to generate either a more reduced cofactor or a more open protein structure, *i.e.*, a structure that allows substrates and products to enter or leave easily the cofactor pocket.

Early studies have shown that, under turnover conditions in the absence of substrate, the resting-state EPR signals of both the MoFe-protein and Fe-protein significantly decrease. This occurs because during turnover the Fe-protein donates an electron, converting its metal cluster to the diamagnetic $[4Fe\text{-}4S]^{2+}$ state, while the MoFe- protein receives the electron, converting the M-center to the integer spin, EPR-silent M^R state. While low-electron-flux experiments result in an approximately 50% reduction of M^N, high-flux experiments often produce $\geq$ 90% reduction of this signal intensity. This extent of reduction is unexpected since, according to the Lowe/Thorneley mechanistic scheme, the MoFe-protein should alternate between half-integer and integer states, such that in the steady state about half of the M-centers are half-integer and EPR-active.

Rapid quench experiments on Kp1 detected several S = 1/2 and 3/2 signals of unknown origin [107]. One of these signals, at g = [2.139, 2.001, 1.977] was proposed to be associated with H_2 evolution. Of recent interest is a study which shows a series of new substrate-independent S = 3/2 signals elicited during turnover with Av1 [108]. In this study the intensity of the resting-state M^N signal (called signal 1a with g = 4.32) rapidly decreases following initiation of turnover, as described above. During this decrease, there is a parallel increase of a second signal (1b, g = 4.21). Under conditions of rapid electron flux, this second signal (1b) also decrease,s to produce a third signal (1c, g = 4.69). It has been proposed that signals 1b and 1c arise from reduced states of the Lowe/Thorneley mechanistic scheme and reflect different conformations of the FeMo cofactor with different protonation states.

4.2 Turnover Under CO

In the late 1970's a series of EPR studies investigated the effect of substrates and inhibitors on the EPR spectrum of nitrogenase turnover samples [107,109-111]. While minor signals were detected for several substrates, all of these studies showed the presence of two intense signals when the non-competitive inhibitor CO was present. One of these signals (lo-CO: 0.08 atm) with g = [2.09, 1.97, 1.93] appears under low CO pressures and is replaced by another (hi-CO: 0.5 atm) with g = [2.17, 2.06, 2.06] at high pressures. Because these signals are present only during enzymatic turnover, they obviously represent mechanistic intermediate states of CO inhibition. Recently these

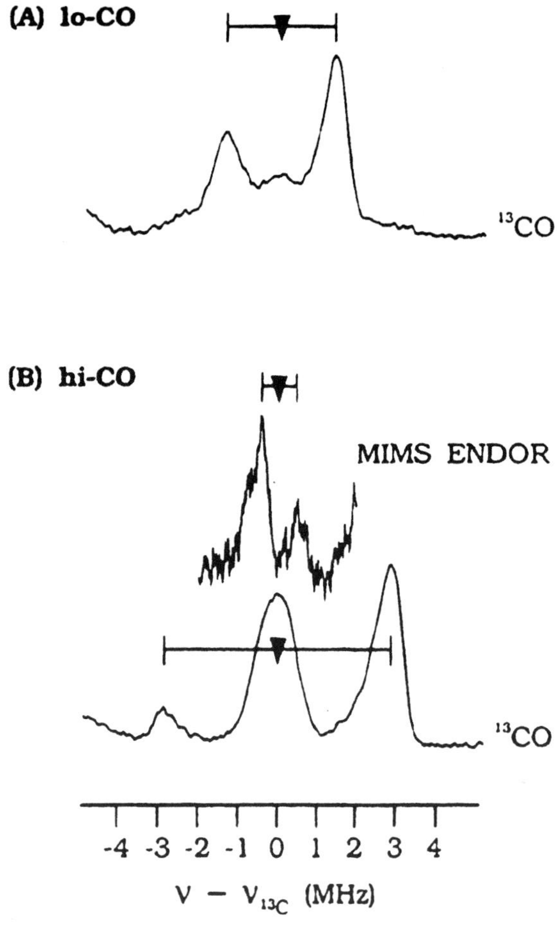

Figure 6. ^{13}C ENDOR spectra of ^{13}CO labeled lo-CO (A) and hi-CO (B). Spectra show the existence of a single CO molecule bound to the M-center of Av1 in lo-CO and two CO molecules bound in hi-CO. Mims ENDOR was used on the hi-CO sample to help resolve the small hyperfine interaction of one of the CO molecules.

signals have been used to probe the enzyme during catalysis [43,112-115]. Using ^{13}C ENDOR spectroscopy, it has been shown that lo-CO corresponds to one CO bound to a metal cluster of Av1 while hi-CO represents the addition of a second CO to the same metal cluster (Figure 6). Similar ENDOR studies on ^{57}Fe-labeled isotopomers (*i.e.*, Av1 with ^{57}Fe labeling specifically in either the M-center or P-cluster) definitively demonstrated that CO binds to the M-center of Av1. Orientation-selective ^{13}C ENDOR spectroscopy also revealed that CO in lo-CO is bound to the Fe atoms in the waist region of the FeMo cofactor in a bridging fashion while both COs in hi-CO exhibit terminal binding. Similarly, orientation-selective ^{57}Fe ENDOR spectroscopy allowed a determination of the Fe metal ion valencies in M^N, lo-CO and hi-CO to be [Mo^{4+}, Fe^{3+}, $6Fe^{2+}{}_6$]. As mentioned above, this assignment is consistent with theoretical studies [95] but differs from that predicted from Mössbauer studies [33].

The initial studies on CO were extremely significant because they clearly demonstrated for the first time the binding of a small substrate or inhibitor to the M-center of Mo-nitrogenase. While this binding had always been assumed in the past, these studies definitively showed it. They also revealed a third, previously undetected, CO-induced EPR signal. Unlike lo-CO and hi-CO, this new signal arises from an S = 3/2 state in the g 5 region [115,116]. It has the same CO concentration dependency as hi-CO but is observed only under high electron flux and in the presence of high MgATP concentrations. Because of these criteria, this signal has been labeled hi(5)-CO. Whether this signal represents another CO-bound state similar to hi-CO has not been determined.

CO binding to nitrogenase also has been investigated with stopped-flow Fourier transform infrared spectroscopy (FTIR), which detected a single absorption under low CO concentrations and 3-4 transitions at high CO concentration [117]. The single absorption at low concentrations may represent the bridging CO of lo-CO. Similar CO binding also has been detected by FTIR on isolated FeMoco electrochemically reduced using cyclic voltammetry in the presence of CO [118].

4.3 Turnover Under CO_2, COS, KSCN or CS_2

Some of the most unusual and interesting substrates of Mo-nitrogenase are COS, KSCN and CS_2 [119]. In the past, compounds containing only C, N and/or O in multiple bonds (along with H) were known to be reduced by Mo-nitrogenase. This new group of substrates is the first to contain sulfur. In the initial paper describing COS reduction, the reduction of CO_2 was also outlined [120]. All of these substrates generate S = 1/2 EPR signals during turnover. The nature of these signals has been investigated only for CS_2. During CS_2 reduction (to generate H_2S and other unknown products), at least three overlapping S = 1/2 species are generated with low-field g-factors at 2.035, 2.111 and 2.211 corresponding to signals "a", "b" and "c", respectively. Using rapid-quench techniques, the time-course for the formation and decay of each has been plotted and shown to be sequential "a" then "b" then "c". ^{13}C ENDOR spectroscopy was used to demonstrate the presence of carbon (originating from $^{13}CS_2$) in each signal, although the exact identity of each adduct could not be determined [121], in part due to the complexity of signal overlapping.

4.4 Turnover Under C_2H_2

Other then N_2, C_2H_2 is undoubtedly the most important nitrogenase substrate. This importance is due, in part, to the major use of it in the determination of specific activity. Furthermore, C_2H_2 has a relatively low K_m and a simple two-electron reduction scheme, which is easy to investigate. Past EPR studies were unable to detect [107,110] an C_2H_2-induced $S = 1/2$ signals similar to those observed with CO or CS_2. However, using a variant MoFe-protein, α-Gln195, investigators recently detected two intense signals [116,122] simultaneously generated under turnover conditions (Figure 7). One of the signals (S_{EPR1}) has $g = [2.123, 1.978, 1.949]$ and obviously arises from the M-center. Using $^{13}C_2H_2$ and C_2D_2 in either H_2O or D_2O, ENDOR spectroscopy has been successful in detecting three ^{13}C-transition pairs associated with two C_2H_2 adducts [123]. One adduct has been identified as a reduced C_2H_2 bound side-on to Fe atoms in the waist region of the M-center. The identity of the second adduct has not yet been determined. The second EPR signal (S_{EPR2}) observed is an unusual sharp, narrow, nearly isotropic

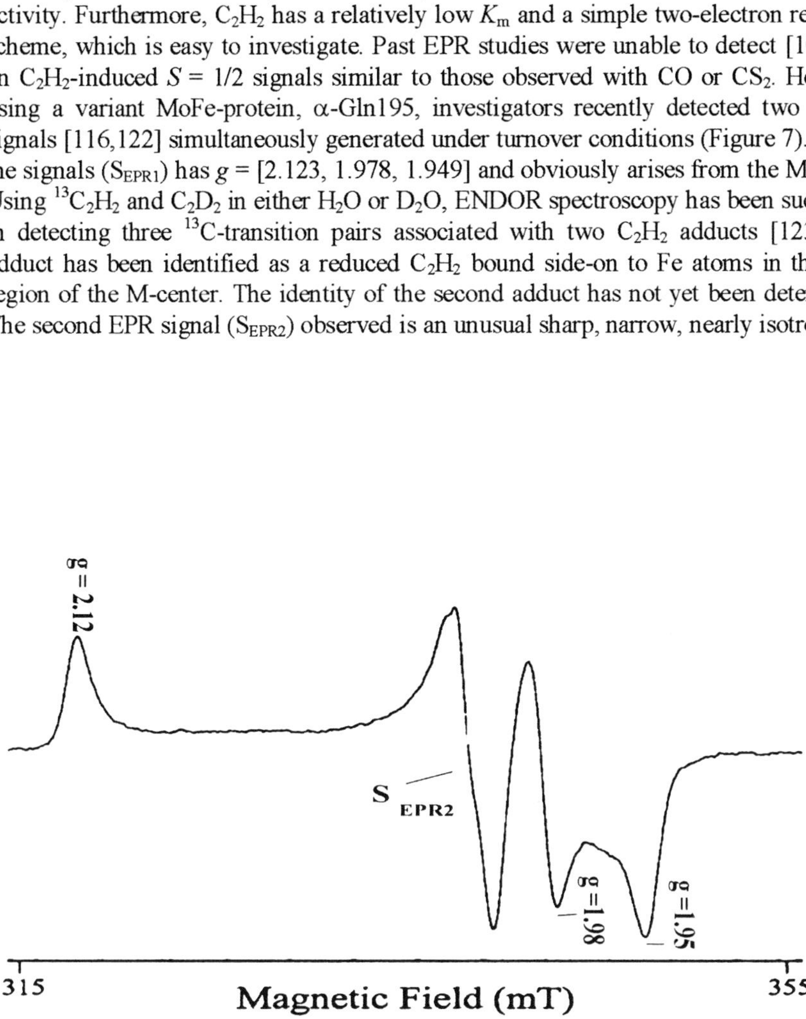

Figure 7. EPR spectrum of acetylene-induced signal in turnover sample using the α-Gln195 variant Av1 showing the S_{EPR1} ($g = [2.12, 1.98, 1.95]$) and S_{EPR2} ($g \sim 2$) signals.

signal with g = [2.007, 2.000, 1.992]. Similar signals have never been reported for nitrogenase but have been observed in O-bridged 2Fe centers [116]. Nitrogenase does not possess this type of cluster. While the origin of this signal is unknown, it has been proposed to arise either from an unusual metal cluster, possibly generate under turnover conditions, or from a radical species such as a homocitrate radical.

Finally, a very interesting set of S = 3/2 EPR signals, similar to M^N, have been reported [124]. These signals are induced when C_2H_2 or HCN interacts with variant α-96 MoFe-proteins. The highly unusual aspect of these signals is that, unlike all the signals described above, these do not require turnover conditions. ^{13}C ENDOR spectroscopy detected hyperfine couple in the HCN sample but could not find similar interaction when C_2H_2 was used. The investigators in this study propose that the wild-type α-Arg96 amino acid acts as a "gatekeeper" for the passage of substrates to and from the M-center. When this amino acid is changed, substrate passage is no longer blocked, allowing it to bind freely to cofactor in the as-isolated (M^N) state.

There are important implications from this work. It had been generally assumed that substrates require turnover conditions in order to bind to the M-center. This study shows that is not the case. Binding can occur in the resting state and does not need turnover conditions. Therefore, even though the initial X-ray structure of the MoFe-protein suggested that there is no direct path to the cofactor, substrates are obviously able to enter the protein. Further studies, similar to the one in this report [124], will have to be undertaken to determine the exact path into the protein to the FeMo-cofactor

It is also interesting to note that all turnover-induced substrate or inhibitor signals, where ENDOR spectroscopy definitively shows binding, arise from S = 1/2 systems. In the recent study using variant α-96 MoFe-proteins, the non-turnover signals are all S = 3/2 spin systems. It appears that binding of a substrate to the resting state cofactor does not induce a spin state change from S = 3/2 to S = 1/2. Turnover conditions are needed for this conversion. The meaning or need for this spin-state conversion will certainly be the focus of future research.

5. REFERENCES

1. G. Palmer, in L. Que Jr. (ed.), Physical Methods in Bioinorganic Chemistry, University Science Books, Sausalito, 2000, p. 121.
2. W. R. Hagen, in A. G. Sykes and R. Cammack (eds.), Advances in Inorganic Chemistry: Iron-Sulfur Proteins, Academic Press, 1992, p. 165.
3. D. N. Chasteen and P. A. Snetsinger, in L. Que Jr. (ed.), Physical Methods in Bioinorganic Chemistry, University Science Books, Sausalito, 2000, p. 187.
4. M. K. Johnson, in L. Que Jr. (ed.), Physical Methods in Bioinorganic Chemistry, University Science Books, Sausalito, 2000, p. 233.

5. E. Münck, in L. Que Jr. (ed.), Physical Methods in Bioinorganic Chemistry, University Science Books, Sausalito, 2000, p. 287.
6. R. Scott, in L. Que Jr. (ed.), Physical Methods in Bioinorganic Chemistry, University Science Books, Sausalito, 2000, p. 465.
7. W. H. Orme-Johnson, N. R. Orme-Johnson, C. Touton, M. Emptage, M. Henzl, J. Rawlings, K. Jacobson, J. P. Smith, W. B. Mims, B. H. Huynh, E. Münck and G. S. Jacob, in W. E. Newton and S. Otsuka (eds.), Molybdenum Chemistry of Biological Significance, Plenum Press, New York, London, 1980, p. 85.
8. T. V. Morgan, J. McCracken, W. H. Orme-Johnson, W. B. Mims, L. E. Mortenson and J. Peisach, Biochemistry 29 (1990) 3077.
9. W. R. Hagen, R. R. Eady, W. R. Dunham and H. Haaker, FEBS Lett. 189 (1985) 250.
10. G. D. Watt and J. W. McDonald, Biochemistry 24 (1985) 7226.
11. M. J. Ryle, W. N. Lanzilotta, L. E. Mortenson, G. D. Watt and L. C. Seefeldt, J. Biol. Chem. 270 (1995) 13112.
12. W. H. Orme-Johnson, W. D. Hamilton, T. L. Jones, M.-Y. W. Tso, R. H. Burris, V. K. Shah and W. J. Brill, Proc. Natl. Acad. Sci. U. S. A. 69 (1972) 3142.
13. W. R. Dunham, W. R. Hagen, A. Braaksma, H. Haaker, S. Gheller, W. E. Newton and B. Smith, in H. J. Evans, P. J. Bottomley and W.E. Newton (eds.), Nitrogen Fixation; Research Progress, Martinus Nijhoff, Dordrecht, 1985, p. 591.
14. P. A. Lindahl, N. J. Gorelick, E. Münck and W. H. Orme-Johnson, J. Biol. Chem. 262 (1987) 14945.
15. P. A. Lindahl, E. P. Day, T. A. Kent, W. H. Orme-Johnson and E. Münck, J. Biol. Chem. 260 (1985) 11160.
16. W. N. Lanzilotta, R. C. Holz and L. C. Seefeldt, Biochemistry 34 (1995) 15646.
17. P. J. Stephens, C. E. McKenna, B. E. Smith, H. T. Nguyen, M.-C. McKenna, A. J. Thomson, F. Devlin and J. B. Jones, Proc. Natl. Acad. Sci. U. S. A. 76 (1979) 2585.
18. M. J. Ryle and L. C. Seefeldt, Biochemistry 35 (1996) 4766.
19. M. J. Ryle, W. N. Lanzilotta, L. C. Seefeldt, R. C. Scarrow and G. M. Jensen, J. Biol. Chem. 271 (1996) 1551.
20. W. N. Lanzilotta, K. Fisher and L. C. Seefeldt, J. Biol. Chem. 272 (1997) 4157.
21. J. M. Chan, W. Wu, D. R. Dean and L. C. Seefeldt, Biochemistry 39 (2000) 7221.
22. P. A. Lindahl, B.-K. Teo and W. H. Orme-Johnson, Inorg. Chem. 26 (1987) 3912.
23. L. Chen, N. Gavini, H. Tsuruta, D. Eliezer, B. K. Burgess, S. Doniach and K. O. Hodgson, J. Biol. Chem. 269 (1994) 3290.
24. Y. A. Onate, M. G. Finnegan, B. J. Hales and M. K. Johnson, Biochim. Biophys. Acta 1164 (1993) 113.
25. J. Meyer, J. Gaillard and J.-M. Moulis, J. Am. Chem. Soc. (1988).
26. E. Münck, H. Rhodes, W. H. Orme-Johnson, L. C. Davis, W. J. Brill and V. K. Shah, Biochim. Biophys. Acta 400 (1975) 32.
27. B. E. Smith and G. Lang, Biochem. J. 137 (1974) 169.
28. R. Zimmermann, E. Münck, W. J. Brill, V. K. Shah, M. T. Henzl, J. Rawlings and W. H. Orme-Johnson, Biochim. Biophys. Acta 536 (1978) 185.

29. J. Rawlings, V. K. Shah, J. R. Chisnell, W. J. Brill, R. Zimmermann, E. Münck and W. H. Orme-Johnson, J. Biol. Chem. 253 (1978) 1001.
30. B. H. Huynh, E. Münck and W. H. Orme-Johnson, Biochim. Biophys. Acta 527 (1979) 192.
31. B. H. Huynh, M. T. Henzl, J. A. Christner, R. Zimmermann, W. H. Orme-Johnson and E. Münck, Biochim. Biophys. Acta 623 (1980) 124.
32. B. E. Smith, M. J. O'Donnel, G. Lang and S. Kevos, Biochem. J. 191 (1980) 449.
33. S. J. Yoo, H. C. Angove, V. Papafthymiou, B. K. Burgess and E. Münck, J. Am. Chem. Soc. 122 (2000) 4926.
34. R. H. Burris and W. H. Orme-Johnson, in W. E. Newton and C. J. Nyman (eds.), Proceedings of the First International Symposium on N_2 Fixation, WSU Press, Pullman, WN, 1976, p. 208.
35. M. J. O'Donnel and B. E. Smith, Biochem. J. 173 (1978) 831.
36. G. D. Watt, A. Burns, S. Lough and D. L. Tennent, Biochemistry 19 (1980) 4926.
37. D. J. Lowe and B. E. Smith, Biochem. Soc. Trans. 13 (1985) 579.
38. T. V. Morgan, L. E. Mortenson, J. W. McDonald and G. D. Watt, J. Inorg. Biochem. 33 (1988) 111.
39. L. C. Davis, V. K. Shah, W. J. Brill and W. H. Orme-Johnson, Biochim. Biophys. Acta 256 (1972) 512.
40. B. M. Hoffman, R. A. Venters and J. E. Roberts, J. Am. Chem. Soc. 104 (1982) 4711.
41. R. A. Venters, M. J. Nelson, P. A. McLean, A. E. True, M. A. Levy, B. M. Hoffman and W. H. Orme-Johnson, J. Am. Chem. Soc. 108 (1986) 3487.
42. A. E. True, M. J. Nelson, R. A. Venters, W. H. Orme-Johnson and B. M. Hoffman, J. Am. Chem. Soc. 110 (1988) 1935.
43. H.-I. Lee, B. J. Hales and B. M. Hoffman, J. Am. Chem. Soc. 119 (1997) 11395.
44. P. A. Lindahl, V. Papaefthymiou, W. H. Orme-Johnson and E. Münck, J. Biol. Chem. 263 (1988) 19412.
45. W. R. Hagen, H. Wassink, R. R. Eady, B. E. Smith and H. Haaker, Eur. J. Biochem. 169 (1987) 457.
46. K. K. Surerus, M. P. Hendrich, P. D. Christie, D. Rottgardt, W. H. Orme-Johnson and E. Münck, J. Am. Chem. Soc. 114 (1992) 8579.
47. J. T. Bolin, A. E. Ronco, L. E. Mortenson, T. V. Morgan, M. Williamson and N.-H. Xuong, in P. M. Gresshoff, L. E. Roth, G. Stacey and W. E. Newton (eds.), Nitrogen Fixation: Achievements and Objectives, Chapman and Hall, New York, 1990, p. 117.
48. J. Kim and D. C. Rees, Nature 360 (1992) 553.
49. J. Kim and D. C. Rees, Science 257 (1992) 1677.
50. M. K. Chan, J. Kim and D. C. Rees, Science 260 (1993) 792.
51. D. C. Rees, J. Kim and D. Woo, Biochemistry 32 (1993) 7104.
52. J. T. Bolin, N. Campobasso, S. W. Muchmore, T. V. Morgan and L. E. Mortenso, in E. I. Stiefel, D. Coucouvanis and W. E. Newton (eds.), Molybdenum Enzymes, Cofactors, and Model Systems, American Chemical Society, Washington, D.C., 1993, p. 186.

53. H. Thomann, T. V. Morgan, H. Jin, S. J. N. Burgmayer, R. E. Bare and E. I. Steifel, J. Am. Chem. Soc. 109 (1987) 7913.
54. H. Thomann, M. Bernardo, W. E. Newton and D. R. Dean, Proc. Natl. Acad. Sci. U. S. A. 88 (1991) 6620.
55. V. J. DeRose, C.-H. Kim, W. E. Newton, D. R. Dean and B. M. Hoffman, Biochemistry 34 (1995) 2809.
56. H.-I. Lee, T. K. S., D. R. Dean, W. E. Newton and B. M. Hoffman, Biochemistry 37 (1998) 13370.
57. M. K. Johnson, A. J. Thomson, A. E. Robinson and B. E. Smith, Biochim. Biophys. Acta 671 (1981) 61.
58. E. Robinson, A., A. J. M. Richards, A. j. Thomson, T. R. Hawkes and B. E. Smith, Biochem. J. 219 (1984) 495.
59. L. Ma, N. Gavini, H. I. Liu, B. Hedman, K. O. Hodgson and B. K. Burgess, J. Biol. Chem. 269 (1994) 18007.
60. S. P. Cramer, T. K. Eccles, F. W. Kutzler, K. O. Hodgson and L. E. Mortenson, J. Am. Chem. Soc. 98 (1976) 1287.
61. S. P. Cramer, K. O. Hodgson, W. O. Gillum and L. E. Mortenson, J. Am. Chem. Soc. 100 (1978) 3398.
62. S. P. Cramer, W. O. Gillum, K. O. Hodgson, L. E. Mortenson, E. I. Stiefel, J. R. Chisnell, W. J. Brill and V. K. Shah, J. Am. Chem. Soc. 100 (1978) 3814.
63. K. O. Hodgson, in W. E. Newton and W. H. Orme-Johnson (eds.), Nitrogen Fixation, University Park Press, Baltimore, 1980, p. 263.
64. A. M. Flank, M. Weininger, L. E. Mortenson and S. P. Cramer, J. Am. Chem. Soc. 108 (1986) 1049.
65. S. D. Conradson, B. K. Burgess, W. E. Newton, L. E. Mortenson and K. O. Hodgson, J. Am. Chem. Soc. 109 (1987) 7507.
66. S. D. Conradson, B. K. Burgess, W. E. Newton, K. O. Hodgson, J. W. McDonald, J. F. Rubinson, S. F. Gheller, L. E. Mortenson, M. W. W. Adams, P. K. Mascharak, W. A. Armstrong and R. H. Holm, J. Am. Chem. Soc. 107 (1985) 7935.
67. M. K. Eidsness, A. M. Flank, B. E. Smith, A. C. Flood, C. D. Garner and S. P. Cramer, J. Am. Chem. Soc. 108 (1986) 2746.
68. B. Hedman, P. Frank, S. F. Gheller, A. L. Roe, W. E. Newton and K. O. Hodgson, J. Am. Chem. Soc. 110 (1988) 3798.
69. J. Chen, J. Christiansen, R. C. Tittsworth, B. J. Hales, S. J. George, D. Coucouvanis and S. P. Cramer, J. Am. Chem. Soc. 115 (1993) 5509.
70. J. Chen, J. Christiansen, S. J. George, J. van Elp, R. Tittsworth, B. J. Hales, S. Al-Ahmad, D. Coucouvanis, J. T. Bolin and S. P. Cramer, in E. I. Stiefel, D. Coucouvanis and W. E. Newton (eds.), Molybdenum Enzymes, Cofactors, and Model Systems, American Chemical Society, Washington D.C.,, 1993, p. 231.
71. J. Chen, J. Christiansen, N. Campobasso, J. T. Bolin, R. C. Tittsworth, B. J. Hales, J. J. Rehr and S. P. Cramer, Angewante Chemie International 32 (1993) 1592.
72. J. Christiansen, R. C. Tittsworth, B. J. Hales and S. P. Cramer, J. Am. Chem. Soc. 117 (1995) 10017.

73. V. K. Shah and W. J. Brill, Proc. Natl. Acad. Sci. U. S. A. 74 (1977) 3249.
74. R. L. Robson, R. R. Eady, T. H. Richardson, R. W. Miller, M. Hawkins and J. R. Postgate, Nature 322 (1986) 388.
75. R. R. Eady, R. L. Robson, T. H. Richarson, R. W. Miller and M. Hawkins, Biochem. J. 244 (1987) 197.
76. B. J. Hales, E. E. Case, J. E. Morningstar, M. F. Dzeda and L. A. Mauterer, Biochemistry 25 (1986) 7251.
77. W. G. Zumft, G. Palmer and L. E. Mortenson, Biochim. Biophys. Acta 292 (1973) 413.
78. R. R. Eady, R. Pau, D. J. Lowe and F. J. Luque, in P. M. Gresshoff, L. E. Roth, G. Stacey and W. E. Newton (eds.), Nitrogen Fixation: Achievements and Objectives, Chapman and Hall, New York, 1990, p. 125.
79. J. E. Morningstar and B. J. Hales, J. Am. Chem. Soc. 109 (1987) 6854.
80. C. Z. Blanchard and B. J. Hales, Biochemistry 35 (1996) 472.
81. D. R. Dean and L. Zheng, J. Biol. Chem. 269 (1994) 18723.
82. J. E. Morningstar, M. K. Johnson, E. E. Case and B. J. Hales, Biochemistry 26 (1987) 1795.
83. N. Ravi, V. Moore, S. Lloyd, B. J. Hales and B. H. Huynh, J. Biol. Chem. 269 (1994) 20920.
84. J. M. Arber, B. R. Dobson, R. R. Eady, P. Stevens, S. S. Hasnain, C. D. Garner and B. E. Smith, Nature 325 (1987) 372.
85. G. N. George, C. L. Coyle, B. J. Hales and S. P. Cramer, J. Am. Chem. Soc. 110 (1988) 4057.
86. J. M. Arber, B. R. Dobson, R. R. Eady, S. S. Hasnain, C. D. Garner, H. Matsushita, M. Nomura and B. E. Smith, Biochem. J. 258 (1989) 733.
87. I. Harvey, J. M. Arber, R. R. Eady, B. E. Smith, D. C. Garner and S. S. Hasnain, Biochem. J. 266 (1990) 929.
88. J. R. Chisnell, R. Premakumar and P. E. Bishop, J. Bacteriol. 170 (1988) 27.
89. K. Schneider, A. Müller, U. Schramm and W. Klipp, Eur. J. Biochem. 195 (1991) 653.
90. K. Schneider, U. Gollan, M. Dröttboom, S. Selsemeier-Voigt and A. Müller, Eur. J. Biochem. 244 (1997) 789.
91. L. Lehman and G. P. Roberts, J. Bacteriol. 173 (1991) 5705.
92. R. Davis, L. Lehman, R. Petrovich, V. K. Shah, G. P. Roberts and P. W. Ludden, J. Bacteriol. 178 (1996) 1145.
93. A. Müller, K. Schneider, K. Knüttel and W. R. Hagen, FEBS Lett. 303 (1992) 36.
94. E. Krahn, J. R. Weiss, M. Kröckel, J. Groppe, G. Henkel, S. P. Cramer, A. X. Trautwein, K. Schneider and A. Müller, J. of Biol. Inorganic Chem. (2001).
95. T. Lovell, J. Li, T. Liu, D. A. Case and L. Noodleman, J. Am. Chem. Soc. 123 (2001) 12392.
96. G. D. Watt and K. R. N. Reddy, J. Inorg. Biochem. 53 (1994) 281.
97. H. C. Angove, S. J. Yoo, B. K. Burgess and E. Münck, J. Am. Chem. Soc. 119 (1997) 8730.

98. S. J. Yoo, H. C. Angove, B. K. Burgess, M. P. Hendrich and E. Münck, J. Am. Chem. Soc. 121 (1999) 2534.
99. K. B. Musgrave, H. C. Angove, B. K. Burgess, B. Hedman and K. O. Hodgson, J. Am. Chem. Soc. 120 (1998) 5325.
100. S. J. Yoo, H. C. Angove, B. K. Burgess, E. Münck and J. Peterson, J. Am. Chem. Soc. 120 (1998) 9704.
101. P. Strop, P. M. Takahara, H.-C. Chiu, H. C. Angove, B. K. Burgess and D. C. Rees, Biochemistry 40 (2001) 651.
102. R. J. Gurbiel, J. T. Bolin, A. E. Ronco, L. Mortenson and B. M. Hoffman, J. Magn. Reson. 91 (1991) 227.
103. M. K. Johnson, A. J. Thomson, A. E. Robinson, K. K. Rao and D. O. Hall, Biochim. Biophys. Acta 667 (1981) 433.
104. R. C. Tittsworth and B. J. Hales, J. Am. Chem. Soc. 115 (1993) 9763.
105. R. C. Tittsworth and B. J. Hales, Biochemistry 35 (1996) 479.
106. K. B. Musgrave, H. I. Liu, M. Li, B. K. Burgess, G. Watt, B. Hedman and K. O. Hodgson, J. .Biol. Inorganic Chem. 3 (1998) 344.
107. D. J. Lowe, R. R. Eady and R. N. F. Thorneley, Biochem. J. 173 (1978) 277.
108. K. Fisher, W. E. Newton and D. J. Lowe, Biochemistry 40 (2001) 3333.
109. M. G. Yates and D. J. Lowe, FEBS Lett. 72 (1976) 121.
110. L. C. Davis, M. T. Henzl, R. H. Burris and W. H. Orme-Johnson, Biochemistry 18 (1979) 4860.
111. W. H. Orme-Johnson and L. C. Davis, in W. Lovenberg (ed.), Iron-Sulfur Proteins, Academic Press, New York, 1978, p. 15.
112. R. C. Pollock, H.-I. Lee, L. M. Cameron, V. J. DeRose, B. J. Hales, W. H. Orme-Johnson and B. M. Hoffman, J. Am. Chem. Soc. 117 (1995) 8686.
113. P. D. Christie, H.-I. Lee, L. M. Cameron, B. J. Hales, W. H. Orme-Johnson and B. M. Hoffman, J. Am. Chem. Soc. 118 (1996) 8707.
114. H.-I. Lee, L. M. Cameron, B. J. Hales and B. M. Hoffman, J. Am. Chem. Soc. 119 (1997) 10121.
115. L. M. Cameron and B. J. Hales, Biochemistry 37 (1998) 9449.
116. M. Sørlie, J. Christiansen, B. Lemon, J. W. Peters, D. R. Dean and B. J. Hales, Biochemistry 40 (2001) 1540.
117. S. J. George, G. A. Ashby, C. W. Wharton and R. N. F. Thorneley, J. Am. Chem. Soc. 119 (1997) 6450.
118. S. K. Ibrahim, K. Vincent, C. A. Gormal, B. E. Smith, S. P. Best and C. J. Pickett, Chem. Commun. (1999) 1019.
119. M. E. Rasche and L. C. Seefeldt, Biochemistry 36 (1997) 8574.
120. L. C. Seefeldt, M. E. Rasche and S. A. Ensign, Biochemistry 34 (1995) 5382.
121. M. J. Ryle, H.-I. Lee, L. C. Seefeldt and B. M. Hoffman, Biochemistry 39 (2000) 1114.
122. M. Sørlie, J. Christiansen, D. R. Dean and B. J. Hales, J. Am. Chem. Soc. 121 (1999) 9457.

123. H.-I. Lee, M. Sorlie, J. Christiansen, R. Song, D. R. Dean, B. J. Hales and B. M. Hoffman, J. Am. Chem. Soc. 122 (2000) 5582.
124. P. M. C. Benton, S. M. Mayer, J. Shao, B. M. Hoffman, D. R. Dean and L. C. Seefeldt, Biochemistry 40 (2001) 13816.

Nitrogen Fixation at the Millennium.
G. Jeffery Leigh, Editor.

Chapter 4

The Gene Products of the *nif* Regulon

Luis M. Rubio and Paul W. Ludden
Department of Biochemistry, University of Wisconsin, Madison, WI 53705 USA

1. INTRODUCTION

The relatively simple substrate (N_2) and product (NH_3) for the nitrogenase enzyme system of bacteria and archaea belie the complexity of the enzymatic mechanism, the number of gene products required and the regulation of expression of nitrogenase. This Chapter deals with the gene products required for molybdenum-dependent nitrogen fixation in the model organisms *Azotobacter vinelandii* and *Klebsiella pneumoniae* with references to a few other systems. Some gene products will be described in great detail in other Chapters in this volume, for example the structural proteins of the nitrogenase enzyme system, dinitrogenase reductase (also referred to as the Fe protein) and dinitrogenase (also known as the MoFe protein). Other gene products are hardlybe mentioned elsewhere, as their roles are incompletely defined.

The designation "*nif*" (for ***ni***trogen ***f***ixation) was first employed by Valentine and coworkers [1] to indicate genes required for nitrogen fixation, and was subsequently adopted by other workers [2]. In these early studies, little was known of the complexity that lay ahead. For the purposes of this review, we will use the operational definition for *nif* genes as *nifA* and those genes under the control of NifA that are responsible for the production of a functional molybdenum nitrogenase. Subsequent to the description of *nif* genes, sets of genes required for the production of an analogous vanadium nitrogenase (*vnf* genes; ***v***anadium ***n***itrogen ***f***ixation) and an iron-only nitrogenase (*anf* genes; ***a***lternative ***n***itrogen ***f***ixation) ([3] and references therein) were defined. The picture is complicated because some *nif* genes (at least *nifB*, *nifM, nifU, nifS*, and *nifV* in *A. vinelandii*) are required for production of functional *vnf*- and *anf*-encoded nitrogenases [4,5].

Before the development of genetic tools to address questions regarding the identities, roles and regulation of gene products involved in nitrogen fixation, there were few insights into the process. It was recognized that nitrogen fixation is an "adaptive" process even in pre-Jacob-Monod days [6] and subsequent work explored the relationship of growth conditions on nitrogen fixation [7-10]. A study of nitrogen fixation by the phototrophic bacterium *Rhodospirillum rubrum* grown in continuous culture in the absence of N_2 demonstrated that the substrate, N_2, is not necessary for the formation of nitrogenase [11].

A number of advances allowed workers to begin the thorough analysis of the gene products required for nitrogen fixation. The development of methods for extraction of active nitrogenase from *Clostridium pasteurianum* by Mortenson and co-workers [12] and the adaptation of these methods to obtain active extracts from other organisms initiated the biochemical analysis of nitrogenase. This analysis led to the discovery by Bulen and Lecomte that two proteins were required for N_2 reduction [13]. The development of the acetylene reduction assay [14,15] based on the discovery of ability of nitrogenase to reduce acetylene to ethylene [16,17] provided a rapid assay to assess the activity of mutant strains. Finally, powerful genetic tools developed in other organisms, especially *Escherichia coli*, were adapted for use in the genetic and biochemical dissection of the *nif* system.

At the time of the first reports of genetic analysis the *nif* system, the model organisms for investigation of the biochemistry of nitrogen fixation were *C. pasteurianum* and *A. vinelandii*. Because the best and most reproducible activities for nitrogenase were obtained with *C. pasteurianum*, much of our knowledge of the enzyme came from that source. The Kettering labs at Yellow Springs, Ohio, had adopted *A. vinelandii* for study and a number of advances in our understanding came from those investigations. However, good genetic tools were unavailable for either organism at that time and subsequent advances relied on the use *of K. pneumoniae*, a close relative of *E. coli*. *K. pneumoniae* had been adopted for genetic studies by the ARC Unit at the University of Sussex. After attempts to develop facile genetic tools in *A. vinelandii* failed, *K. pneumoniae* was employed by the Wisconsin group as well. In the intervening period, investigation of *C. pasteurianum* nitrogenase has diminished and the investigation of *A. vinelandii* has expanded because of the development of genetic tools and the tractability of biochemical analysis of that organism.

The power of the genetic/biochemical analysis for understanding complex systems such as those involved in nitrogen fixation can be seen in the rapid progress that was made between the first and the third major international gathering of scientists studying nitrogen fixation. At the first meeting in Pullman, Washington, in 1974, a few genes and their linkages were reported. By the time of the third such gathering in 1978, an extensive map of the *K. pneumoniae nif* regulon was available [18]. The development of new tools also allowed the sequencing of the *nif* regulons from *K. pneumoniae* [19] and *A. vinelandii* [20,27], which revealed a number of previously unknown reading frames; the roles of the products of some of these have become known while others remain obscure. The advent of the age of genomics and proteomics will now allow the *nif* regulons of nitrogen-fixing organisms to be investigated in the context of the entire genome or proteome of the organism. The genomes of several nitrogen-fixing organisms are known, but not those of *K. pneumoniae* or *A. vinelandii*, although both should be disclosed soon. Undoubtedly, these will provide some surprises.

Much of our information regarding the gene products required for nitrogen fixation has been obtained from analysis of *K. pneumoniae* and *A. vinelandii*. Although these organisms have significant physiological differences, both are members of the gamma division of the proteobacteria and thus may not provide the full range of gene products involved in production and function of a molybdenum nitrogenase. This may be especially true for the gene products required for electron-donation to nitrogenase and regulation of expression of *nif* gene products. Significant insight has been obtained from analysis of gene products from *C. pasteurianum, Rhodobacter capsulatus, Anabaena, Azospirillum brasilense, R. rubrum* and *Methanosarcina barkeri* among other organisms. The focus of this review on gene products from *K. pneumoniae* and *A. vinelandii* is not intended to detract from the value of information provided by analysis of those systems.

2. ORGANIZATION OF NITROGEN-FIXATION (*nif*) GENES IN *Azotobacter vinelandii* AND *Klebsiella pneumoniae*

The products of the structural genes of the Mo-containing nitrogenase are not initially competent for substrate reduction. Rather, the products of a number of other nitrogen-fixation (*nif*)-specific genes are required for maturation of the nitrogenase components to their active forms, for electron transport to nitrogenase, and for activation or inhibition of *nif*-gene expression under the appropriate physiological conditions (Table 1). The regulation of *nif*-gene expression in *A. vinelandii* and other diazotrophs is discussed in detail in Chapter 6. The genes required for electron transport to nitrogenase in *K. pneumoniae* are *nifF* and *nifJ*, which encode a flavodoxin and a pyruvate:flavodoxin oxidoreductase, respectively; these genes will be briefly discussed in Section 5.5. The pathways for electron transport to nitrogenase in *A. vinelandii* and other diazotrophs are not yet well defined; nevertheless, current knowledge on the matter will be presented in Sections 5.5 and 5.6. The *nif*-genes required for the biosynthesis of the iron-molybdenum cofactor of nitrogenase (FeMo-co) will be discussed thoroughly in Section 4, while other *nif* genes known to be required for the maturation of the nitrogenase component proteins will be analyzed in Section 5.

The anaerobe facultative *K. pneumoniae* was the first diazotroph whose nitrogen-fixation genes were thoroughly analyzed. A combination of genetic and biochemical techniques, together with DNA sequencing, was used to identify the *nif* genes and *nif*-gene products of *K. pneumoniae* and, finally, to develop the first genetic map of nitrogen fixation [18,19,21-26]. In *K. pneumoniae*, 20 *nif*-genes, *nifJHDKTYENXUSVWZMFLABQ*, are clustered in a single 23-kb region in the chromosome and organized in 8 transcriptional units (Figure 1A).

The *nif* genes of the free-living bacterium *A. vinelandii* were subsequently identified and analyzed in comparison to the *nif* genes of *K. pneumoniae* ([20,27], and references therein). In *A. vinelandii*, the *nif*-specific genes are clustered into two different linkage groups: (1) the so-called major *nif* cluster, containing the *nifHDKTYENXUSVWZMF* genes and several ORFs interspersed among the *nif* genes [27]; and (2) the *nifABQ* region [4]. The subsequent identification of additional *nif* or *nif*-related genes (*nifL* [28], *fdxN* and *nifO* [29], *rnfABCDGEH* and *nafY* [30,31]) within and upstream from the *nifABQ* region has increased the size of the minor cluster. Eleven potential *nif*-specific promoters have been identified in the major *nif* cluster, and only three in the minor *nif* cluster. The two clusters of *nif* genes from *A. vinelandii* are presented in Figure 1B.

Although there are many *nif*-specific genes identified in free-living and symbiotic diazotrophs (*e.g.*, those from *R. capsulatus* or *Bradyrhizobium japonicum*), our discussions will focus on current knowledge of the biochemical genetics of nitrogen fixation in *A. vinelandii* and *K. pneumoniae*.

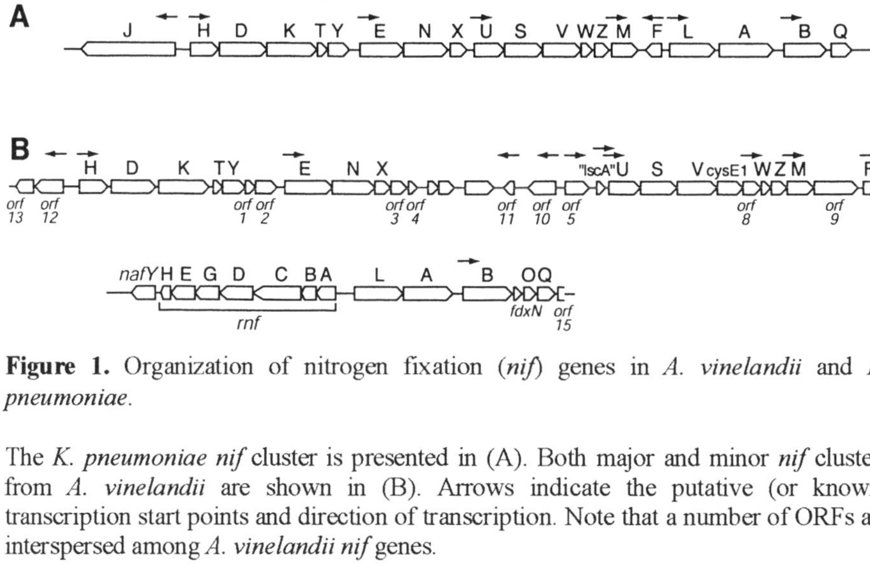

Figure 1. Organization of nitrogen fixation (*nif*) genes in *A. vinelandii* and *K. pneumoniae*.

The *K. pneumoniae nif* cluster is presented in (A). Both major and minor *nif* clusters from *A. vinelandii* are shown in (B). Arrows indicate the putative (or known) transcription start points and direction of transcription. Note that a number of ORFs are interspersed among *A. vinelandii nif* genes.

3. STRUCTURAL GENES FOR DINITROGENASE AND DINITROGENASE REDUCTASE

3.1 NifD and NifK Gene Products

The nitrogenase enzyme is composed of two oxygen-labile metalloproteins: dinitrogenase (also referred to as MoFe protein or component I) and dinitrogenase reductase (also referred to as Fe protein or component II) [13,32]. Dinitrogenase is a 230-kDa $\alpha_2\beta_2$ tetramer of the *nifD* and *nifK* gene products that presents a two-fold symmetrical structure between pairs of αβ dimers. Although the α and β subunits have only minimal sequence similarity, they have remarkably similar folding patterns, in which each subunit has three internal folding domains [33].

Dinitrogenase contains two types of metallocluster, known as the iron-molybdenum cofactor (FeMo-co) and the P cluster [34-40]. Each αβ dimer contains one FeMo-co paired with a P cluster separated by a distance of 19 Å. FeMo-co has an inorganic composition of 7Fe: 1Mo: 9S, arranged as a [Mo-3Fe-3S] cluster and a [4Fe-3S] cluster connected by three sulfurs; it also contains one molecule of (*R*)-homocitrate with the C-2 carboxyl and hydroxyl groups coordinated to the Mo atom (Figure 2).

FeMo-co is buried 10 Å below the protein surface and is almost completely encompassed by the α subunit of dinitrogenase, although some residues from the β subunit are linked to the homocitrate by water molecules [36]. Dinitrogenase has only two protein ligands to FeMo-co: α-His^{442}, which binds to the Mo atom at one end of

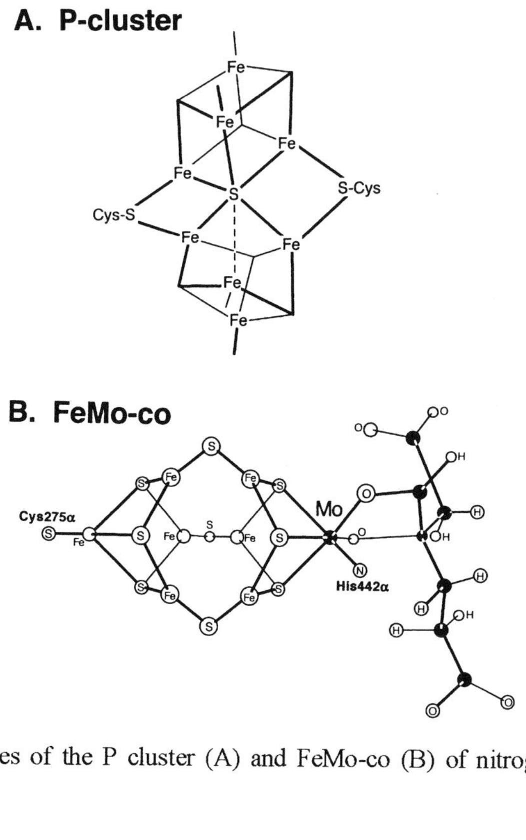

Figure 2. Structures of the P cluster (A) and FeMo-co (B) of nitrogenase. Adapted from [33,37].

the molecule, and α-Cys^{275}, which binds to the Fe atom located at the opposite site of the cofactor. The P clusters are located at the interface between the α and β subunits at 12-15 Å below the protein surface. Each P cluster is composed of two [4Fe-4S] clusters that are bridged by two thiol groups provided by residues α-Cys^{88} and β- Cys^{95} and by a sulfur atom [37,40] (Figure 2). In addition, each αβ dinitrogenase dimer donates another four cysteine residues to coordinate the remaining Fe sites in the P cluster (α-Cys^{62}, α-Cys^{154}, β-Cys^{70} and β-Cys^{153}) [36,38].

Table 1. Genes involved in nitrogen fixation and roles of their products.

Gene	Nif phenotype	Identity / role
nifH	-	Dinitrogenase reductase. Obligate electron donor to dinitrogenase. Also required for FeMo-co biosynthesis and apo-dinitrogenase maturation
nifD	-	α-subunit of dinitrogenase. FeMo-co binding subunit. Forms an $\alpha_2\beta_2$ tetramer with NifK
nifK	-	β-subunit of dinitrogenase
nifT	+	Unknown
nifY	+	In *K. pneumoniae,* aids in the insertion of FeMo-co into apo-dinitrogenase
nifE	-	Required as scaffold for FeMo-co synthesis. Forms an $\alpha_2\beta_2$ tetramer with NifN
nifN	-	Required as scaffold for FeMo-co synthesis. Forms an $\alpha_2\beta_2$ tetramer with NifE
nifX	+	Involved in FeMo-co synthesis and, possibly, in stability of apo-dinitrogenase
nifU	-	Mobilizes Fe for [Fe-S] cluster synthesis and repair
nifS	±	Mobilizes S for [Fe-S] cluster synthesis and repair
nifV	±	Homocitrate synthase. Involved in FeMo-co synthesis
nifW	±	Involved in maturation of dinitrogenase. Also proposed to protect dinitrogenase from O_2 inactivation
nifZ	+	Unknown. May form a complex with NifW
nifM	-	Required for the maturation of NifH
nifF	+	Flavodoxin. Physiological electron donor to NifH
nifL	+	Sensor protein. Acts as NifA anti-activator. Forms a two-component regulatory system with NifA
nifA	-	Transcriptional activator. Forms a two-component regulatory system with NifL
nifB	-	Required for FeMo-co synthesis. Its product, NifB-co serves as specific Fe and S donor to FeMo-co
fdxN	±	[4Fe-4S] Ferredoxin. Electron donor to nitrogenase
nifO	+	Unique to *A. vinelandii.* Unknown role
nifQ	-	Required for FeMo-co biosynthesis. Processes Mo prior to its incorporation into FeMo-co

rnfA	-	Integral membrane protein. Part of the Rnf complex, that may be involved in electron transfer to nitrogenase
rnfB	-	[Fe-S] protein. Part of the Rnf complex
rnfC	-	[Fe-S] protein. Part of the Rnf complex
rnfD	-	Membrane protein. Part of the Rnf complex
rnfG	-	Secretory membrane protein. Part of the Rnf complex
rnfE	-	Integral membrane protein. Part of the Rnf complex
rnfH	?	Soluble protein. Part of the *rnf* gene cluster
nafY	+	In *A. vinelandii*, γ subunit of apo-dinitrogenase. Stabilizes apo-dinitrogenase prior to FeMo-co insertion. Also binds FeMo-co specifically

+ Diazotrophic growth, ± slow diazotrophic growth, - no diazotrophic growth, ? unknown.

Table 1 (Cont.) Genes involved in nitrogen fixation and roles of their products.

3.2 NifH

Dinitrogenase reductase is a symmetrical 64-kDa α_2 dimer of the *nifH* gene product that contains a single [4Fe-4S] cluster bridging the subunits [41]. The [4Fe-4S] cluster is coordinated by two cysteinyl residues from each subunit (Cys^{97} and Cys^{132} in the *A. vinelandii* enzyme), and is positioned just beneath the surface. Dinitrogenase reductase also holds one MgATP-binding site per subunit, located in the interface between subunits and separated from the [4Fe-4S] cluster by a distance of 20 Å. The most notable structural features in each monomer are a β-sheet core surrounded by α-helices, and the switch loops I and II that interact with the nucleotide γ-phosphate and the Mg^{2+} [33].

NifH has at least three roles in the nitrogenase enzyme system [42]: first, it specifically serves as electron donor to nitrogenase; second, it participates in the biosynthesis of FeMo-co (see Section 4.4); and third, it is required for maturation of apo-dinitrogenase to a FeMo-co-activatable form (see Section 6.2). The electron-transfer role of NifH is well understood, but the specific roles of NifH in the other two processes are obscure. The reduction of dinitrogenase by dinitrogenase reductase occurs one electron at a time, in a process that involves the association and dissociation of the component proteins and the hydrolysis of two MgATPs by NifH for each electron transferred [32,43-46]. The electrons are transferred from the [4Fe-4S] cluster in dinitrogenase reductase to the P cluster of dinitrogenase and then from the P cluster to FeMo-co, the site of substrate reduction [47,48]. Recently, an all-ferrous form of the [4Fe-4S] cluster of NifH has been obtained *in vitro* [49,50], and the crystallographic structure of the all-ferrous NifH has been determined [51]. These observations have led to the speculation that NifH may serve as a two-electron donor to nitrogenase.

4. GENE PRODUCTS REQUIRED FOR THE BIOSYNTHESIS OF THE IRON-MOLYBDENUM COFACTOR (FeMo-co) OF NITROGENASE

The genes encoding the dinitrogenase component polypeptides, *nifD* and *nifK*, are not required for FeMo-co biosynthesis, suggesting that FeMo-co is assembled elsewhere in the cell and is then inserted into a semi-maturated apo-dinitrogenase that already contains the P clusters [52,53]. A whole set of *nif*-specific gene products is known to be involved in the biosynthesis of FeMo-co. The development of the *in vitro* FeMo-co synthesis assay was a milestone in the study of FeMo-co biosynthesis, as it provided an assay for the purification of gene products involved in the biosynthesis of FeMo-co [54]. As a result, nearly all components known to be required for FeMo-co synthesis have been isolated in purified form. The *in vitro* FeMo-co synthesis system involves the mixing either of purified components or of extracts from mutant strains defective in the synthesis of FeMo-co to reconstitute nitrogenase activity by biochemical complementation. The minimum requirements for *in vitro* FeMo-co synthesis are NifNE, NifB, NifH, molybdate, homocitrate, and ATP. The function of NifQ is hypothesized to be in the processing of molybdenum, and NifV is responsible for the generation of homocitrate for FeMo-co biosynthesis [55]. Therefore, molybdate and homocitrate can replace NifQ and NifV in the *in vitro* system. In addition, although it is not absolutely required, NifX stimulates the *in vitro* FeMo-co synthesis three-fold. Thus, at least seven *nif* gene products, *nifB*, *nifE*, *nifN*, *nifH*, *nifQ*, *nifV*, and *nifX*, are involved in the synthesis of FeMo-co in *K. pneumoniae* and *A. vinelandii* [39,42,56,57]. The addition of exogenous Fe and S to the *in vitro* FeMo-co synthesis reaction has no effect on the yield of the reaction. It appears that the Fe and S required for FeMo-co synthesis is already present in protein-bound forms and is readily incorporated into FeMo-co (V. K. Shah, unpublished results).

Figure 3 depicts the current model for FeMo-co biosynthesis. The model is based on the results obtained from a biochemical-genetic approach to the problem of FeMo-co biosynthesis and can be summarized as follows: (1) NifB-co is assembled on NifB. It is likely that NifU and NifS are involved in the mobilization of Fe and S for the biosynthesis of NifB-co, but there is no experimental evidence supporting this hypothesis. (2) NifB-co is transferred from NifB to NifNE, a scaffold protein upon which FeMo-co is synthesized. It is not known if NifNE catalyzes any modification on NifB-co, but it is clear that NifB-co will serve as specific Fe and S donor to FeMo-co. (3) Mo is then added to NifB-co (or to a modified NifB-co) in a NifH-dependent reaction. Presumably, Mo has been previously processed by the activity of NifQ. (4) Homocitrate is incorporated into the FeMo-co precursor in a reaction possibly catalyzed by NifX. (5) The finished cofactor is then bound to NafY, the product of a non-*nif* gene that also stabilizes the FeMo-co-deficient apo-dinitrogenase. (6) Finally,

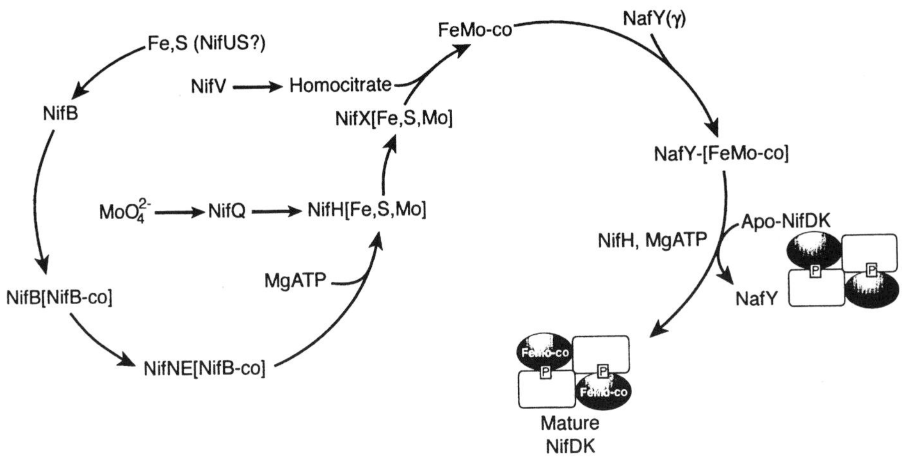

Figure 3. Proposed model for FeMo-co biosynthesis.

FeMo-co is inserted into apo-dinitrogenase to reconstitute the holoenzyme. To be competent for FeMo-co activation, apo-dinitrogenase must be previously maturated in a reaction that requires NifH. While presented here as a linear biosynthetic pathway composed of individual players that transfer the intermediates in FeMo-co biosynthesis to each other sequentially, it is likely that most of the reactions catalyzed may occur in a single multi-protein complex that we denominate the "FeMo-co biosynthesis factory". Figure 4 illustrates the organization of the hypothetical "factory". NifNE would constitute the core of the complex, to which NifB, NifH and NifX component polypeptides would most probably associate in a spatial or temporal arrangement.

The specific roles of the *nif* gene products involved in the biosynthesis of FeMo-co are discussed in detail below.

4.1. NifQ

NifQ is involved in the processing of Mo prior to its incorporation into FeMo-co [58]. The *nifQ* gene was first defined in *K. pneumoniae* by MacNeil *et al.* [22] as the *nif* complementation group closest to the *his locus*, forming an operon with *nifB*. All point mutations, Mu insertions and deletion mutations in *nifQ* yielded a very leaky phenotype, making the genetic analysis difficult. NifQ mutants from *K. pneumoniae* are deficient in FeMo-co biosynthesis, and thus present a clear Nif$^-$ phenotype only when low concentrations of extra-cellular molybdate (nanomolar levels) are available. High concentrations of molybdate in the medium (micromolar levels) produce a phenotypic reversion in *nifQ* mutant strains [58]. A similar effect of Mo in the suppression of the *nifQ* phenotype has been observed in *A. vinelandii* [4,29] and *R. capsulatus* [59], suggesting that NifQ has an essential role for nitrogen fixation only when performed under conditions of low levels of available Mo, conditions found in many natural environments. Interestingly, substitution of cystine for sulfate as the sulfur source in the medium also suppresses the effect of the *nifQ* mutation. This has been interpreted as indication of a reaction between Mo and a sulfur donor at an early step in the biosynthesis of FeMo-co [60]. Although normally NifQ would catalyze this reaction, such a transformation might also occur non-enzymatically when the levels of

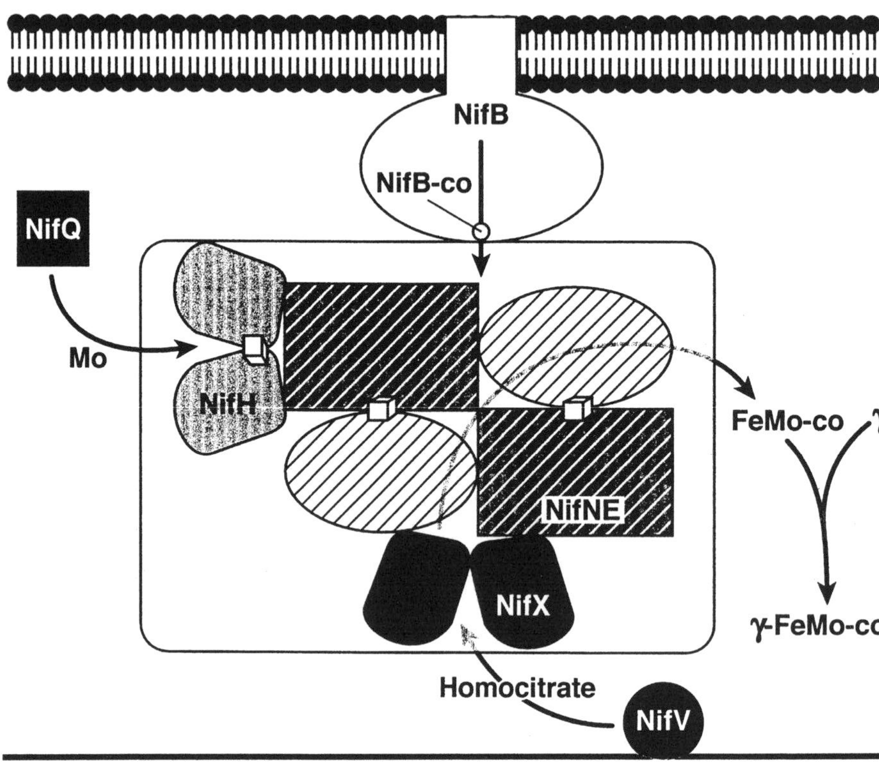

Figure 4. The "FeMo-co biosynthetic factory" model.

the reactants are high. Consistent with the above results, while *nifQ* mutants accumulate lower levels of Mo than the wild type in normal growth conditions, the inclusion of cystine or a high concentration of molybdate in the growth medium increases the accumulation of intracellular Mo in *nifQ* mutants to wild-type levels and restores maximum nitrogenase activity.

It is significant that *nifQ* mutants are not defective in the biosynthesis of the molybdenum cofactor (Mo-co) that is present in molybdoenzymes other than nitrogenase, such as nitrate reductase. Therefore, the *nifQ* gene seems to be specifically involved in the biosynthesis of FeMo-co [61].

The sequence of the *nifQ* gene from *K. pneumoniae* implies a molecular weight for the NifQ polypeptide of 20 kDa; however the product of *K. pneumoniae nifQ* has not yet been visualized on a gel. On the other hand, *A. vinelandii* NifQ has been detected in cell extracts resolved by two-dimensional gel electrophoresis as a protein of 22 kDa and a pI of 6.1 [62], behaving exactly as predicted from the *nifQ* DNA sequence [4]. NifQ is present in *A. vinelandii* cells grown in the presence of molybdenum and the absence of ammonium. The primary amino acid sequence of NifQ reveals a cluster of four cysteines near the C-terminus that is located in a very hydrophilic region of the protein and would likely be at its surface [63]. The presence of these putative metal-binding thiol moieties is consistent with NifQ being involved in the synthesis of a Mo-S intermediate. It is also possible that NifQ is assisting in the assembly of FeMo-co precursors, such as the $MoFe_3S_3$ unit of FeMo-co. The fact that both cysteine, the likely sulfur source for Fe-S cluster biosynthesis, and Mo can suppress the *nifQ* phenotype [60] is consistent with both hypotheses. Until NifQ is available in a purified form, the details of Mo processing before its incorporation into FeMo-co will remain speculative.

4.2. NifB

NifB activity is absolutely required for FeMo-co biosynthesis. *K. pneumoniae* and *A. vinelandii* strains with mutations in *nifB* accumulate a FeMo-co-deficient dinitrogenase, referred to as apo-dinitrogenase, that can be activated *in vitro* by purified FeMo-co [4,18,64,65]. The product of the *A. vinelandii nifB* gene has a predicted molecular weight of 54 kDa and a pI of 6.1 [4]; it contains a cluster of cysteines near the N-terminus that is conserved among NifB homologs, consistent with the proposed role of NifB in FeMo-co biosynthesis (see below). There are a number of NifB homologs identified in nitrogen-fixing organisms, including *A. chroococcum* [66], *R. capsulatus* [67,68], *Rhizobium leguminosarum* [69], *Rhizobium meliloti* [70], and the cyanobacteria *Anabaena* 7120 [71] and *Anabaena variabilis* [72]. Furthermore, the NifB family of proteins has been proposed to belong to the "Radical SAM" protein superfamily that contains more than 600 proteins [73]. Members of the Radical SAM superfamily catalyze a great diversity of reactions that may share a common mechanism: the generation of radical species by reductive cleavage of S-adenosylmethionine (SAM) through an unusual [Fe-S] center.

In *A. vinelandii*, the *nifB* gene is required for the activity of all three nitrogenases (Mo nitrogenase, V nitrogenase, and Fe-only nitrogenase). Consequently, *nifB* mutants are unable to grow under any of the three diazotrophic conditions (nitrogen limitation and: Mo sufficiency, V sufficiency together with Mo deficiency, and Mo and V deficiencies) [4,29,74,75]. It is not surprising, then, that nitrogenase-derepressed cells show *nifB* expression [4,29,76] and NifB accumulation [62] under all of the conditions mentioned above. In all cases, the presence of ammonium in the medium represses the expression of *nifB*. There are potential binding sites for the transcriptional activators responsible for the expression of genes involved in the Mo and V nitrogenase systems (NifA and VnfA, respectively) in the region upstream of *nifB* transcription start site, and it has been shown that NifA, VnfA and, to some extent, AnfA (the iron-only nitrogenase transcriptional activator) are able to activate *nifB* transcription [77].

NifB has recently been identified in *A. vinelandii* cell extracts resolved by two-dimensional gel electrophoresis as a protein of 54 kDa and a pI of 6.5 [62]. Although an active and purified form of NifB is not yet available, the metabolic product of NifB, termed NifB-co, has been isolated from membranes of *K. pneumoniae* using the detergent *N*-lauryl sarcosine [78]. Solutions of NifB-co thus obtained contain no detectable protein or Mo. NifB-co seems to be a small [Fe-S] cluster exhibiting many of the characteristics of FeMo-co: it is greenish-brown, stable in *N*-methylformamide, and extremely oxygen-labile. Multiple lines of evidence led Shah *et al.* [78] to propose that NifB-co is a small [Fe-S] cluster synthesized by NifB as a precursor of FeMo-co. First, the requirement for NifB activity in the *in vitro* FeMo-co synthesis assay is satisfied by the simple addition of NifB-co, and the amount of FeMo-co synthesized *in vitro* is stoichiometric to the amount of added NifB-co when all other components are in excess. Second, NifB-co-dependent *in vitro* FeMo-co synthesis is also absolutely dependent on the presence of molybdate, homocitrate and NifNE. Third, the only metal detected in purified NifB-co preparations is Fe. This hypothesis was later confirmed by showing the specific incorporation into FeMo-co of Fe and S from ^{55}Fe- and ^{35}S-labeled NifB-co [79].

In this context, it has been proposed that the role of NifB could be that of a molecular scaffold for the assembly of a portion of FeMo-co (that would be NifB-co) that must be donated to the NifNE complex [80]. There are a number of observations supporting this hypothesis: First, NifB contains a cluster of cysteines near the N-terminus, that is conserved among NifB homologs, and is located in a region of the protein that is rich in polar and charged amino acid residues [69,70]. It seems a good site to bind an [Fe-S] cluster (NifB-co) that has to be further transferred to an interacting protein (NifNE). Second, the interaction between NifB and NifN domains has been demonstrated in *C. pasteurianum*, since *nifB* and *nifN* genes are fused in this organism [81]. Third, the formation of a NifNE•NifB-co complex has been well demonstrated *in vitro* by following the changes in the electrophoretic mobility of NifNE upon NifB-co binding [82] and the incorporation into NifNE of Fe and S from ^{55}Fe- and ^{35}S-labeled NifB-co [79].

4.3. NifNE

The *nifN* and *nifE* genes are also essential for the biosynthesis of FeMo-co. Like *nifB* mutant strains, *nifN* or *nifE* mutant strains produce an apo-dinitrogenase that can be activated *in vitro* by purified FeMo-co [18]. NifNE has been purified from a *nifB* mutant strain, applying the *in vitro* FeMo-co synthesis assay, as an $\alpha_2\beta_2$ tetramer of about 200 kDa of molecular weight [83]. The UV-visible spectrum of NifNE and its oxygen lability suggested the presence of an [Fe-S] cluster in the complex. Later on, the purified NifNE complex was shown to contain two identical [4Fe-4S] clusters having an S=1/2 ground state in the reduced form and a reduction potential of −350 mV [84]. NifNE does not function catalytically in the *in vitro* FeMo-co synthesis system. It accumulates a FeMo-co precursor that is further utilized for the synthesis of the cofactor through the consumption of some moiety of NifNE. This led Paustian *et al.* [83] to suggest that the consumed moiety in NifNE was probably Fe or S from the [Fe-S] clusters.

Roll *et al.* [82] reported the accumulation of different forms of NifNE in *A. vinelandii* cell extracts depending on the presence of a functional *nifB* gene in the cell. The NifNE species accumulated in a *nifHDK*$^-$ *nifB*$^+$ background is associated with NifB activity. A second NifNE species, with different electrophoretic mobility, accumulates in a *nifB*$^-$ background and can be converted to the first one by the simple addition of NifB-co. It seems that, in the absence of NifH, NifB-co is bound to NifNE but is not further processed. Further *in vitro* work with ^{55}Fe-labeled NifB-co showed that NifNE indeed binds specifically NifB-co [79]. Recent studies show that a FeMo-co precursor that has incorporated Mo also accumulates on NifNE when NifH, NifB-co and ATP are provided to the system (P. Rangaraj and P. W. Ludden, unpublished results).

Figure 5 illustrates the high degree of sequence similarity that exists between NifE and NifD (encoding the α subunit of dinitrogenase) and between NifN and NifK polypeptides (encoding the β subunit of dinitrogenase) [85-87]. Note that the NifE-NifD and NifK-NifN subsets of the alignment are also similar to each other. It has been proposed that *nifD*, *nifK*, *nifE*, and *nifN* constitute a paralogous family of genes that arose from an ancestor gene by two consecutive events of paralogous duplication, followed by divergence [88]. The first duplication and divergence event produced an ancestral bicistronic operon that, in turn, underwent duplication and divergence generating the *nifDK* and *nifNE* clusters (Figure 6).

The extent of the sequence conservation, the mutual stability requirement for NifE and NifN [18], and the fact that dinitrogenase is not required for the biosynthesis of FeMo-co led to speculation that NifNE forms a complex that provides a molecular scaffold for the synthesis of FeMo-co [85]. Muchmore *et al.* modeled the NifNE complex [80] based on the crystal structure of dinitrogenase [89]. These authors suggested that each of the two [4Fe-4S] clusters in the NifNE tetramer is bridged between a NifE-NifN subunit interface at a position analogous to that occupied by the P clusters in dinitrogenase. In contrast to the P clusters in dinitrogenase, the [4Fe-4S]

```
NifD    1  ...............MTGMSREEVESLIQEVLEVYPEKARKDRNKHLAVNDPAVTQSKKC
NifE    1  .......................................MKAKDIAELLDEPACSHNKKE

NifK    1  MSQQVDKIKASYPLFLDQDYKDMLAKKRDGFEEKYPQDKIDEVFQWTTTKEYQELNFQRE
NifN    1  ...................................................MAEIINRNK

NifD   46  IISNKKSQPGLMTIRGCAYAGSKGVVWGPIKDMIHISHGPVGCGQYSRAGRRNYYIGTTG
NifE   22  KSGCAKPKPGATDGR.CSFDG.AQIALLPVADVAHIVHGPIACAGSSWDNRGTRSSGPD.

NifK   61  ALTVNPAKACQPLG........AVLCALGFEKTMPYVHGSQGCVAYFRSYFNRHFREPVS
NifN   10  ALAVSPLKASQTMG........AALAILGLALSMPLFHGSQGCTAFAKVFFVRHFREPVP

NifD  106  VNAFVTMNFTSDFQEKDIVFGGDKK..LAKLIDEVETLFPLNKGISVQ..SECPIGLIGD
NifE   78  ...LYRIGMTTDLTENDVIMGRAEKRLFHAIRQAVESYLPP..AVFVY..NTCVPALIGD

NifK  112  .......CVSDSMTEDAAVFGGQQN.MKDGLQNCKATYKPDMIAVSTTCMAEVIGDDLNA
NifN   61  .......LQTTAMDQVSSVMGADEN.VVEALKTICERQNPSVIGLLTTGLSETQGCDLHT

NifD  162  DIESVSKVKGAELSKTIVPVRCEGFRGVSQSLGHHIANDAVRDWVLGKRDEDTTFASTP.
NifE  132  DVDAVCKAAAERFGTPVIPVDSAGFYGTKN.LGNRIAGEAMLKYVIGTREPDPLPVGSER

NifK  165  FINNSKKEGFIPDEFPVPFAHTPSFVGSHV.TGWDNMFEGIARYFTLKSMDDKVVGSNK.
NifN  114  ALHEFRTQYEEYKDVPIVPVNTPDFSGCFE.SGFAAAVKAIVETLVPERRDQVGKRPRQ.

NifD  220  .....YDVAIIGDYNIG.GDAWSSRILLEEMGLR..CVAQWSG..DGSISEIELTPKVKL
NifE  191  PGIRVHDVNLIGEYNIA.GEFWHVLPLLDELGLR..VLCTLAG..DARYREVQTMHRAEV

NifK  222  ......KINIVPGFETYLGNFRVIKRMLSEMGVGYSLLSDPEEVLDTPADG..QFRMYAG
NifN  171  .......VNVLCSANLTPGDLEYIAESIESFGLRPLLIPDLSGSLDGHLDENRFNALTTG

NifD  271  NLVHCYRSMNYISRHMEEKYGIPWMEYNFFGPTKTIESLRAIAAKFDES.IQKKCEEVIA
NifE  246  NMMVCSKAMLNVARKLQETYGTPWFEGSFYGITDTSQALRDFARLLDDPDLTARTEALIA

NifK  275  GTTQEEMKDAPNALNTVLLQPWHLEKTKKFVEGTWKHEVPKLNIPMGLDWTDEFLMKVSE
NifN  225  GLSVAELATAGQSVATLVVGQSLAGAADALAERTGVPD.RRFGMLYGLDAVDAWLMALAE

NifD  330  KYKPEWEAVVAKYRPRLEGKRVMLYIGGLRPRHVIGAYEDLGM....EVVGTGYEFAHND
NifE  306  REEAKVRAALEPWRARLEGKRVLLYTGGVKSWSVVSPLQDLGM....KVVATGTKKSTEE

NifK  335  ISGQPIPASLTKERGRLVDMMTDSHTWLHGKRFALWGDPDFVMGLVKFLLELGCEPVHIL
NifN  284  ISGNPVPDRYKRQRAQLQDAMLDTHFMLSSARTAIAADPDLLLGFDALLRSMGAHTVAAV

NifD  386  DYDR............TMKEMGDSTLLYDDVTGYEFEEFVKRIKPDLIGSGIK.......
NifE  362  DKAR............IRELMGDDVKMLDEGNARVLLKTVDEYQADILIAGGR.......

NifK  395  CHNGNKRWKKAVDAILAASPYGKNATVYIGKDLWHLRSLVFTDKPDFMIGNSYGKFIQRD
NifN  344  VPAR............AAALVDSPLPSVRVGDLEDLEHAARAGQAQLVIGNSH.......

NifD  427  EKFIFQKMGIPFRQMHSWDYSGPYHGFDGFAIFAR...DMDMTLNNPCWKKLQ..APWEA
NifE  403  NMYTALKGRVPFLDINQEREFG.YGGYDRMLELVR...HVCITLECPVWEAVRRPAPWDI

NifK  455  TLHKGKEFEVPLIRIG.FPIFDRHHLHRSTTLGYEGAMQILTTLVNSILERLDEETRGMQ
NifN  385  ALASARRLGVPLLRAG.FPQYDLLGGFQRCWSGYRGSSQVLFDLANLLVEHHQGIQPYHS
```

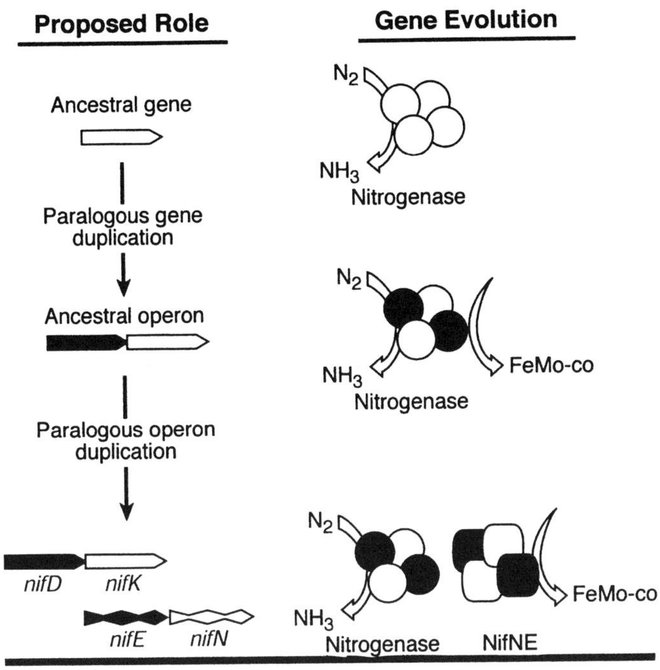

Figure 6. Model for the evolution of the *nifD, nifK, nifE,* and *nifN* genes. Adapted from Fani *et al.* [88].

clusters in NifNE would be asymmetrically coordinated to the NifN and NifE subunits (three cysteine residues would be donated by NifE and only one cysteine by NifN). This model places the [4Fe-4S] clusters in NifNE far from the proposed FeMo-co assembly site. So, in contrast to Paustian *et al.*, it is suggested that the clusters may have a redox or structural role rather than becoming incorporated into FeMo-co during cofactor biosynthesis. In this model, the pocket for the assembly of FeMo-co seems to be large enough to fit homocitrate in and contains Cys^{250} as a candidate for FeMo-co binding.

Figure 5. Amino acid sequence alignment of NifD, NifK, NifN, and NifE proteins from *A. vinelandii*.
Symbols above NifD and below NifN indicate amino acid residues identical (circles) or similar (asterisks) in the pairs NifD-NifE and NifK-NifN, respectively. Identical (circles) and similar (asterisks) amino acid residues in three or four of the aligned proteins are depicted between NifE and NifK sequences. The last 15 amino acids of the alignment have been removed for clarity.

The *vnf* regulon contains an analogous set of genes that have been designated *vnfEN* [90]. However, the products of *vnfEN* have not been demonstrated to be functionally similar to NifNE. The *anf* system contains no homologous set of *anfEN* genes. An extensive search of the entire genome of *R. capsulatus*, which produces a highly active Fe-only nitrogenase (encoded by the *anf* system), reveals no *anfEN* genes. Klipp *et al.* have demonstrated that functional *nifNE* genes are essential for molybdenum-dependent diazotrophic growth of *R. capsulatus* [91], while *nifNE* mutant strains grow well diazotrophically in the absence of Mo. This has led to the hypothesis that the structural genes of the iron-only nitrogenase (*anfDK*) play this role for the iron-only system (S. Nordlund, unpublished work). To further complicate the picture, a recent paper by Klipp and co-workers [92] indicates that NifNE is not required absolutely for the biosynthesis of FeMo-co in *R. capsulatus*. They propose that, in the absence of NifNE, a less efficient assembly of FeMo-co takes place in a γ-like protein.

4.4. NifH

Apart from the role in electron transfer to dinitrogenase, NifH is required for FeMo-co biosynthesis and apo-dinitrogenase maturation [93-96]. It is well known that *nifH* mutants of both *A. vinelandii* and *K. pneumoniae* fail to synthesize FeMo-co [93,94]. An absolute requirement for NifH has also been demonstrated using the *in vitro* FeMo-co synthesis system [97]. Since FeMo-co is probably assembled into the NifNE complex, which is structurally similar to dinitrogenase, it was originally thought that the role of NifH in FeMo-co biosynthesis was similar to its role in catalysis— that is, NifH would serve as electron donor to NifNE. However, it was shown later that the role of NifH in FeMo-co synthesis is independent of its function during nitrogenase turnover [98,99]. Moreover, neither the presence of the [4Fe-4S] cluster in NifH [100] nor the hydrolysis of Mg-ATP by NifH [101], which are processes known to be required for electron transfer to dinitrogenase, are necessary for the function of NifH in FeMo-co biosynthesis. Another example of the independence of the roles played by NifH is provided by the phenotype of *nifM* mutants [102]. Strains with mutations in *nifM* accumulate an immature form of NifH that is catalytically inactive as dinitrogenase reductase but remains competent for FeMo-co biosynthesis or apo-dinitrogenase maturation. It is also unlikely that the role of NifH in FeMo-co biosynthesis is that of Fe or S donor to FeMo-co because, as we noted above, the apo-form of NifH (whose [4Fe-4S] cluster has been removed by chelation) still functions in the biosynthesis of the cofactor [100].

One possibility could be that NifH ensures that Mo and no other heterometal is incorporated into FeMo-co. The rationale for this proposal is that each nitrogenase system (Mo-, V- and Fe-only-nitrogenase) has an associated dinitrogenase reductase. The highly similar proteins NifH, VnfH and AnfH could be responsible for the specific incorporation of Mo, V or Fe into each individual type of cofactor. Although there is some experimental evidence suggesting a role for NifH in the addition of Mo to a precursor of FeMo-co (P. Rangaraj and P. W. Ludden, unpublished results), it has been demonstrated that dinitrogenase reductase from the vanadium system (VnfH) can replace NifH function in the processes of *in vitro* biosynthesis of FeMo-co and apo-dinitrogenase maturation [103]. It is therefore unlikely that NifH, VnfH and AnfH alone can specify the metal that will be incorporated into the cofactor.

4.5. NifX and the family of X-proteins

The product of *nifX* is not required for diazotrophic growth in *K. pneumoniae*, *A. vinelandii* or *R. capsulatus* and, therefore, mutations in *nifX* render a Nif^+ phenotype [27,91,104]. However, Shah *et al.* [56] have shown that when only purified components are used for the *in vitro* synthesis of FeMo-co, the addition of NifX stimulates FeMo-co synthesis three-fold. This result strongly suggests that NifX may be a required factor for FeMo-co synthesis under certain *in vivo* conditions. Although the exact nature of requirement for NifX in FeMo-co biosynthesis remains unknown, there are reports that attribute at least three different possible roles to NifX in nitrogen fixation, as listed below.

(1) When the amino acid sequence of NifX is compared to the amino acid sequences deposited in protein databases, it becomes evident that *nifX* encodes a protein that is similar to NifY, NafY, VnfX, and the C-terminal half of NifB from *A. vinelandii* and other bacterial sources (Figure 7). That there was significant sequence similarity between the products of *nifX* and *nifY* was first reported by Dean *et al.* [27]. The actual role of NifY in the *A. vinelandii nif* system is not known. However, NifY in *K. pneumoniae* appears to be involved in apo-dinitrogenase maturation [105,106]. Sequence identity between NifY/NifX and NifB proteins was also noted, and a role for NifY/NifX proteins in FeMo-co maturation or stabilization was suggested, based on the known role of NifB in FeMo-co biosynthesis [91]. NifB is required for the three nitrogenase systems (Mo-, V- and Fe-only-nitrogenases) [74], and its role is the synthesis of NifB-co, an [Fe-S] cluster that serves as precursor for FeMo-co, FeV-co and FeFe-co biosyntheses [78]. Another homolog to the *nifX* gene-product has been identified in the product of *vnfX* in the vanadium-dependent nitrogenase system [90]. There is evidence showing that VnfX binds a V-Fe cluster that serves as precursor for the synthesis of FeV-co and showing the homocitrate-dependent transfer of the FeV-co precursor from VnfX to apo-dinitrogenase [107,108]. Finally, the non-*nif* gene encoding the apo-dinitrogenase γ subunit of *A. vinelandii* (*nafY*) has been found to encode a NifY/NifX-like protein [31]. In *A. vinelandii*, γ binds specifically to free FeMo-co or to apo-dinitrogenase; therefore, a role of γ in the insertion of FeMo-co into apo-dinitrogenase was initially suggested [109].

As we describe above, consistent with the sequence similarity between γ (NafY), NifY, NifX, NifB and VnfX polypeptides, *in vitro* studies provide evidence supporting the conclusion that they are functionally related, as they are able to interact with some intermediates of the FeMo-co or FeV-co biosynthetic pathways (Table 2). First, NafY

```
Av NafY   1  MVT.PVN....MSRETALRIALAARALPGTTVGQLLEILHQRIEGPLTEESLQGVSVTDL
Av NifY   1  MTAQPPFGQAPLPAHLALRIALAARSLKGVDTAHLLRALIAAVGEPITEARLRKLRASRL
Kp NifY   1  MSD.........NDTLFWRMLALFQSLPDLQPAQIVDWLAQESGETLTPERLATLTQPQL
Av NifX      ............................................................
Av VnfX   1  ..........................................................MI
Kp NifB      ............................................................
                                                                        *

Av NafY  56  KIGLAGS..EEDVDMLDTPMSALKDAVRILWGEAE...VDSLPQPVKLER.VPEGSIRVA
Av NifY  61  RTRLLETCGEGVQSTLTD..RQLHSALGLLKGRGVRMPEDPLPIPEPYRNGEFQDSVRIA
Kp NifY  52  AASFPSATAVMSPARWSRVMASLQGAL...........PAHLRIVRPAQR...TPQLLAA
Av NifX   1  .....................................MSSPTRQLQVLDSEDDGTLLKVA
Av VnfX   3  KVAFASNDRVNVNLHFGA....................ADTLMVYDISPG..YAELLGVG
Kp NifB 331  .............................................ATRGESEADDACLVA
                                                      *              * **

Av NafY 110  IASNNGEQLDGHFGSCLRFLVYQVSAKDASLVDIRST.......LDVALA..EDKNAWRV
Av NifY 119  CASDNGERLDGIFSNCTRFLIYQISPRESRLIDLREP.......GPCRED..EDRHARRA
Kp NifY  98  FCSQDGLVINGHFGQGRLFFIYAFDEQGGWLYDLRRY.......PSAPHQ..QEANEVRA
Av NifX  24  FASSDRELVDQHFGSSRSFAIYGVNPERSQLLSVVEF.......GELEQDGNEDKLARKI
Av VnfX  41  EFVQANMKGENRFKSLSDGQTNIIDLQ....LSAEEL.......ERLAAKPPEDKVIAKL
Kp NifB 346  VASSRGDVIDCHFGHADRFYIYSLSAAGMVLVNERFTPKYCQGRDDCEPQDNAARFAAIL
              **  ** ** *•*    * ** *      ** **                 *** * *

Av NafY 161  EQIQDCQVLYVVSIGGPAAAKVVRAGIHPLKKPKGCAAQEAIAELQTVMAGSPPPWLAKL
Av NifY 170  ELLADCQLLYTLSIGGPAAAKVVRAGVHPVRLARARPAREIVEELQRVLATAPPPWLAKA
Kp NifY 149  RLIEDCQLLFCQEIGGPAAARLIRHRIHPMKAQPGTTIQAQCEAINTLLAGRLPPWLAKG
Av NifX  77  DLLDGCVAVYCCACGASAVRQLMAIGVQPIKVSEGARIAELIEALQVELREGPSAWLAKA
Av VnfX  90  KFLEGCSAIYAASIGTSSIKRLIMAGIQPIIVGTGQTIEDLLNEVSLALHCGGLSWVERA
Kp NifB 406  ELLADVKAVFCVRIGHTPWQQLEQEGIEPCVDGAWRPVSEVLPAWWQQRRGSWPAALPHK
             *** **  **   *•* *  ***  ** •**   *    ***  **  *    * ****

Av NafY 221  VGVSAEERVRFSVSDDEDEAARA..........
Av NifY 230  MGAEPDQRIRFTQ....................
Kp NifY 209  LTGITLWKNAFFNPCFVLVAR............
Av NifX 137  IQRTRGPDMRRFDAMAAEGWDE...........
Av VnfX 150  KAKAERAAAQNPAAPHTPSGNGLRLITSIEELE
Kp NifB 466  GVA..............................
```

Figure 7. Amino acid sequence alignment of members of the NafY/NifY/NifX/VnfX family of iron and molybdenum (or vanadium) cluster-binding proteins. The amino acid sequences are indicated by the single letter code. Gaps were introduced for optimal alignments. Asterisks indicate amino acid residues identical or similar in four or more of the aligned proteins. Circles indicate residues identical in all the aligned proteins. Only residues 331-468 in the *K. pneumoniae* NifB sequence are aligned. The entire sequence of all other polypeptides is shown. Av, *A. vinelandii*; Kp, *K. pneumoniae*.

binds FeMo-co [109] and NifB-co (L. M. Rubio and P. W. Ludden, unpublished results). Second, NifX from *A. vinelandii* is also able to bind FeMo-co, NifB-co, and a ^{55}Fe-labeled FeMo-co precursor [110]. Recent data imply that the NifX-bound form of NifB-co can be transferred to NifNE (P. Rangaraj and P. W. Ludden, unpublished data). NifX may interact with the NifNE complex, and the fact that the *nifNEX* cluster of genes is widespread among bacteria strongly suggests that the products of the three genes indeed have some kind of interaction [111]. Third, VnfX binds FeMo-co, NifB-co and a FeV-co precursor (vanadium-containing Fe-S cluster) lacking homocitrate ([107]; C. Rüttimann-Johnson and P. W. Ludden, unpublished results). Fourth, NifB is the protein where NifB-co is assembled and it is therefore evident that a NifB-co-binding domain must exist in NifB. Collectively considered these findings, we propose the existence of a new family of iron and molybdenum (or vanadium) cluster-binding proteins that includes NifX, VnfX, NifY, NafY and the C-terminal domain of NifB. We have designated this family as the X-family of proteins of the nitrogenase system.

Protein	NifB-co	FeMo-co	FeMo-co / FeV-co precursor
NafY (γ)	Yes	Yes	?
NifY	?	?	?
NifX	Yes	Yes	Yes
VnfX	Yes	Yes	Yes
NifB	Yes	?	?

? Unknown binding capability.

Table 2. Cluster-binding capabilities of members of the X-proteins family from *A. vinelandii.*

Interestingly, the NifX domain has recently been detected using methods for sequence-profile analysis and protein-structure comparison that identify it as a small-molecule-binding domain of ancient origin [112].

(2) A clear Nif⁻ phenotype appears when an *A. vinelandii nifXnafY* double mutant is derepressed for nitrogen fixation under Mo-limiting conditions. When the levels of Mo in the growth medium are sufficient to repress the expression of the alternative nitrogenase systems, but low enough to stress the cells, *nifXnafY* mutants exhibit severely affected diazotrophic growth rates and very low levels of dinitrogenase activity in crude extracts when compared to the wild-type under the same conditions [31]. The authors, Rubio *et al.*, using their results on *in vitro* FeMo-co insertion experiments, suggest a role for NifX and NafY in the stabilization of the form of apo-dinitrogenase that is competent for FeMo-co activation. Since there is sequence similarity between NifX and NafY and between NifNE and NifDK, it is also possible that both NifX and NafY bind to (and stabilize) NifNE and the apo-dinitrogenase.

(3) Other results suggest a role for NifX as negative regulator of *nif*-gene expression in response to NH_4^+ and O_2 [104]. The over-expression of NifX in *K. pneumoniae* blocked *nif*-protein synthesis and accumulation of *nif*-specific mRNAs. Conversely, a *nifX* mutant strain accumulates more *nif*-specific mRNAs than does the wild-type upon NH_4^+ or O_2 addition. These results provide evidence for the involvement of NifX in balancing *nif*-gene expression.

4.6. NifV

The product of *nifV* is homocitrate synthase [113]. While *A. vinelandii* and *K. pneumoniae* strains with mutations in *nifV* reduce effectively protons and acetylene, they are unable to reduce dinitrogen effectively, and show a slow diazotrophic growth [22,102,114,115]. The defect in *nifV* strains lies in the FeMo-co itself. Hawkes *et al.* [47] presented the first direct evidence that FeMo-co constitutes the substrate reduction site when they activated apo-dinitrogenase with FeMo-co isolated from dinitrogenase of a *nifV* mutant strain, and the resulting holo-dinitrogenase exhibited the altered catalytic properties associated with the *nifV* phenotype. This suggested that *nifV* strains were somehow defective in FeMo-co biosynthesis. The hypothesis was confirmed when Hoover *et al.* [116] showed that extracts from *K. pneumoniae nifV* mutants were not able to direct *in vitro* FeMo-co synthesis.

In *K. pneumoniae* and *A. vinelandii*, the accumulation of homocitrate in the medium during diazotrophic cell growth correlates with the presence of a functional *nifV* [116]. Moreover, the addition of homocitrate to the growth medium of a *nifV* mutant during nitrogenase derepression results in a phenotypic reversion, i.e. synthesis of a dinitrogenase identical to the wild-type dinitrogenase [117]. Finally, the *A. vinelandii* NifV protein has been overproduced in a recombinant *E. coli* strain, purified and shown to be the only gene product required to catalyze the condensation of acetyl coenzyme A and α-ketoglutarate to form homocitrate [113].

The identification of homocitrate as an integral part of FeMo-co [118] and the correlation of homocitrate accumulation with a functional *nifV* gene led to the hypothesis that, in *nifV* mutant strains, homocitrate in FeMo-co was being replaced by another organic acid, and that such a replacement would account for the altered catalytic properties exhibited by dinitrogenase of those mutants. Purified dinitrogenase from a *K. pneumoniae nifV* mutant was subsequently shown to contain citrate instead of homocitrate [119]. The low stoichiometric yield of citrate suggested that other analogs might be replacing homocitrate in *nifV* mutants. When analogs of homocitrate were used in the *in vitro* FeMo-co synthesis system, the resulting dinitrogenases exhibited altered substrate specificities and inhibitor susceptibilities. In a meticulous study, Imperial *et al.* [120] determined that the following structural features of homocitrate are required for the synthesis of the proper FeMo-co: (1) the 1- and 2-carboxyl groups; (2) the hydroxyl group; (3) the *R* configuration of the chiral center; and (4) a 4 to 6 carbon chain length containing two terminal carboxyl groups.

It is still not known how Mo becomes coordinated to homocitrate or why homocitrate and not another organic acid is incorporated into FeMo-co *in vivo*. It is possible that the NifV activity produces such a large intracellular concentration of homocitrate that it avoids the incorporation of other analogs into FeMo-co.

5. OTHER *nif*-GENE PRODUCTS

5.1. NifU and NifS

The products of the *nifU* and *nifS* genes from *A. vinelandii* are required for the full activation of nitrogenase because they are involved in the maturation of the two nitrogenase component proteins. NifU and NifS function together in the synthesis of an [Fe-S] cluster, precursor to the [Fe-S] clusters harbored by both components of the nitrogenase [121]. The role of NifU will be discussed first.

The initial indication that implied that NifU might be involved in the maturation or stability of the nitrogenase component proteins came from experiments designed by Jacobson *et al.* [102]. These showed that a deletion of *nifU* in *A. vinelandii* resulted in a 20-fold reduction of dinitrogenase reductase activity and a 3-fold reduction in dinitrogenase activity. It was later shown that *nifU* is required for activity of all three nitrogenase systems in *A. vinelandii* [5]. The initial characterization of NifU was achieved after its over-expression and purification from recombinant *E. coli* cells. NifU is a homo-dimer of 33 kDa subunits that contains one stable [2Fe-2S] cluster per subunit, with a potential of about -260 mV, and coordinated to the polypeptide entirely by cysteinyl thiolate ligands [122].

An alignment of the deduced amino acid sequences of the *K. pneumoniae*, *A. vinelandii* and *Anabaena* 7120 *nifU* gene products shows nine conserved cysteinyl residues contained in regions of high sequence similarity [63]. NifU proteins are modular proteins consisting of at least two domains. The first domain is contained in the N-terminal third of NifU and has three cysteines that provide a rubredoxin-like Fe-binding site. At low temperature, this site can bind one labile ferric iron per monomer. The second domain includes the central portion of NifU and contains four cysteines, which are ligands to the stable [2Fe-2S] cluster described above. All seven cysteines, those involved in the ligation of the mononuclear Fe in the rubredoxin-like site and in the ligation of the [2Fe-2S] cluster, are required for the function of NifU [123]. NifU has two other cysteines in the C-terminal end of the protein that are not necessary for iron binding, nor are they required for its full physiological function. However, it is likely that the C-terminal region of NifU constitutes a third domain of unidentified function. Agar *et al.* [123] proposed a model in which the iron bound to the labile rubredoxin-like sites within the N-terminal domain of each NifU subunit is used for [Fe-S] cluster formation, whereas the stable [2Fe-2S] clusters contained in the central domains would have a redox function involving the release of Fe or an [Fe-S] cluster precursor from the rubredoxin-like binding sites. This model has been largely confirmed by the observation of the assembly of a transient [2Fe-2S] cluster within NifU through the interaction with NifS and the formation of an $\alpha_2\beta_2$ NifUS hetero-tetrameric complex [121]. The same three cysteine residues that comprise the rubredoxin-like Fe-binding site are also required for the NifS-dependent assembly of the transient [2Fe-2S] cluster in NifU. This second [2Fe-2S] cluster species is labile, bridged between the two subunits of NifU, and rapidly released from NifU upon reduction.

Similar to *nifU*, the *nifS* mutants of *A. vinelandii* and *K. pneumoniae* are characterized by reduced dinitrogenase and dinitrogenase reductase activities [18,102]. Again, a functional *nifS* gene is required for activity of all three nitrogenase systems in *A. vinelandii* [5]. NifS is a pyridoxal phosphate (PLP) homo-dimer that catalyzes the desulfurylation of L-cysteine to yield L-alanine and elemental sulfur [124]. In mechanistic studies, the catalytic activity of NifS was found to be extremely sensitive to alkylating reagents due to the reactivity of NifS-Cys^{325}. This residue was shown to be catalytically important and to accept the sulfur from the substrate, L-cysteine, yielding an enzyme-bound cysteinyl persulfide intermediate [125]. The NifS-bound persulfide so formed is proposed to be the physiological sulfur donor for the formation of [Fe-S] clusters for the nitrogenase component proteins. It was further demonstrated that NifS was indeed able to catalyze the *in vitro* reconstitution of an apo-dinitrogenase reductase whose [4Fe-4S] cluster had been removed by chelation [126]. The presence of Cys^{325} in NifS is essential for this reconstitution process, since a NifS-Ala^{325} mutant variant lacked desulfurase activity and was unable to catalyze the reactivation of the apo-dinitrogenase reductase. Recently, the crystal structure of a NifS-like protein from *Thermotoga maritima* has been resolved at 2.0 Å resolution [127]. NifS is structured in two domains, the larger bearing the PLP-binding active site and the smaller having the active site cysteine in the middle of a highly flexible loop. Based on the crystal structure analysis, it is suggested that, once charged with sulfur, the loop could deliver S^0 directly to other sites far from NifS, *e.g.*, to an already folded apo-dinitrogenase reductase dimer.

The series of researches by Dean *et al.* on NifS and NifU has implications significantly beyond the field of nitrogen fixation. Homologs of NifS and NifU, designated as IscS and IscU respectively (to indicate their role in ***i***ron-***s***ulfur-***c***luster formation), are found to be widely conserved in nature [128]. Thus, the reactions catalyzed by NifU and NifS may represent a specialized way to boost the mobilization of Fe and S necessary for nitrogenase maturation, and IscU and IscS proteins possibly have housekeeping functions related to a general [Fe-S] cluster formation [39,128]. While IscS and NifS proteins are highly similar in their entire amino acid sequences, IscU proteins are much smaller than NifU proteins and exhibit high sequence-identity when compared to the N-terminal domain of NifU alone. Hence, IscU proteins contain only the features found in the first module of NifU proteins. The high degree of sequence similarity between IscU proteins from species as divergent as humans and *Haemophilus influenzae* suggests that IscU is among the most highly conserved proteins known [129]. Also in common with NifU and NifS, IscU and IscS have been shown to form an $\alpha_2\beta_2$ hetero-tetrameric complex [130].

A mechanism for the IscS-dependent assembly of [Fe-S] clusters within IscU has been demonstrated [130-132]. The [Fe-S] cluster assembly within IscU proceeds in sequential steps. The transfer of S^0 from the cysteine persulfide on IscS to one or more of the three cysteine residues on IscU is the initial step in the [Fe-S] cluster biosynthesis. The resultant cysteine persulfides on IscU are then reduced by two ferrous ions, concomitant with the oxidation of two cysteine thiols to yield a [2Fe-2S] cluster per IscU dimer [132]. This IscU form is then converted to a form containing two [2Fe-2S] clusters per dimer and finally to a form that contains one [4Fe-4S] cluster per dimer. It is very possible that the same general mechanism operates in NifUS during the assembly of [Fe-S] clusters for the nitrogenase components.

5.2. NifW and NifZ

Mutant strains of *A. vinelandii* deleted for *nifW*, *nifZ* or *nifW* and *nifZ* exhibit lowered diazotrophic growth rates and a modest reduction in dinitrogenase activity, but no apparent reduction in dinitrogenase reductase activity [102,133]. A similar phenotype was observed in *K. pneumoniae* [134]. These data suggest that NifW and NifZ are involved in the formation or accumulation of an active dinitrogenase. *nifW* and *nifZ* homologs have been identified in several diazotrophs, including *A. vinelandii*, *K. pneumoniae*, *A. chroococcum* and *R. capsulatus* [63].

The addition of purified FeMo-co to *nifW* or *nifZ* mutants does not stimulate dinitrogenase activity [102]. However, it is still possible that *nifW* and *nifZ* are involved in the biosynthesis of FeMo-co but accumulate an altered form of FeMo-co that cannot be displaced from dinitrogenase by the native cofactor. Kim and Burgess [133] purified both components of nitrogenase from a strain deleted in *nifW* and found that, while dinitrogenase reductase from the *nifW* strain was indistinguishable from that of the wild-type strain, the mutant dinitrogenase presented CD spectroscopy and specific activities different from the wild-type. It was then suggested that dinitrogenase from a *nifW* mutant has an alteration near the FeMo-co site, possibly at homocitrate, which causes differential inhibition with different substrates. In this context, higher concentrations of homocitrate are found to be required for the phenotypic reversion of a *R. capsulatus nifVW* double mutant when compared to a *nifV* mutant [135], indicating that NifW may be involved in homocitrate transport or in incorporation into FeMo-co. In contrast, Kim and Burgess [136] later suggested that NifW is part of a system that protects dinitrogenase from oxygen. They observed that NifW from *A. vinelandii* behaves as a trimer when anaerobic conditions are maintained, but becomes part of a very high-molecular-weight complex that also includes dinitrogenase when cell extracts are exposed to oxygen.

The similar diazotrophic growth rates and component-protein activities exhibited by *nifW* and *nifZ* mutant strains, and the apparent translational coupling of *nifW* and *nifZ*, suggest that NifW and NifZ may be components of the same enzyme complex [102]. More recently, the interaction of NifW with itself *in vivo* and also with NifZ has been demonstrated by the use of the yeast two-hybrid system [137]. It would be interesting to test the putative interaction of NifW with dinitrogenase *in vivo* using the same system.

5.3. NifM

A.vinelandii and *K. pneumoniae* strains with mutations in *nifM* have a clear Nif$^-$ phenotype [18,102]. When *nifM* mutants were analyzed for nitrogenase component protein activities, it was found that such strains are primarily deficient in dinitrogenase reductase activity [18,102]. In spite of being thus deficient, the cell extracts from *nifM* mutant strains exhibited substantial accumulation of dinitrogenase reductase protein, suggesting that the *nifM* gene product is involved in the activation of dinitrogenase reductase.

By using a binary plasmid system in *E. coli* that directed the heterologous expression of *K. pneumoniae nifH* in combination with other *nif*-specific products, Howard *et al.* [138] showed that NifM and NifH are the only *nif*-specific gene products required for the expression of a catalytically active dinitrogenase reductase. Expression of *nifH* in the absence of *nifM* expression resulted in a poor accumulation of dinitrogenase reductase, and additionally inactive. Paul and Merrick [134] confirmed Howard's conclusions unambiguously by using more specific plasmid constructs that lacked the previously unnoticed, *nifZ* gene upstream from *nifM*.

The function of NifM in dinitrogenase reductase maturation is not yet very clear. The sequence of the C-terminal domain of NifM is similar to the sequence of peptidyl-prolyl cis-trans isomerases (PPIases). Recently, it has been shown that the activity of NifM changes the configuration of the apo-dinitrogenase reductase and allows the subsequent insertion of the [4Fe-4S] cluster [139]. In this process, the N-terminal domain of NifM appears to mediate the interaction with dinitrogenase reductase, and the C-terminal domain would be responsible for the PPIase activity. Consistently, it has been possible to demonstrate the interaction between NifM and dinitrogenase reductase *in vivo* using the yeast two-hybrid system [140].

5.4. NifY

The *nifY* gene product is not essential for nitrogen fixation, at least under the physiological conditions tested so far on *A. vinelandii* and *K. pneumoniae nifY* mutants. Mutations in *nifY* slightly affect the nitrogenase activity and render a Nif$^+$ phenotype [27,105,141]. It is proposed that the *nifY* gene product from *K. pneumoniae* plays the same role in apo-dinitrogenase maturation as does γ (NafY) in *A. vinelandii*, *i.e.*, the stabilization of an apo-dinitrogenase conformation that can be activated by FeMo-co (see Section 6.2). NifY has been found to form a stable complex with apo-dinitrogenase from *nifB* or *nifN* mutant strains, and to dissociate from the complex when apo-dinitrogenase is activated by the addition of FeMo-co [105,106].

The over-expression of *nifY* in *K. pneumoniae* blocks the accumulation of *nifHDKTY* mRNA in a regulatory mechanism that is likely to be exerted through interaction with the NifLA regulatory proteins [141]. It is possible that, as part of the maturation process, NifY has a role in sensing and signaling the status of nitrogenase and that it provides a feedback mechanism to regulate nitrogenase production.

5.5. NifF and NifJ

The products of the *nifF* and *nifJ* genes are involved in the transfer of electrons to nitrogenase in *K. pneumoniae*. The electron donor to nitrogenase is known definitively only in the case of *K. pneumoniae*, where biochemical and genetic evidence has shown that a specific flavodoxin, encoded by the *nifF* gene, uniquely donates electrons to nitrogenase *in vivo* [142]. The NifF flavodoxin is, in turn, reduced by a pyruvate:flavodoxin oxidoreductase, the product of the *nifJ* gene. It is established that pyruvate:flavodoxin oxidoreductase, flavodoxin, and both nitrogenase component proteins are the only proteins required to obtain pyruvate-coupled nitrogenase activity *in vitro* [143]. Pyruvate is thus the physiological source of electrons to nitrogenase in *K. pneumoniae*.

The pathway for electron transport to nitrogenase in other diazotrophs is uncertain. In *R. rubrum*, a *nifJ* gene has been identified and its product purified, but a *nifJ* mutant does not exhibit a Nif$^-$ phenotype [144,145]. In the case of *A. vinelandii*, a NifF homologue is present; but the corresponding flavodoxin is not essential for nitrogen fixation, although it is required for maximum *in vivo* nitrogenase activity [146]. Moreover, no gene product homologous to *K. pneumoniae* NifJ has yet been detected in *A. vinelandii*. The product of the *fdxA* gene (ferredoxin I) of *A. vinelandii* has also been implicated in electron transport to nitrogenase. However, double mutant strains lacking both flavodoxin and ferredoxin I are still able to fix nitrogen, suggesting that a third protein, at least, may serve as an electron donor to nitrogenase [147]. In this context, the recently reported presence of *rnf*-gene homologs in *A. vinelandii* is very suggestive (see below).

In *R. capsulatus*, NifF contributes to some extent to the *in vivo* nitrogenase activity under iron-replete nitrogen-fixation conditions; however, it is absolutely required for nitrogen fixation under iron limitation [148]. As in the case of *K. pneumoniae*, a pyruvate:flavodoxin oxidoreductase is able to couple the oxidation of pyruvate to nitrogenase activity *via* the reduction of NifF [149]. In contrast, pyruvate:flavodoxin oxidoreductase from *R. capsulatus* has little specificity for NifF and can also reduce ferredoxins *in vitro* (including the product of the *fdxN* gene, that serves as electron donor to nitrogenase). Under iron-replete conditions, the electron transfer to nitrogenase seems to be mediated by the products of the *rnf* and the *fdxN* genes [150]. The Rnf proteins form a membrane complex that couples the proton motive force generated by the photosynthetic apparatus to the reduction of ferredoxin I (FdxN) (see below). Therefore, in *R. capsulatus*, donation of electrons to nitrogenase may occur *via* a NifJ-NifF system, as in *K. pneumoniae,* or *via* the Rnf-FdxN system, depending on the availability of iron [151].

5.6. The Rnf complex

Upstream from *nifE, R. capsulatus* possesses a set of *nif*-regulated genes designated *rnf* genes (for ***R***hodobacter ***n***itrogen ***f***ixation). The Rnf polypeptides are proposed to constitute a membrane complex that would couple the membrane potential to the nitrogenase activity *via* reduction of ferredoxins [150]. The *rnf* genes are arranged in two transcriptional units. The first unit contains the *rnfF* gene and the ferredoxin-encoding genes *fdxC* and *fdxN*, and is transcribed in the same orientation as *nifE*. The second unit consists of the *rnfABCDGEH* genes and is transcribed in the opposite sense to *nifE* [150,152]. Homologs to the *rnf* genes have been found in the nitrogen-fixing bacteria *P. stutzeri* [153] and *A. vinelandii* [30,31] and in several non-fixing bacteria (*e.g., H. influenzae, E. coli,* and *Vibrio alginolyticus*). In *A. vinelandii*, the *rnf* cluster consists of seven genes (*rnfA, –B, –C, –D, –G, –E,* and *-H* genes) located between the *nifLA* gene cluster and the *nafY* gene (that encodes the apo-dinitrogenase γ subunit), and is transcribed in the opposite direction to *nifL* [30,31].

R. capsulatus strains carrying mutations in any one of the *rnf* genes are unable to grow diazotrophically, either in the light or in the dark [150,154]. *rnf* mutants exhibit very low nitrogenase activity *in vivo* (<2% of the values observed in the wild-type strain). They also exhibit reduced levels of *in vitro* nitrogenase activity (<25% of the values from the wild-type); but this phenotype correlates well with the lower levels of accumulation of the nitrogenase component proteins found in the *rnf* mutant strains when compared to the wild-type. On the other hand, *R. capsulatus* strains that over-express the *rnf* genes exhibit higher *in vivo* nitrogenase activity than the wild-type, apparently without increasing the level of accumulation of nitrogenase proteins [155].

The presence of the *rnf* genes in *A. vinelandii* is particularly important because double mutants lacking both putative electron donors to nitrogenase (NifF and FdxA) are not impaired in nitrogen fixation [147]. There must be at least a third way of donating electrons to nitrogenase, and such a way could involve the activity of the Rnf protein complex. Klugkist *et al.* [156] reported the coupling of the electron flux through the respiratory chain in the cytoplasmic membrane to the *in vivo* nitrogenase activity in *A. vinelandii*. It is possible that the two membrane-associated *nif*-specific polypeptides of 29 and 30 kDa of molecular weight they observed are the products of the *rnfG* and *rnfE* genes (with predicted molecular weights of 26 and 25 kDa, respectively).

On the basis of similarity searches and topological experiments, it has been proposed that the Rnf polypeptides constitute a membrane complex of chimeric structure that may have evolved to function specifically in nitrogen fixation [152,157]. The first sub-complex would be composed of the RnfA, RnfD and RnfE polypeptides.

They are integral membrane proteins, similar to components of the Na^+-dependent NADH:quinone oxidoreductase, and therefore are proposed to form a sodium channel. A second sub-complex would be integrated by the RnfB and RnfC polypeptides, which are peripheral membrane Fe-S proteins similar to the flavoprotein fraction of the H^+-dependent NADH:quinone oxidoreductase [158]. The RnfBC subcomplex would be the site where ferredoxin I is reduced. RnfB contains one [2Fe-2S] cluster, and RnfC contains at least one [4Fe-4S] cluster and putative binding sites for NADH and FMN. It is also predicted that the soluble protein, RnfG, would be associated primarily with the RnfBC subcomplex. Finally, the function of the soluble protein RnfH has not yet been inferred.

6. NON-*nif* PROTEIN REQUIREMENTS

6.1. The molybdenum storage protein (Mo-sto)

The capacity of *A. vinelandii* to scavenge Mo from the medium is impressive; it can accumulate up to 25 times more Mo than it actually requires for maximum nitrogenase activity [159]. *A. vinelandii* has a molybdenum storage protein (Mo-sto) that is responsible for the high levels of Mo accumulation observed in this bacterium [160]. The purified Mo-sto protein has a molecular weight of 90 kDa and is a hetero-tetramer of 21 and 24 kDa subunits; it does not contain iron and binds a minimum of 15 atoms of Mo per tetramer, although the Mo content of the Mo-sto decreases during purification. By using ^{99}Mo-labeled molybdate, Allen *et al.* [161] have confirmed that Mo accumulates in Mo-sto and that the binding of Mo to Mo-sto is a nucleotide-dependent process. The form of Mo bound to Mo-sto is not known, but it is probably not molybdate because it fails to accumulate in Mo-sto when molybdate is added to an *in vitro* reaction mixture.

Even though *A. vinelandii* does not require Mo while growing with ammonium as a nitrogen source, the synthesis of Mo-sto is not repressed by ammonium [160]. Nevertheless, the level of incorporation of Mo into Mo-sto increases when cells are derepressed for nitrogenase synthesis (P. Rangaraj and P. W. Ludden, unpublished work), consistent with the proposed role for Mo-sto as a reservoir of Mo to be utilized when cells are actively fixing nitrogen.

6.2. The γ protein (NafY)

Two roles in the nitrogenase system have been proposed for the γ protein: (1) as a "molecular prop" whose function is to keep the conformation of the FeMo-co-deficient apo-dinitrogenase competent for FeMo-co insertion [105,106,109,162], and (2) as a specific FeMo-co insertase [109]. We will discuss each role separately.

Gamma was first identified in the analysis of purified apo-dinitrogenase from an *A. vinelandii nifB* mutant strain [65]. In contrast to the $\alpha_2\beta_2$ subunit composition of the mature dinitrogenase, apo-dinitrogenase from strains of *K. pneumoniae* or *A. vinelandii* with mutations in *nifB*, *nifN* or *nifE* has a $\alpha_2\beta_2\gamma_2$ hexameric composition. The hexameric apo-dinitrogenase can be activated *in vitro* by the simple addition of purified FeMo-co [65,105]. In *K. pneumoniae*, the γ subunit has been identified as the product of the *nifY* gene [105,106], while in *A. vinelandii* it is the product of the non-*nif* gene termed *nafY* (***n**itrogenase **a**ccessory **f**actor Y*) [31].

On the other hand, apo-dinitrogenase accumulated in extracts of mutant strains deleted in *nifH* has a tetrameric composition ($\alpha_2\beta_2$) and requires some type of NifH- and MgATP-dependent maturation that, in turn, promotes the association of the γ subunit leading to the form that is competent for FeMo-co activation [95,96]. The maturation of apo-dinitrogenase occurs in several steps. First, the $\alpha_2\beta_2$ form of apo-dinitrogenase is synthesized with the FeMo-co site inaccessible for FeMo-co insertion. Second, the $\alpha_2\beta_2$ form is then converted to a form with the FeMo-co site accessible, in a reaction that requires NifH and MgATP. Magnuson *et al.* [162] have followed the structural changes in the $\alpha_2\beta_2$ apo-dinitrogenase form due to this reaction by using chemically-modified NifD-Cys^{275} as a probe. NifD-Cys^{275} is one of the two protein ligands to FeMo-co in the mature dinitrogenase (the other protein ligand to FeMo-co is NifD-His^{442}), and has been shown to be necessary for the incorporation of FeMo-co into apo-dinitrogenase. NifD-Cys^{275}, which is buried in the $\alpha_2\beta_2$ form of apo-dinitrogenase, is exposed to solvent after the NifH- and MgATP-dependent maturation of the $\alpha_2\beta_2$ apo-dinitrogenase. Third, the activated apo-dinitrogenase readily binds to γ and is converted to the $\alpha_2\beta_2\gamma_2$ form. Therefore it is believed that the binding of γ stabilizes a conformation that leaves Cys^{275} exposed for binding FeMo-co. Finally, FeMo-co is inserted and the γ subunit dissociates from the mature dinitrogenase.

The hypothesis that the *A. vinelandii* γ protein (the product of the *nafY* gene) plays a role in the stabilization of the apo-dinitrogenase conformation that is competent for FeMo-co activation has also found support in recent *in vitro* FeMo-co insertion experiments. These show that, even though apo-dinitrogenase is accumulated to the same extent in *nifB* strains (FeMo-co-less mutants) as in *nafYnifB* mutants, the mutation in *nafY* reduces by 50% the amount of apo-dinitrogenase that can be activated by purified FeMo-co [31].

The second proposed role for γ is that of a FeMo-co insertase [109]. The proposal was based on *in vitro* studies with crude extracts of an *A. vinelandii nifB* mutant that showed that γ is able to bind either to free FeMo-co or to apo-dinitrogenase, which are expected properties of a FeMo-co insertase. However, the FeMo-co insertase model has not been generally accepted because Christiansen *et al.* [163] later showed that γ is not essential for the activation by FeMo-co of a purified His-tagged apo-dinitrogenase.

This His-apo-dinitrogenase lacked the γ subunit (possibly due to the purification procedure or to interference of the His-tag with the binding site for γ in the apo-dinitrogenase) but was still almost fully activated by the addition of FeMo-co.

Nevertheless, the presence of the His-tag in the $\alpha_2\beta_2$ apo-dinitrogenase made less certain the conclusions drawn by the authors. Finally, the results obtained by the analysis of *A. vinelandii* mutant strains deleted in *nafY* did not provide evidence supporting a role for γ as a FeMo-co insertase [31]. Briefly, it was shown that the presence of γ in an *in vitro* FeMo-co-insertion reaction mix did not increase the rate of FeMo-co insertion into apo-dinitrogenase. It is possible that the experimental conditions necessary to determine whether γ is an insertase have not yet been properly defined.

7. FUTURE STUDIES

In the last two decades, there has been much progress in our knowledge of the individual roles played by many of the *nif*-gene products and in our understanding of the biochemistry of the nitrogenase system as a whole. The field of FeMo-co biosynthesis, in particular, has advanced enormously, thanks to the application of a genetic-biochemical approach and to the development of an *in vitro* FeMo-co synthesis assay. Yet a number of very important issues remain unclarified. While some of them will be easy to solve, others will require great effort and skill, but every step forward will be rewarding. For example, although it is probably the case, it is not known whether NifU and NifS are directly involved in the initial mobilization of Fe and S for FeMo-co biosynthesis. The structure of NifB-co is another enigma. Since we do not know the structure of any intermediate cluster in the biosynthesis of FeMo-co, it is very difficult to imagine the specific reactions that lead to the assembly of the final cofactor. Which protein(s) specifies the heterometal that will be incorporated into the cofactor? What are the specific roles of NifQ and NifH in the biosynthesis of FeMo-co?

A structural approach should also help us to understand how the Nif proteins interact with each other in order to perform their functions. The three-dimensional structure of apo-dinitrogenase and the determination of the chemical nature of the NifH-dependent maturation of the apoenzyme are important unsolved issues. Likewise, the nitrogenase system provides a perfect model to investigate protein domains involved in handling and assembling complex metalloclusters. The crystal structure of any of the X-protein family members will be a very valuable piece of data. Finally, one of the most exciting advances to us would be the physical identification of the putative "FeMo-co biosynthetic factory".

8. ACKNOWLEDGMENTS

We thank Gail Stirr for reading critically the manuscript and Laura Vanderploeg for help with the Figures.

9. REFERENCES

1. S. Streicher, E. Gurney and R. Valentine, Proc. Natl. Acad. Sci. USA 68 (1971) 1174.
2. R. Dixon and J. Postgate, Nature 234 (1971) 47.
3. P. E. Bishop and R. Premakumar in: Biological Nitrogen Fixation, G. Stacey, R. H. Burris, and H. J. Evans (eds.), Chapman & Hall, New York, 1992, p. 736.
4. R. D. Joerger and P. E. Bishop, J. Bacteriol. 170 (1988) 1475.
5. C. Kennedy and D. R. Dean, Mol. Gen. Genet. 231 (1992) 494.
6. R. H. Burris, A. Phelps and J. Wilson, Soil Sci. Soc. Am. Proc. 7 (1942) 272.
7. I. Zelitch, E. Rosenblum, R. H. Burris and P. Wilson, J. Bacteriol. 62 (1951) 747.
8. G. Strandberg and P. Wilson, Can. J. Microbiol. 14 (1967) 25.
9. H. Dalton and J. Postgate, J. Gen. Microbiol. 56 (1969) 307.
10. G. Daesch and L. Mortenson, J. Bacteriol. 110 (1972) 103.
11. T. O. Munson and R. H. Burris, J. Bacteriol. 97 (1969) 1093.
12. J. E. Carnahan, L. E. Mortenson, H. F. Mower and J. E. Castle, Biochim. Biophys. Acta 44 (1960) 520.
13. W. A. Bule, and J. R. LeComte, Proc. Natl. Acad. Sci. USA 56 (1966) 979.
14. W. P. D. Stewart, G. P. Fitzgerald and R. H. Burris, Proc. Natl. Acad. Sci. USA 58 (1967) 2071.
15. R. H. Burris in: Nitrogen Fixation by Free-living Micro-organisms, W. D. P. Stewart (ed.), Cambridge University Press, 1975, p. 249.
16. R. Schollhorn and R. H. Burris, Proc. Natl. Acad. Sci. USA 58 (1967) 213.
17. M. J. Dilworth, Biochim. Biophys. Acta 127 (1966) 285.
18. G. P. Roberts, T. MacNeil, D. MacNeil and W. J. Brill, J. Bacteriol. 136 (1978) 267.
19. W. Arnold, A. Rump, W. Klipp, U. B. Priefer and A. Pühler, J. Mol. Biol. 203 (1988) 715.
20. K. E. Brigle, W. E. Newton and D. R. Dean, Gene 37 (1985) 37.
21. C. Elmerich, J. Houmard, L. Sibold, I. Manheimer and N. Charpin, Mol. Gen. Genet. 165 (1978) 181.
22. T. MacNeil, D. MacNeil, G. P. Roberts, M. A. Supiano and W. J. Brill, J. Bacteriol. 136 (1978) 253.
23. M. Merrick, M. Filser, R. Dixon, C. Elmerich, L. Sibold and J. Houmard, J. Gen. Microbiol. 117 (1980) 509.
24. G. E. Riedel, F. M. Ausubel and F. C. Cannon, Proc. Natl. Acad. Sci. USA 76 (1979) 2866.
25. G. P. Roberts and W. J. Brill, J. Bacteriol. 144 (1980) 210.
26. G. P. Roberts and W. J. Brill, Ann. Rev. Microbiol. 35 (1981) 207.
27. M. R. Jacobson, K. E. Brigle, L. T. Bennett, R. A. Setterquist, M. S. Wilson, V. L. Cash, J. Beynon, W. E. Newton and D. R. Dean, J. Bacteriol. 171 (1989) 1017.

28. G. Blanco, M. Drummond, P. Woodley and C. Kennedy, Mol. Microbiol. 9 (1993) 869.
29. F. Rodríguez-Quiñones, R. Bosch and J. Imperial, J. Bacteriol. 175 (1993) 2926.
30. L. M. Rubio, C. S. Brown and P. W. Ludden, GeneBank AF450501.
31. L. M. Rubio, P. Rangaraj, M. J. Homer, G. P. Roberts, and P. W. Ludden, J. Biol. Chem. 277 (2002) 14299.
32. R. V. Hageman and R. H. Burris, Proc. Natl. Acad. Sci. USA 75 (1978) 2699.
33. J. B. Howard and D. C. Rees in: Prokaryotic Nitrogen Fixation: a model system for analysis of a biochemical process, E. W. Triplett (ed.), Horizon Scientific Press, Wymondham, UK, 2000, p. 43.
34. V. K. Shah and W. J. Brill, Proc. Natl. Acad. Sci. USA 74 (1977) 3249.
35. M. K. Chan, J. Kim and D. C. Rees, Science 260 (1993) 792.
36. J. Kim and D. C. Rees, Nature 360 (1992) 553.
37. J. Kim and D. C. Rees, Science 257 (1992) 1677.
38. J. T. Bolin, N. Campobasso, S. W. Muchmore, T. V. Morgan and L. E. Mortenson in: Molybdenum Enzymes, Cofactors and Models, E. I. Stiefel, D. Coucouvanis and W. E. Newton (eds.), American Chemical Society, Washington, DC, 1993, p. 186.
39. D. R. Dean, J. T. Bolin and L. Zheng, J. Bacteriol. 175 (1993) 6737.
40. J. W. Peters, M. H. B. Stowell, S. M. Soltis, M. G. Finnegan, M. K. Johnson and D. C. Rees, Biochemistry 36 (1997) 1181.
41. M. M. Georgiadis, H. Komiya, P. Chakrabarti, D. Woo, J. J. Kornuc and D. C. Rees, Science 257 (1992) 1653.
42. R. M. Allen, R. Chatterjee, M. Madden, P. W. Ludden and V. K. Shah, Crit. Rev. Biotech. 14 (1994) 225.
43. T. Ljones and R. H. Burris, Biochemistry 17 (1978) 1866.
44. R. N. F. Thorneley, Phil. Trans. R. Soc. Lond. B 336 (1992) 73.
45. L. C. Seefeldt, T. V. Morgan, D. R. Dean and L. E. Mortenson, J. Biol. Chem. 267 (1992) 6680.
46. L. C. Seefeldt, J. M. Chan, W. Wu, and J. E. Huyett in: Nitrogen Fixation: from Molecules to Crop Productivity. Proceedings of the 12th International Congress on Nitrogen Fixation, F. O. Pedrosa, M. Hungria, G. Yates and W. E. Newton (eds.), Kluwer Academic Publishers, Dordrecht, 1999, p. 19.
47. T. R. Hawkes, P. A. McLean and B. E. Smith, Biochem. J. 217 (1984) 317.
48. P. A. McLean, A. E. True M. J. Nelson, S. Chapman, M. R. Godfrey, B. K. Teo, W. H. Orme-Johnson and B. M. Hoffman, J. Am. Chem. Soc. 109 (1987) 943.
49. G. D. Watt and K. R. N. Reddy, J. Inorg. Biochem. 53 (1994) 281.
50. S. J. Yoo, H. C. Angove, B. K. Burgess, M. P. Hendrich and E. Münck, J. Inorg. Biochem. 53 (1999) 281.
51. P. Strop, P. M. Takahara, H. Chiu, H. C. Angove, B. K. Burgess and D. C. Rees, Biochemistry 40 (2001) 651.
52. A. C. Robinson, B. K. Burgess and D. R. Dean, J. Bacteriol. 166 (1986) 180.
53. R. A. Ugalde, J. Imperial, V. K. Shah and W. J. Brill, J. Bacteriol. 159 (1984) 888.

54. V. K. Shah, J. Imperial, R. A. Ugalde, P. W. Ludden and W. J. Brill, Proc. Natl. Acad. Sci. USA 83 (1986) 1636.
55. T. R. Hoover, A. D. Robertson, R. L. Cerny, R. N. Hayes, J. Imperial, V. K. Shah and P. W. Ludden, Nature 329 (1987) 855.
56. V. K. Shah, P. Rangaraj, R. Chatterjee, R. M. Allen, J. T. Roll, G. P. Roberts and P. W. Ludden, J. Bacteriol. 181 (1999) 2797.
57. P. Rangaraj, C. Rüttimann-Johnson, V. K. Shah and P. W. Ludden in: Prokaryotic Nitrogen Fixation: a Model System for Analysis of a Biochemical Process, E. W. Triplett (ed.), Horizon Scientific Press, Wymondham, UK, 2000, p. 55.
58. J. Imperial, R. A. Ugalde, V. K. Shah and W. J. Brill, J. Bacteriol. 158 (1984) 187.
59. C. Moreno-Vivián, S. Hennecke, A. Pühler, and W. Klipp, J. Bacteriol. 171 (1989) 2591.
60. R. A. Ugalde, J. Imperial, V. K. Shah and W. J. Brill, J. Bacteriol. 164 (1985) 1081.
61. J. Imperial, R. A. Ulgalde, V. K. Shah and W. J. Brill, J. Bacteriol. 163 (1985) 1285.
62. R. Bosch, F. Rodríguez-Quiñones and J. Imperial, FEMS Microbiol. Lett. 157 (1997) 19.
63. D. R. Dean and J. M. R. in: Biological Nitrogen Fixation, G. Stacey, R. H. Burris, and H. J. Evans (eds.), Chapman and Hall, New York, 1992, p. 763.
64. T. R. Hawkes and B. E. Smith, Biochem. J. 209 (1983) 43.
65. T. D. Paustian, V. K. Shah and G. P. Roberts, Biochemistry 29 (1990) 3515.
66. D. Evans, R. Jones, P. Woodley and R. Robson, J. Gen. Microbiol. 134 (1988) 931.
67. B. Masepohl, W. Klipp and A. Pühler, Mol. Gen. Genet. 212 (1988) 27.
68. W. Klipp, B. Masepohl and A. Pühler, J. Bacteriol. 170 (1988) 693.
69. L. Rossen, Q. S. Ma, E. A. Mudd, A. W. Johnston and J. A. Downie, Nucleic Acids Res. 12 (1984) 7123.
70. W. J. Buikema, J. A. Klingensmith, S. L. Gibbons and F. M. Ausubel, J. Bacteriol. 169 (1987) 1120.
71. M. E. Mulligan and R. Haselkorn, J. Biol. Chem. 264 (1989) 19200.
72. E. M. Lyons and T. Thiel, J. Bacteriol. 177 (1995) 1570.
73. H. J. Sofia, G. Chen, B. G. Hetzler, J. F. Reyes-Spindola and N. E. Miller, Nucleic Acids Res. 29 (2001) 1097.
74. P. E. Bishop and R. D. Joerger, Annu. Rev. Plant Physiol. Plant Mol. Biol. 41 (1990) 109.
75. R. D. Joerger, R. Premakumar and P. E. Bishop, J. Bacteriol. 168 (1986) 673.
76. R. Bosch and J. Imperial, Recent Res. Devel. Microbiology 4 (2000) 131.
77. M. Drummond, J. Walmsley and C. Kennedy, J. Bacteriol. 178 (1996) 788.
78. V. K. Shah, J. R. Allen, N. J. Spangler and P. W. Ludden, J. Biol. Chem. 269 (1994) 1154.
79. R. M. Allen, R. Chatterjee, P. W. Ludden and V. K. Shah, J. Biol. Chem. 270 (1995) 26890.

80. S. W. Muchmore, R. F. Jack and D. R. Dean in: Mechanisms of Metallocenter Assembly, R. P. Hausinger, G. L. Eichhorn and L. G. Marzilli, (eds.) VCH Publishers, Inc., New York, 1996, p. 111.
81. J.-S. Chen, S.-Z. Wang and J. L. Johnson in: Proceedings 8th International Congress on Nitrogen Fixation, P. M. Gresshoff, L. E. Roth, G. Stacey and W. E. Newton (eds.), Chapman & Hall, Knoxville, USA, 1990.
82. J. T. Roll, V. K. Shah, D. R. Dean and G. P. Roberts, J. Biol. Chem. 270 (1995) 4432.
83. T. D. Paustian, V. K. Shah and G. P. Roberts, Proc. Natl. Acad. Sci. USA 86 (1989) 6082.
84. P. J. Goodwin, J. N. Agar, J. T. Roll, G. P. Roberts, M. K. Johnson and D. R. Dean, Biochemistry 37 (1998) 10420.
85. K. E. Brigle, M. C. Weiss, W. E. Newton and D. R. Dean, J. Bacteriol. 169 (1987) 1547.
86. K. E. Brigle and D. R. Dean, Nucleic Acids Res. 16 (1988) 5214.
87. R. Setterquist, K. E. Brigle, J. Beynon, M. Cannon, A. Ally, F. Cannon and D. R. Dean, Nucleic Acids Res. 16 (1988) 5215.
88. R. Fani, R. Gallo and P. Lio, J. Mol. Evol. 51 (2000) 1.
89. J. B. Howard and D. C. Rees, Ann. Rev. Biochem. 63 (1994) 235.
90. E. D. Wolfinger and P. E. Bishop, J. Bacteriol. 173 (1991) 7565.
91. C. Moreno-Vivián, M. Schmehl, B. Masepohl, W. Arnold and W. Klipp, Mol. Gen. Genet. 216 (1989) 353.
92. S. Siemann, K. Schneider, K. Behrens, A. Knochel, W. Klipp and A. Muller, Eur. J. Biochem. 268 (2001) 1940.
93. W. A. Filler, R. M. Kemp, J. C. Ng, T. R. Hawkes, R. A. Dixon and B. E. Smith, Eur. J. Biochem. 160 (1986) 371.
94. A. C. Robinson, D. R. Dean and B. K. Burgess, J. Biol. Chem. 262 (1987) 14327.
95. R. M. Allen, M. J. Homer, R. Chatterjee, P. W. Ludden, G. P. Roberts and V. K. Shah, J. Biol. Chem. 268 (1993) 23670.
96. A. C. Robinson, T. W. Chun, J.-G. Li and B. K. Burgess, J. Biol. Chem. 264 (1989) 10088.
97. V. K. Shah, T. R. Hoover, J. Imperial, T. D. Paustian, G. P. Roberts and P. W. Ludden in: Nitrogen Fixation: Hundred Years After, H. Bothe, F. J. de Bruijn and W. E. Newton (eds.), Gustav Fischer, Cologne, 1988.
98. N. Gavini, and B. K. Burgess, J. Biol. Chem. 267 (1992) 21179.
99. D. Wolle, D. R. Dean and J. B. Howard, Science 258 (1992) 992.
100. P. Rangaraj, V. K. Shah and P. W. Ludden, Proc. Natl. Acad. Sci. USA 94 (1997) 11250.
101. P. Rangaraj, M. J. Ryle, W. N. Lanzilotta, P. W. Ludden and V. K. Shah, J. Biol. Chem. 274 (1999) 19778.
102. M. R. Jacobson, V. L. Cash, M. C. Weiss, N. F. Laird, W. E. Newton and D. R. Dean, Mol. Gen. Genet. 219 (1989) 49.

103. R. Chatterjee, R. M. Allen, P. W. Ludden and V. K. Shah, J. Biol. Chem. 272 (1997) 21604.
104. M. M. Gosink, N. M. Franklin and G. P. Roberts, J. Bacteriol. 172 (1990) 1441.
105. M. J. Homer, T. D. Paustian, V. K. Shah and G. P. Roberts, J. Bacteriol. 175 (1993) 4907.
106. T. C. White, G. S. Harris and W. H. Orme-Johnson, J. Biol. Chem. 267 (1992) 24007.
107. C. Rüttimann-Johnson, C. R. Staples, P. Rangaraj, V. K. Shah and P. W. Ludden, J. Biol. Chem. 274 (1999) 18087.
108. C. Rüttimann-Johnson, P. Rangaraj, V. K. Shah and P. W. Ludden, J. Biol. Chem. 276 (2001) 4522.
109. M. J. Homer, D. R. Dean and G. P. Roberts, J. Biol. Chem. 270 (1995) 24745.
110. P. Rangaraj, C. Rüttimann-Johnson, V. K. Shah and P. W. Ludden, J. Biol. Chem. 276 (2001) 15968.
111. A. R. Mushegian and E. V. Koonin, Trends Genet. 12 (1996) 289.
112. V. Anantharaman, E. V. Koonin and L. Aravind, J. Mol. Biol. 307 (2001) 1271.
113. L. Zheng, R. H. White and D. R. Dean, J. Bacteriol. 179 (1997) 5963.
114. P. A. McLean and R. A. Dixon, Nature 292 (1981) 655.
115. P. A. McLean, B. E. Smith and R. A. Dixon, Biochem. J. 211 (1983) 589.
116. T. R. Hoover, V. K. Shah, G. P. Roberts and P. W. Ludden, J. Bacteriol. 167 (1986) 999.
117. T. R. Hoover, J. Imperial, P. W. Ludden and V. K. Shah, J. Bacteriol. 170 (1988) 1978.
118. T. R. Hoover, J. Imperial, P. W. Ludden and V. K. Shah, Biochemistry 28 (1989) 2768.
119. J. Liang, M. Madden, V. K. Shah and R. H. Burris, Biochemistry 29 (1990) 8577.
120. J. Imperial, T. R. Hoover, M. S. Madden, P. W. Ludden and V. K. Shah, Biochemistry 28 (1989) 7796.
121. P. Yuvaniyama, J. N. Agar, V. L. Cash, M. K. Johnson and D. R. Dean, Proc. Natl. Acad. Sci. USA 97 (2000) 599.
122. W. Fu, R. F. Jack, T. V. Morgan, D. R. Dean and M. K. Johnson, Biochemistry 33 (1994) 13455.
123. J. N. Agar, P. Yuvaniyama, R. F. Jack, V. L. Cash, A. D. Smith, D. R. Dean and M. K. Johnson, J. Biol. Inorg. Chem. 5 (2000) 167.
124. L. Zheng, R. H. White, V. L. Cash, R. F. Jack and D. R. Dean, Proc. Natl. Acad. Sci. USA 90 (1993) 2754.
125. L. Zheng, R. H. White, V. L. Cash and D. R. Dean, Biochemistry 33 (1994) 4714.
126. L. Zheng, and D. R. Dean, J. Biol. Chem. 269 (1994) 18723.
127. J. T. Kaiser, T. Clausen, G. P. Bourenkow, H. D. Bartunik, S. Steinbacher and R. Huber, J. Mol. Biol. 297 (2000) 451.
128. L. Zheng, V. L. Cash, D. H. Flint and D. R. Dean, J. Biol. Chem. 273 (1998) 13264.

129. D. M. Hwang, A. Dempsey, K. T. Tan and C. C. Liew, J. Mol. Evol. 43 (1996) 536.
130. J. N. Agar, L. Zheng, V. Cash, D. R. Dean and M. K. Johnson, J. Am. Chem. Soc. 122 (2000) 2136.
131. J. N. Agar, C. Krebs, J. Frazzon, B. H. Huynh, D. R. Dean and M. K. Johnson, Biochemistry 39 (2000) 7856.
132. A. D. Smith, J. N. Agar, K. A. Johnson, J. Frazzon, I. J. Amster, D. R. Dean and M. K. Johnson, J. Am. Chem. Soc. 123 (2001) 11103.
133. S. Kim and B. K. Burgess, J. Biol. Chem. 269 (1994) 4215.
134. W. Paul and M. Merrick, Eur. J. Biochem. 178 (1989) 675.
135. B. Masepohl, S. Angermuller, S. Hennecke, P. Hubner, C. Moreno-Vivián and W. Klipp, Mol. Gen. Genet. 238 (1993) 369.
136. S. Kimand B. K. Burgess, J. Biol. Chem. 271 (1996) 9764.
137. S. H. Lee, L. Pulakat, K. C. Parker and N. Gavini, Biochem. Biophys. Res. Commun. 244 (1998) 498.
138. K. S. Howard, P. A. McLean, F. B. Hansen, P. V. Lemley, K. S. Koblan and W. H. Orme-Johnson, J. Biol. Chem. 261 (1986) 772.
139. N. Gavini and L. Pulakat in: Proceedings 101st General Meeting of the American Society for Microbiology, Orlando, Florida (2001), p. 411.
140. N. Petrova, L. Gigova, and P. Venkov, Biochem. Biophys. Res. Commun. 270 (2000) 863.
141. H. M. Simon, M. M. Gosink and G. P. Roberts, J. Bacteriol. 181 (1999) 3751.
142. D. Nieva-Gomez, G. P. Roberts, S. Klevickis and W. J. Brill, Proc. Natl. Acad. Sci. USA 77 (1980) 2555.
143. V. K. Shah, G. Stacey and W. J. Brill, J. Biol. Chem. 258 (1983) 12064.
144. A. Lindblad, J. Jansson, E. Brostedt, M. Johansson, U. Hellman and S. Nordlund, Mol. Microbiol. 20 (1996) 559.
145. E. Brostedt and S. Nordlund, Biochem. J. 279 (1991) 155.
146. L. T. Bennett, M. R. Jacobson and D. R. Dean, J. Biol. Chem. 263 (1988) 1364.
147. A. E. Martin, B. K. Burgess, S. E. Iismaa, C. T. Smartt, M. R. Jacobson and D. R. Dean, J. Bacteriol. 171 (1989) 3162.
148. G. Gennaro, P. Hubner, U. Sandmeier, A. F. Yakunin and P. C. Hallenbeck, J. Bacteriol. 178 (1996) 3949.
149. A. F. Yakunin and P. C. Hallenbeck, Biochim. Biophys. Acta 1409 (1998) 39.
150. M. Schmehl, A. Jahn, A. Meyer zu Vilsendorf, S. Hennecke, B. Masepohl, M. Schuppler, M. Marxer, J. Oelze and W. Klipp, Mol. Gen. Genet. 241 (1993) 602.
151. B. Masepohl and W. Klipp, Arch. Microbiol. 165 (1996) 80.
152. Y. Jouanneau, H. S. Jeong, N. Hugo, C. Meyer and J. C. Willison, Eur. J. Biochem. 251 (1998) 54.
153. N. Desnoues, M. Lin and C. Elmerich in GenBank AJ297529 (2000).
154. K. Saeki, and H. Kumagai, Arch. Microbiol. 169 (1998) 464.
155. H. S. Jeong, and Y. Jouanneau, J. Bacteriol. 182 (2000) 1208.
156. J. Klugkist, H. Haaker and C. Veeger, Eur. J. Biochem. 155 (1986) 41.

157. H. Kumagai, T. Fujiwara, H. Matsubara and K. Saeki, Biochemistry 36 (1997) 5509.
158. Y. Hatefi, Annu. Rev. Biochem. 54 (1985) 1015.
159. V. K. Shah, R. A. Ugalde, J. Imperial and W. J. Brill, Annu. Rev. Biochem. 53 (1984) 231.
160. P. T. Pienkos and W. J. Brill, J. Bacteriol. 145 (1981) 743.
161. R. M. Allen, J. T. Roll, P. Rangaraj, V. K. Shah, G. P. Roberts and P. W. Ludden, J. Biol. Chem. 274 (1999) 15869.
162. J. K. Magnuson, T. D. Paustian, V. K. Shah, D. R. Dean, G. P. Roberts, D. C. Rees and J. B. Howard, Tetrahedron 53 (1997) 11971.
163. J. Christiansen, P. J. Goodwin, W. N. Lanzilotta, L. C. Seefeldt and D. R. Dean, Biochemistry 37 (1998) 12611.

Nitrogen Fixation at the Millennium.
G. Jeffery Leigh, Editor.

CHAPTER 5

Use of Short-Chain Alkynes to Locate the Nitrogenase Catalytic Site

Suzanne M. Mayer[a], Patricia dos Santos[a], Lance C. Seefeldt[b], Dennis R. Dean[a]
[a]Department of Biochemistry, Virginia Tech, Blacksburg, Virginia 24061, USA;
[b]Department of Chemistry and Biochemistry, Utah State University, Logan, Utah 84322, USA

1. INTRODUCTION

The mechanism by which molybdenum nitrogenase catalyzes the MgATP-dependent reduction of dinitrogen to ammonia has challenged enzymologists for more than four decades [1,2]. Because the quantification of ammonia has proven to be difficult and laborious, acetylene is frequently used as an alternative substrate for probing the nitrogenase catalytic mechanism [3]. Here we summarize some of the problems that have denied a complete understanding of this mechanism and describe how acetylene and other short-chain alkynes have recently been used in a combination of biochemical and genetic studies aimed at resolving one of the remaining unsolved questions – where exactly on the complex metallocluster known as FeMo-cofactor do nitrogenase substrates interact and become reduced?

2. BACKGROUND

The basic features concerning nitrogenase enzymology have been known for many years. The enzyme is comprised of two separate component proteins that contain metal-sulfur clusters. The smaller of the two component proteins is the Fe protein, which serves to deliver reducing equivalents [4, 5] to the other component protein, referred to as MoFe protein, which provides the site for substrate-binding [6-8] and reduction[‡] [9-11]. The Fe protein is so named as it is a 60-kDa homodimer that contains a single [4Fe-4S] cluster [12]. During catalysis, electrons are delivered one at a time from the Fe protein to the MoFe protein [4] and each electron-transfer requires

---

[‡] The structures and organization of both the Fe protein and the MoFe protein have been crystallographically determined and are described in detail in Chapter 2. The biophysical properties associated with nitrogenase are described in Chapter 3 and the catalytic process is discussed in several Chapters. The reader's attention is also directed to several recent reviews [9-12].

the hydrolysis of at least 2 MgATP. The Fe protein [4Fe-4S] cluster cycles between the 2+ and 1+ oxidation levels as it transfers the electrons sequentially. In certain nitrogen-fixing organisms, the physiological reductant of the Fe protein is a flavodoxin [13-15], which itself cycles between a reduced (hydroquinone) and oxidized (semiquinone) state. However, for *in vitro* enzymatic studies using the purified components, dithionite is often used as an artificial electron donor to the Fe protein [16]. A possible complication regarding the use of dithionite has recently surfaced since investigators have discovered it is possible to induce an *in vitro* all-ferrous Fe protein [17-24]. This finding raises the possibility that the Fe protein [4Fe-4S] cluster might actually cycle between the 2+/0 oxidation levels during physiological nitrogenase catalysis [23]. This also suggests *in vitro* enzymatic studies using dithionite as the artificial source of reducing equivalents for the Fe protein may be flawed since dithionite is only capable of inducing the 2+/1+ oxidation levels of the [4Fe-4S] cluster of the Fe protein during catalysis. Consequently, the current *in vitro* experimental system most frequently used may not necessarily faithfully duplicate *in vivo* events. As yet, however, there is no compelling experimental evidence that the all-ferrous (iron(II)) Fe protein actually exists *in vivo*. Therefore, the possible mechanistic implications of two-electron transfers from the Fe protein to the MoFe protein will not be considered further in this Chapter.

The 240-kDa $\alpha_2\beta_2$ MoFe protein contains two types of unique metal-sulfur clusters known as the P-cluster [8Fe-7S] and FeMo-cofactor [7Fe-9S-Mo-homocitrate], of which there are two copies of each cluster type per tetramer [25-27]. The P-clusters are located at an $\alpha\beta$ interface while the FeMo-cofactors are located in each α-subunit and therefore each $\alpha\beta$ half contains one cluster of each type. Each P-cluster/FeMo-cofactor pair is often considered to encompass one independently operating catalytic unit, although recent work has indicated that this assumption might not be entirely correct [28]. Nevertheless, the available evidence indicates that any communication between the separate "halves" of the MoFe protein contributes only to catalytic efficiency rather than being a critical aspect of the overall catalytic mechanism, because, MoFe proteins that have only one FeMo-cofactor per MoFe protein tetramer remain relatively active [29,30].

The precise role of the P-cluster has never been fully defined but current evidence suggests that it serves as an intermediate in electron transfer from the Fe protein [4Fe-4S] cluster to the site of substrate-binding/reduction at the FeMo-cofactor [19,31-34]. Its location, exactly midway between these two clusters, further emphasizes this possibility [27]. It is known that electrons are transferred from the Fe protein to the MoFe protein one at a time [4] but six electrons are required to reduce one molecule of dinitrogen to yield two molecules of ammonia. A question that emerges is: where do these electrons accumulate during the catalytic process? One possibility is that they are sequestered as semi-reduced dinitrogen intermediates. Alternatively, electrons could be accumulated within the MoFe protein prior to delivery to the substrate. Because both the P-cluster and FeMo-cofactor contains a large number of Fe atoms it was conceivable that they might jointly store a sufficient number of electrons necessary to achieve complete reduction of dinitrogen in a single concerted step. However, this does not appear to be what happens because acid quenching of the enzyme under turnover conditions results in the production of hydrazine, indicating that at least one semi-reduced substrate intermediate is formed under catalytic conditions [35-37]. Nevertheless, careful kinetic analyses have shown that the MoFe protein must receive three or four electrons from the Fe protein before dinitrogen is able to bind [38]. Thus, it appears that the catalytic mechanism with regard to electron transfer is complicated and likely to include both the storage and distribution of electrons within the metallocluster types as well as proceeding through discrete semi-reduced intermediates. It is important to emphasize that the Fe protein is the obligate source of electrons used for substrate reduction and that MgATP hydrolysis, which accompanies intercomponent electron transfer, is also essential for productive electron transfer.

Why multiple electrons must be delivered to the MoFe protein before it is competent to bind dinitrogen is not entirely clear but this could be related to the activation of the metal-sulfur substrate-binding site by a reduction and protonation event. This suggestion arises from two observations. First, dinitrogen reduction at the active site appears to require eight electrons and eight protons rather than the minimum theoretical values of six electrons and six protons [39]. The extra electrons and protons appear in the overall stoichiometry of the reaction products as one dihydrogen molecule evolved for each dinitrogen reduced, indicating that hydrides might be eliminated from an activated metal-sulfur surface upon dinitrogen-binding [40, 41]. Second, when the enzyme is incubated under turnover conditions in the absence of dinitrogen, electrons and protons are consumed and produced as hydrogen gas at about the same rate as if dinitrogen was being used as the substrate [39]. It is not yet known whether or not the hydrogen evolved as an obligate aspect of dinitrogen reduction is also formed at the same site and by the same mechanism as the hydrogen evolved when the enzyme turns over in the absence of dinitrogen. These features highlight one of the major complicating issues associated with nitrogenase enzymology, namely, that the MoFe protein must be partially reduced to achieve the capacity to bind dinitrogen yet it cannot be stably prepared in this state because, in the absence of dinitrogen, hydrogen is evolved, which continuously returns the enzyme to the original resting state.

Experimental measurement of the naturally occurring nitrogenase products, ammonia and hydrogen gas, in order to analyze the nitrogenase catalytic activity, also presents difficulties [38]. For example, measuring the rate of dinitrogen reduction is troublesome because quantitation of ammonia is a relatively difficult task [42]. Also, nitrogenase-directed proton reduction, although easy to measure by gas chromatography, is not suitable for most mechanistic studies because the concentration of protons cannot be experimentally controlled. Because of these issues other multiply-bonded substrates have been used as mechanistic probes [43] and among these, acetylene has proved to be the most useful [3]. Nevertheless, there also are several complications associated with using acetylene reduction as a model for analysis of dinitrogen-binding and reduction. First, acetylene is reduced by only two-electrons, to yield one molecule of ethylene, whereas, dinitrogen is reduced by six-electrons to yield two molecules of ammonia. Second, dinitrogen-binding/reduction apparently involves the obligate evolution of hydrogen gas [39,41] whereas acetylene does not. Finally, hydrogen gas is a competitive inhibitor of dinitrogen-binding but it is not an inhibitor of acetylene-binding. All of these features point to fundamental mechanistic differences between dinitrogen reduction and acetylene reduction. Nonetheless, considering that both acetylene and dinitrogen are small triply bonded molecules, it is reasonable to expect that acetylene should be useful for determining the exact location of substrate interaction at the FeMo-cofactor.

The nature of the relationship between dinitrogen-binding and reduction and acetylene-binding and reduction catalyzed by nitrogenase was evaluated many years ago by determining the mutual inhibition properties of the different substrates. This relationship turned out to be complex, as acetylene was found to be a non-competitive inhibitor of dinitrogen reduction whereas dinitrogen was found to be a weak competitive inhibitor of acetylene reduction [3,40]. Clues towards the understanding of this non-reciprocal behavior were provided by a series of simple "electron flux" experiments [44,45,46,]. Electron flux is defined as the rate at which electrons flow through the nitrogenase complex to achieve substrate reduction. As already noted, the nitrogenase Fe protein is the obligate electron donor to the MoFe protein, which provides the substrate reduction site. Electrons can only accumulate within the MoFe protein by several single-electron transfer events, which requires several Fe protein:MoFe protein association and disassociation cycles [4]. Therefore, the rate at which any substrate can be reduced by nitrogenase depends upon the rate of electron-delivery from the Fe protein to the MoFe protein. Because electrons are delivered from the Fe protein to the MoFe protein in one-electron steps, and because at least two electrons are required for the reduction of any substrate, overall substrate reduction can be controlled by adjusting the relative concentrations of the Fe protein and the MoFe protein. Therefore, a high Fe protein/MoFe protein ratio maximizes electron delivery and this condition is referred to as high-flux. Low-flux refers to a low Fe protein/MoFe protein ratio and under this condition the rate of substrate reduction is limited by the rate of accumulation of the several electrons required for substrate reduction. In early experiments, it was shown that, for low-flux conditions under a dinitrogen atmosphere, a disproportionately higher level of electrons were allocated to proton reduction relative to dinitrogen reduction [44,46,47]. The related key experiment with regard to acetylene reduction is depicted in Figure 1 [44]. This figure illustrates that under low-flux conditions acetylene is a very effective inhibitor of dinitrogen reduction but is a significantly less effective inhibitor of dinitrogen reduction under high-flux conditions.

How variations in flux might affect the relative ability of acetylene to inhibit dinitrogen reduction, as well as the non-reciprocity for the patterns of mutual inhibition

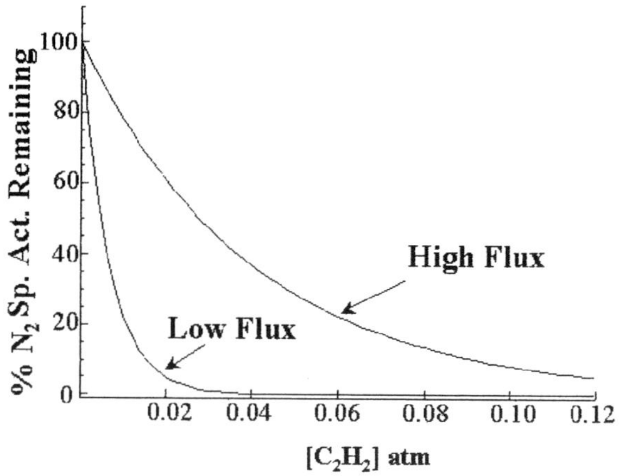

Figure 1. Effect of electron flux upon the inhibition of dinitrogen reduction by acetylene. Under low-flux conditions (low Fe protein:MoFe protein ratio) acetylene is significantly better at inhibiting dinitrogen reduction compared to high-flux conditions (high Fe protein:MoFe protein ratios).

of acetylene and dinitrogen reduction, can be explained using the model shown in Figure 2. In Figure 2 the MoFe protein is depicted as sequentially progressing through increasingly reduced states as it accumulates electrons delivered from the Fe protein [37]. The resting state, with no accumulated electrons, is designated E_0 and is incapable of binding substrates at this redox level. Progressively more reduced states are formed as a consequence of multiple electron transfers from the Fe protein and are depicted as E_2, E_4 and E_6.

Because acetylene is able to bind to a less reduced state of MoFe protein than does dinitrogen [47], acetylene is expected to be a more effective inhibitor of dinitrogen reduction under low-flux conditions than under high-flux conditions. This is because under low-flux conditions a population of MoFe protein accumulates that is available for acetylene-binding but not dinitrogen-binding. Thus, acetylene has favorable access to the FeMo-cofactor active site before there is any opportunity for dinitrogen-binding. Upon acetylene reduction, a large fraction of the available semi-reduced MoFe protein is returned to the original resting state, which removes the potential of this fraction of protein for interaction with dinitrogen. This explains why acetylene exhibits a non-competitive inhibition pattern on dinitrogen reduction. As described above, the presence of acetylene effectively lowers the pool of enzyme available for dinitrogen-binding, which is how classical non-competitive inhibition is defined. However, in this particular case, the inhibitor (acetylene) does not lower the pool of competent active sites by binding at a site separate from that which binds the substrate (dinitrogen) but rather by binding at a different redox state.

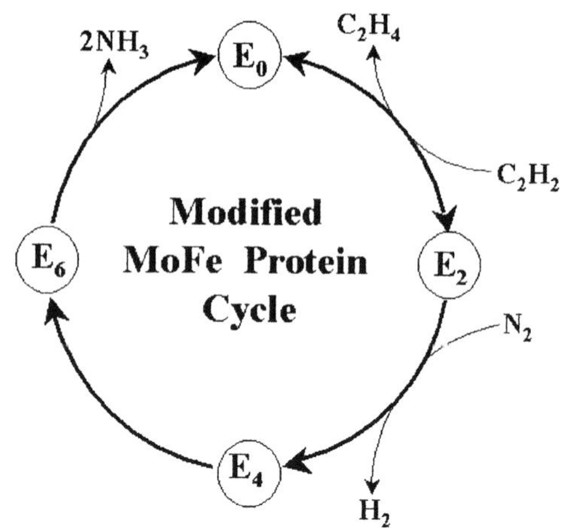

Figure 2. A simplified representation of the MoFe protein catalytic cycle. E_0 represents the resting state of the MoFe protein where no electrons have yet been transferred by the Fe protein. All states represented have accumulated multiples of two-electrons from the sequential delivery from the Fe protein. Therefore, E_2 represents a two-electron reduced state, E_4 a four-electron reduced state, and E_6 a six-electron reduced state. Redox states of the enzyme having odd numbers of electrons accumulated are not depicted for the sake of simplicity and because all substrates are reduced by multiples of two-electrons. Acetylene binds and is reduced at the E_2 state; dinitrogen binds at some time after the E_2 state has been reached and likely evolves a hydrogen as a consequence of dinitrogen binding.

For dinitrogen to exhibit a competitive pattern of inhibition on acetylene reduction it is expected that acetylene should also be able to bind to the enzyme when it is in the more reduced state, which is the state required for dinitrogen binding. This explains why dinitrogen is only a very weak competitive inhibitor of acetylene reduction, namely, only a small fraction of the total enzyme capable of binding acetylene is also subjected to competitive inhibition by dinitrogen-binding. In other words, acetylene-binding is expected to occur at a redox state that is not yet competent for dinitrogen-binding and therefore not susceptible to inhibition by dinitrogen.

3. INSIGHT INTO THE EFFECTS OF AMINO ACID SUBSTITUTIONS AROUND THE FeMo-COFACTOR ON SUBSTRATE REDUCTION

Among the central questions that remain in nitrogenase enzymology is this: where and how do substrates interact with FeMo-cofactor? Inspection of the FeMo-cofactor structure is not particularly informative on this point, as it is a symmetrical molecule [48]. As shown in Figure 3, the metal-sulfur core of the FeMo-cofactor is constructed from [4Fe-3S] and [4Fe-Mo-3S] substructures, geometrically analogous to

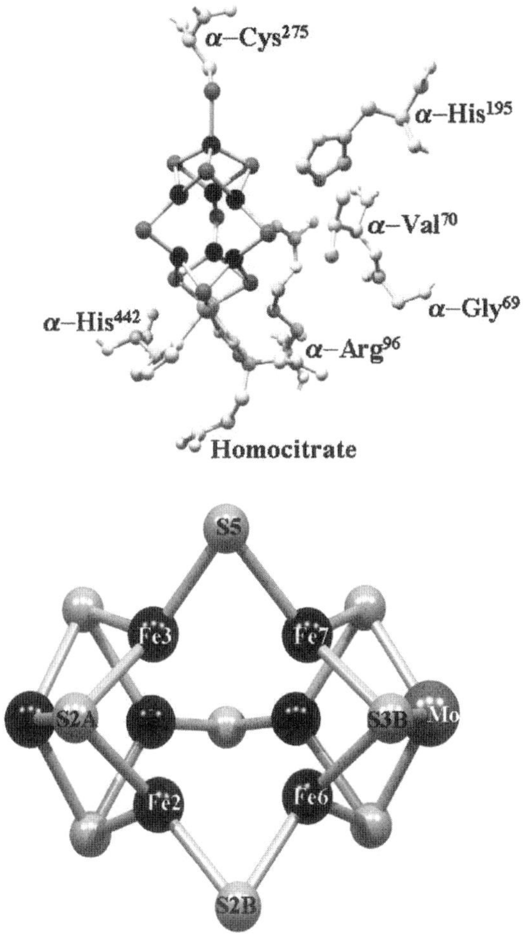

Figure 3. The structure and immediate environment of the FeMo-cofactor is shown in the upper figure while the central metal-sulfur core with the specific 4Fe-4S face capped by the α-Val^{70}, α-Arg^{96} and α-His^{195} residues is labeled in the lower figure. The color scheme for the FeMo-cofactor atoms is as follows: Fe = black, S = light gray and Mo = dark gray.

[4Fe-4S] cubanes and linked by three bridging sulfides. The central region of FeMo-cofactor can be considered as having three identical 4Fe-4S faces, with each face sharing Fe atoms and bridging sulfides with one of the two other faces. Thus, there is no geometrically unique site within the metal-sulfur core that is an obvious site for substrate interaction. However, in the context of the polypeptide environment, which almost certainly controls or affects substrate access and reactivity, FeMo-cofactor is not symmetrical. Inspection of the MoFe protein crystal structure reveals there is a limited number of amino acids located within the first shell of non-covalent interactions with the FeMo-cofactor. These facts shaped our original experimental approach and work in our laboratories has focused on attempts to determine the location of the substrate interaction site by placing amino acid substitutions within the FeMo-cofactor environment and determining the effect of these substitutions on dinitrogen- and acetylene-binding and reduction.

The effect of one such substitution, MoFe protein α-subunit His^{195} substituted by α-Gln^{195}, proved to be the most informative [49-54]. In the wild-type enzyme, the ε-nitrogen of the imidazole group of the α-His^{195} residue is within hydrogen-bonding distance of one of the three bridging sulfides, which connect the two FeMo-cofactor subclusters. It was assumed that replacing this histidine with a glutamine would eliminate the hydrogen bond to the FeMo-cofactor and perhaps provide some insight about substrate interaction with the FeMo-cofactor. Characterization of the α-Gln^{195} MoFe protein revealed that, while it retained the ability to reduce protons and acetylene at approximately the same rate as the wild-type α-His^{195} protein [49], it could reduce dinitrogen at only approximately 1-2% of the wild-type rate [50]. Furthermore, even though dinitrogen was only a very poor substrate for the α-Gln^{195} MoFe protein, it still inhibited proton and acetylene reduction as effectively as wild-type α-His^{195} MoFe protein [49]. This observation indicates that the α-Gln^{195} MoFe protein is able to bind dinitrogen in the normal way but is very ineffective in its capacity to reduce dinitrogen. The three dimensional structure of the α-Gln^{195} MoFe protein was also solved crystallographically [54] and the overall structure was found to remain intact, except that the hydrogen bond normally provided by the imidazole group of the wild-type α-His^{195} was replaced by a similar interaction with the amide nitrogen of the α-Gln^{195} side-chain. There were no other structural perturbations within the altered MoFe protein that could explain its substrate reducing abilities. The simplest explanation for the inhibition of proton and acetylene reduction caused by dinitrogen-binding to the α-Gln^{195} MoFe protein is that all three potential substrates (protons, acetylene, dinitrogen) bind at the same or overlapping sites. Furthermore, an inability to reduce dinitrogen could be explained if it is assumed that the α-His^{195} is a proton donor required for effective dinitrogen reduction but which is not essential for the reduction of certain other substrates, such as acetylene, as these require fewer protons for their reduction.

While the work with the α-Gln^{195} substituted-MoFe protein was unable to clearly establish where substrates bind, it did provide some new insights about substrate-binding at the FeMo-cofactor. More importantly, it demonstrated that manipulating the polypeptide environment of the FeMo-cofactor is a reasonable approach for the analysis of substrate-FeMo-cofactor interactions.

4. RATIONALE AND STRATEGY FOR ISOLATION OF AN ACETYLENE-RESISTANT MoFe PROTEIN

Even though the results the α-Gln^{195}-substituted MoFe protein were informative, it was soon realized that the approach of randomly substituting amino acids proximal to the FeMo-cofactor needed to be reconsidered. Because the MoFe protein must receive several electrons before it is competent to bind any substrate, and different substrates are able to bind at different redox states, it is difficult to assign uniquely the basis for any particular effect to any particular parameter. As one example, many altered MoFe proteins that retain the ability to reduce acetylene and protons but are not able to reduce dinitrogen, in addition to the one described in Section 3, have been isolated [55]. This phenotype can be reasonably ascribed as resulting from a combination of any or all three of the following changes: (i) an alteration in the electron flux so that the enzyme is unable to achieve the redox state necessary for dinitrogen-binding, (ii) an alteration in the chemical reactivity of FeMo-cofactor towards dinitrogen-binding, (iii) a structural perturbation that specifically affects dinitrogen accessibility. Because each of these parameters (flux, reactivity, and accessibility) is likely to impact the others, determination of the specific origin of an amino acid substitution that affects dinitrogen reduction while leaving acetylene reduction intact is not possible.

An alternative approach to identifying the substrate-interaction site involves searching for an altered MoFe protein that is affected in its ability to reduce acetylene without also being affected in its ability to reduce dinitrogen. The binding and reduction of dinitrogen is chemically and electronically more demanding than acetylene-binding and reduction. Therefore, any alteration yielding a MoFe protein that retains a normal capacity for dinitrogen reduction but which is compromised in acetylene reduction cannot be interpreted as arising from alterations in either electron flux or the chemical reactivity of the FeMo-cofactor. Rather, such a phenotype can only be assigned to a structural perturbation that specifically alters the capacity of acetylene to interact with the active site. The expectation that such an altered MoFe protein might exist, even if dinitrogen and acetylene recognize essentially the same site, seemed reasonable given that acetylene is larger than dinitrogen. The challenge was to devise a strategy for the isolation of a strain that would produce such a protein.

As acetylene is known to be both an *in vitro* and a physiological inhibitor of dinitrogen reduction, the initial objective was to isolate a mutant strain that was physiologically resistant to the inhibitory effect of acetylene. However, this approach presented a technical challenge because the organism used for these studies, *Azotobacter vinelandii*, is an obligate aerobe. Consequently, the largest hurdle to overcome was to discover a way to achieve complete growth inhibition of *A. vinelandii* cells using the necessary concentrations of acetylene, which would not be explosive when mixed with the requisite concentration of dioxygen essential for aerobic growth. [56].

Previous work regarding the *in vitro* effect of electron flux on the capacity of acetylene to inhibit dinitrogen reduction provided the inspiration to circumvent this problem [44]. These studies indicated that a strain defective in its ability to achieve maximum electron flux should be physiologically more sensitive to lower concentrations of acetylene than is the wild-type strain. As previously defined, electron flux is the rate at which electrons are delivered to the substrate-reduction site. This rate can be controlled during *in vitro* experiments by altering the Fe protein:MoFe protein ratio, to give high- or low-flux rates. The way to control *in vivo* electron flux is to use mutant strains that are defective in their capacity for inter- or intra-component electron transfer [32,57]. Any mutant that is altered in its ability to transfer electrons from the Fe protein to the MoFe protein or in its capacity for electron-transfer from the P-cluster to the FeMo-cofactor is, by definition, unable to achieve high-flux.

Such mutants of both types were isolated previously and were subsequently shown to be unable to reduce dinitrogen at a sufficient level to support growth when cultured under a 0.025 acetylene atmosphere [32,56,57]. However, after prolonged incubation under these conditions, mutants resistant to acetylene were obtained [56]. Such acetylene-resistant mutants might have arisen, either as a result of a correction in the original flux defect or because the MoFe protein had become altered so that acetylene was no longer able to effectively interact with the MoFe protein. Genetic and sequence analyses showed that, for the particular acetylene-resistant strains isolated, the original flux defect was left intact [56]. It was also shown that the mutation leading to acetylene-resistance could effectively suppress the acetylene-sensitive phenotype when placed in combination with a variety of other individual mutations, each of which lowers flux in a different way. Thus, the acetylene-resistance phenotype is the result of an alteration in the ability of the altered MoFe protein to use acetylene as an effective substrate.

This conclusion was confirmed by the biochemical analysis of an acetylene-resistant MoFe protein [56,58]. For these studies, the mutation conferring resistance to acetylene was first moved to a strain with a wild-type genetic background. Thus, the new strain from which the altered MoFe protein was isolated and characterized carried the mutation leading to the acetylene-resistant phenotype but it did not carry the mutation leading to the original flux defect. This altered MoFe protein was found tohave about the same K_m and V_{max} for dinitrogen reduction as for the wild-type protein. In contrast, the acetylene K_m for the altered acetylene-resistant MoFe protein increased twenty fold when compared to the wild- type MoFe protein. However, the V_{max} for acetylene reduction was the same for both the altered and wild-type MoFe proteins. These results provide an explanation for the physiological phenotype of the acetylene-resistant strain, namely, that at low concentrations acetylene it is not able to effectively inhibit dinitrogen reduction. A second important biochemical feature of the acetylene-resistant MoFe protein is that acetylene was converted from a non-competitive to a competitive inhibitor of dinitrogen reduction [59].

There are at least two different explanations for the catalytic properties of the acetylene-resistant MoFe protein. The first is a "two-site" model that suggests there are two separate acetylene-binding sites, a high-affinity site and a low-affinity site, located within the MoFe protein. In this case, the high-affinity site is eliminated in the altered acetylene-resistant MoFe protein leaving the low-affinity acetylene-binding site intact. In this model, because acetylene was converted from a non-competitive inhibitor to a competitive inhibitor of dinitrogen reduction, the low-affinity acetylene-binding site must be the same as, or overlap with, the dinitrogen-binding site.

The second explanation uses a "two-state" model, which suggests that both the high-affinity acetylene-binding site and the low-affinity acetylene-binding site are essentially the same. However, in this case the capacity for acetylene-binding changes as the MoFe protein proceeds through different redox states. In the two-state model, the high-affinity site is the one that becomes available at the less reduced redox level and is responsible for non-competitive inhibition by acetylene on dinitrogen reduction in the wild-type enzyme. As the MoFe protein accumulates more electrons and becomes more reduced, the high-affinity acetylene-binding site is converted to a low-affinity acetylene-binding site, thus enabling dinitrogen to compete for the same binding site. This model also demands that structural transitions occur within the FeMo-cofactor and/or its environment and that these transitions are manifested in changes in the binding affinity for acetylene as the protein progresses through more reduced redox levels. In the case of the acetylene-resistant MoFe protein, these structural transitions must perturb the high-affinity acetylene-binding site so that it is no longer available at the more oxidized redox level. Therefore, the only acetylene-binding site that remains is the mutually competitive dinitrogen-binding site that is only available at the more reduced redox level.

So what changes have occurred in the nitrogenase enzyme to induce an acetylene-resistant strain and MoFe protein? DNA sequence analysis of the *nifD* gene encoding the MoFe protein α-subunit revealed that all of the isolated mutant strains have the α-Gly^{69} residue substituted by α-Ser^{69} [56]. It was later demonstrated that other amino acid substitutions other than serine at the α-69-position could also induce the acetylene-resistant phenotype. For example, substitution of α-Gly^{69} by α-Ala^{69} or α-Asp^{69} also results in an acetylene-resistant strain. This presumably means that the effect is not the consequence of a specific feature of a particular side-chain (*i.e.*, size or charge). Rather, it is likely that acetylene-resistance arises from a regional inflexibility that is a consequence of replacing the glycine residue, which does not have a side-chain, with one that does possess a side-chain. Inspection of the MoFe protein crystal structure reveals many interesting features of the α-Gly^{69} environment. First, the α-Gly^{69} residue is located on a short helix that connects the P-cluster to the FeMo-cofactor. This helix begins at the P-cluster ligand α-Cys^{61} and continues toward the FeMo-cofactor. It ends at the α-Gly^{69} neighboring residue α-Val^{70}, which is positioned 3.5 Å from the central waist of one of the three equivalent 4Fe-4S faces of the FeMo-cofactor (Figure 3).

Closer inspection reveals that the carbonyl groups of both α-Gly^{69} and α-Val^{70} are within hydrogen-bonding distance of a guanidine group provided by the side-chain of α-Arg^{96}. This α-Arg^{96} side-chain also approaches the same 4Fe-4S face as does α-Gly^{69} and α-Val^{70} and it also hydrogen bonds to one of the central bridging sulfurs of the FeMo-cofactor. Thus, the α-Ser^{69} substitution could affect the accessibility of acetylene to this 4Fe-4S face, due either to the movement of the α-Val^{70} side-chain or alternatively to the movement of the main polypeptide chain, which could perturb the interaction with α-Arg^{96} and ultimately the FeMo-cofactor. Several other structural features and considerations also make these possibilities attractive.

As mentioned earlier, α-Val^{70} is located on a short helix that connects the P-cluster to the FeMo-cofactor. In addition, α-Arg^{96} is also connected to the P-cluster in an analogous way. The α-Arg^{96} is also located on a short helix that begins at the P-cluster ligand α-Cys^{88} and ends with the α-Arg^{96} side-chain hydrogen bonding to the FeMo-cofactor. Because the P-cluster is believed to be an electron-transfer agent to FeMo-cofactor, and because the P-cluster is known to undergo redox-dependent structural changes [60,61], communication between the P-cluster and FeMo-cofactor during catalysis seems mechanistically essential. These structural rearrangements in the P-cluster could therefore result in the rearrangement of the polypeptide environment surrounding the 4Fe-4S face approached by α-Arg^{96} and α-Val^{70}, communicated through these connecting helices.

There is one final consideration to be made for the location of these residues in relation to this specific 4Fe-4S face. When compared to the other two 4Fe-4S faces, the environment of this particular face is uniquely rich in potential proton donors. These potential proton donors include a pool of water molecules surrounding homocitrate, the organic constituent of FeMo-cofactor. When considered collectively, the structural features of the position that induces the acetylene-resistance (*i.e.*, the α-69 position) and the experimental results from the acetylene-resistant α-Ser^{69} MoFe protein implicate the 4Fe-4S face of the FeMo-cofactor approached by α-Val^{70} and α-Arg^{96} as providing a likely acetylene interaction site.

5. EXPANDING THE CAPACITY OF NITROGENASE TO REDUCE LONGER-CHAIN ALKYNES

Ordinarily, nitrogenase is not able to reduce short-chain alkynes other than acetylene, presumably because such molecules are too big to gain access to the active site. As a way to test whether or not the 4Fe-4S face of the FeMo-cofactor approached by α-Val^{70} and α-Arg^{96} comprises the surface for substrate interaction, these residues were substituted by amino acids having shorter side-chains. In doing so, it was expected that the substrate selectivity of nitrogenase might be expanded to include alkynes larger than acetylene. For these experiments, mutant strains having individual amino acid substitutions for either α-Val^{70} or α-Arg^{96} were evaluated for growth inhibition by propargyl alcohol ($HC{\equiv}CCH_2OH$) [58]. Propargyl alcohol was selected for this evaluation because it is the smallest substituted-acetylene molecule that is also water-soluble at ambient temperature.

Addition of propargyl alcohol to the growth medium of strains having various amino acid substitutions at the α-Arg^{96} position had no affect on the cell's ability to grow and fix dinitrogen. However, the diazotrophic growth of a strain having the α-Val^{70} residue replaced by an alanine was completely inhibited when 6 mM propargyl was present in the growth medium. The addition of 6 mM propargyl alcohol had no effect on the growth of this strain if a source of fixed nitrogen was also added to the growth medium, indicating that the effect was specific to physiological dinitrogen reduction.

Kinetic analysis of the purified α-Ala^{70}-substituted-MoFe protein showed that propargyl alcohol is an effective inhibitor of both proton- and dinitrogen-reduction for this altered MoFe protein [58]. Although difficult to quantify, allyl alcohol ($CH_2{=}CHCH_2OH$) was shown to be the two-electron reduction product of propargyl alcohol. Propargyl alcohol was also shown to be a competitive inhibitor ($K_i \approx 4.0$ mM) of acetylene reduction catalyzed by the α-Ala^{70} MoFe protein. In contrast, when propargyl alcohol was tested as a substrate for the wild-type α-Val^{70} MoFe protein, no reduction product could be identified and no inhibition of proton-, dinitrogen-, or acetylene-reduction activities occurred in its presence. The difficulties incurred in quantifying allyl alcohol made determining accurate kinetic parameters for its formation catalyzed by the α-Ala^{70} MoFe impossible. Therefore, propyne, an alternative short-chain alkyne of equivalent size to propargyl alcohol, and whose two-electron reduction product, propene, can be accurately determined by gas chromatography, was used as a substrate.

Propyne was found to be an effective inhibitor of proton reduction catalyzed by the α-Ala^{70} MoFe protein, as well as an effective substrate. The K_m for propyne reduction for the α-Ala^{70} MoFe protein was 0.02 atm propyne, which is quite close to the K_m for acetylene reduction (0.005 atm) catalyzed by the wild-type α-Val^{70} MoFe protein. Propyne is also a competitive inhibitor of acetylene reduction catalyzed by the α-Ala^{70} MoFe protein. In contrast, the wild-type α-Val^{70} MoFe is unable to effectively reduce propyne or to inhibit reduction of protons in its presence. It is unlikely that the ability of the altered α-Ala^{70} MoFe protein to reduce propyne (and propargyl alcohol) is the result of the fortuitous formation of a new alkyne-binding site within the MoFe protein since both are competitive inhibitors of acetylene reduction. Rather, these results indicate that a property of the α-Val^{70} residue in the wild-type MoFe protein is to place structural constraints at the substrate-binding site so that alkynes larger than acetylene are not effective substrates. More recently, preliminary results have shown that shortening the α-Val^{70} even more, using an α-Gly^{70} substitution, further relaxes these constraints suggesting that this interpretation is correct.

6. ACETYLENE INTERACTION WITH THE MoFe PROTEIN IN THE RESTING STATE

Examination of the MoFe protein resting state crystal structure indicates that, when van der Waals forces are taken into consideration, there does not appear to be enough space to accommodate acetylene-binding at the 4Fe-4S face of the FeMo-cofactor approached by residues α-Val^{70} and α-Arg^{96}. This is especially so if the expected side-on binding of acetylene to metal or metals occurs. However, as previously suggested, substrate-binding could require the dynamic redox-dependent movement of the polypeptide environment surrounding the 4Fe-4S face of the FeMo-cofactor including either or both the α-Val^{70} and α-Arg^{96} side-chains. Such movement might be necessary to relax constraints imposed by these residues that deny access of substrates to the FeMo-cofactor in the resting state of the enzyme. Indeed, as already discussed, the α-Val^{70} side-chain acts in this way to restrict access of alkynes larger than acetylene to the FeMo-cofactor under catalytic conditions. Based on these considerations, the ability of altered MoFe proteins having substitutions for either the α-Val^{70} or α-Arg^{96} position were evaluated for their ability to interact with acetylene in the resting state.

Replacement of the α-Val^{70} residue by amino acids having shorter side-chains (Ala or Gly) did not endow the altered MoFe proteins with the ability to interact with acetylene in the resting state. In contrast, substitution of the α-Arg^{96} residue by leucine, glutamine, alanine, or histidine all resulted in an alteration in the MoFe protein so that it is able to interact with acetylene in the resting state [62]. The acetylene interaction was evaluated by perturbations in the normal resting-state $S=3/2$ electron paramagnetic resonance (EPR) signal arising from the FeMo-cofactor. This result is consistent with the idea that the α-Arg^{96} side-chain becomes repositioned under turnover conditions to accommodate substrate binding. However, it is unlikely that such a movement is the only feature that controls substrate access to the FeMo-cofactor when other variables are considered.

First, the amount of acetylene required to cause a significant perturbation in the resting state $S=3/2$ EPR signal is concentration-dependent and is much greater than the amount required to detect acetylene reduction under turnover conditions. This feature could be an indication that the redox-dependent repositioning of other amino acid side-chains may also occur during turnover, in order to facilitate effective substrate interaction at the FeMo-cofactor. Likewise, structural changes that are remote from the substrate-binding site might be required in order to permit substrate diffusion from the polypeptide surface to the FeMo-cofactor. Second, the addition of dinitrogen to the altered α-Arg^{96} MoFe proteins did not cause any perturbation of the resting state S=3/2 EPR signal. It is unlikely that the substitutions at the α-Arg^{96} position would permit acetylene-binding, yet deny dinitrogen-binding. In this context, it seems more likely that both the reactivity and the accessibility of the FeMo-cofactor towards substrates change as the MoFe protein proceeds through more reduced states. Along these lines, it is emphasized that the premise for examining the altered α-Arg^{96} MoFe proteins was the expectation that this side-chain moves to accommodate substrate accessibility to the FeMo-cofactor. However, the results obtained do not rule out the possibility that a function of α-Arg^{96} movement is to control the chemical reactivity of the FeMo-cofactor through changes in hydrogen-bonding interactions between its guanidine group and the FeMo-cofactor during catalysis, rather than to provide space for substrate-binding.

7. SUMMARY AND OUTLOOK

The evidence summarized in this Chapter suggests that the metals and sulfurs involved in providing the acetylene-binding site are likely to be located within a particular 4Fe-4S face of the FeMo-cofactor. Nevertheless, the question of which individual atoms are specifically involved in binding as well as the geometric features of substrate-binding remains a challenging problem. We favor a model where acetylene- and dinitrogen-binding occur at essentially the same site.

However, because a more reduced state and more protons are required for dinitrogen-binding and reduction than for acetylene-binding and reduction, all of the atoms that participate in dinitrogen-binding might not also be required for acetylene-binding.

It should also be noted that, even though the work described here provides strong support for the idea that acetylene is able to interact with a specific 4Fe-4S face of the FeMo-cofactor, it does not eliminate the possibility that such interactions also occur at other sites. In our opinion this is unlikely, at least for the wild-type MoFe protein, but this issue remains open, and does not preclude the possibility that Mo might become directly involved in acetylene- or dinitrogen-binding at a later stage in the catalytic cycle. In spite of these reservations, we believe that approaches using various alkynes as mechanistic probes will continue to be profitable. In particular, the ability to use substituted short-chain alkynes as effective substrates raises the possibility of monitoring the electronic features of FeMo-cofactor during catalysis or, perhaps, even the development of mechanism-based suicide inhibitors. Finally, the availability of altered MoFe proteins having the capacity to interact with acetylene in the resting state presents the unprecedented opportunity to determine crystallographically the structure of a MoFe protein in at least some stage of substrate-binding.

8. ACKNOWLEDGEMENTS

We thank the National Institutes of Health Grant R01-GM59087 (D.R.D. and L.C.S) and the National Science Foundation Grants MCB-9630127 (D.R.D.) and MCB-0090187 (L.C.S.) for sponsoring our research.

9. REFERENCES

1. J. E. Carnahan, L. E. Mortenson, H. F. Mower and J. E. Castle, Biochem. Biophys. Acta. 127 (1960) 520.
2. R. H. Burris, J. Biol. Chem. 266 (1991) 9339.
3. M. J. Dilworth, Biochim. Biophys. Acta 127 (1966) 285.
4. R. V. Hageman and R. H. Burris, Proc. Natl. Acad. Sci. USA 75 (1978) 2699.
5. T. Ljones and R. H. Burris, Biochemistry 17 (1978) 1866.
6. V. K. Shah, L. C. Davis, J. K. Gordon, W. H. Orme-Johnson and W. J. Brill, Biochim. Biophys. Acta 292 (1973) 246.
7. R. V. Hageman and R. H. Burris, J. Biol. Chem. 254 (1979) 11189.
8. T. R. Hawkes, P. A. McLean and B. E. Smith, Biochem. J. 217 (1984) 317.
9. B. K. Burgess and D. J. Lowe, Chem. Rev. 96 (1996) 2983.
10. D. C. Rees and J. B. Howard, Curr. Opin. Chem. Biol. 4 (2000) 559.
11. J. Christiansen, D. R. Dean and L. C. Seefeldt, Annu. Rev. Plant Physiol. Plant Mol. Biol. 52 (2001) 269.
12. M. M. Georgiadis, H. Komiya, P. Chakrabarti, D. Woo, J. J. Kornuc and D. C. Rees, Science 257 (1992) 1653.
13. D. Nieva-Gomez, G. P. Roberts, S. Klevickis and W. J. Brill, Proc. Natl. Acad. Sci. USA 77 (1980) 2555.
14. V. K. Shah, G. Stacey and W. J. Brill, J. Biol. Chem. 258 (1983) 12064.
15. R. N. F. Thorneley and J. Deistung, Biochem. J. 253 (1988) 587.
16. W. A. Bulen, R. C. Burns and J. R. LeComte, Proc. Natl. Acad. Sci. USA 53 (1965) 532.
17. G. D. Watt and K. R. N. Reddy, J. Inorg. Biochem. 53 (1994) 281.
18. H. C. Angove, S. J. Yoo, B. K. Burgess and E. Münck, J. Am. Chem. Soc. 119 (1997) 8730.
19. H. C. Angove, S. J. Yoo, E. Münck and B. K. Burgess, J. Biol. Chem. 273 (1998) 26330.
20. K. B. Musgrave, H. C. Angove, B. K. Burgess, B. Hedman and K. O. Hodgson, J. Am. Chem. Soc. 120 (1998) 5325.
21. S. J. Yoo, H. C. Angove, B. K. Burgess, E. Münck and J. Peterson, J. Am. Chem. Soc. 120 (1998) 9704.
22. J. A. Erickson, A. C. Nyborg, J. L. Johnson, S. M. Truscott, A. Gunn, F. R. Nordmeyer and G. D. Watt, Biochemistry 38 (1999) 14279.

23. A. C. Nyborg, J. L. Johnson, A. Gunn and G. D. Watt, J. Biol. Chem. 275 (2000) 39307.
24. P. Strop, P. M. Takahara, H. Chiu, H. C. Angove, B. K. Burgess and D. C. Rees, Biochemistry 40 (2001) 651.
25. J. T. Bolin, N. Campobasso, S. W. Muchmore, W. Minor, T. V. Morgan and L. E. Mortenson, in Molybdenum Enzymes, Cofactors and Model Systems (E. I. Stiefel, D. Coucouvanis, and W. E.Newton eds.) American Chemical Society, Washington, D.C., 1993, p 186.
26. J. T. Bolin, A. E. Ronco, T. V. Morgan, L. E. Mortenson and N. H. Xuong, Proc. Natl. Acad. Sci. USA 90 (1993) 1078.
27. J. Kim and D. C. Rees, Nature 360 (1992) 553.
28. S. Maritano, S. A. Fairhurst and R. R. Eady, J. Biol. Inorg. Chem. 6 (2001) 590.
29. T. A. Clarke, S. Maritano and R. R. Eady, Biochemistry 39 (2000) 11434.
30. J. G. Grossmann, S. S. Hasnain, F. K. Yousafzai and R. R. Eady, J. Biol. Chem. 276 (2001) 6582.
31. D. J. Lowe, K. Fisher and R. N. F. Thorneley, Biochem. J. 292 (1993) 93.
32. J. W. Peters, K. Fisher, W. E. Newton and D. R. Dean, J. Biol. Chem. 270 (1995) 27007.
33. L. Ma, M. A. Brosius and B. K. Burgess, J. Biol. Chem. 271 (1996) 10528.
34. J. M. Chan, J. Christiansen, D. R. Dean and L. C. Seefeldt, Biochemistry 38 (1999) 5779.
35. R. N. F. Thorneley, R. R. Eady and D. J. Lowe, Nature 272 (1978) 557.
36. M. J. Dilworth and R. N. F Thorneley, Biochem. J. 193 (1981) 971.
37. R. N. F. Thorneley and D. J. Lowe, in Molybdenum Enzymes (T. G. Spiro ed.), Wiley, New York, 1985 p 221.
38. D. J. Lowe and R. N. F. Thorneley, Biochem. J. 224(1984) 877.
39. K. L. Hadfield and W. A. Bulen, Biochemistry 8 (1969) 5103.
40. J. M. Rivera-Ortiz and R. H. Burris, J. Bact. 123 (1975) 537.
41. F. B. Simpson and R. H. Burris, Science 224 (1984) 1095.
42. M. J. Dilworth, M. E. Eldridge and R. R. Eady, Analyt. Biochem. 207 (1992) 6.
43. B. K. Burgess, in Molybdenum Enzymes (T. J. Spiro ed.), Wiley-Interscience Publication, New York, 1985. p 161.
44. L. C. Davis, V. K. Shah and W. J. Brill, Biochim. Biophys. Acta 403 (1975) 67.
45. S. Wherland, B. K. Burgess, E. I. Stiefel and W. E. Newton, Biochemistry 20 (1981) 5132.
46. A. P. Sadkov and G. I. Likhtenshtein, Biochemistry-Russia 55 (1990) 446.
47. D. J. Lowe, K. Fisher and R. N. F. Thorneley, Biochem. J. 272 (1990) 621.
48. J. Kim and D. C. Rees, Science 257 (1992) 1677.
49. C. H. Kim, W. E. Newton and D. R. Dean, Biochemistry 34 (1995) 2798.

50. M. J. Dilworth, K. Fisher, C. H. Kim and W. E. Newton, Biochemistry 37 (1998) 17495.
51. K. Fisher, M. J. Dilworth and W. E. Newton, Biochemistry 39 (2000) 15570.
52. K. Fisher, M. J. Dilworth, C. H. Kim and W. E. Newton, Biochemistry 39 (2000) 10855.
53. K. Fisher, M. J. Dilworth, C. H. Kim and W. E. Newton, Biochemistry 39 (2000) 2970.
54. M. Sorlie, J. Christiansen, B. J. Lemon, J. W. Peters, D. R. Dean and B. J. Hales, Biochemistry 40 (2001) 1540.
55. J. Shen, D. R. Dean and W. E. Newton, Biochemistry 36 (1997) 4884.
56. J. Christiansen, V. L. Cash, L. C. Seefeldt and D. R. Dean, J. Biol. Chem. 275 (2000) 11459.
57. D. Wolle, C. H. Kim, D. R. Dean and J. B. Howard, J. Biol. Chem. 267 (1992) 3667.
58. S. M. Mayer, W. G. Niehaus and D. R. Dean, J. Chem. Soc., Dalton Trans. 5 (2002) 802.
59. J. Christiansen, L. C. Seefeldt and D. R. Dean, J. Biol. Chem. 275 (2000) 36104.
60. J. W. Peters, M. H. B. Stowell, S. M. Soltis, M. G. Finnegan, M. K. Johnson and D. C. Rees, Biochemistry 36 (1997) 1181.
61. S. M. Mayer, D. M. Lawson, C. A. Gormal, S. M. Roe and B. E. Smith, J. Mol. Biol. 292 (1999) 871.
62. P. M. C. Benton, S. M. Mayer, J. Shao, B. M. Hoffman, D. R. Dean and L. C. Seefeldt, Biochemistry 40 (2001) 13816.

Nitrogen Fixation at the Millennium.
G. Jeffery Leigh, Editor.

CHAPTER 6

Regulation of Mo Nitrogenases

P. Rudnick and C. Kennedy
Department of Plant Pathology, The University of Arizona, Tucson, AZ, 85721, USA

1. INTRODUCTION

Diazotrophic bacteria have access to an essentially unlimited supply of nitrogen in the form of atmospheric dinitrogen gas (N_2). The ability to fix N_2 to ammonium (NH_4^+), which is readily assimilated, is dependent on these organisms' ability to express effectively, protect, and utilize the enzyme nitrogenase. Diazotrophs have evolved this ability in a variety of ecologically diverse zones, including the oceans, fresh water, soils, the root systems of leguminous plants, and even the guts of termites. Nitrogen fixation supplements the requirements for combined nitrogen of these bacteria, and in some cases, provides fixed nitrogen into the metabolic pool of a higher organism through a symbiotic or associative relationship. The ability to use N_2 efficiently as a nitrogen source is often, however, limited by the heavy energy expenditure required for the reduction of the strong triple bond between the N atoms. In the case of the classic Mo-dependent nitrogenase (Chapter 2), complete reduction of a single mole of dinitrogen to form 2 moles of NH_4^+ requires a minimum input of 16 ATP (Fig. 1). Thus, it is hardly surprising that most aspects of nitrogen fixation are subject to one or more levels of metabolic and environmental regulation.

$$N_2 + 8e^- + 16\,H^+ + 16\,MgATP^{2-} + 18\,H_2O \xrightarrow{\text{nitrogenase}} 2NH_4^+ + 2OH^- + 16\,MgADP^- + 16\,H_2PO_4^- + H_2$$

Figure 1. The stoichiometry of dinitrogen reduction by molybdenum nitrogenases.

The mechanism and structure of nitrogenase have been discussed in previous chapters (Chapters 2 and 5) as have the genes and gene products of the *nif* regulon (Chapter 4.); when discussing regulation, one should be familiar with these. Of special importance with respect to regulation is thc *nifA*, gene found in all diazotrophic species of the Proteobacteria phylum of Eubacteria, many of which are discussed below. The *nifA* gene encodes the NifA protein, a *nif*-specific transcriptional activator, similar to the response regulator of two-component regulatory systems. Gene clusters for alternative nitrogenase systems often contain genes encoding NifA homologs. These systems, as well as a description of the regulators, can be found in Chapter 8. NifA is central to a discussion of regulation of nitrogen fixation and will serve as a starting point. Very often, control of NifA expression or activity provides good control of nitrogen fixation, which can be sensitive to the two most common environmental stimuli limiting the efficiency of nitrogen fixation: **oxygen** and **fixed nitrogen (usually as NH_4^+)**. Oxygen inhibits nitrogen fixation because it inactivates nitrogenase; hence regulatory mechanisms that protect nitrogenase from oxygen have evolved. Sources of fixed nitrogen also have a negative effect on nitrogen fixation because these molecules often require less energy input to transport and metabolize them than does the reduction of N_2, and therefore the presence of these molecules often precludes a necessity to fix atmospheric nitrogen.

Regulation of NifA affects the level of transcription from other *nif* gene promoters; there are also post-translational mechanisms at work which affect nitrogenase activity in some organisms, involving ADP-ribosylation of dinitrogenase reductase, or rapid switch-off of nitrogenase activity, the enzymes and mechanisms for which are discussed in Chapter 4. Together with transcriptional regulation, these two mechanisms provide the necessary tight regulatory control required to maintain the efficiency of nitrogen fixation. The detailed mechanisms by which these occur differ among diazotrophs and there are often subtleties and even novel components in each organism. Regulation of nitrogen fixation should therefore be examined on a per organism basis, although common themes exist and are discussed next.

2. COMPONENTS OF THE *nif*-PROMOTER REGION

In diazotrophs belonging to the eubacterial phylum Proteobacteria, the components of the promoters just upstream from *nif* genes or *nif* operons are similar. The promoter regions are typified by the occurrence of an NifA-binding site, located between 30-225 base pairs upstream from the promoter and its consensus sequence is TGT-N_{10}-ACA [1]. The second component is the promoter itself, which is recognized by the alternate sigma factor, (RpoN) σ^{54}- (or σ^N)-containing form of RNA polymerase. The consensus binding-site for this form of the polymerase is 5'-T<u>GG</u>CCACN$_4$TTT<u>GC</u>(A/T)-3'; the –24 and –12 (underlined) are highly conserved and are required for promoter recognition by σ^{54} [2].

The exceptional genes with respect to regulation by NifA appear to be the *nifA* genes themselves, or operons containing *nifA*. In some cases, activation of *nifA* transcription occurs only in the absence of fixed nitrogen, by the phosphorylated form of NtrC, a NifA homolog bearing a different consensus binding-site [3,4]. NtrC belongs to the general nitrogen-sensing cascade, which is well studied for the enteric organisms [5].

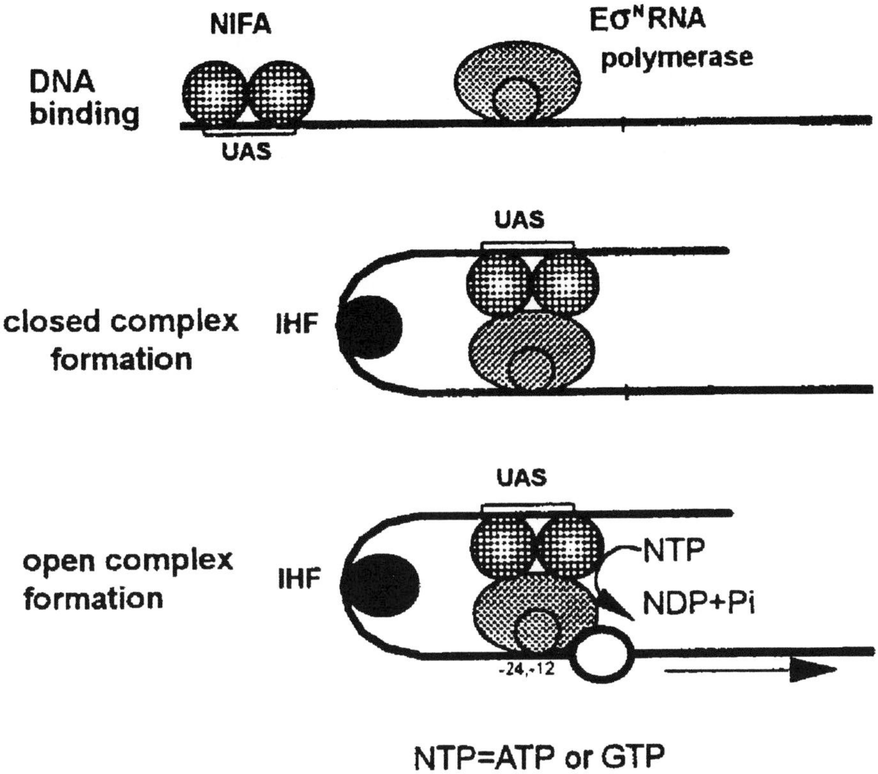

Figure 2. Model for the activation of Nif A (from reference 10).

In other cases, such as in *Azotobacter vinelandii*, the expression of *nifA* appears to be unregulated and its promoter is probably recognized by the housekeeping form of the RNAP holoenzyme containing σ^{70} [6].

3. MECHANISMS OF TRANSCRIPTIONAL ACTIVATION

As mentioned above, the first step in transcriptional activation of a *nif* promoter is recognition, achieved by the σ^{54}-bound form of the polymerase. The second step, as with other promoters, is isomerization or melting, to form an open complex. However, unlike other promoters recognized by RNAP-σ^{70} holoenzyme, forms of RNAP, melting at σ^{54} -dependent promoters does not occur in the absence of an activator protein, such as NifA or NtrC. There is really only a single model of how this occurs: the activator protein contacts the polymerase and catalyzes isomerization [7] by a mechanism involving DNA-looping, as occurs with other similar activators [1]. The latter is supported by more experimental data, such as activation of the p*glnA* promoter *in trans* [5], and, convincingly, structures of DNA looping have been visualized [8,9]. Supporting this model is the specific role of IHF, a DNA-bending protein, which enhances activation *in vitro* [7]. A current model for activation by NifA is depicted in Figure 2.

Transition to the open complex requires not only contact with an activator such as NifA but also hydrolysis of ATP or other nucleoside triphosphate [10]; NifA has a preference for ATP or GTP and does not hydrolyze CTP [11]. Nucleotide hydrolysis is apparently a key step in transcriptional activation, as mutant forms of the activator, deficient in ATP hydrolysis, also fail to activate transcription, and non-hydrolyzable analogs of ATP cannot substitute for ATP *in vitro*. Transcriptional initiation follows promoter melting, which also requires ATP hydrolysis [12,13].

For the enteric diazotroph *Klebsiella pneumoniae*, control of mRNA stability might offer another level of regulation. In this organism, the half-life of *nif* transcripts is reduced upon exposure to fixed nitrogen or molecular oxygen, a function that is dependent on the presence of the *nifL* gene [14]. The *nifL* gene, common to members of the γ-subgroup of the Proteobacteria but seldom outside this group, encodes a negative regulator of NifA activity. A role for NifL in destabilizing transcripts might also suggest a possible role for NifA activity in stabilizing *nif* mRNAs [14]. This model was recently challenged; Simon *et al.* found that NifA in trans is not sufficient to stabilize *nif* transcripts, and instead, they report that active nitrogenase and the *nifY* gene may be responsible for sensing those conditions conducive to nitrogen fixation and *nif* mRNA stability [15]. These experiments provide insight into the earliest steps in regulation. Mechanisms controlling transcriptional activation by the regulatory proteins and the conditions affecting their transcription and activity will be discussed next.

4. *Trans*-ACTING FACTORS INFLUENCING NITROGENASE EXPRESSION

4.1 NifA

Before discussion of the regulatory features of several of the relevant organisms, it is essential to cover the components governing each regulatory step. In general terms, there are two different types of regulatory mechanisms affecting transcriptional regulation: (i) *nif* specific, and (ii) global. The *nif*-specific factors under consideration here are called, NifA and NifL. As mentioned above, NifA is ubiquitous among Proteobacterial diazotrophs; it is composed of three domains: (i) an N-terminal domain that is essentially of unknown function, although it has been proposed to be a regulatory domain, (ii) a central catalytic domain, which hydrolyzes ATP and potentially interacts with σ^{54}-bound RNAP, and (iii) a C-terminal domain that shares homology to other two-component response regulator proteins and contains a conserved HTH motif for binding DNA at the conserved UAS (upstream activating site). In some organisms, NifA contains an additional motif located between the central domain and the C-terminal domain. This domain contains the sequence motif CXXXXC, which, together with other C residues in this region of the protein, has been proposed to form a metal-binding site [16]. Consistent with this hypothesis is the observation that removal of trace metals from the culture medium eliminates NifA activity. However the metal content and the mechanism by which this may occur remains elusive since these proteins are difficult to manipulate *in vitro*. Thus far, this feature of NifA is found only within the α- and β-subgroups of the Proteobacteria, hence more is known for the NifA proteins of the γ-subgroup, although sensing oxygen in these systems requires the presence of a second protein, NifL, discussed below. Diazotrophs also respond to the level of fixed nitrogen within the cell by modulating the expression and activity of NifA.

4.2 NifL

In the absence of an oxygen-sensing mechanism intrinsic to NifA, γ-subgroup members, species of *Klebsiella* (or *Enterobacter*) and *Azotobacter*, encode a second protein, NifL. *nifL* was originally identified on the basis of a mutation located very near *nifA* in *K. pneumoniae* that had a slightly leaky Nif $^-$ phenotype and gave rise to 'pseudorevertants' in which oxygen and ammonium no longer prevented *nifHDK* gene expression [17-19]. The *nifL* gene of *A. vinelandii* was identified by sequencing the region just downstream of *nifA* [6]. In both organisms, *nifL* is located downstream of, and cotranscribed with, the *nifA* gene. A *nifL* was also recently identified in *Azoarcus* sp BH72, a member of the β subgroup of Proteobacteria [20]. The NifL protein acts as a specific inhibitor of NifA activity. The NifL and NifA proteins interact stoichiometrically to function as an activator/anti-activator system. The *nifL* and *nifA* genes are translationally coupled in at least one organism, *K. pneumoniae* [21]. NifL also has a modular structure, with some features similar to the sensor-kinase component of two-component regulatory systems. NifL contains two domains, an N-terminal sensor domain and a C-terminal kinase-like domain, separated by a glutamine-rich linker (Figure 3) [22]. The N-terminal domain contains the oxygen-sensing region of the protein. Oxygen-sensing by NifL proteins is dependent on the presence of a covalently attached FAD moiety, bound at the N-terminal PAS domain, found in several other oxygen-sensitive proteins from all three domains of life [23]. This moiety can be oxidized or reduced; the reduced form favors the non-inhibitory conformation of NifL, or one which does not antagonize NifA activity. In contrast, when the FAD becomes oxidized, NifL specifically and rapidly inhibits NifA activity [24].

The NifL protein of *A. vinelandii* has been well characterized, that of *K. pneumoniae* to a lesser extent [25,26]. For *A. vinelandii*, evidence indicates that NifL is also sensitive to ADP *in vitro* [24] and to fixed nitrogen *in vivo* [27]. These features make NifL a sensor of potentially three separate signals. The ADP response is strong *in*

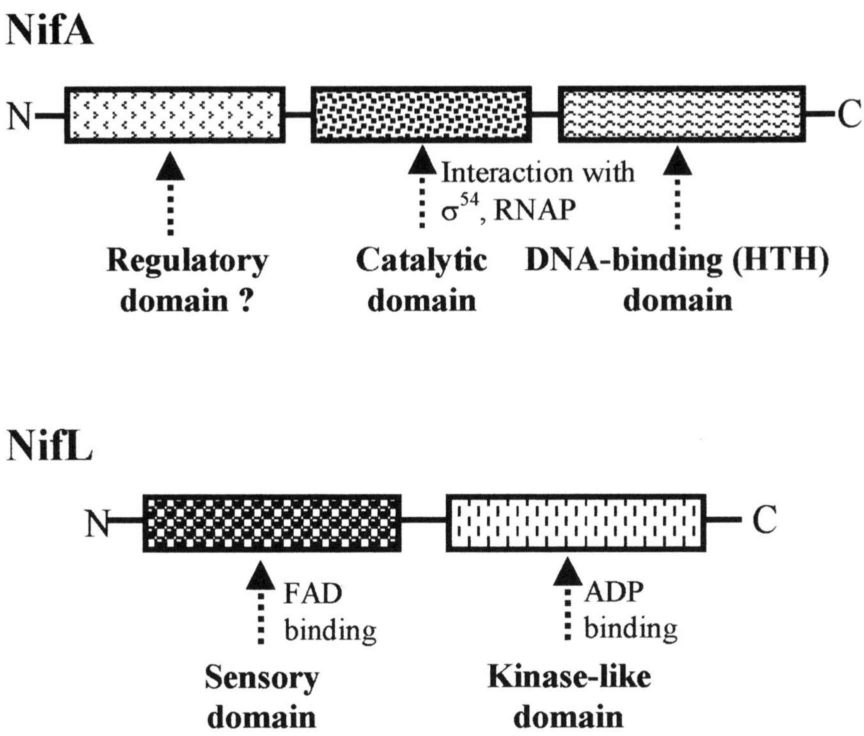

Figure 3. Modular structure of NifA and NifL, adapted from reference 10.

vitro and overrides the oxygen response [24], although the relevance of this response has not been validated *in vivo*. ADP was found to protect the C-terminal domain from proteolytic cleavage, and an N-terminal deletion of 146 amino acids of NifL does not disrupt either the ADP response or the fixed N response, but entirely eliminates the oxygen response, indicating that ADP-binding and the oxygen or fixed N response are separate and discrete functions [25]. The mechanism by which fixed nitrogen is sensed by NifL was found to be indirect, and has only recently been elucidated. This mechanism involves interplay of the signal-transmitter protein GlnK, a member of the class of PII-like signal-transduction proteins and is the general nitrogen-sensing pathway common to most bacteria, Archaea and some higher organisms. Since many of the mechanistic details by which diazotrophs utilize these proteins and sense and respond to the level of fixed nitrogen does not appear to be *nif* specific, this response is discussed in the next section.

5. NITROGEN-SENSING

As described above, oxygen and/or energy (ADP)-sensing is thought to occur as a function of either NifA or NifL. On the other hand, sensing of fixed nitrogen involves several other components, most of which are highly conserved with homologous components in non-diazotrophic species. At the center of the general nitrogen-sensing cascade, is the GlnD protein, a uridylyl-transferase/uridylyl-removing enzyme, specific for the attachment of a UMP moiety to PII-like proteins. PII proteins are highly conserved across genera and species and have been found in some higher organisms. These proteins are small ~12 kDa proteins that form trimers, and have a conserved T-loop region which carries the conserved site of uridylylation, Tyr51. The T-loop region of the protein is the site of interaction with receptor enzymes and proteins. Several recent and very comprehensive reviews are available on the subject of PII signal transduction proteins [28,29].

In response to nitrogen limitation, the uridylyltransferase function of GlnD predominates and PII-UMP is formed (Figure 4). In the model organism, *Escherichia coli*, this form of the protein has at least two functions. The first is to signal indirectly the auto-phosphorylation of the nitrogen-responsive kinase, NtrB. Apparently PII-UMP does not interact with NtrB, as does the unmodified form, and it does not influence the NtrC-phosphatase activity of NtrB. The phosphate group of NtrB is transferred specifically to NtrC. In *E. coli*, NtrC-P is responsible for the activation of several genes, such as *glnA*, *ntrBC*, *put* (proline utilization) *etc*., important for growth under nitrogen-limiting conditions or in the presence of less preferred sources of fixed nitrogen. Another operon activated by NtrC-P is *glnK-amtB* [30]. The *glnK* gene in this operon encodes a second PII-like protein, which can also be uridylylated by GlnD. This protein differs at only two residues within the T-loop and is 68% identical at the amino acid level, but differs in some of its functions. For example, the rate of deuridylylation by GlnD is much slower for GlnK than for PII, encoded by the *glnB* gene. It has also been proposed that these two proteins form heterotrimers with functions that differ from either of the two homotrimers [31,32]. This mechanism is thought to allow fine-tuning with respect to receptor interactions.

The second function of PII, and under N-deficient conditions GlnK, is to signal the adenylylation state of glutamine synthetase (GS). This enzyme, encoded by *glnA*, is a key for the assimilation of NH_4^+ into glutamine, a process which consumes ATP and is therefore post-translationally as well as transcriptionally regulated. GS is a dodecameric enzyme and each of its subunits is susceptible to adenylylation by GlnE (ATase/AR) in response to high nitrogen conditions, signaled by the unmodified form of PII. The process of adenylylation is reversible; PII-UMP or GlnK-UMP promote AMP removal from GS by GlnE and the deadenylylated enzyme is active [29].

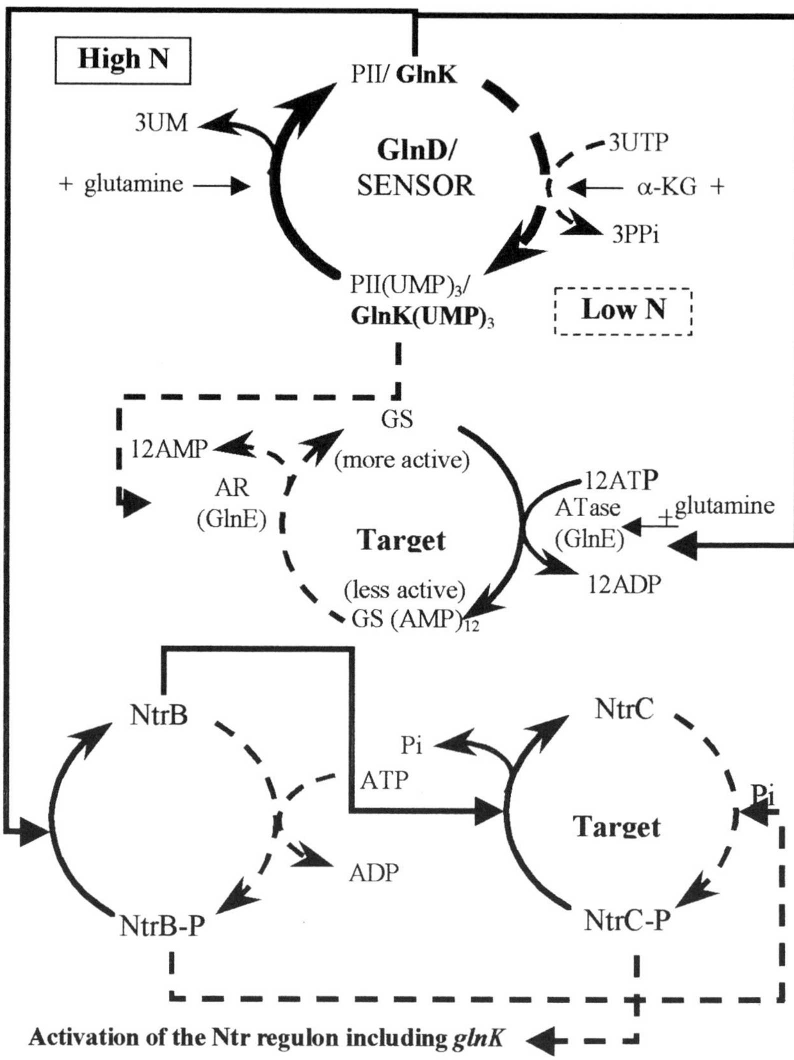

Figure 4. Nitrogen sensing and signal transduction in *E. coli*.
Dashed arrows indicate favoured reactions during N-sufficiency (high glutamine), solid arrows N-limitation.

E. coli has also been used as a heterologous system for the study of nitrogenase expression [25,27,33,34]. These studies have been fruitful and have uncovered a role for GlnK in the regulation of NifA activity in *K. pneumoniae* [34]. Using a p*nifH-lacZ* reporter on the chromosome and a plasmid which expresses *nifLA* from p*tac*, it was found. that *glnK* expression is required to relieve NifL inhibition of NifA activity [34]. This function is apparently specific for GlnK, as GlnB (PII) cannot substitute for it. An interesting aside from these experiments is that the role of GlnK in the relief of NifL inhibition occurs irrespective of the uridylylation state of GlnK, as relief of NifL inhibition occurred under N-limiting conditions in a *glnD* background and also in a strain carrying a *glnK* Y50N allele, encoding an unuridylylatable form of the protein [34]. These studies have been validated in the host organism, *K. pneumoniae*; it was also found that increasing the concentration of GlnB within the cell can partially relieve NifL inhibition of NifA activity [35]. Also, for *K. pneumoniae*, a combination of two point mutations in the T-loop region of PII (the *glnB* gene product), to make it resemble the T-loop of GlnK, conferred the ability to relieve NifL inhibition in response to N-limitation [36].

In contrast to what occurs in *K. pneumoniae*, *A. vinelandii* encodes only a single PII-like protein, GlnK. The gene encoding this protein is linked to *amtB* like other *glnK* genes but more closely resembles *E. coli* GlnB at key residues within the T-loop of the protein [36,37]. Consistent with this observation, in a heterolgous *E. coli* system carrying *A. vinelandii* NifL and NifA, *E. coli* GlnB and *A. vinelandii* GlnK influence NifL activity [25]. However, in contrast to what occurs in *K. pneumoniae*, the presence of these proteins negatively affects the activity of NifA by increasing the affinity of NifL for NifA [38]. Also in contrast to the *K. pneumoniae* system, the uridylylation state of GlnK is key in determining the influence of GlnK on the NifL-NifA system, as a mutation in *glnK* rendering the protein unuridylylatable prevents maximal nitrogenase expression in response to nitrogen limitation [39]. It has also been determined that GlnK exerts its effects directly on NifL, shown by two-hybrid interaction with these proteins and also in vitro [39,40]. Why these systems have developed two separate pathways to achieve the same regulatory response is unknown but is evidenced by their evolutionary divergence.

6. INVOLVEMENT OF PII IN CONTROL OF NifA ACTIVITY IN OTHER PROTEOBACTERIA

PII proteins have also been shown to influence the activity of NifA in *Azospirillum brasilense*, *Herbaspirillum seropedicae*, and *Rhodopseudomonas capsulatus*, and also *Rhodospirillum rubrum*, four α group members of the phylum Proteobacteria, none of which have a NifL homolog. This is supported largely by genetic evidence. In *A. brasilense*, *glnB* mutants, NifA is synthesized but is inactive [31]. Deletion of the N-terminal domain resulted in restoration of NifA activity in the absence of PII (GlnB) protein and in the presence or absence of ammonium [41]. Substitution of Phe for Tyr at residue 18 in NifA also resulted in restoration of NifA activity in cultures grown with or without ammonium [42], suggesting that NifA itself might be covalently modified by PII or by another unidentified protein which is influenced by PII. Similarly, truncation of the N-terminal domain of *H. seropedicae* NifA leads to its being constitutively active, supporting a model for the N-terminal domain being responsible for regulation of NifA activity in response to levels of fixed N [43]. *R. capsulatus* has two copies of genes encoding NifA proteins which differ from each other in the 19 or 22 N-terminal amino acids. Each can substitute for the other as demonstrated by analysis of single and double *nifA1* and *nifA2* mutants [44]. As with the other NifAs, ammonium inhibits activity of both NifA proteins of *R. capsulatus* by a mechanism dependent on PII proteins [45,46]. Mutations in NifA leading to loss of inhibition by ammonium were found to map both within the N-terminal domain and also in the interdomain linker found between the central catalytic domain and the C-terminal DNA binding domain [47]. In contrast to the NifAs in *H. seropedicae* and *A. brasilense*, truncation of the entire N-terminus in *R. capsulatus* NifA led to it being inactive. Finally, PII appears to be required for NifA activity in the related *R. rubrum* [48].

7. CONCLUDING REMARKS

There are many other separate mechanisms controlling transcription and translation of nitrogenase in diazotrophic bacteria, the scope of which is beyond this review. However, the cases presented here give us a glimpse at the level of mechanistic details at work to effectively express nitrogenase while conserving valuable metabolites during times of nitrogen availability. In diazotrophic organisms in the kingdom Archeae, there are many similarities with prokaryotic relatives. It will be interesting to learn more about these organisms, as well as the role of nitrogen-sensing proteins in higher organisms. Even in the most advanced systems studied thus far there are new levels of detail being illuminated, and it will be interesting to learn more about how these organisms adjust to a changing environment with respect to nitrogen, oxygen and other stimuli.

8. ACKNOWLEDGEMENTS

Work in the Department of Plant Pathology, The University of Arizona, related to this review was supported by USDA NRICG grant 95-37305-2067.

9. REFERENCES

1. M. Buck, W. Cannon and J. Woodcock, Nucl. Acids Res. 15 (1987) 9945.
2. H. Barrios, B. Valderrama and E. Morett, Nucl. Acids Res. 27 (1999) 4305.
3. E. Santero, T. Hoover, J. Keener and S. Kustu, Proc. Natl. Acad. Sci. USA 86 (1989) 7346.
4. S. Austin, N. Henderson and R. Dixon, Mol. Microbiol. 1 (1987) 92.

5. I. Rombel, A. North, I. Hwang, C. Wyman and S. Kustu, Cold Spring Harb. Symp. Quant. Biol. 63 (1998) 157.
6. G. Blanco, M. Drummond, P. Woodley and C. Kennedy, Mol. Microbiol. 9 (1993) 869.
7. T. R. Hoover, E. Santero, S. Porter and S. Kustu, Cell 63 (1990) 11.
8. C. Wyman, I. Rombel, A. K. North, C. Bustamante and S. Kustu, Science 275 (1997) 1658.
9. W. Su, S. Porter, S. Kustu and H. Echols, Proc. Natl. Acad. Sci. USA 87 (1990) 5504.
10. R. Dixon, Arch. Microbiol. 169 (1998) 371.
11. H. S. Lee, D. K. Berger and S. Kustu, Proc. Natl. Acad. Sci. USA 90 (1993) 2266.
12. A. B. Wedel and S. Kustu. Genes Dev. 9 (1995) 2042.
13. W. Cannon, M. T. Gallegos, P. Casaz and M. Buck, Genes Dev. 13 (1999) 357.
14. J. Collins, G. P. Roberts and W. J. Brill, J Bacteriol. 168 (1986) 173.
15. H. M. Simon, M. M. Gosink and G. P. Roberts, J Bacteriol. 181 (1999) 3751.
16. H.-M. Fischer, T. Bruderer and H. Hennecke, Nucl. Acids Res. 16 (1988) 2207.
17. C. Kennedy, Mol. Gen. Genet. 157 (1977) 199.
18. S. Hill, C. Kennedy, E. Kavanagh, R. B. Goldberg and R. Hanau, Nature 290 (1981) 424.
19. M. Merrick, S. Hill, H. Hennecke, R. Hahn, R. Dixon and C. Kennedy, Mol. Gen. Genet. 185 (1982) 75.
20. B. Rheinhold-Hurek, personal communication.
21. F. Govantes, E. Andujar and E. Santero, EMBO J. 17 (1998) 2368.
22. J. C. Wootton and M. H. Drummond, Protein Eng. 22 (1989) 535.
23. I. Zhulin and R. Dixon, Trends in Biological Sciences (1997) 331.
24. S. Hill, S. Austin, T. Eydmann, T. Jones and R. Dixon, Proc. Natl. Acad. Sci. USA 93 (1996) 2143.
25. E. Soderback, F. Reyes-Ramirez, T. Eydmann, S. Austin, S. Hill and R. Dixon, Mol. Microbiol. 28 (1998) 179.
26. R. A. Schmitz, K. Klopprogge and R. Grabbe, J. Mol. Microbiol. Biotechnol. 4 (2002) 235.
27. F. Reyes-Ramirez, R. Little and R. Dixon, J. Bacteriol. 183 (2001) 3076.
28. A. J. Ninfa and M. R. Atkinson, Trends Microbiol. 8 (2000) 172.
29. T. Arcondeguy, R. Jack and M. Merrick, Microbiol. Mol. Biol. Rev. 65 (2001) 80.
30. W. C. van Heeswijk, S. Hoving, D. Molenaar, B. Stegeman, D. Kahn and H. V. Westerhoff, Mol. Microbiol. 21 (1996) 133.
31. M. de Zamaroczy, Mol. Microbiol. 29 (1998) 449.
32. W. C. van Heeswijk, D. Wen, P. Clancy, R. Jaggi, D. L. Ollis, H. V. Westerhoff and S. G. Vasudevan, Proc. Natl. Acad. Sci. USA 97 (2000) 3942.
33. L. He, E. Soupene and S. Kustu, J. Bacteriol. 179 (1997) 7446.
34. L. He, E. Soupene, A. Ninfa and S. Kustu, J. Bacteriol. 180 (1998) 6661.

35. T. Arcondeguy, W. C. van Heeswijk and M. Merrick, FEMS Microbiol. Lett. 180 (1999) 263.
36. T. Arcondeguy, D. Lawson and M. Merrick, J. Biol. Chem. 275 (2000) 38452.
37. D. Meletzus, P. Rudnick, N. Doetsch, A. Green and C. Kennedy, J. Bacteriol. 180 (1998) 3260.
38. R. Little, F. Reyes-Ramirez, Y. Zhang, W. C. van Heeswijk and R. Dixon, EMBO J. 19 (2000) 6041.
39. P. Rudnick, C. Kunz, M. K. Gunatilaka, E. R. Hines and C. Kennedy, J. Bacteriol. 184 (2002) 812.
40. R. Little, V. Colombo, A. Leech and R. Dixon, J. Biol. Chem. 277 (2002) 15472.
41. F. Arsene, P. A. Kaminski and C. Elmerich, J. Bacteriol. 178 (1996) 4830.
42. F. Arsene, P. A. Kaminski and C. Elmerich, FEMS Microbiol. Lett. 179 (1999) 339.
43. F. O. Pedrosa, E. M. Benelli, M. G. Yates, R. Wassem, R. A. Monteiro, G. Klassen, M. B. Steffens, E. M. Souza, L. S. Chubatsu and L. U. Rigo, J. Biotechnol. 91 (2001) 189.
44. W. Klipp, B. Masepohl and A. Puhler, J. Bacteriol. 170 (1988) 693.
45. P. Hubner, B. Masepohl, W. Klipp and T. A. Bickle, Mol. Microbiol. 10 (1993) 123.
46. B. Masepohl, T. Drepper, A. Paschen, S. Gross, A. Pawlowski, K. Raabe, K. U. Riedel and W. Klipp, J. Mol. Microbiol. Biotechnol. 4 (2002) 243.
47. A. Paschen, T. Drepper, B. Masepohl and W. Klipp, FEMS Microbiol. Lett. 200 (2001) 207.
48. Y. Zhang, E. L. Pohlmann, P. W. Ludden and G. P. Roberts, J. Bacteriol. 182 (2000) 983.

Nitrogen Fixation at the Millennium.
G. Jeffery Leigh, Editor.

CHAPTER 7

Actinorhizal symbioses

K. Pawlowski
Albrecht von Haller Institute, Department of Plant Biochemistry, Göttingen University, 37073 Göttingen, Germany

1. PREFACE

Actinorhizal root nodules are nitrogen-fixing symbioses between the actinomycete *Frankia* and roots of dicotyledonous plants belonging to 25 genera from eight different plant families (Table 1). The microsymbionts fix nitrogen while being hosted within nodule cells and export the products of nitrogen fixation to the host plant, thereby rendering it independent of soil nitrogen sources. With one exception, *Datisca glomerata,* actinorhizal plants are trees or woody shrubs. Due to their symbiosis, actinorhizal plants mostly represent pioneer plants and are often used in reforestation or soil reclamation. The use of actinorhizal plants, particularly alders, in forestry in Europe can be traced back to the Roman empire [1]. Mature actinorhizal nodules are perennial organs consisting of multiple lobes, each of which represents a modified lateral root without root cap and with infected cells in the expanded cortex. Although actinorhizal nodules were first described in 1829 [2], only in 1895 was it shown that they contributed to the nitrogen nutrition of the plant [3], and the identification of the microsymbiont as an actinomycete did not take place before the 1930s [4,5]. The first isolation of an actinorhizal microsymbiont took place in 1959 [6], but the cultures were subsequently lost. Hence, the precise systematic position of the actinorhizal microsymbionts remained controversial until the microsymbiont of *Comptonia peregrina* was isolated and used successfully for reinfection [7].

2. MICROSYMBIONTS

The genus *Frankia* comprises gram-positive and gram-variable actinomycetes that grow in hyphal form. Under nitrogen limitation and aerobic conditions they form special organs for nitrogen fixation, spherical vesicles, at the ends of hyphae or short side hyphae (Figure 1). Vesicles are surrounded by envelopes of multiple layers containing hopanoids, bacterial steroid lipids [8]. The number of layers is correlated with the oxygen tension, indicating that they act as a gas-diffusion barrier [9]. The hopanoid composition also seems to play a role in modulating oxygen permeability, since it is dependent on the depth of soil where *Frankia* is growing, and *in planta* on the external oxygen concentration [10,11]. Under microaerobic conditions and nitrogen starvation, *Frankia* expresses nitrogenase in hyphae [12]. *In planta,* shape, septation and

plant family	actinorhizal genera	infection mechanism	*Frankia* isolate	Nodule aeration	shape of *Frankia* vesicles
Betulaceae	*Alnus*	intracellular	+	lenticels	spherical
Casuarinaceae	*Casuarina*	intracellular	+	nodule roots	no vesicles
	Allocasuarina	n.d.		nodule roots	no vesicles
	Gymnostoma	intracellular		nodule roots	spherical
Myricaceae	*Comptonia*	intracellular	+	nodule roots	club-shaped
	Myrica	intracellular		nodule roots	small spherical
Elaeagnaceae	*Elaeagnus*	intercellular	+	no nodule	spherical
	Hippophae	intercellular		roots	spherical
	Shepherdia	intercellular			spherical
Rhamnaceae	*Ceanothus*	intercellular	+	no nodule	spherical
	Colletia	n.d.	(exception:	roots	spherical
	Discaria	intercellular	*Ceanothus*)		spherical
	Kentrothamnus	n.d.			
	Retanilla	n.d.			spherical
	Talguenea	n.d.			spherical
	Trevoa	n.d.			
Rosaceae	*Cercocarpus*	intercellular	-	no nodule	pear-shaped
	Chamaebatia	n.d.		roots	elliptical
	Cowania	n.d.			elliptical
	Dryas	n.d.			elliptical
	Purshia	intercellular?			elliptical
Datiscaceae	*Datisca*	intercellular?	-	lenticels and nodule roots	lanceolate
Coriariaceae	*Coriaria*	intercellular?	-	one large lenticel	lanceolate

Infection mechanisms are given as far as examined (for references see text). "Intercellular?" infection is given when only indirekt evidence is available regarding the infection pathway. "*Frankia* isolate" is positive when members of an actinorhizal plant family can be infected by cultured *Frankia* strains. Data on lenticels and nodule roots are taken from [112] and [186]. Data on shapes of vesicles are taken from [13] and [14]. n.d., no data available.

Table 1. List of actinorhizal plant families/genera.

subcellular localization of *Frankia* vesicles depends on the host plant (Table 1; reviewed in [13,14]). *Frankia* can also form multilocular sporangia [15,16; for review see 17]. On the basis of the presence or absence of sporangia within root nodules, *Frankia* strains can be classified as spore$^+$ or spore$^-$, respectively [17]. Spore$^+$ strains seem to be more infective than spore$^-$ strains [18]. Because, like cyanobacteria and in contrast to rhizobia, *Frankia* strains can also fix nitrogen in the free-living state, it is not surprising that it is found in soil in the absence of its host plants. Using trap plants, symbiotic *Frankia* bacteria have been detected in soils that had been devoid of actinorhizal plants for decades [19,20], but were rare in nitrogen-fertilized plots [21]. However, it should be considered that these results might be due to the persistance of *Frankia* spores in the soil. In circumpolar soils outside the normal area of distribution of *Alnus, Frankia* was detected only in very few locations [22], implying that it has not been spread to these places. Thus, the presence of the host plant may still play a role in the distribution of *Frankia*..

Up to now, it has not been possible to isolate *Frankia* strains from nodules of Rosaceae, *Datisca, Coriaria* or *Ceanothus* species, nor to infect these plants with cultured *Frankia* strains (Table 1). Fortunately, several methods are available to identify *Frankia* strains without cultivation, in nodules as well as in soil (for review, see 23). This way, it could be shown that actinorhizal nodules can contain multiple bacterial strains [24], not only different *Frankia* strains but also, in the outer cortex, non-nitrogen-fixing related actinomycetes that cannot reinfect the host plant on their own [25].

The lack of a genetic system for *Frankia* has been a major obstacle in the study of actinorhizal symbioses. Meanwhile, the basis for the development of cloning vectors has been laid in that studies of antibiotic resistance patterns of *Frankia* strains are available [26], and *Frankia* plasmids have been analysed [27].

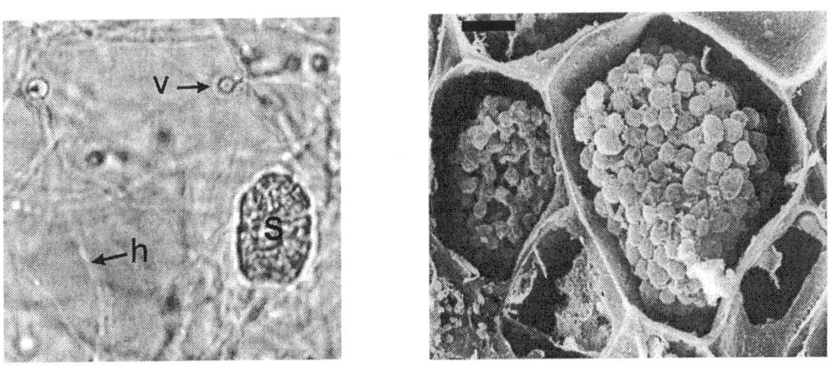

(A) *Frankia* strain Thr culture grown under normal oxygen tension and nitrogen-limiting conditions. h, hypha; v, vesicle; s, sporangium. **(B)** Electron micrograph of an infected cell of an *Alnus glutinosa* nodule (kindly provided by A.D.L. Akkermans). *Frankia* has formed spherical vesicles. The bar represents 10 µm.

Figure 1. *Frankia* in the free-living state and *in planta.*

2.1. Methods of *Frankia* systematics

Frankia strains were catalogued previously according to host specificity, morphological criteria (*e.g,*. shape of sporangia), cell chemistry and physiology and DNA homology (reviewed in 28). The development of molecular methods for strain identification has revolutionized the analysis of *Frankia* populations in soil and in nodules, since such analyses are not affected either by culturability or by infectivity of different *Frankia* strains. The application of the polymerase chain reaction (PCR) has allowed the retrieval of sequence information from uncultured bacteria and their use for phylogenetic analysis. For obvious reasons, rRNA genes have been targeted for such analyses, but nitrogenase genes or the intergenic spacers between two nitrogenase genes (*nifH-D*) [29,30] or the glutamine synthetase gene *glnII* [31] have also been investigated. 16S rDNA sequencing has led to the emendation of the genus *Frankia*, containing four main subdivisions, namely (i) a large group mainly comprised of *Frankia alni* and other strains of the *Alnus*-infective and the *Casuarina*-infective host specificity groups, (ii) uncultured endophytes of *Dryas, Coriaria, Datisca, Ceanothus* and *Purshia* [32, 33], (iii) strains of the *Elaeagnus*-infective host specificity group and (iv) atypical non-nitrogen-fixing strains [34]. A more detailed picture involving subgroups of (i) was based on the comparison of an actinomycete-specific 23S rDNA insertion of several *Frankia* strains [35]. A group of uncultured non-nitrogen-fixing endophytes in nodules of *Alnus glutinosa* were shown to belong to a formerly undescribed subdivision within the genus *Frankia* [36]. Nodules induced by *Frankia* strains that cannot fix nitrogen are called "ineffective"; in them, *Frankia* exhibits the characteristics of a mild pathogen [37].

At a lower taxonomic level, *e.g.*, for studies on the diversity of *Frankia* within single nodules, restriction fragment length polymorphism (RFLP) has often been used to distinguish between strains [24]. Nowadays RFLP analyses are usually performed on PCR-amplified DNA fragments such as the 16S rDNA [31,38,39]. Using different amplicons and different sets of restriction enzymes, single isolates can be distinguished and genetic groups of strains can be identified [40, 41]

2.2. *Frankia* physiology

Originally, two physiological groups of *Frankia* strains, groups A and B, were defined [42] on the basis of DNA homology, serology and physiology. Group A strains (corresponding roughly to the *Frankia elaeagni* group as proposed by [43]) are rather diverse and can utilize several monosaccharides and disaccharides, while group B strains (mostly isolated from *Alnus, Myrica* or *Comptonia* nodules) are more homogenous and, with a few exceptions [44,45], do not grow on carbohydrates. All *Frankia* strains analysed to date grow well on short chain fatty acids such as acetate or propionate (with one exception, the group A strain $EAN1_{pec}$ which was isolated from *Elaeagnus* nodules [46] and cannot grow on propionate) and variably on Tween or dicarboxylic acids or pyruvate. Altogether, it is surprising that although glycogen and trehalose are major storage compounds of *Frankia* strains [47-49], so many strains seem unable to grow on carbohydrates. Accordingly, it has been suggested that carbohydrates might be used to a greater extent than implied by these early studies [50]. At any rate, it should always be considered that many symbiotic *Frankia* strains have not yet been cultured and thus any grouping criteria are preliminary.

As in most other diazotrophs, ammonium assimilation in *Frankia* follows the glutamine synthetase (GS)/glutamate synthase (GOGAT) pathway [51]. Glutamine dehydrogenase and alanine dehydrogenase have similar activities in *Frankia* strain CpI1 grown on ammonia-rich and nitrogen-deficient medium, respectively, indicating that at least in the free-living state, neither enzyme plays a role in ammonium assimilation [52]. Like rhizobia, *Frankia* contains two types of GS, GSI and GSII. GSI is regulated by adenylylation [53] and GSII is synthesized in response to nitrogen starvation and is the predominant GS under nitrogen-fixing conditions [52,54]. GSII is present in hyphae and vesicles, while GOGAT is present only in hyphae [52]. This implies that glutamine is transfered to the hyphae to be converted to glutamate [52, 55], comparable to the situation in cyanobacterial heterocysts [56]. Several polyamines – putrescine, cadaverine, spermidine and spermine – have been found in cultured *Frankia* strains [57], while polyamines are not abundant in symbiotic *Frankia in planta* [58], probably due to the fact that here, the fixed nitrogen is exported to the plant cells.

Nitrogen fixation requires high amounts of iron for the Fe- and MoFe-proteins of nitrogenase and the prosthetic hemes of cytochromes. Interestingly, although hydroxamate siderophore-mediated iron uptake does occur in selected *Frankia* isolates (*e.g*., strain 52065), it is not uniformly distributed among all strains [59,60]. *Frankia* also releases compounds with antimicrobial and calcium-transport inhibiting activities [61, 62]. It remains to be established whether these so-called frankiamides also have a function in symbiosis.

3. MACROSYMBIONTS

More than 200 species of dicotyledonous plants, distributed in 25 genera belonging to eight different families can enter a root nodule symbiosis with *Frankia* (Table 1). With one exception, *Datisca glomerata*, actinorhizal plants represent trees or woody shrubs. Interestingly, not all species of a genus have to be actinorhizal; for instance, in *Dryas* only one actinorhizal species is known [63]. Thanks to their symbiosis, actinorhizal plants can grow on marginal soils and have been used in soil reclamation, erosion control, agroforestry, dune stabilization. They also have been used in fuel wood, pulp and timber production [for reviews, see 64-67]. *Hippophae rhamnoides* is currently being domesticated since its fruits are very nutritious, rich in vitamin C and carotenes, and the seed oil is highly unsaturated and has light absorption and emollient properties that make it a promising ingredient in cosmetics and phytopharmaceuticals [68-70].

4. INFECTION PROCESS AND NODULE DEVELOPMENT

In actinorhizal symbioses, two ways of infection are known, either intracellular *via* root hairs or intercellular by penetration between epidermal cells. The infection route is determined solely by the host plant [71]. Members of the Betulaceae, Casuarinaceae and Myricaceae families are infected intracellularly (Table 1), while members of the other families are infected intercellularly. Strictly speaking, infection mechanisms of *Datisca,*

Coriaria, rhamnaceous species except for *Ceanothus* and *Discaria,* and rosaceous species except for *Cercocarpus* have not been analysed yet. However, since intracellular infection involves the formation of a prenodule and leads to the formation of files of infected cells in the mature nodule lobes (see below), the morphology of a mature nodule lobe allows conclusions with regard to the infection pathway.

4.1. Intracellular infection

During intracellular infection, the first response of the plant to the presence of the microsymbiont is the deformation and branching of growing root hairs [72-75], probably due to blockage and re-initiation of tip growth as in legumes [76]. This is due to factors excreted by the microsymbiont [77] the chemical nature of which has not yet been analysed [78]. In contrast to legumes, root hair deformation can also be induced by non-symbiotic soil bacteria [79,80]. Only one or two root hairs become infected: a *Frankia* hypha associated with a deformed root hair penetrates the root hair cell wall, while the root hair plasma membrane invaginates and an infection thread-like structure develops (Figure 2). Within this structure, the *Frankia* hypha is embedded within a plant-derived cell-wall-like pectin-rich matrix, the so-called encapsulation [74,75,80-82]. Concomitantly, cell divisions are induced in the root cortex close to the infected root hair, giving rise to the formation of a so-called prenodule. The infection thread-like structures grow toward the prenodule by cell-to-cell passage, and infect some, but not all, prenodule cells by extensive branching within these cells, filling them from the center outward [83]. The infected prenodule cells become hypertrophic, while uninfected prenodule cells accumulate starch [73]. *Frankia* starts fixing nitrogen in the infected prenodule cells [84,85]. Studies on cell differentation based on gene expression patterns indicate that, at least in *Casuarina glauca,* prenodules are primitive symbiotic organs that consist of three cell types with unique differentiation features equivalent to their counterparts in mature nodule lobes. These cell types are (i) infected cells containing nitrogen-fixing *Frankia,* (ii) uninfected cells showing the same features as uninfected cortical cells in the mature nodule lobe while differing from root cortical cells [85], and (iii) polyphenol-containing cells the gene expression pattern of which differs from that of polyphenol-containing root cortical cells while resembling those of polyphenol-containing cortical cells of mature nodule lobes [86]. Nevertheless, the prenodule is only an intermediate stage in nodule development. While it is developing, the primordium of the nodule lobe is initiated in the root pericycle near the infection site, and *Frankia* hyphae in infection thread-like structures grow from the prenodule to the nodule primordium, again by cell-to-cell passage, and infect primordium cells.

Infection does not always lead to an effective symbiosis. Some strains can induce nitrogen-fixing nodules on one plant species but only ineffective nodules on another, *i.e.*, they are incompatible with the second host plant [87]. Furthermore, non-nitrogen-fixing *Frankia* strains have been found that could induce ineffective nodules on alder (so-called ineffective strains; see *e.g.*, [88]). The ability of an alder tree to discrimiate between ineffective and effective *Frankia* strains, *i.e.*, its resistance against ineffective strains, seems to be determined genetically [37,88,89]. There is even evidence that the intracellular infection pathway relies on signals of the microsymbiont that can be imitated by other organisms. The fungus *Penicillium nodositatum* can induce so-called myconodules on the alder roots [90-93]. The nodule induction process used by the fungus - intracellular infection via root hairs - resembles that employed by *Frankia* [92]. Structurally, myconodules resemble ineffective, *i.e.*, non-nitrogen-fixing actinorhizal nodules [92]. As in ineffective nodules, the infection of the host plant by the fungus does not elicit a strong resistance response [93], and like ineffective nodules, myconodules remain small (mostly single-lobed) and contain large amounts of polyphenolics [92].

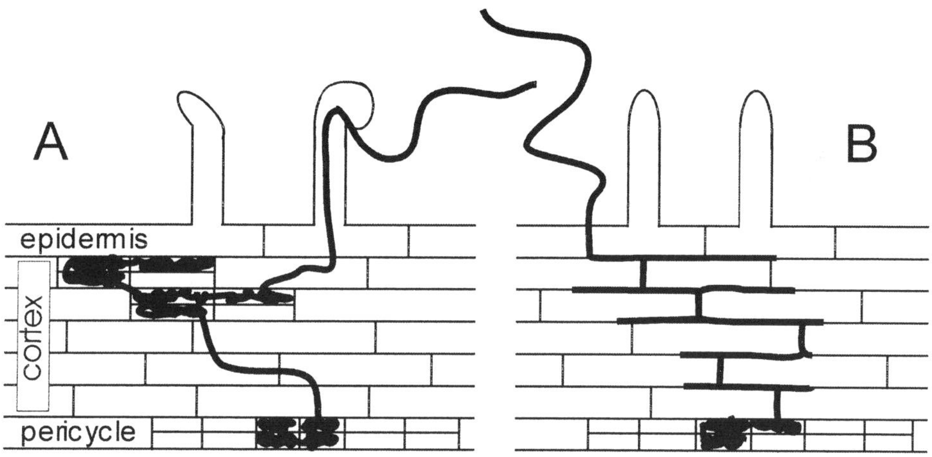

(A) Intracellular infection. Root hairs deform and branch in response to the presence of *Frankia* in the rhizosphere. A *Frankia* hypha enters the plant through an infection thread-like structure formed in a curled root hair. Concomitantly, cell divisions are induced in the outer cortex and the infection thread-like structure grows toward these dividing cells and infects, *i.e.*, fills them with branching encapsulated hyphae. The dividing cortical cells represent the prenodule. Now, the formation of the nodule primordium is induced in the pericycle opposite to a protoxylem pole, and hyphae grow in infection thread-like structures from the prenodule to the nodule primordium and infect primordium cells. **(B)** Intercellular infection. *Frankia* hyphae penetrate the root epidermis between epidermal cells and colonize the root cortex intercellularly. Cell divisions are induced in the root pericycle and lead to the formation of a nodule primordium. Cells of the nodule primordium are infected by *Frankia* hyphae.

Figure 2. Infection mechanisms.

4.2. Intercellular infection

No root hair deformation takes place during this type of infection, but *Frankia* hyphae penetrate the middle lamella between adjacent cells of the root epidermis and progressively colonize the intercellular spaces of the root cortex [71,94-97]. The penetration of the epidermis is probably achieved by pectolytic activity that has been shown for several *Frankia* strains [98]. Epidermal and cortical sells secrete pectin-rich material into the intercellular spaces [99] which might represent the equivalent of the cell-wall-like material encapsulating the *Frankia* hyphae in infection thread-like structures. Concomitantly, the nodule lobe primordium is initiated in the root pericycle, and *Frankia* hyphae infect primordium cells from the intercellular spaces. During infection, the plasma membrane of the infected cells invaginates and the hyphae are embedded in cell-wall-like material such as in infection thread-like structures.

4.3. Autoregulation of nodulation

The autoregulation of nodulation by the host plant has been extensively studied in legumes [100]. The plant can regulate the number of effective root nodules produced, depending on light, water, nitrogen and phosphorus availability, soil pH *etc.* [reviewed in 14]. Nitrogen inhibition of nodulation [101-104] and nitrogen inhibition of nodule nitrogen fixation [105] seem similar in actinorhizal plants and in legumes. However, nitrogen inhibition of nodulation seems to work on different levels of nodule development in intracellularly infected, as opposed to intercellularly infected, actinorhizal plants [reviewed in 106]. Recent studies have shown that nitrogen inhibition of nodulation depends on the N/P ratio in legumes and in intracellulary as well as in intercelularly infected actinorhizal plants [107-110]. Furthermore, low nitrogen concentrations may stimulate nodulation, and threshold nitrate concentrations for inhibition of nodule formation differ between plant species [102].

5. MATURE ACTINORHIZAL NODULES

5.1. Nodule morphology

As mentioned above, actinorhizal nodules consist of lobes which represent modified lateral roots (Figure 3A) without root caps. Due to the activity of the apical meristem, the infected cells in the cortex of a nodule lobe are arranged in a

developmental gradient (Figure 3B). In intracellularly infected plants, infected cells often occur in files, since infection thread-like structures grow from one cell to the next. In intercellularly infected plants, where infection proceeds from the intercellular spaces, such files of infected cells are not found. It is not understood which features decide whether a particular cortical cell will be infected by the microsymbiont. In most symbioses, infected cells are interspersed with uninfected cells. In nodules formed on the roots of *Datisca* or *Coriaria* species, the pattern of infected cells is unusual in that they form a continuous patch on one side of the acentric stele, not interspersed with uninfected cells (Figure 3C) [111-112].

5.2. Oxygen regulation in nodules

Although nitrogenase is irreversibly denatured by oxygen, N_2 fixation has a high energy demand requiring oxygen for respiratory processes in the nodules. Therefore, the supply of O_2 to the microsymbiont has to be tightly controlled. In the different actinorhizal plant families, diverse strategies are applied to achieve this goal, involving nodule morphology, vesicle structure, plant and bacterial metabolism and the formation of oxygen-binding proteins in infected cells.

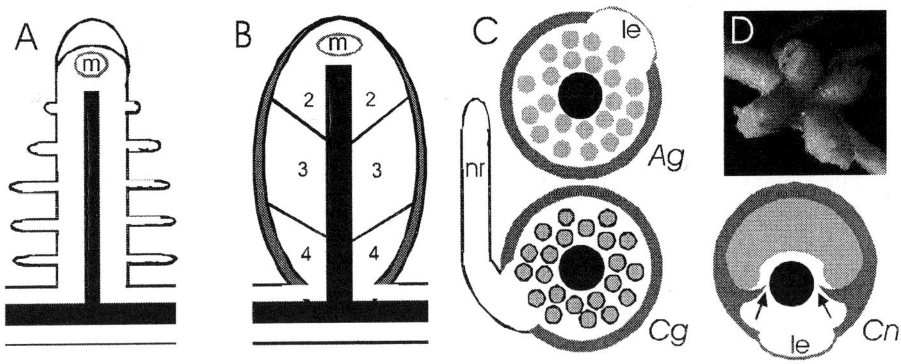

(A) Scheme of a lateral root for comparison. The vascular system is labeled in black. Calyptra and root hairs are shown. m, meristem. **(B)** Scheme of an actinorhizal nodule lobe. Due to the activity of the apical meristem (m), the nodule cortex can be divided into the infection zone (2), where cortical cells become gradually filled with *Frankia* hyphae, and afterwards, vesicles differentiate (with the exception of *(Allo-)Casuarina* nodules), the fixation zone (3) where nitrogen is fixed by *Frankia* in the infected cells, and the senescence zone (4), where *Frankia* hyphae and vesicles are degraded. Nodule lobes are surrounded by a periderm which is shown in dark gray. **(C)** Cross sections of nodule lobes from different actinorhizal plants (*Ag, Alnus glutinosa; Cg, Casuarina glauca; Cn, Coriaria nepalensis*). The infected cells are labeled in light grey. For aeration, the periderm may be interupted by lenticels (le) or by agraviotropically growing nodule roots (nr). For *Casuarina*, the lignification of the walls of infected cells is indicated by thick black lines. Arrows mark the control points of air access to the infected cells in *Coriaria*. **(D)** Photograph of a mature four-lobed nodule from *Datisca glomerata, ca.* eight weeks after infection.

Figure 3. Actinorhizal nodule structure.

Actinorhizal plants that grow in wet or waterlogged soils have developed mechanisms for gas transport to the nodules. *Casuarina*, *Gymnostoma, Myrica* and *Comptonia* species that often grow in wetlands form nodules surrounded by a periderm that is more-or-less impermeable to gas, and oxygen is provided *via* air spaces in the nodule roots [113]. Nodule roots are formed at the tips of nodule lobes and grow upwards; their length is negatively correlated with the aeration of the root substrate [114]. Nodules of *Alnus glutinosa*, which often grows in waterlogged soil, do not form nodule roots, but solve the problem of oxygen access *via* thermo-osmotically mediated gas transport from the aerial parts to the roots [115]. In well-drained soils, *Alnus* nodules are well aerated since their periderm is interrupted by lenticels and their cortex contains large intercellular spaces, particularly in the outer cortex [116]. Nodules of *Coriaria* and *Datisca*, in which the infected cells are only found on one side of the acentric stele, are surrounded by a dense periderm with suberised cell walls [112,117]. The thickness of the periderm in *Coriaria* increases at elevated O_2 concentrations [118].

In *Coriaria* nodules, a single, large lenticel is present on the uninfected side of the nodule lobe, limiting the gas-diffusion pathway to the infected cells to the narrow gap between the inner periderm and the stele (Figure 3C) [118]. Nodules of *Datisca cannabina* form nodule roots [113]), while nodules of *Datisca glomerata* can form nodule roots in liquid culture or water-logged soil and lenticels in well-drained soil [A.M. Berry and K. Pawlowski, unpublished observation].

Mechanisms for oxygen protection of nitrogenase in actinorhizal nodules are complex and differ between host plant families. For instance, there is evidence that in *Alnus* nodules, *Frankia* contributes to oxygen protection of nitrogenase. The number of layers making up the vesicle envelope which is thought to serve as an oxygen-diffusion barrier were shown to increase with increasing oxygen tension in symbiotic *Frankia* in *Alnus* nodules [10,119]. The fact that the increase in number of layers relative to oxygen tension is lower *in planta* than in the free-living state indicates that the plant must be contributing to oxygen protection.

In nodules of *Casuarina*, however, the plant seems to be solely responsible for oxygen protection of nitrogenase. Here, the zone of infected cells contains very few intercellular spaces [120] and the walls of infected cells are lignified, which is supposed to restrict O_2 diffusion [121]. *Frankia* does not form vesicles, indicating that nitrogenase is formed in hyphae which are kept in an environment with low O_2 tension [122], which is consistent with the fact that high concentrations of the oxygen transport protein hemoglobin are found specifically in the infected cells [123]. This protein is encoded by a nodule-specific gene as in legumes. Large amounts of hemoglobin, and an oxygen-diffusion barrier around the infected cells, were also found in nodules of *Myrica gale*, although here, *Frankia* does form vesicles [124,125]. In *Alnus glutinosa*, hemoglobin concentration in nodules is elevated compared to roots, but not enough to account for a role of hemoglobin in nitrogenase protection [126].

No hemoglobin has been found in nodules of *Datisca* and *Coriaria* [113]. In both plants, *Frankia* forms finger-shaped vesicles with rather thin envelopes which point at the central vacuole, forming a ring around it [112,117]. The major diffusion-resistance found in *Datisca glomerata* nodules was in the infected cells, but the vesicle envelopes are much thinner here than in other actinorhizal nodules [127,128]. However, since the vesicles are very tightly packed, leading to low surface exposure, perhapseven the thin vesicle envelopes could serve as the major O_2 diffusion barrier. A continuous blanket of mitochondria was found around the bases of the vesicles [127]. It is proposed that this blanket acts as a gas-diffusion barrier, and that the short diffusion path required for O_2 transport into the vesicles precludes the need for an O_2 transport system like hemoglobin. In contrast, in *Coriaria* nodules the only O_2 diffusion pathway from the lenticel to the infected cells is the two-cell-layer gap between the endodermis and the internal periderm, which would open the possibility of regulating O_2 concentration *via* minimal changes in the turgor of these cells [118]. Interestingly, *Coriaria* is unique among actinorhizal plants in that the nodules can perform short-term adaption to varying pO_2 levels, as do legume nodules, which also regulate O_2 access to the infected cells *via* turgor changes [129].

5.3. Nodule cell differentiation

One regulatory feature that determines infectability of cortical cells may be cell-specific flavan biosynthesis. During nodule formation in *Alnus glutinosa* and *Casuarina glauca,* cell-specific flavan accumulation delimits cortical compartments containing *Frankia*-infected cells and might restrict endophyte invasion [130; K. Pawlowski, unpublished observation]. In plant-pathogen interactions, layers or tissues of polyphenolic-filled cells seem to serve as pathogen barriers [131-133]. Ineffective, *i.e.*, non-nitrogen-fixing *A. glutinosa* nodules that are induced by incompatible *Frankia* strains [88,134], remain small (mostly single-lobed) and contain large amounts of polyphenolics. In such nodules, polyphenolic-filled vacuoles are often found within infected cells, a phenomenon rarely observed in effective nodules [135], suggesting that these polyphenolics may be part of a defense response of the plant. This interpretation is supported by the fact that in these nodules, endosymbiont degradation starts soon after the infected cells have been filled with *Frankia* hyphae [135]. However, in nodules of *Coriaria nepalensis*, the central vacuole of infected cells is filled with polyphenolics [117], implying that polyphenolic-filled vacuoles do not necessarily affect the viability of intracellular *Frankia.* In myconodules of *Alnus glutinosa*, the distribution of fungus-infected cells in the cortex is different from that of *Frankia*-infected cells in effective and ineffective nodules. Infected areas of the cortex tend to contain very few uninfected cells and are mostly surrounded by, andnot interspersed with, areas of cells containing polyphenolic-filled vacuoles [136]. It seems that in effective interactions, *Frankia* only infects particular cortical cells, while in ineffective interactions *Frankia* also invades cortical cells of other differentiation types, *i.e.*, those destined to accumulate polyphenolics. This would imply that the infection of a particular cortical cell is the result of a signal exchange between this cell and the microsymbiont, and that this communication is disturbed in ineffective symbioses.

Altogether, nodules contain several cell types that are not present in roots, the three cell types also found in prenodules, *i.e.*, infected cortical cells, uninfected cortical cells that contain starch grains, and polyphenol-containing cortical cells [85,86], and additionally the cells of the multilayered pericycle of the nodule vascular system. At least two nodule-specific genes are expressed exclusively in the infected cells and in the nodule pericycle [C.H. Guan and K. Pawlowski, unpublished]. There is evidence that signals controlling the differentiation of infected cells are conserved between legumes and actinorhizal plants. The β-glucuronidase (GUS) fusion of a nodule-infected cell-specific symbiotic hemoglobin promoter from an actinorhizal plant, *Casuarina glauca,* is expressed specifically in infected cells of legume nodules, and a leghemoglobin promoter-GUS fusion from a legume, soybean, retains its infected cell-specific expression in actinorhizal nodules [123,137]. Thus, at least one transcription factor responsible for infected cell-specific gene expression is conserved in both legumes and intracellularly infected actinorhizal plants. Interestingly, a hemoglobin promoter-GUS fusion from *Parasponia,* the only non-legume infected by rhizobia, retained its expression pattern in the actinorhizal plant *Casuarina glauca,* but not in a legume [137,138]. In spite of being induced by rhizobia, *Parasponia* nodules developmentally and morphologically resemble actinorhizal nodules [139,140], and phylogenetically, *Parasponia* is more closely related to actinorhizal plants than to legumes [141].

Apart from the gene for symbiotic hemoglobin [123], some other genes that are induced specifically upon infection of cortical cells of *Alnus glutinosa* [142,143] have been characterized. They encode a serine protease of the subtilisin family, a member of which is involved in the differentiation of stomatal guard cells [142,144] and a metal-binding protein that might form a part of the cell wall-like matrix surrounding *Frankia* in infected cells [143]. Homologs of these genes have been found in *Casuarina glauca* [145; B.C. Mullin, personal communication], indicating the similarity of the infection process in both plants. However, no homologs of these genes have yet been found to be expressed in legume nodules, or in actinorhizal nodules of *Datisca glomerata* [K. Pawlowski, unpublished]. A nodule-specific gene from *Datisca* nodules has a homolog in legume nodules, but not in *Alnus/Casuarina* [146]. Altogether, at the present time not enough data are available to draw conclusions on the similarities and differences in the process of differentiation of infected cells in actinorhizal nodules from different plant familes and from legume nodules, far less on the process of infected-cell differentiation *per se*. However, one interesting item should be mentioned here, namely that in the process of infection of cortical cells from *Datisca glomerata* nodules, the transcription of rubisco activase is induced [147]. The mRNA is not spliced correctly and does not leave the nucleus, but the fact of its synthesis indicates interaction of the signal transduction pathways.

6. NODULE PHYSIOLOGY

Actinorhizal nodule metabolism has been analysed physiologically and with molecular methods. Physiological examinations have sometimes been impeded by the fact that it is not possible in all systems to isolate symbiotic *Frankia* without contamination of plant cell organelles [reviewed in 14]. Molecular methods have allowed the analysis of the physiological specialization of different tissues [148,149].

6.1. Carbon metabolism

Nodules represent strong carbon sinks, since they need assimilates for growth and maintenance, N_2 fixation as well as ammonium assimilation and transport of nitrogenous solutes, and starch biosynthesis. Uninfected cortical cells of nodules (with the exception of *Casuarina* sp.; E. Duhoux, unpublished observation) tend to contain large starch grains, but the function of nodule starch is not known. It is used neither for nitrogen fixation during the dark period, nor for spring growth [150,151]. Nodule starch formation may simply represent an overflow valve for superfluous photosynthates delivered to the nodule *via* the root phloem. In that case, we have to wonder what storage compounds are formed in *Casuarina* nodules instead of starch, or whether *Casuarina* nodules are exceptional in not receiving excess photosynthates.

In most plants, sucrose is the main carbon-transport form in the phloem. The phloem unloading activity in sink organs is determined either by the distribution of energy-dependent transmembrane sugar transporters in case of apoplastic sugar transport, or by the activity of sucrose-degrading enzymes – sucrose synthase or invertases – in the sink cells in case of symplastic sugar transport. *Alnus* nodules contain a multilayered pericycle consisting of small cells with dense cytoplasm and large mitochrondria, indicating high metabolic activity and a role in post-phloem transport of sugars [152]. Sucrose synthase has been shown to be expressed at enhanced levels in *Alnus* nodules compared to roots, particularly in the infected cortical cells and in the nodule pericycle [148], but the role of invertases in nodule-carbon metabolism has not yet been addressed. Preliminary results indicate that symplastic carbon transport is favoured over apoplastic carbon transport in nodules compared to roots [M. Schubert and K. Pawlowski, unpublished]. Since the microsymbiont is located in the apoplast and thus has access to apoplastic sugars, this may reflect the necessity for the plant to control the flow of carbon to *Frankia*.

In order to determine which carbon metabolites are transferred from the host to *Frankia,* vesicle clusters have been isolated from nodules [153-155] and checked for consumption of exogenously added sugars. For instance, sucrose, trehalose, maltose, glucose and fructose stimulated respiration in vesicle clusters from *Alnus rubra* [156], but these observations may simply reflect the fact that *Frankia* can degrade its own carbon-storage forms [48] and do not necessarily prove that hexoses are the carbon source provided by the plant. In legume symbioses, the plant is transferring dicarboxylic acids to the rhizobial bacteroids [157]. Physiological studies on *Frankia* isolated from actinorhizal nodules seem to point toward an analogous situation: succinate, malate and glutamate are able to stimulate respiration in vesicle clusters of *Alnus* [158-160], and the tricarboxylic acid cycle enzymes required for the consumption of dicarboxylic acids were shown to be active in these clusters [158]. However, all these results could be explained by passive uptake of the substrates.

Recent results on the expression patterns of sugar transporters and sucrose degrading enzymes [M. Schubert and K. Pawlowski, unpublished] indicate that, in *Datisca glomerata* nodules, hexoses may in fact be transferred to endosymbiotic *Frankia*. At the very least, a glucose transporter is expressed specifically in infected cells, and due to the absence of significant activities of apoplastic invertase in the nodules, it is unlikely to be involved in uptake of glucose from the apoplast. Since subcellular localization of the transporter has not yet been shown, this can only be considered circumstantial evidence. At any rate, without bacterial mutants involved in the uptake of particular carbon sources, it is impossible to state whether (a) hexoses are the only carbon sources transferred to endosymbiotic *Frankia* and whether (b) this is the case for all actinorhizal symbioses, or only for *Datisca* nodules.

6.2. Nitrogen metabolism

The pattern of ^{15}N labelling of amino acids indicates that ammonia derived from bacterial nitrogen fixation is assimilated *via* glutamine synthetase (GS) in both *Alnus glutinosa* and *Myrica gale* [105,161]. No GS activity was found in vesicle clusters isolated from nodules [162], and bacterial GS protein could not be found in *Frankia* prepared from *Alnus incana* nodules, while plant GS was found in nodule extracts [163], indicating that ammonia/ammonium is exported by the bacteria and assimilated by the plant. However, while it had been generally for legumes accepted, on the basis of ^{15}N labelling studies, that ammonium is the sole secretion product of N_2-fixation by the bacteroids, detailed studies have revealed only recently that rhizobial bacteroids can assimilate ammonia via alanine dehydrogenase and that export of alanine by bacteroids plays a role in symbiosis [164,165]. Hence, the fate of fixed nitrogen in actinorhizal nodules is not yet solved.

6.3. Uptake hydrogenase

Nitrogenase forms H_2 concurrently with the reduction of N_2 to ammonium. Since H_2 is an inhibitor of nitrogenase, the oxidation of H_2 by uptake hydrogenase is important to prevent H_2 accumulation and to recover reductant and ATP that was used in H_2 production. *Frankia* forms hydrogenase in hyphae as well as in vesicles under N_2-fixing conditions, thus indicating that the role of hydrogenase may not be restricted to preventing the loss of energy by nitrogenase [166,167]. Although in legume nodules many rhizobial strains lack hydrogenase activity, all actinorhizal symbioses studied except for one have shown hydrogenase activity [168, 169]. Up to now, it has. Not. been possible to elucidate the benefits of uptake hydrogenase for nodule metabolism, since the exceptional Hup^- inoculant forms spores in nodules, while the strains used for comparison do not [170], and spore formation is supposed to reduce the efficiency of nodule N_2 fixation [17].

7. PHYLOGENY

7.1. Actinorhizal plants and legumes

Molecular phylogenetic analysis of seed plants based on rbcL gene sequences has revealed that all plants able to enter a nitrogen-fixing root nodule symbiosis belong to the rosid I clade [141,171]. These data imply that the ability to form such a symbiosis relies on a predisposition that arose only once during angiosperm evolution. Nodulated plants within the rosid I clade can be grouped into four major lineages. One includes the intracellularly-infected actinorhizal plants of the Hamamelid families Myricaceae, Betulaceae and Casuarinaceae (branch (a) in Figure 4). The second group comprises the symbiotic legumes. The third group includes actinorhizal plants of the Coriariaceae and Datiscaceae families (branch (b) in Figure 4), and the fourth group includes intercellularly infected actinorhizal plants of the Elaeagnaceae, Rhamnaceae and Rosaceae families as well as the only non-legume infected by rhizobia, *Parasponia* (Ulmaceae) (branch (c) in Figure 4). This grouping is supported by the conservation of morphological features of actinorhizal nodules in the different lineages [171,172], as well as by the fossil record and the geographical distribution of actinorhizal species [173].

7.2. Macro- and Microsymbiont phylogeny

Molecular phylogenetic studies on *Frankia* have been performed using 16S RNA sequences [32,34,174,175], *nifD* gene sequences [176] and recently also *recA* and *glnII* gene sequences [177]. Based on the 16S rDNA sequences, three major groups were distinguished (Figure 4), as well as a fourth clade of "*Frankia*-like" strains comprising actinomycetes that can be isolated from nodules but that are not able to reinfect their host plant and cannot fix nitrogen (*e.g.*,see 25). This grouping is also supported by morphological, physiological and host-specificity features of *Frankia* strains (reviewed in 106). One of the three groups does not contain any isolates, but is based solely on *Frankia* gene sequences amplified from nodules of Rosaceae, *Datisca, Coriaria* and *Ceanothus.*

Plant and bacterial groupings are correlated [106,173] (Figure 4). Detailed analysis suggests that the actinorhizal symbiosis evolved independently at least three or four times [172,174], followed by co-evolution with limited lateral transfer [173]. For instance, differences in the host range of actinorhizal plants allow conclusions with regard to the evolution of host specificity. According to the fossil record, the Hamamelid families Myricaceae, Betulaceae and Casuarinaceae contain the oldest extant actinorhizal genera, while Coriariaceae are the youngest [173,178]. Among the Casuarinaceae, *Gymnostoma* has diverged first [179]. Within the Hamamelid actinorhizal plants, there is a clear difference in host range. *Myrica* and *Gymnostoma* which are readily nodulated by *Frankia* strains from both Clades I and II in the greenhouse as well as in the field, rank highest, followed by *Alnus*, while *Allocasuarina* and *Casuarina* are infected mostly by a specialized lineage within Clade I [179,180, reviewed in 173]. Hence, a narrow host range seems to characterize the more recent actinorhizal genera. It is also interesting that there seems to be no overlap in host range of *Frankia* Clade III with the other clades, except for one instance where Clade II strains have been found in *Ceanothus* nodules from Southern California [173,181], which raises the question about the origins of Clade III.

As mentioned above, there are several instances among closely-related plant species, such as different *Dryas* species or *Betula* and *Alnus*, where one may be actinorhizal while the another is not. This fact implies that the symbiotic ability was lost at several points during evolution of the rosid I clade. Hence, it can be concluded that the ability to enter a nitrogen-fixing symbiosis does not always confer an evolutionary advantage to the host plant. For instance, to support bacterial nitrogen fixation, symbiotic plants need iron, molybdenum and cobalt (summarized for legumes in 182). Under conditions when nitrogen supply is not the only or major growth-limiting factor, a nitrogen-fixing root nodule symbiosis may turn out to be a competitive disadvantage to the host plant.

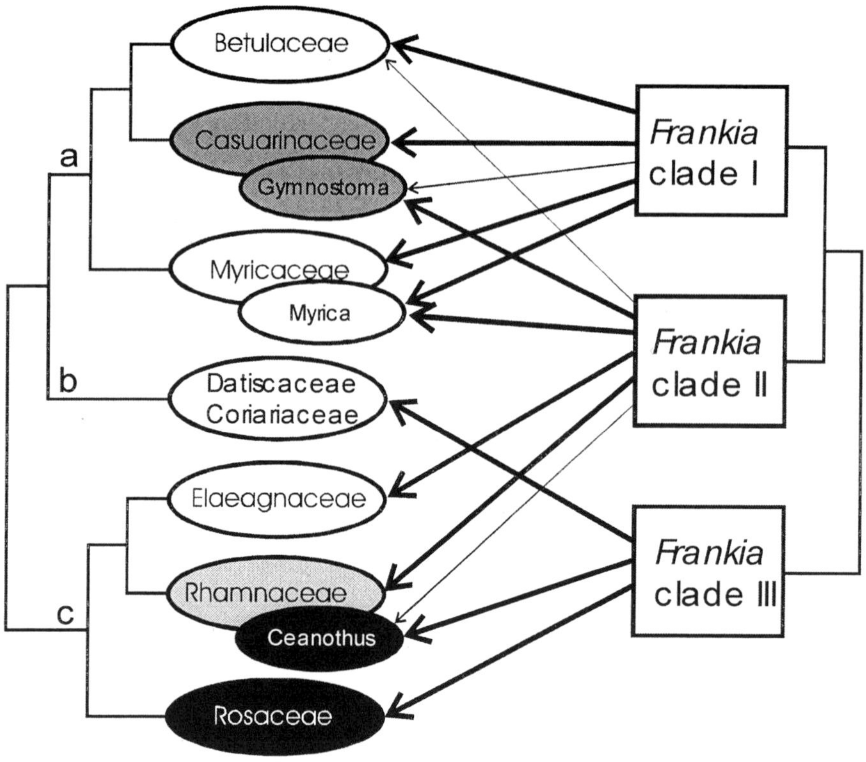

The relationships between groups of actinorhizal plants and *Frankia* are shown in a simplified scheme based on [106] und [173]. Thick arrows connecting *Frankia* clades with the plant families/genera indicate that members of that clades are commonly associated with the plants. Thin arrows indicate that members of that clades have been isolated from or detected in an effective or ineffective nodule of a member of the plant genus/family at least once. Some genera (*Gymnostoma, Myrica, Ceanothus*) differ in microsymbiont specificity from the rest of the plant family they belong to. Plant groups that show a particular native geographical distribution are labeled (dark grey, Australia and western Pacific; light gray, South America and southern New Zealand; black, western North America). Actinorhizal Elaeagnaceae and Betulaceae occur on most continents, while Coriariaceae and Datiscaceae are found in temperate zones of both hemispheres.
No members of *Frankia* Clade III have been isolated yet. Host specificity exists within the *Frankia* clades, *i.e.*, not all members of a clade can nodulate all plants associated with that clade. For instance, all members of Clade I tested can nodulate *Myrica,* but only some of them can nodulate either *Alnus* or *Casuarina.*

Figure 4. Phylogeny of *Frankia* and actinorhizal plants.

8. OUTLOOK

Looking at root nodule symbioses from the phylogenetic point of view, the question arises is: which unique feature was acquired by the progenitor of the rosid I clade that enabled plants of this clade to develop root nodule symbioses ?[141,171]. The analysis of legume, and particularly of actinorhizal, symbioses has shown that for most practical requirements such as oxygen supply, oxygen protection, sugar partitioning *etc.*, multiple solutions are possible and have been adopted in different symbiotic genera. So what is it that makes the symbiosis possible? Basically, three features come to mind, (i) the ability to integrate stably a bacterial microsymbiont into a plant cell, (ii) the ability to form lateral- root structures in response to microsymbiont signals, and (iii) the ability to repress the defense response against bacterial entry to the root. One might argue that (i) is realized in most terrestrial plants since arbuscular mycorrhizal fungi are stably internalized [183]. However, the fact that, in root nodule symbioses, only cells that have been formed after signal exchange with the microsymbiont can internalize stably rhizobia or *Frankia* implies that this process requires a specific differentiation of the plant cell that is not present in non-symbiotically grown plants. With regard to (ii), it is interesting that several symbiotic trees, namely some legume trees and the actinorhizal *Gymnostoma*, have been shown to form symbiotic lateral-root structures also in response to infection by arbuscular mycorrhizal fungi ("mycorrhizal nodules") [184,185], indicating that the formation of lateral-root structures is triggered very easily in these plants, perhaps due to some peculiarities of the root phytohormone balance. However, such mycorrhizal nodules were also found on non-symbiotic tree species [184]. Regarding the suppression of the defense response, it is interesting that in actinorhizal plants, the entry into the root seems to be less strictly controlled than in legumes. First, there is *Alnus* where *Penicillium* fungi can also use the *Frankia*'s intracellular infection pathway [92], and secondly, several actinorhizal nodules have been found to contain multiple actinomycete strains, not all of which are able to reinfect the plant on their own [25,186].

Further research is required to understand these phenomena. At present, the most urgent objective in actinorhizal research should be to develop a transformation method for *Frankia,* and to find culture conditions for members of the *Frankia* Clade III.

9. REFERENCES

1. R. F. Tarrant and J. M. Trappe, Plant Soil, special volume (1971) 335.
2. J. Meyen, Flora (Jena) 12 (1829) 49.
3. L. Hiltner, Landwirt. Verständnisstudien 46 (1895) 153.
4. O. Krebber, Arch. Microbiol. 2 (1932) 588.
5. R. Schaede, Planta 19 (1933) 389.
6. E. H. Pommer, Ber. Deut. Bot. Ges. 72 (1959) 138.
7. D. Callaham, P. Del Tredici and J. G. Torrey, Science 199 (1978) 899.

8. A. M. Berry, O. T. Harriott, R. A. Moreau, S. F. Osman, D. R. Benson and A. D. Jones, Proc. Natl. Acad. Sci. U.S.A. 90 (1993) 6091.
9. R. Parsons, W. B. Silvester, S. Harris, W. T. M. Gruijters and S. Bullivant, Plant Physiol. 83 (1987) 728.
10. G. Kleemann, G. Alskog, A. M. Berry AM and K. Huss-Danell, Protoplasma 183 (1994) 107.
11. R. Nalin, A.-M. Domenach and A. M. Berry in Biological Nitrogen Fixation for the 21st Century, C. Elmerich, A. Kondorosi and W. E. Newton (eds.), Kluwer Academic Publishers, Dordrecht, 1998, p. 355.
12. M. A. Murry, Z. Zhang and J. G. Torrey, Can. J. Microbiol. 31 (1985) 804.
13. W. R. Newcomb and S. Wood, Internat. Rev. Cytol. 109 (1987) 1.
14. K. Huss-Danell, New Phytol. 136 (1997) 375.
15. S. Burleigh and J. G. Torrey, Appl. Env. Microbiol. 56 (1990) 2565.
16. J. G. Torrey, Physiol. Plant.70 (1987) 279.
17. C. R. Schwintzer in The Biology of *Frankia* and Actinorhizal Plants, C. R. Schwintzer and J. D. Tjepkema (eds.), Academic Press, San Diego, 1990, p. 178.
18. P. Normand and L. Lalonde, Can. J. Microbiol. 28 (1982) 1133.
19. R. Nalin, P. Normand and A.-M. Domenach, Physiol. Plant. 99 (1997) 732.
20. L. Maunuksela, K. Zepp, T. Koivula, J. Zeyer, K. Haahtela and D. Hahn, FEMS Microbiol. Ecol. 28 (1999) 11.
21. S. Elo, L. Maunuksela, M. Salkinoja-Salonen, A. Smolander and K. Haahtela, FEMS Microbiol. Ecol. 31 (2000) 143.
22. K. Huss-Danell, H. Sverrisson, A.-S. Hahlin and K. Danell, Arct. Antarct. Alp. Res. 31 (1999) 400.
23. D. Hahn, A. Nickel and J. Dawson, FEMS Microbiol. Ecol. 29 (1999) 215.
24. D. D. Baker and B. C. Mullin, Soil Biol. Biochem. 26 (1994) 547.
25. H. Ramirez-Saad, J. D. Janse and A. D. L. Akkermans, Can. J. Microbiol. 44 (1998) 140.
26. L. S. Tisa, M. S. Chéval, G. D. Krumholz and J. Richards, Can. J. Bot. 77 (1999) 1257.
27. C. Lavire, D. Louis, G. Perriere, J. Briolay, P. Normand and B. Cournoyer, FEMS Microbiol. Lett. 197 (2001) 111.
28. M. P. Lechevalier and H. A. Lechevalier in The Biology of *Frankia* and Actinorhizal Plants, C. R. Schwintzer and J. D. Tjepkema (eds.), Academic Press, San Diego, 1990, p. 35.
29. P. Simonet, M. C. Grosjean, A. K. Misra, S. Nazaret, B. Cournoyer and P. Normand, Appl. Env. Microbiol. 57 (1991) 3278.
30. S. Jamann, M. P. Fernandez and P. Normand, Mol. Ecol. 2 (1993) 17.
31. B. Cournoyer and P. Normand, Soil Biol. Biochem 26 (1994) 553.
32. D. R. Benson, D. W. Stephens, M. L. Clawson and W. B. Silvester, Appl. Env. Microbiol. 62 (1996) 2904.
33. M. L. Clawson, M. Caru and D. R. Benson, Appl. Env. Microbiol. 64 (1998) 3539.
34. P. Normand, S. Orso, B. Cournoyer, P. Jeannin, C. Chapelon, J. Dawson, L. Evtushenko and A. Misra, Int. J. Syst. Bacteriol. 46 (1996) 1.

35. W. Hönerlage, D. Hahn, K. Zepp, J. Zeyer and P. Normand, Syst. Appl. Microbiol. 17 (1994) 433.
36. D. J. Wolters, C. van Dijk, E. G. Zoetendal and A. D. L. Akkermans, Mol. Ecol. 6 (1997) 971.
37. C. van Dijk and A. Sluimer, New Phytol. 128 (1994) 497.
38. N. R. McEwan, C. T. Wheeler and J. J. Milner, Soil Biol. Biochem. 26 (1994) 541.
39. L. Maggia and J. Bousquet, Mol. Ecol. 3 (1994) 459.
40. E. Lumini, M. Bosco and M. P. Fernandez, FEMS Microbiol. Ecol. 21 (1996) 303.
41. C. Rouvier, J. Schwencke, Y. Prin, E. Navarro, P. Benoist, A. Muller, M. Girgis, S. Selim, P. Reddell, D. Gauthier, G. Rinaudo, P. Normand and P. Simonet, Acta Bot. Gallica 143 (1996) 567.
42. M. P. Lechevalier and H. A. Lechevalier in Bergey's Manual of Systematic Bacteriology Vol. 4, S. T. Williams, M. E. Sharpe and J. G. Holt (eds.), Williams & Wilkins, Baltimore, 1989, p. 2405.
43. M. Lalonde, L. Simon, J. Bousquet and A. Séguin in Nitrogen Fixation: Hundred Years After, H. Bothe, F. J. de Bruijn and W. E. Newton, eds, Fischer, Stuttgart, 1988, p. 671.
44. W. A. Shipton and A. J. P. Burggraaf, Can. J. Bot. 61 (1983) 2783.
45. M. F. Lopez and J. G. Torrey, Arch. Microbiol. 143 (1985) 209.
46. L. Tisa, M. McBride and J. C. Ensign, Can. J. Bot. 61 (1983) 2768.
47. M. F. Lopez, C. S. Whaling and J. G. Torrey, Can. J. Bot. 61 (1983) 2834.
48. M. F. Lopez, M. S. Fontaine, J. G. Torrey, Can. J. Microbiol. 30 (1984) 209.
49. S. Burleigh and J. O. Dawson, Soil Biol. Biochem. 26 (1994) 593.
50. D. R. Benson and N. A. Schultz in The Biology of *Frankia* and Actinorhizal Plants, C. R. Schwintzer and J. D. Tjepkema (eds.), Academic Press, San Diego, 1990, p. 107.
51. K. R. Schubert, Annu. Rev. Plant Physiol. 37 (1986) 539.
52. N. A. Schultz and D. R. Benson, J. Bacteriol. 172 (1990) 1380.
53. J. Edmands, N. A. Noridge and D. R. Benson, Proc. Natl. Acad. Sci. U.S.A. 84 (1987) 6126.
54. Y.-L. Tsai and D. R. Benson, Arch. Microbiol. 152 (1989) 382.
55. X. Zhang and D. R. Benson, Arch. Microbiol. 158 (1992) 256.
56. R. Haselkorn, Annu. Rev. Plant Physiol. 29 (1994) 319.
57. C. T. Wheeler, G. S. Tonin and A. Sutcliffe, Soil Biol. Biochem. 26 (1994) 577.
58. S. Fujihara, H. Abe, Y. Minakawa, S. Akao and T. Yoneyama, Plant Cell Physiol. 35 (1994) 1127.
59. D. G. Aronson and G. L. Boyer, Soil Biol. Biochem. 26 (1994) 561.
60. G. L. Boyer, S. A. Kane, J. A. Alexander and D. B. Aronson, Can. J. Bot. 77 (1999) 1316.
61. P. Haansuu, P. Vuorela and K. Haahtela, Pharm. Pharmacol. Lett. 9 (1999) 1.
62. K. D. Klika, J. P. Haansuu, V. V. Ovcharenko, K. K. Haahtela, P. M. Vuorela and K. Pihlaja, J. Org. Chem. 66 (2001) 4065.

63. D. D. Baker and C. R. Schwintzer in The Biology of *Frankia* and Actinorhizal Plants, C. R. Schwintzer and J. D. Tjepkema (eds.), Academic Press, San Diego, 1990, p. 1.
64. H. G. Diem and Y. R. Dommergues in The Biology of *Frankia* and Actinorhizal Plants, C. R. Schwintzer and J. D. Tjepkema (eds.), Academic Press, San Diego, 1990, p. 317.
65. L. F. Benoit and A. M. Berry in The Biology of *Frankia* and Actinorhizal Plants, C. R. Schwintzer and J. D. Tjepkema eds., Academic Press, San Diego, 1990, p. 281.
66. J. I. Sprent and R. Parsons, Field Crops Res. 65 (2000) 183.
67. J. Schwencke and M. Carú, Arid Land Res Manage 15 (2001) 285.
68. R. Beveridge, T. S. C. Li, B. D. Oomah and A. Smith, J. Agr. Food Chem. 47 (1999) 3480.
69. S. Jana and W. R. Schroeder, J. Sustain. Forest. 10 (2000) 335.
70. N. Jeppsson, Fauna Flora (Stockholm) 95 (2000) 61.
71. S. Racette and J. G. Torrey, Can. J. Bot. 67 (1989) 2873.
72. J. G. Torrey, Am. J. Bot. 63 (1976) 335.
73. D. Callaham and J. G. Torrey, Can. J. Bot. 51 (1977) 2306.
74. D. Callaham, W. Newcomb, J. G. Torrey and R. L. Peterson, Bot. Gaz. (Suppl.) 140 (1979) S1.
75. A. M. Berry, L. McIntyre and M. E. McCully, Can. J. Bot. 64 (1986) 292.
76. R. Heidstra, R. Geurts, H. Franssen, H. P. Spaink, A. van Kammen and T. Bisseling, Plant Physiol. 105 (1994) 787.
77. M. van Ghelue, E. Lovaas, E. Ringo and B. Solheim, Physiol. Plant. 99 (1997) 579.
78. H. Céremonie, F. Debelle and M. P. Fernandez, Can. J. Bot. 77 (1999) 1293.
79. S. Knowlton, A. M. Berry and J. G. Torrey, Can. J. Microbiol. 26 (1980) 971.
80. A. M. Berry and J. G. Torrey, Can. J. Bot. 61 (1983) 2863.
81. M. Lalonde and R. Knowles, Can. J. Bot. 53 (1975) 1951.
82. R. H. Berg, Protoplasma 159 (1990) 35.
83. C. R. Schwintzer, A. M. Berry and L. D. Disney, Can. J. Bot. 60 (1982) 746.
84. A. F. Angulo Carmona, Acta Bot. Neerl. 23 (1974) 257.
85. L. Laplaze, E. Duhoux, C. Franche, T. Frutz, S. Svistoonoff, T. Bisseling, D. Bogusz and K. Pawlowski, Mol. Plant-Microbe Interact. 13 (2000) 107.
86. L. Laplaze, M. C. Bon, M. O. Sy, A. Smouni, C. Allonneau, F. Auguy, T. Frutz, M. Rio, F. Guermache, E. Duhoux, C. Franche and D. Bogusz in Molecular Biology of Woody Plants, Vol 1, S. M. Jain and S. C. Minocha (eds.), Kluwer Academic Publishers, Dordrecht, 2000, p. 269.
87. K. A. VandenBosch and J. G. Torrey, Can. J. Bot. 61 (1983) 2898.
88. D. Hahn, M. J. C. Starrenburg and A. D. L. Akkermans, Plant Soil 107 (1988) 233.
89. D. J. Wolters, C. van Dijk, A. D. L. Akkermans and J. W. Woldendorp, Acta Oecol. 20 (1999) 71.
90. E. H. Pommer, Flora (Jena) 143 (1956) 603.

91. C. van Dijk, PhD thesis, Rijksuniversiteit Leiden, Leiden, 1984.
92. J. Sequerra, A. Capellano, M. Faure-Raynaud and A. Moiroud, Can. J. Bot. 72 (1994) 955.
93. J. Sequerra, A. Capellano, V. Gianinazzi-Pearson and A. Moiroud, New Phytol. 130 (1995) 545.
94. I. M. Miller and D. D. Baker, Protoplasma 128 (1985) 107.
95. A. M. Berry and L. A. Sunell in The Biology of *Frankia* and Actinorhizal Plants, C. R. Schwintzer and J. D. Tjepkema (eds.), Academic Press, San Diego, 1990, p. 61.
96. Q. Liu and A. M. Berry, Protoplasma 163 (1991) 82.
97. C. Valverde and L. G. Wall, New Phytol. 141 (1999) 345.
98. A. Séguin and M. Lalonde, Plant Soil 118 (1989) 221.
99. Q. Liu and A. M. Berry, Protoplasma 164 (1991) 93.
100. G. Caetano-Anolles, Field Crops Res. 53 (1997) 47.
101. J. A. Arnone, S. J. Kohls and D. D. Baker, Soil Biol. Biochem. 26 (1994) 599.
102. S. J. Kohls and D. D. Baker, Plant Soil 118 (1989) 171.
103. K. A. Thomas and A. M. Berry, Plant Soil 118 (1989) 181.
104. L. G. Wall, K. Huss-Danell, Physiol. Plant. 99 (1997) 594.
105. A. Baker and R. Parsons, Physiol. Plant. 99 (1997) 640.
106. L. G. Wall, J. Plant Growth Regul. 19 (2000) 167.
107. Y. Yang, Plant Soil 176 (1995) 161.
108. Y. Yang, W. A. Shipton and P. Reddel, Plant Soil 189 (1997) 75.
109. C. Valverde, L. G. Wall and K. Huss-Danell, Symbiosis 28 (2000) 49.
110. L. G. Wall, A. Hellsten and K. Huss-Danell, Symbiosis 29 (2000) 91.
111. W. Newcomb and C. E. Pankhurst, New Zeal. J. Bot. 20 (1982) 93.
112. F. Hafeez, A. D. L. Akkermans and A. H. Chaudhary, Plant Soil 79 (1984) 383.
113. W. B. Silvester, S. L. Harris and J. D. Tjepkema in The Biology of *Frankia* and Actinorhizal Plants, C. R. Schwintzer and J. D. Tjepkema (eds.), Academic Press, San Diego, 1990, p. 157.
114. J. D. Tjepkema, Am. J. Bot. 70 (1978) 59.
115. P. Schröder, Ann. Sci. Forest. 46 (1989) 310.
116. C. T. Wheeler, J. C. Gordon and T. M. Ching, New Phytol. 82 (1979) 449.
117. S. Mirza, K. Pawlowski, F. Y. Hafeez, A. H. Chaudhary and A. D. L. Akkermans, New Phytol. 126 (1994) 131.
118. W. B. Silvester and S. L. Harris, Plant Soil 118 (1989) 97.
119. W. B. Silvester, J. Whitbeck, J. K. Silvester and J. G. Torrey, Can. J. Bot. 66 (1988) 1762.
120. S. Zeng, J. D. Tjepkema and R. H. Berg, Plant Soil 118 (1989) 119.
121. R. H. Berg and L. McDowell, Can. J. Bot. 66 (1987) 2038.
122. R. H. Berg and L. McDowell, Protoplasma 136 (1987) 104.
123. K. Jacobsen-Lyon, E. Østergaard-Jensen, J.-E. Jørgensen, K. A. Marcker, J. Peacock and E. Dennis, Plant Cell 7 (1995) 213.
124. S. Zeng and J. D. Tjepkema, Soil Biol. Biochem. 5 (1994) 633.
125. S. M. Pathirana and J. D. Tjepkema, Plant Physiol. 107 (1995) 827.

126. U. K. J. Suharjo and J. D. Tjepkema, Physiol. Plant. 95 (1995) 247.
127. W. B. Silvester, B. Langenstein and R. H. Berg, Can. J. Bot. 77 (1999) 1358.
128. J. D. Tjepkema, G. Y. Du and C. R. Schwintzer, Can. J. Bot. 77 (1999) 1367.
129. F. R. Minchin, Soil Biol. Biochem 29 (1997) 881.
130. L. Laplaze, H. Gherbi, T. Frutz, K. Pawlowski, C. Franche, J.-J. Macheix, F. Auguy, D. Bogusz and E. Duhoux, Plant Physiol. 121 (1999) 113.
131. E. de Neergaard, Can. J. Plant Pathol. 11 (1989) 28.
132. M. Lummerzheim, D. D. Oliveira, C. Casresana, F. C. Miguens, E. Louzada, D. Roby, M. van Montagu and B. Timmerman, Mol. Plant-Microbe Interact. 6 (1993) 532.
133. K. A. Spanos and S. Woodward, Eur. J. Forest Pathol. 27 (1997) 13.
134. C. van Dijk, A. Sluimer-Stolk and F. Weber, Physiol. Plant. 72 (1988) 349.
135. C. H. Guan, D. J. Wolters, C. van Dijk, A. D. L. Akkermans, A. van Kammen, T. Bisseling and K. Pawlowski, Acta Bot. Gallica 143 (1996) 613.
136. K. Pawlowski, D. J. Wolters, C. van Dijk and T. Bisseling in From Symbiosis to Eukaryotism: ENDOCYTOBIOLOGY VII, E. Wagner, J. Normann, H. Greppin, J. H. P. Hackstein, K. V. Kowallik, H. E. A. Schenk and H. Seckbach (eds.), University of Freiburg/University of Geneva, 1999, p. 241.
137. C. Franche, D. Diouf, L. Laplaze, F. Auguy, T. Frutz, M. Rio, E. Duhoux and D. Bogusz, Mol. Plant-Microbe Interact. 11 (1998) 887.
138. C. R. Andersson, D. J. Llewellyn, W. J. Peacock and E. S. Dennis, Plant Physiol. 113 (1997) 45.
139. S. A. Lancelle and J. G. Torrey, Protoplasma 123 (1984) 26.
140. S. A. Lancelle and J. G. Torrey, Can. J. Bot. 63 (1984) 25.
141. D. E. Soltis, P. S. Soltis, D. R. Morgan, S. M. Swensen, B. C. Mullin, J. M. Dowd and P. G. Martin, Proc. Natl. Acad. Sci. U.S.A. 92 (1995) 2647.
142. A. Ribeiro, A. D. L. Akkermans, A. van Kammen, T. Bisseling and K. Pawlowski, Plant Cell 7 (1995) 785.
143. K. Pawlowski, P. Twigg, S. Dobritsa, C. Guan and B. C. Mullin, Mol. Plant-Microbe Interact. 10 (1997) 656.
144. D. Berger and T. A. Altmann, Genes Dev. 14 (2000) 1119.
145. L. Laplaze, A. Ribeiro, C. Franche, E. Duhoux, F. Auguy, D. Bogusz and K. Pawlowski Mol. Plant-Microbe Interact. 13 (2000) 113.
146. P. A. Okubara, N. A. Fujishige, A. M. Hirsch and A. M. Berry, Plant Physiol. 122 (2000) 1073.
147. P. A. Okubara, K. Pawlowski, T. M. Murphy and A. M. Berry, Plant Physiol. 120 (1999) 411.
148. M. van Ghelue, A. Ribeiro, B. Solheim, A. D. L. Akkermans, T. Bisseling and K. Pawlowski, Mol. Gen. Genet. 250 (1996) 437.
149. C. H. Guan, A. Ribeiro, A. D. L. Akkermans, Y. X. Jing, A. van Kammen, T. Bisseling and K. Pawlowski, Plant Mol. Biol. 32 (1996) 1177.
150. C. T. Wheeler and A. C. Lawrie in Symbiotic Nitrogen Fixation in Plants, P. S. Nutman (ed.), Cambridge University Press, Cambridge, 1976, p. 497.
151. C. R. Wheeler, S. H. Watts and J. R. Hillman, New Phytol. 95 (1983) 209.

152. D. Burgess and R. L. Peterson, Can. J. Bot. 65 (1987) 1647.
153. J. van Straten, A. D. L. Akkermans and W. Roelofsen, Nature 266 (1977) 437.
154. P.-Å. Vikman and K. Huss-Danell, Physiol. Plant. 70 (1987) 349.
155. P.-O. Lundquist and K. Huss-Danell, New Phytol.119 (1991) 121.
156. M. F. Lopez, P. Young and J. G. Torrey, Can. J. Microbiol. 32 (1986) 353.
157. J. G. Streeter, Adv. Bot. Res. 18 (1991) 129.
158. A. D. L. Akkermans, K. Huss-Danell and W. Roelofsen, Physiol. Plant. 53 (1981) 289.
159. K. Huss-Danell, W. Roelofsen, A. D. L. Akkermans and P. Meijer, Physiol. Plant. 54 (1982) 461.
160. P.-Å. Vikman and K. Huss-Danell, J. Exp. Bot. 42 (1991) 221.
161. A. Baker, G. F. Hill and R. Parsons, L. J. Exp. Bot. 48 (1997) 67.
162. A. D. L. Akkermans, W. Roelofsen, J. Blom, K. Huss-Danell and R. Harkink, Can. J. Bot. 61 (1983) 2793.
163. P.-O. Lundquist and K. Huss-Danell, FEMS Microbiol. Lett. 91 (1992) 141.
164. J. K. Waters, B. L. Hughes, L. C. Purcell, K. O. Gerhardt, T. P. Mawhinney and D. W. Emerich, Proc. Natl. Acad. Sci. U.S.A. 95 (1998) 12038.
165. D. Allaway, E. M. Lodwig, L. A. Crompton, M. Wood, R. Parsons, T. R. Wheeler and P. S. Poole, Mol. Microbiol. 36 (2000) 508.
166. A. Sellstedt and P. Lindblad, Plant Physiol. 92 (1990) 809.
167. U. Mattsson, L. Johansson, G. Sandstrom and A. Sellstedt, Curr. Microbiol. 42 (2001) 438.
168. A. Sellstedt, Physiol. Plant. 75 (1989) 304.
169. S. F. Zitzer and J. O. Dawson, Tree Physiol. 5 (1989) 185.
170. A. Sellstedt, K. Huss-Danell and A. S. Ahlqvist, Physiol. 66 (1986) 99.
171. S. M. Swensen and B. C. Mullin, Physiol. Plant. 99 (1997) 565.
172. S. M. Swensen, Am. J. Bot. 83 (1996) 1503.
173. D. R. Benson and M. L. Clawson in Prokaryotic Nitrogen Fixation: A Model System for Analysis of a Biological Process, E. W. Triplett (ed.), Horizon Scientific Press, Symondham, 2000, p. 207.
174. S. C. Jeong, N. J. Ritchie and D. D. Myrold, Mol. Phylogenet. Evol. 13 (1999) 493.
175. N. J. Ritchie and D. D. Myrold, Can. J. Bot. 77 (1999) 1208.
176. P. Normand, B. Cournoyer, P. Simonet and S. Nazaret, Gene 111 (1992) 119.
177. B. Cournoyer and C. Lavire, FEMS Microbiol. Lett.. 177 (1999) 29.
178. J. Muller, Bot. Rev. 47 (1981) 1.
179. L. Maggia, S. Nazaret and P. Simonet, Acta Oecol. 13 (1992) 453.
180. S. Nazaret, B. Cournoyer, P. Normand and P. Simonet, J. Bacteriol. 173 (1991) 4072.
181. M. A. Murry, A. S. Konopka, S. D. Pratt and T. L. Vandergon, Physiol. Plant. 99 (1997) 714.
182. G. W. O'Hara, Aust. J. Exp. Agr. 41 (2001) 417.
183. M. Parniske, Curr. Opin. Plant Biol. 3 (2000) 320.
184. M. Béreau and J. Garbaye, Ann. Sci. Forest. 51 (1994) 407.

Nitrogen Fixation at the Millennium.
G. Jeffery Leigh, Editor.

CHAPTER 8

Alternative Nitrogenases

B. Masepohl [a], K. Schneider [b], T. Drepper [a], A. Müller [b], and W. Klipp [a]

[a] Lehrstuhl für Biologie der Mikroorganismen, Ruhr-Universität Bochum, D-44780 Bochum, Germany.

[b] Lehrstuhl für Anorganische Chemie I der Universität Bielefeld, D-33501 Bielefeld, Germany.

1. INTRODUCTION

It was like a change in paradigm when it became evident in the late 1980s that three distinct types of nitrogenases, and not only the long known molybdenum (Mo)-containing system (Mo nitrogenase), exist in nature. As long ago as in 1936, Bortels [1] showed that low concentrations of vanadium (V) stimulate diazotrophic growth of *Azotobacter* species. The possibility that V could substitute for Mo was not excluded, however other hypotheses were favoured, *e.g.*, that binding of V to the conventional Mo nitrogenase would lead to a stabilization of the enzyme and thus to a more effective utilization of the small amounts of Mo found in Mo-starved cells [2]. Slow diazotrophic growth in media lacking both Mo and V were usually attributed to the incorporation of trace amounts of contaminating Mo into nitrogenase. Stronger evidence for the existence of at least one Mo-independent nitrogenase system was presented in 1980 by Bishop and coworkers [3], who found that *Azotobacter vinelandii* Nif^- point mutants (defective in synthesis of the *nif*-encoded Mo nitrogenase) are able to fix nitrogen in the absence of Mo. Although this observation was confirmed by other publications (*e.g.*, [4]), the proposal that a Mo-independent route of biological N_2 fixation should exist, was received with great scepticism, because it challenged the long-held dogma that Mo was absolutely required for N_2 fixation. The possibility that the observed capacity to fix nitrogen in the absence of Mo resulted from the leakiness of the point mutants used in the intial studies rather than from derepression of an alternative nitrogenase, was ruled out when mutant strains carrying stable deletions in the structural genes of the Mo nitrogenase (*nifHDK*) were shown to undergo the same phenotypic reversal [5].

Alternative nitrogenases containing vanadium instead of molybdenum (V nitrogenase) were first (1986) isolated from *Azotobacter chroococcum* [6] and *A. vinelandii* [7]. Two years later a second type of alternative nitrogenase, lacking both Mo and V (Fe nitrogenase or Fe-only nitrogenase), was isolated from *A. vinelandii* as well [8]. The latter enzyme system has afterwards also been isolated from phototrophic bacteria, *i.e., Rhodobacter capsulatus* [9] and *Rhodospirillum rubrum* [10]. Detailed biochemical studies on isolated alternative nitrogenases have been so far restricted to these few enzyme systems. However, molecular genetic studies revealed that Mo-independent nitrogenases are widely distributed among diazothrophic eubacteria and archaea ([11,12]; Table 1; see below).

Species	Characteristics	Mo nitrogenase (*nif*-encoded)	V nitrogenase (*vnf*-encoded)	Fe nitrogenase (*anf*-encoded)
Klebsiella pneumoniae	enterobacterium (γ proteobacterium)	+	–	–
Azotobacter vinelandii	soil bacterium (γ proteobacterium)	+	+	+
Azotobacter chroococcum	soil bacterium (γ proteobacterium)	+	+	–
Rhodobacter capsulatus	phototrophic α proteobacterium	+	–	+
Anabaena variabilis	heterocyst-forming cyanobacterium	+ / + (*nif1* / *nif2*) [a]	+	–
Clostridium pasteurianum	Gram-positive firmibacterium	+	–	+
Methanosarcina barkeri	archaea	+ (*nif2*) [b]	+ (*nif1*) [b]	–

[a] *A. variabilis* harbors two genetically distinct sets of *nif* genes coding for Mo nitrogenases.
[b] The *M. barkeri nif2* and *nif1* genes most likely encode a Mo and a V nitrogenase, respectively.

Table 1. Occurence of Mo-containing and alternative Mo-independent nitrogenases.

DNA sequencing established that all three nitrogenase systems – although clearly related – are based on genetically distinct systems and do not simply arise from the substitution of metals. Whereas the conventional Mo nitrogenase is encoded by the *nif* (**ni**trogen **f**ixation) genes, the genes encoding the V nitrogenase are called *vnf* (vanadium-dependent **n**itrogen **f**ixation) genes [13], and the genes encoding the Fe nitrogenase have been designated as *anf* (**a**lternative **n**itrogen **f**ixation) genes [14]. It should be emphasized at this point that both the V and the Fe nitrogenases are "alternative" nitrogenases in the sense that they are Mo-independent.

Like the conventional Mo nitrogenase, both the V and the Fe nitrogenases consist of two dissociable metalloprotein components called "component 1" or "dinitrogenase" (MoFe protein, VFe protein, FeFe protein) and "component 2" or "dinitrogenase reductase" (Fe protein). The apo-MoFe protein (the *nifDK* product, NifDK) exhibits a tetrameric structure ($\alpha_2\beta_2$). Both the VFe protein (VnfDGK) and the FeFe protein (AnfDGK) contain additional subunits (δ, encoded by the structural genes *vnfG* and *anfG*, respectively) resulting in hexameric structures ($\alpha_2\beta_2\delta_2$). The two unique metal clusters, the cofactor (also called "M cluster") and the electron-transfering "P cluster", are both located within the dinitrogenase component of each system.

The dinitrogenase reductase (NifH, VnfH, AnfH), responsible for the ATP-dependent electron transfer to the corresponding dinitrogenase, has a comparable structure in all nitrogenase systems. It is a dimer of two identical subunits (γ_2), and contains a conventional $[Fe_4S_4]$ cluster bridging the two subunits.

The prokaryotic species shown in Table 1 represent only a selected few of well-characterized diazotrophic organisms. Even closely related species such as *A. vinelandii* and *A. chroococcum* may differ in their capacity to synthesize alternative nitrogenases. In addition to the bacteria shown in Table 1, many other species containing alternative nitrogenases have been identified, as inferred from the presence of the *vnfG* and *anfG* genes unique to the alternative nitrogenase systems [12]. It is worth noting that until now no bacterium has been described containing an alternative nitrogenase but not containing a conventional Mo nitrogenase.

A novel type of nitrogenase unrelated to the Nif, Vnf and Anf systems, which involves a heterotrimeric dinitrogenase, has recently been reported to be present in *Streptomyces thermoautotrophicus* [15]. Electrons for N_2 reduction are derived from superoxide oxidation coupled to CO oxidation. In contrast to the conventional and the alternative nitrogenases, all components of the superoxide-dependent nitrogenase are completely insensitive to O_2. The *S. thermoautotrophicus* enzyme system will be described in detail in Chapter 10 of this book.

2. ORGANIZATION OF NITROGEN FIXATION GENES (*nif*, *vnf* AND *anf* GENES)

2.1. *Klebsiella pneumoniae nif* genes and their function in nitrogen fixation

K. pneumoniae was the first diazotroph in which the nitrogen fixation process was analyzed at the molecular level (starting in the early 1970s), primarily because most of the genetic tools developed for *Escherichia coli* could be used in this closely related enterobacterium. With respect to the topic of this chapter, *K. pneumoniae* is a "poor" diazotroph, since it is able to synthesize only a conventional Mo-containing (but no alternative) nitrogenase. However, *K. pneumoniae* served for a long time as a paradigm, and the nomenclature of nitrogen-fixation genes is based on its *nif* genes. Therefore, we will briefly summarize the genetic organization of the *K. pneumoniae nif* genes and the functions of their products (also see Chapter 4).

The *K. pneumoniae nif* regulon consists of 20 genes arranged in seven transcriptional units, namely *nifJ*, *nifHDKTY*, *nifENX*, *nifUSVWZM*, *nifF*, *nifLA*, and *nifBQ* (Figure 1; [16,17]). As mentioned above, the proteins that comprise nitrogenase are encoded by the *nifH*, *nifD* and *nifK* genes. The products of several genes, including *nifE*, *nifN*, *nifX*, *nifB*, and *nifQ*, are involved in biosynthesis of the iron-molybdenum cofactor (FeMoco) of nitrogenase. In addition, the *nifV*-encoded homocitrate synthase provides homocitrate, which is an essential component of FeMoco. NifY may have a role in the insertion of FeMoco into apo-dinitrogenase. The *nifU* and *nifS* gene products are involved in the synthesis of iron-sulfur clusters, whereas the *nifM* gene product is required for maturation of dinitrogenase reductase. In *K. pneumoniae*, electron transfer from pyruvate to nitrogenase reductase is mediated through a pyruvate:flavodoxin oxidoreductase (NifJ), and a flavodoxin (NifF). The product of *nifA* is a transcriptional activator that is required for the expression of all other *nif* genes, whereas the product of *nifL* negatively regulates the activity of NifA in response to oxygen and fixed nitrogen.

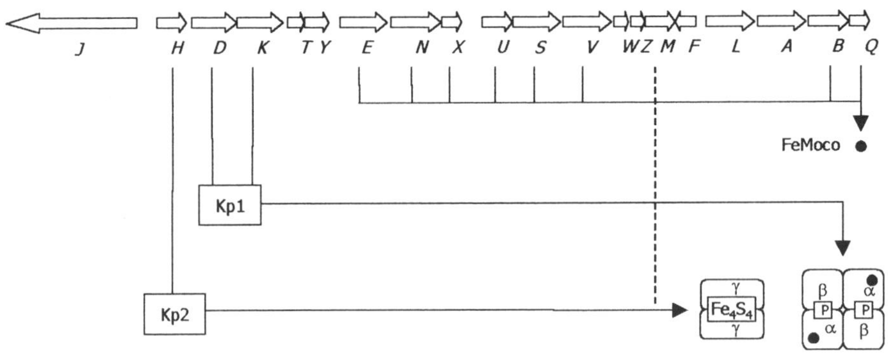

Figure 1. The *K. pneumoniae nif* gene cluster. Arrows indicate sizes and directions of transcription of the *K. pneumoniae nif* genes. The roles of the corresponding products are indicated in the lower part of the figure. The [Fe_4S_4] cluster bridging the subunits of component 2 (Kp2), the localizations of the P clusters at the αβ interfaces of component 1 (Kp1) and the localizations of the FeMo cofactors within the α subunit of Kp1 are shown schematically. For details, see text.

2.2 Organization of *vnf* and *anf* genes in selected bacteria

A. vinelandii harbors three genetically distinct nitrogenases encoded by the *nif*, *vnf* and *anf* genes [18,19]. With the exception of *nifJ*, homologs of the 19 remaining *K. pneumoniae nif* genes have been identified in two clusters in the chromosome of

A..vinelandii [20-22]. The *A. vinelandii nif* genes have the same sequential arrangement as the corresponding *nif* genes from *K. pneumoniae*. In addition, twelve other open reading frames (*orfs*) whose expression is subject to *nif*-specific regulation are interspersed among the *A. vinelandii nif* genes. Although at least ten of these *orfs* seem to be dispensable for diazotrophic growth in *A. vinelandii* [21], several have highly conserved counterparts in other diazotrophic bacteria.

Compared to the number of *nif* genes, relatively few *vnf* genes have been identified in *A. vinelandii* (Figure 2; [18,19]), including the structural genes of the V nitrogenase *vnfH*, *vnfD*, *vnfG*, and *vnfK*. Mutational analysis revealed that (beside *vnfH*, *vnfD* and *vnfK*) the *vnfG* gene (coding for the additional subunit δ of the V dinitrogenase) is required for Mo-independent diazotrophic growth *via* the V nitrogenase [23]. In contrast to the structural genes of the Mo nitrogenase (*nifHDK*) and the Fe nitrogenase (*anfHDGK*, see below), which are cotranscribed from promoters located upstream of *nifH* and *anfH*, respectively, the *A. vinelandii vnfH* gene does not form a transcriptional unit with *vnfDGK*. The *vnfH* gene forms an operon with the *vnfFd* gene which codes for a ferredoxin-like protein [24]. A *vnfFd* mutant is unable to fix nitrogen under conditions in which the V nitrogenase operates, and therefore, the possibility that the product of this gene might be involved in transfer of electrons to the *vnfH*-encoded dinitrogenase reductase has been discussed [24].

The *A. vinelandii vnfENX* genes appear to be organized as a single transcriptional unit [25]. Mutants defective for *vnfE* and *vnfN* exhibit Vnf $^+$ and Anf $^+$ phenotypes, respectively, as indicated by diazotrophic growth in both molybdenum-deficient medium containing vanadium or in medium lacking both Mo and V. However, a double mutant which contains a *nifEN* deletion and an insertion in *vnfE* is unable to grow diazotrophically in Mo-deficient medium, with or without V. This suggests that NifE and NifN substitute for VnfE and VnfN when the *vnfEN* genes are inactivated.

The *A. vinelandii vnfA* gene is located upstream of the *vnfENX* operon (Figure 2; [26]). The *vnfA* gene codes for a NifA-like transcriptional activator which is not only required for the synthesis of the V nitrogenase but also for expression of several *nif* genes under conditions in which the *vnf* pathway operates (see below).

The *A. vinelandii anfHDGKOR* operon codes for the apo-proteins of dinitrogenase reductase (AnfH) and dinitrogenase (AnfDGK) of the Fe nitrogenase and for two proteins of unknown function (AnfO, AnfR; Figure 2; [14]). *A. vinelandii* strains carrying mutations in either of these six genes are unable to grow with N_2 under conditions in which the Fe nitrogenase operates [14,23]. Mutants carrying in-frame deletions in *anfO* and *anfR* accumulate the subunits of the Fe nitrogenase, and these mutants, which obviously cannot reduce N_2 to ammonia, are still able to reduce the artificial substrate acetylene *via* the heterometal-free system [27]. Similar observations have been obtained for *anfG* and *vnfG* mutants, which reduce acetylene to different degrees *via* the Fe or the V nitrogenase [23]. Although these phenotypes somehow resemble that of *K. pneumoniae nifV* mutants, which are unable to synthesize homocitrate and thus produce a modified form of FeMoco [28], the molecular mechanisms leading to these changes in catalytic features are different.

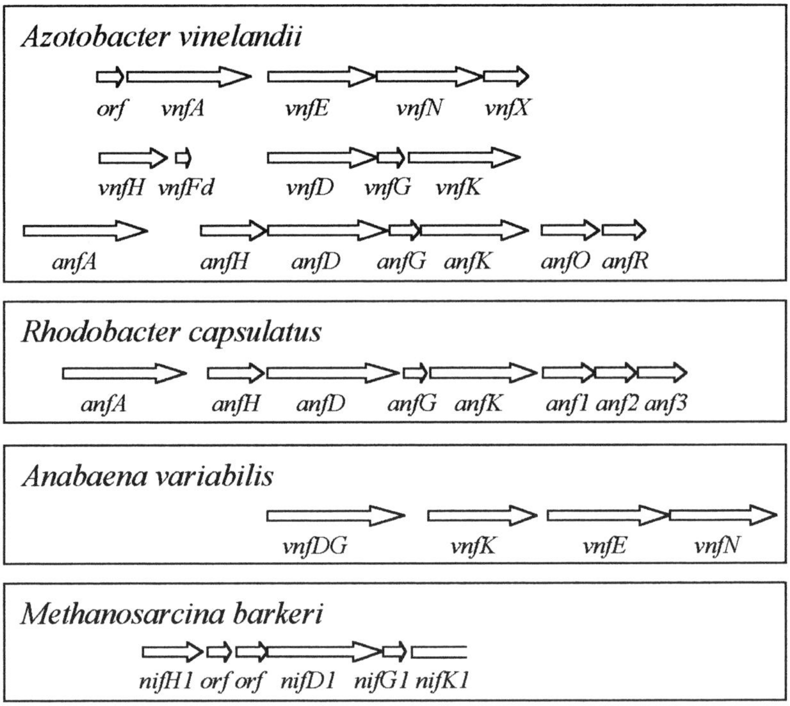

Figure 2. Comparion of *vnf* and *anf* gene regions in selected bacteria. Mo-independent nitrogenases are characterized by an additional subunit (δ, encoded by *vnfG* and *anfG*, respectively) not present in Mo nitrogenases. In the cyanobacterium *A. variabilis*, the *vnfG* gene is fused to *vnfD*. The *orf* upstream of *A. vinelandii vnfA* codes for a protein exhibiting similarity to the C-terminal domain of NifU proteins. Genes most likely coding for a V-containing nitrogenase in *M. barkeri* are designated *nif1* genes. The presence of two *glnB*-like *orfs* separating *nifH1* and *nifD1* seems to be a characteristic feature of archaea.

The *A. vinelandii anfA* gene is located upstream of the *anfHDGKOR* operon (Figure 2; [26]). It codes for a NifA-like transcriptional activator required for expression of the Fe nitrogenase and several *nif* genes under conditions in which the *anf* pathway operates (see below).

Beside the alternative nitrogenase systems of *A. vinelandii*, the *anf*-encoded nitrogenase of *R. capsulatus* (an organism devoid of a V nitrogenase) is well-characterized both genetically and biochemically. In contrast to *A. vinelandii*, the genomic sequence of *R. capsulatus* is almost complete [29]. The *R. capsulatus anf*

genes are also organized in two operons, namely *anfA* and *anfHDGK-1-2-3* (Figure 2; [30]). The *R. capsulatus anf1* and *anf2* genes are homologous to the *A. vinelandii anfO* and *anfR* genes, respectively. Mutants defective for the *anf3* gene, which seems to have no counterpart in *A. vinelandii*, exhibit the same phenotype as *anf1* (*anfO*) and *anf2* (*anfR*) mutant strains (K.-U. Riedel, A. Lindackers, and W. Klipp, unpublished results).

The filamentous heterocyst-forming cyanobacterium *Anabaena variabilis* harbors two Mo nitrogenases (encoded by the so-called *nif1* and *nif2* genes, respectively; see below) and a *vnf*-encoded V nitrogenase (Table 1; [31-33]). One Mo nitrogenase (*nif1*-encoded) as well as the V nitrogenase functions only in heterocysts under either aerobic or anaerobic growth conditions, whereas the second Mo nitrogenase (*nif2*-encoded) functions only under anaerobic conditions in vegetative cells or heterocysts. In contrast to *A. vinelandii*, the *A. variabilis vnfEN* genes are located in close proximity to *vnfK* (Figure 2). It should be emphasized that the *A. variabilis vnfG* gene is fused to *vnfD* [31]. This may have implications on the role of VnfG concerning insertion of the FeV cofactor into apo-dinitrogenase (see below).

Within the archaea, nitrogen fixation has been found only in methanogenic species [34]. *Methanosarcina barkeri* harbors a Mo nitrogenase (*nif2*-encoded) and a V nitrogenase (*nif1*-encoded) [11]. Therefore, the *nif1/nif2* nomenclature of *M. barkeri* differs from that of *A. variabilis*. As described above for eubacterial species, *M. barkeri* also contains a *vnfG*-like gene (*nifG1*) indicative for alternative nitrogenases (Figure 2). However, the *nifH1* and *nifD1* genes are separated by two *glnB*-like *orfs* coding for putative signal transduction proteins. Similarly, two *glnB*-like genes (designated *nifI1* and *nifI2*) are located between the *nifH* and *nifD* genes in *Methanococcus maripaludis* [35]. Mutational analysis revealed that both NifI1 and NifI2 are required for ammonia switch-off of nitrogen fixation.

2.3 Products of nif genes required for alternative nitrogenases

In addition to the *vnf*- and *anf*-specific genes described above, five *nif* genes (*nifB*, *nifU*, *nifS*, *nifV*, and *nifM*) are required for the function of all three nitrogenases suggesting that neither *vnf* nor *anf* analogoues of these genes are present in *A. vinelandii* [36-38]. Since *nifB* (involved in synthesis of NifBco, the iron- and sulfur-containing precursor of FeMoco) is essential for activity of all three nitrogenases, it seems likely that NifBco is the FeS precursor of the cofactors of both the V dinitrogenase (VFeco) and the Fe dinitrogenase (FeFeco). Furthermore, the requirement of the *nifV* gene product (homocitrate synthase) for activity of all three nitrogenases suggests that homocitrate is not only present in FeMoco but also in FeVco and FeFeco (see below). The products of *nifS* (cysteine desulfurase supplying inorganic sulfide) and *nifU* (involved in mobilization of Fe) are required to supply the enlarged demand of [Fe_2S_2] clusters for formation of the FeS clusters present in the nitrogenase component proteins

[39]. Housekeeping homologs of both NifU and NifS designated IscU and IscS, respectively, are found not only in *A. vinelandii* and other diazotrophs but also in non-nitrogen fixing organisms.

As in *A. vinelandii*, homologs of most of the *K. pneumoniae nif* genes have been identified in *R. capsulatus* (Figure 3; [40]). *R. capsulatus nif* genes are also clustered, but the structural genes of the Mo nitrogenase *nifHDK* are separated in the chromosome from the majority of the other *nif* genes. A peculiarity of *R. capsulatus* is the duplication of several *nif* genes (*nifA*, *nifB*, *nifU*). Only strains carrying mutations in both copies of either *nifA* or *nifB* exhibit a Nif $^{-}$ phenotype demonstrating that both copies are functional and the corresponding gene products can substitute for each other.

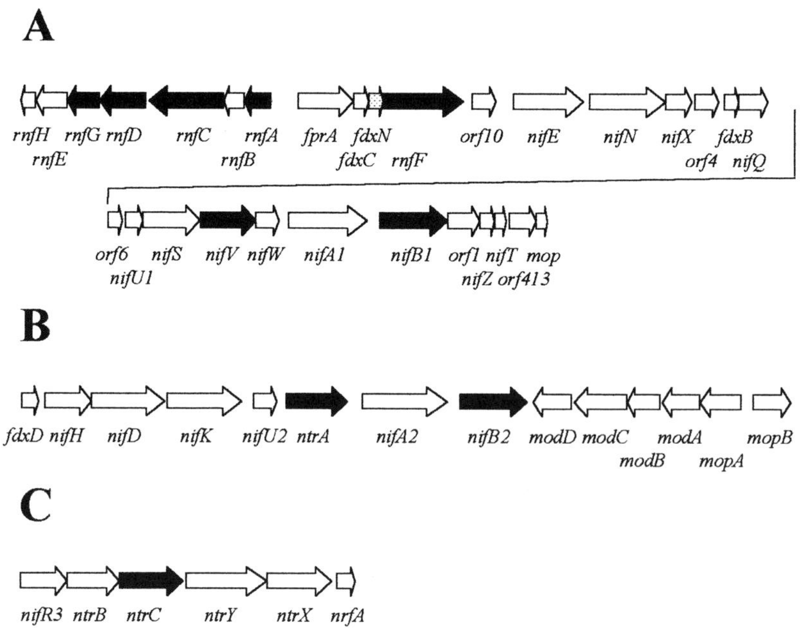

Figure 3. *R. capsulatus nif* genes required for N_2 fixation by the alternative nitrogenase. *R. capsulatus* contains three *nif* gene clusters (A-C) separated in the chromosome. Filled arrows indicate genes essential for synthesis or activity of the alternative Fe nitrogenase. Mutations in *fdxN* result in drastic reduction of diazotrophic growth mediated by either the Mo or the Fe nitrogenase in *R. capsulatus*.

Several *orfs* are found interspersed among the *R. capsulatus nif* genes, some of which have homologs in *A. vinelandii* but not in *K. pneumoniae*. The *rnf* (***Rhodobacter*** **n**itrogen **f**ixation) genes, which belong to the nitrogen fixation regulon, encode a membrane-bound complex involved in electron transport to nitrogenase [41-43]. The product of the *fdxN* gene, a ferredoxin carrying two [Fe_4S_4] clusters is thought to be the electron donor for nitrogenase reductase. In addition, *R. capsulatus* harbors a flavodoxin-encoding *nifF*-like gene, which is not part of the *nif* gene clusters shown in Figure 3 [44]. The *nifF* gene is expressed under nitrogen-limiting conditions both in the presence and absence of molybdenum, and therefore, it has been suggested that NifF may be able to donate electrons not only to the Mo but also to the Fe nitrogenase. NifF contributes significantly to Mo nitrogenase activity under normal nitrogen-fixing conditions (in the presence of iron), but it is absolutely required for nitrogen fixation under iron limitation.

Several of the *R. capsulatus nif* genes (marked in Figure 3) are also required for synthesis and/or activity of the alternative Fe nitrogenase [30]. Among these are the *nifB* and *nifV* genes known to be involved in FeMoco biosynthesis, confirming that NifBco is likely to be either identical to the Fe cofactor (FeFeco) or the precursor of FeFeco, and that FeFeco contains homocitrate. In contrast to *A. vinelandii*, *nifS* is not essential for activity of the Fe nitrogenase in *R. capsulatus* [30]. This can perhaps be explained by the presence of three other *nifS*-like genes in *R. capsulatus* which may substitute for *nifS* under conditions in which the Anf system operates.

Furthermore, the gene products of *rnfA*, *rnfC*, *rnfD*, *rnfG*, and *rnfF* are involved in electron transport to the *anfH*-encoded dinitrogenase reductase [30]. Interestingly, RnfB, which is essential for activity of the Mo nitrogenase, is not required for the alternative system. In an *fdxN* mutant the activity of both nitrogenase systems is dramatically reduced.

In addition, both nitrogenase systems share the regulatory genes *ntrC* and *ntrA* [30]. NtrC is the transcriptional activator for *nifA1*, *nifA2* and *anfA* (see below). In turn, NifA activates transcription of all *nif* genes and AnfA activates all *anf* and selected *nif* genes required for alternative nitrogen fixation. The transcriptional activator proteins NifA and AnfA initiate transcription in concert with RNA polymerase containing the sigma factor NtrA (σ^{54} or RpoN).

As mentioned above, *A. vinelandii* strains carrying mutations in either *nifEN* or *vnfEN* are able to grow diazotrophically *via* both alternative nitrogenase systems, whereas corresponding double mutants can no longer grow with N_2 as sole nitrogen source, indicating that NifEN can substitute for VnfEN [25]. Similarly, *nifEN* are dispensable for activity of the Fe nitrogenase of *R. capsulatus* [30]. However, *R. capsulatus* does not seem to harbor other *nifEN*-like genes (as revealed by analysis of the genomic sequence), leaving the open question of how FeFeco is built and/or inserted into the *anfDGK*-encoded apo-dinitrogenase. It is worth noting that, in contrast to other diazotrophic bacteria, *R. capsulatus nifE* mutants are able to accumulate an active Mo nitrogenase containing its FeMoco, albeit at a clearly reduced level compared to the wild-type [45]. Therefore, NifEN may not be essential for biosynthesis of the cofactor, but its presence guarantees formation of a higher content of intact cofactor-containing dinitrogenase.

3. REGULATION OF ALTERNATIVE NITROGENASE SYSTEMS

3.1. Environmental factors regulating synthesis and activity of alternative nitrogenases

Nitrogen fixation is a highly energy-demanding process, since the nitrogenase enzymes (both Mo-containing and Mo-independent systems) have an absolute requirement for ATP and the reduction of 1 mole of N_2 to ammonia utilizes (at least) 16 moles of ATP. Therefore, diazotrophic bacteria tightly regulate synthesis and actitivity of nitrogenase, in response to a number of environmental factors including (i) availability of ammonium as the preferred source of fixed nitrogen, (ii) environmental oxygen partial pressure, (iii) availability of molybdenum and vanadium, (iv) the C/N ratio at which carbon and nitrogen sources are consumed, and (v) in the case of photosynthetic bacteria, light intensity. The influence of the latter two environmental factors will not be discussed in detail in this review. Furthermore, other factors like iron availability [44] and temperature [46-48] also significantly influence nitrogen fixation.

Ammonium control is predominantly exerted at the level of transcription of nitrogen fixation genes. In proteobacteria (*e.g., K. pneumoniae, R. capsulatus*) the cellular nitrogen status is measured by the nitrogen regulation (Ntr) system mediating expression of nitrogen fixation genes *via* transcriptional activation of the regulatory genes *nifA* and *anfA* (see below). However, ammonium may also control the activity of the respective transcriptional activator proteins NifA and AnfA. As mentioned above, in *K. pneumoniae* (and *A. vinelandii*) NifL negatively regulates the activity of NifA in response to fixed nitrogen and oxygen. In other species (*e.g., Azospirillum brasilense, Herbaspirillum seropedicae, R. capsulatus*), which do not contain *nifL*-like genes, NifA activity is inhibited by ammonium *via* different mechanism(s) [49-51]. Several bacteria (*e.g., Rhodospirillum rubrum, R. capsulatus, A. brasilense*) harbor a further regulatory system (DraT/DraG) controlling the activity of nitrogenase *via* reversible ADP-ribosylation of nitrogenase reductase (NifH, AnfH) in response to ammonium and other environmental signals (see below).

In cyanobacteria, which do not contain an Ntr-like system, nitrogen control involves the regulatory protein NtcA, which can act both as a transcriptional activator and a repressor [52]. Archaea (*e.g. Methanococcus maripaludis*) harbor a mechanism for repression of nitrogen fixation genes in the presence of ammonium [34].

Due to the extreme oxygen sensitivity of nitrogenase, the environmental oxygen partial pressure is a major regulatory factor for most diazotrophs. One strategy adopted by these organisms is to synthesize nitrogenase only under anaerobic (or microaerobic) conditions (as realized for example in *K. pneumoniae* and *R. capsulatus*), but some bacteria have developed strategies allowing them to synthesize nitrogenase under high (external) oxygen partial pressure. O_2-protective function of a small $[Fe_2S_2]$ protein [53] and respiratory protection of nitrogenase in *Azotobacter* species [54] have been postulated, but other factors such as the supply of ATP and reducing equivalents may be even more important for stability of nitrogenase in the presence of high external oxygen concentrations [55]. In some filamentous cyanobacteria (*e.g.*, *Anabaena variabilis*) expression of nitrogen fixation genes is a developmentally regulated process leading to synthesis of nitrogenase in specialized cells called heterocysts, which provide an anaerobic environment [31-33]. As mentioned above, a peculiarity of *A. variabilis* is that it contains two genetically distinct Mo nitrogenases and in addition, a V system (Table 1; [32,33]).

As shown in Table 1, the occurence of the three types of nitrogenase is different in different diazotrophic bacteria. *K. pneumoniae* contains only a Mo nitrogenase, *A. chroococcum* contains a Mo and a V nitrogenase, *R. capsulatus* contains a Mo and a Fe (but no V) nitrogenase, and *A. vinelandii* contains all three nitrogenase systems. Consistently, dependent on the nitrogenase content, there are certain differences in the transcriptional regulation of nitrogen fixation genes by metals. However, there is one common feature concerning metal regulation of Mo-independent nitrogenases, namely the synthesis of alternative systems is repressed by molybdenum.

In *K. pneumoniae*, molybdenum does not seem to be essential for expression of the structural genes of nitrogenase *nifHDK*, but Mo is required for maximal expression [56]. In *A. vinelandii*, Mo is essential for expression of *nifHDK* [57]. On the other hand, Mo represses transcription of *vnfA* and *anfA*, and accordingly both alternative nitrogenase systems are not synthesized [57,58]. In the absence of Mo, V represses transcription of *anfA*. The Mo and the V nitrogenases themselves are somehow involved in metal regulation, since deletion of *nifHDK* allows low expression of *anfHDGK* in the presence of Mo, and deletion of *vnfDGK* allows full expression of the Fe nitrogenase structural genes in the presence of V [57]. The dinitrogenase reductase component of V nitrogenase (VnfH) is required for maximal expression of *anfHDGK*, and mutant strains that lack both NifH and VnfH are unable to grow diazotrophically in Mo- and V-deficient medium [59].

In *R. capsulatus*, the Fe nitrogenase is solely repressed by Mo [9,60,61]. About 50 % repression occurs, if 6 nM molybdate is present in the growth medium [60]. V has no effect on the expression of the structural genes *anfHDGK*.

Synthesis of the Fe nitrogenase of *R. rubrum* requires completely different growth conditions and does not appear to be metal-regulated [62]. This alternative nitrogenase is repressed neither by molybdate nor by tungstate. In contrast, the *R. rubrum* Fe nitrogenase is expressed whenever a strain lacks an active Mo nitrogenase because of either physiological or genetic inactivation [62]. However, the genetic basis of this regulatory phenomenon has not yet been further investigated.

3.2. Mechanisms for the regulation of alternative nitrogenase systems

Regulation of nitrogen fixation genes is best-characterized in proteobacterial species, involving similar regulatory elements including the two-component regulatory system NtrB/NtrC, PII-like signal transduction proteins, specific transcriptional activators (NifA, VnfA and AnfA), and the "alternative" sigma factor σ^{54} (also called NtrA, RpoN or GlnF). Despite the high degree of conservation of these regulatory modules among proteobacteria, there are species-specific differences. For example, in most (but not all) species analyzed so far, NtrC is essential for expression of nitrogen fixation genes, and most (but not all) NifA proteins are sensitive to O_2 [63].

Enteric bacteria harbor a nitrogen regulation (Ntr) system that senses and responds to the cellular nitrogen status reflected by the glutamine/2-ketoglutarate ratio [64,65]. This Ntr system comprises five gene products, namely a bifunctional uridylyl-transferase/uridylyl-removing enzyme encoded by *glnD*, a trimeric signal transduction protein (PII) encoded by *glnB*, a PII-paralogue encoded by *glnK*, and a two-component regulatory system encoded by *ntrB-ntrC*. In response to the cellular nitrogen status PII is regulated by reversible GlnD-mediated uridylylation at tyrosine residue 51. When cells are N-limited, PII is uridylylated, and thus is unable to interact with the sensor kinase NtrB. Under these conditions NtrB promotes the phosphorylation of the response regulator NtrC, and in turn, NtrC~P acts as a transcriptional activator of its target genes.

R. capsulatus contains genes homologous to *glnD*, *glnB*, *glnK*, *ntrB*, and *ntrC*, and therefore, regulatory mechanisms similar to those of the enteric Ntr system have been proposed [66-68]. Under conditions of nitrogen starvation *R. capsulatus* NtrB phosphorylates, and thereby activates, NtrC (Figure 4; [69]). In turn, NtrC~P activates transcription of a number of genes including *glnB-glnA*, *glnK-amtB*, *nifA1*, *nifA2*, *anfA*, and *mopA-modABCD* [68,70]. These genes code for PII (GlnB), glutamine synthetase (GlnA), the PII-paralogue GlnK, a putative ammonium transporter (AmtB), the transcriptional activators of the two nitrogenase systems (NifA1, NifA2, AnfA), a Mo-dependent repressor (MopA), and a high-affinity molybdate uptake system (ModABC), respectively. In contrast to all other members of the NtrC family, *R. capsulatus* NtrC does not require σ^{54} but instead activates transcription of its target genes in concert with RNA polymerase (RNAP) containing the σ^{70}-like house-keeping sigma factor [70,71]. In addition to NtrC, RegA can stimulate transcription of *nifA2* by binding to the *nifA2* (but not to the *nifA1*) promoter [72]. The RegB/RegA signal transduction system is a global two-component regulatory system that activates photosynthesis and carbon assimilation in *R. capsulatus*. Thus, although NtrC is absolutely required for *nifA2* expression, RegA acts as a coactivator of *nifA2*, and thereby links regulation of nitrogen fixation and photosynthesis. Another link between these two processes is mediated by *hvrA*, which is cotranscribed with *regA*. HvrA is a histone-like protein homologous to *E. coli* H-NS. *R. capsulatus* HvrA, which is known as a low-light activator of the photosynthetic apparatus [73], binds to the *nifH* promoter and thereby modulates *nif* gene expression [68,74]. In the absence of ammonium *R. capsulatus* NifA1 and NifA2 activate transcription of their target genes in concert with RNAP containing σ^{54} leading to synthesis of the Mo nitrogenase (Figure 4).

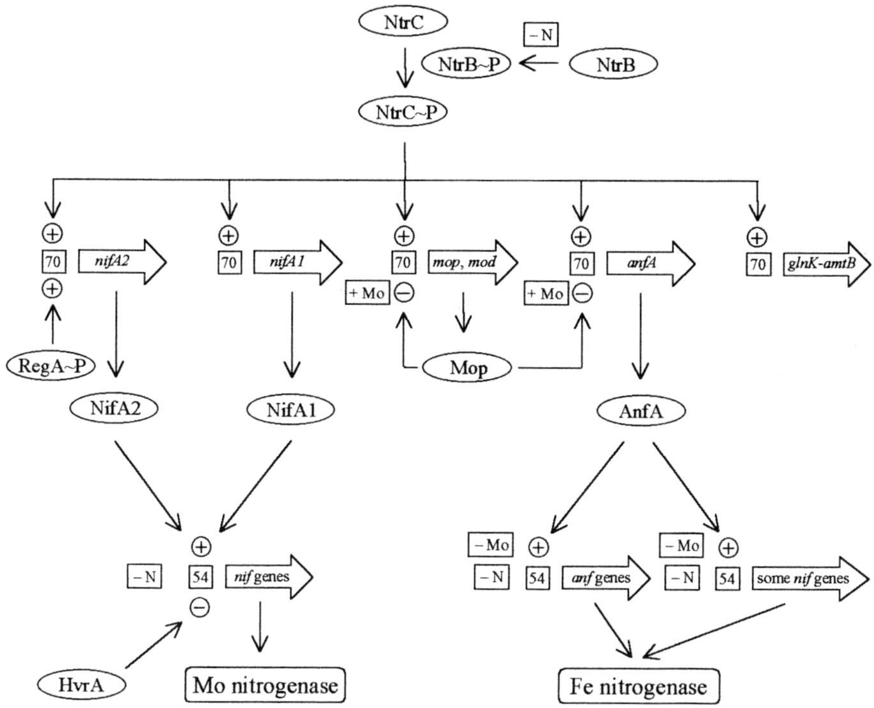

Figure 4. Model of the nitrogen fixation regulon in *R. capsulatus.* The model describes the regulatory cascade controlling *nif* and *anf* gene expression in response to ammonium and molybdenum availability. RNAP-σ^{70}- and RNAP-σ^{54}-dependent promoters are indicated.

Two transcriptional units, namely *mopA-modABCD* and *mopB*, coding for a molybdate transport system and two Mo-dependent repressor proteins are located immediately downstream of *R. capsulatus nifB2* (Figure 3; [61]). In both eubacteria and archaea, molybdate is transported by an ABC-type high-affinity transporter comprising three proteins, ModA (periplasmic binding protein), ModB (membrane protein) and ModC, the ATPase [75]. Both MopA and MopB exhibit clear similarity to *E. coli* ModE, which acts in a Mo-dependent manner as a repressor of the *modABCD* operon [76]. Similarly, *R. capsulatus* MopA and MopB mediate Mo-dependent repression of *mopA-modABCD* and *anfA*, thereby inhibiting expression of the alternative nitrogenase in the presence of molybdenum [77]. This Mo-dependent repression is completely released in mutants carrying a double deletion of *mopA* and *mopB*, and partially released in mutants defective in the high-affinity molybdate uptake system. In contrast, Mo-dependent repression is retained in mutants defective for either *mopA* or *mopB*, indicating that the appropriate repressor proteins can substitute for each other with respect to Mo control of the alternative nitrogenase system. However, MopB (but not MopA) is essential for the Mo-dependent expression of the *R. capsulatus dorCDA* operon (coding for dimethylsulfoxide reductase) [78] demonstrating clear functional differences between the two Mo-responsive proteins.

ModE-like repressor proteins bind to dyad symmetric promotor elements that overlap the transcription start [79]. Binding of the repressor proteins occurs sequence-specific, and a consensus sequence (CGNTATAT-N_7-TATATANCG) has been suggested [75,77]. In addition to Mo repression of *anfA* transcription, activity of the *R. capsulatus* AnfA protein is inhibited by Mo in a MopA/MopB-independent manner [80]. However, in an *nifHDK* deletion strain carrying a *mopA*/*mopB* double mutation, *anfH* transcription is no longer repressed by Mo.

In the absence of both ammonium and molybdenum, *R. capsulatus* AnfA activates transcription of the *anfHDGK-1-2-3* operon in concert with RNAP-σ^{54}. Since *nifA1*/*nifA2* double mutants are still able to grow diazotrophically *via* the Fe nitrogenase [30], it seems likely that AnfA is also responsible for transcriptional activation of *nifB*, *nifV* and the *rnf* genes, which are essential for activity of the alternative nitrogenase (Figure 3). Indeed, the *nifB* promoter of *A. vinelandii* can be activated by NifA, VnfA or AnfA [36].

In addition to transcriptional control of nitrogen fixation genes, the activity of both the Mo-containing and the Fe-only nitrogenase of *R. capsulatus* is controlled by reversible ADP-ribosylation of NifH and AnfH mediated by DraT and DraG in response to ammonium or darkness [81]. Inactivation of dinitrogenase reductase by ADP-ribosylation is catalyzed by DraT (**d**initrogenase **r**eductase **A**DP-ribosyl**t**ransferase), whereas restoration of nitrogenase activity is performed by DraG (**d**initrogenase **r**eductase-**a**ctivating **g**lycohydrolase). Besides *R. capsulatus*, the ADP-ribosylation system has been studied in *Rhodospirillum rubrum* (which serves as the model organism for investigation of post-translational nitrogenase regulation), *Azospirillum brasilense*, *Azospirillum lipoferum*, and *Chromatium vinosum* [82]. In contrast, no such system is present in *K. pneumoniae* or *A. vinelandii*. ADP-ribosylation of dinitrogenase reductase occurs at a specific highly conserved arginine residue (Arg101 in *R. rubrum*). The presence of the ADP-ribose group apparently prevents association of dinitrogenase reductase and dinitrogenase, rather than blocking electron transfer. In addition to the DraTG-mediated system, *R. capsulatus* harbors a second (not yet identified) DraTG-independent mechanism, which leads to nitrogenase "switch-off", when ammonium is added to a nitrogen-fixing culture [83-85].

3.3. Structures of transcriptional activators and cis-regulatory elements

NifA, VnfA and AnfA all have the three-domain structure characteristic for the family of activators acting in concert with RNAP-σ^{54} [63,86]. The N-terminal domains (believed to act as environmental sensors) are not highly homologous, but a conserved cysteine motif is present in VnfA (Cys-X-Cys-X_4-Cys) and AnfA (Ser-X-Cys-X_4-Cys). These motifs are suggested to be involved in the interaction of these proteins with metals. Analysis of *A. vinelandii* AnfA mutants containing amino acid substitutions in this putative metal-binding site indicate that Ser19 is not required for AnfA activity, whereas Cys21 and Cys26 are required [87]. As mentioned above, *A. vinelandii* AnfA appears to require the presence of functional dinitrogenase reductase of either the Mo or the V system for its activity, and it has been suggested that transfer of electrons from dinitrogenase reductase of the V system to AnfA activates AnfA by reduction of a redox site in the N terminus [59]. Deleting the N-terminal domain of AnfA results in fully active protein which is also independent of dinitrogenase reductase [88]. Furthermore, the truncated protein has a constitutive ATPase activity [86].

The central domains of NifA, VnfA and AnfA are highly homologous and contain seven conserved regions (designated C1-C7) including the ATP binding motifs characteristic of other activators acting in concert with RNAP-σ^{54} [17,63]. The central domains are responsible for ATP hydrolysis and transcriptional activation.

The C-terminal domains of NifA, VnfA and AnfA contain a helix-turn-helix motif involved in binding at specific DNA sequences. The DNA binding site of NifA, also referred to as upstream element (UE) or upstream activator-binding sequence (UAS), comprises a highly conserved sequence (typically TGT-N_{10}-ACA [89]). A similar but distinct UAS has been defined for VnfA (GTAC-N_6-GTAC [90]). Both elements are characterized by their palindromic nature. Transcriptional activation from the *A. vinelandii nifB* promoter involves two NifA binding sites each overlapping a VnfA recognition sequence [36]. Furthermore, activation of the *nifB* promoter by AnfA is impaired by overexpressing the DNA-binding domain of NifA, suggesting that binding of NifA and AnfA can be competitive [36]. However, the UAS of AnfA is less certain, but methylase protection experiments using an N-terminally truncated form of AnfA identified the sequence C-N-GG-N_3-GGTA [86].

The upstream activator sequences for binding of NifA, VnfA or AnfA are generally located 100-200 bp upstream of the transcriptional start site. As mentioned above, the members of the NifA family activate transcription in concert with RNAP containing σ^{54}, which binds to a characteristic sequence motif (CTGG-N_8-TTGC) typically located at position -24/-12 upstream of the transcription start site. RNAP-σ^{54} binds to the promoter to form a closed complex which needs the transcriptional activator to undergo isomerization to an open complex that is able to initiate transcription [17]. NifA-mediated transcriptional activation requires contact between NifA and RNAP-σ^{54} bound to distant promoter elements which might be achieved through DNA looping. In some cases the interaction between NifA and RNAP-σ^{54} is stimulated by IHF (integration host factor) [91-93]. IHF is a site-specific DNA-binding protein that induces a sharp bend in the DNA. IHF binds to promoter regions of many *nif* genes from *K. pneumoniae*, *A. vinelandii* and *R. capsulatus* [91]. An IHF binding site is also present in the *R. capsulatus anfH* promoter region suggesting that IHF may also modulate activation of this promoter by AnfA.

4. PROTEIN SIZE, SUBUNIT STRUCTURE AND FEATURES DEDUCED FROM PROTEIN SEQUENCES OF ALTERNATIVE NITROGENASES

4.1. Structural features of alternative dinitrogenase reductases (Fe proteins)

In contrast to conventional Mo nitrogenases, X-ray structures have not been determined for any of the components of alternative nitrogenases. All data available concerning structural features result from genetic, biochemical, spectroscopical and immunological studies. These data suggest that the over-all structures of the alternative nitrogenases and of the Mo nitrogenase are very similar but there are also some distinct differences.

As mentioned above, nitrogenases are composed of a dinitrogenase (MoFe, VFe or FeFe protein) and a dinitrogenase reductase (Fe protein). All dinitrogenase reductases investigated so far (including those of alternative systems) are highly similar as all of them exhibit a homodimeric structure (γ_2), have similar sizes (~ 60 kDa), and display pronounced amino acid sequence similarities. Based on deduced amino acid sequences, high degrees of identity are found, *e.g.*, *A. vinelandii* NifH/VnfH (91 %), *A. vinelandii* VnfH/AnfH (63 %), and *R. capsulatus* NifH/AnfH (62 %) [14, 30, 94].

The three-dimensional crystal structure of the *A. vinelandii* dinitrogenase reductase of the Mo system (NifH protein) confirmed the existence of a conventional [Fe_4S_4] cluster symmetrically coordinated by cysteine residues Cys_{97} and Cys_{132} [95]. The amino acid sequence motifs Gly-Lys-Gly-Gly-Ile-Gly-Lys-Ser (residues 9-16) and Asp-Val-Leu-Gly (residues 125-128) of NifH, which are thought to be involved in nucleotide binding and hydrolysis [96], are also present in VnfH and AnfH [14,30,94]. Based on this great extent of structural similarity, the conclusion can be drawn that the NifH, VnfH and AnfH proteins fulfill an identical function in all three nitrogenase systems, *i.e.* the transfer of a single electron from its [Fe_4S_4] cluster to the P cluster of the respective dinitrogenase component with concomitant hydrolysis of two molecules of ATP [97]. Accordingly, the *R. capsulatus* NifH protein ($Rc2^{Mo}$) can (at least partially) substitute for the AnfH protein ($Rc2^{Fe}$), as shown by *in-vitro* studies measuring acetylene reduction activity of the alternative dinitrogenase ($Rc1^{Fe}$) [98].

4.2. Structural features of alternative dinitrogenases (VFe and FeFe proteins)

In contrast to the Fe proteins of alternative nitrogenases, much research interest has been focused on the VFe and FeFe proteins, since these dinitrogenase components show significant structural deviations from the MoFe protein. Whereas the MoFe protein exhibits a tetrameric structure ($\alpha_2\beta_2$), the VFe and the FeFe proteins form hexameric structures ($\alpha_2\beta_2\delta_2$). α and β subunits have similar sizes in all nitrogenases

(51-59 kDa). The δ subunit sizes range from 13.4 kDa (*R. capsulatus* FeFe protein) to 17.0 kDa (*R. rubrum* FeFe protein) or are fused to VnfD (*A. variabilis* VFe protein). Relative molar masses of the native proteins range around 230,000 for MoFe proteins and 240,000-250,000 for VFe and FeFe proteins. The specific role of the small δ subunit in the alternative dinitrogenases is still not definitely known. Most likely, δ has not simply a structure-stabilizing function but may be involved in cofactor insertion and completion of a functioning enzyme [99].

The protein sequences of the MoFe, VFe and FeFe proteins are less conserved than those of the Fe proteins of the three nitrogenase systems. For example, *A. vinelandii* NifD and VnfD exhibit only 33 % amino acid sequence identity [94]. Higher similarity exists for VnfD and AnfD (54 % identity) [94]. The degree of similarity of *A. vinelandii* NifK, VnfK and AnfK is similar. Despite this relatively low "overall similarity" between the different dinitrogenase proteins, there are some regions and specific amino acid residues functioning as ligands of the metal clusters that are highly conserved in the MoFe, VFe and FeFe proteins. A cysteine and a histidine residue of the α subunit are known to be the only amino acid residues covalently coordinating the FeMo cofactor of the conventional nitrogenase in *A. vinelandii* and *Clostridium pasteurianum* (Cys275 and His442 in *A. vinelandii* [100,101]). In *R. capsulatus* corresponding α subunit residues (Cys290 and His454) are conserved in both the conventional and the Fe nitrogenase [30]. Furthermore, six cysteine residues are suggested to coordinate Fe atoms of the P cluster within the MoFe protein. These cysteine residues are also present at corresponding positions of the *R. capsulatus* FeFe protein, namely at positions 78, 103 and 169 in the α subunit and at positions 70, 95 and 153 in the β subunit [30]. Similarly, all the amino acid residues described to be involved in binding of the dinitrogenase cofactors (FeMoco, FeVco and FeFeco) and the P cluster are highly conserved in all three nitrogenases of *A. vinelandii* as well.

It is worth mentioning that, beside the hexameric form ($\alpha_2\beta_2\delta_2$), the *A. vinelandii* VFe protein has also been purified in a tetrameric ($\alpha_2\beta_2$) and a trimeric $\alpha\beta_2$ form, with the δ subunit associated only in low and variable amounts [102]. Both the tetrameric and the trimeric form exhibit relatively low specific activities, and are considered to represent non-native forms. Similar tetrameric and trimeric forms of the *A. vinelandii* FeFe protein have been isolated, whereas no such forms have been described for the FeFe protein from phototrophic bacteria indicating higher stability of the hexameric form in phototrophs.

5. STRUCTURE AND SPECTROSCOPIC PROPERTIES OF METALLO-SULFUR CLUSTERS

5.1. The [Fe_4S_4] cluster of Fe proteins

The FeS clusters in VnfH and AnfH proteins correspond to the [Fe_4S_4] cluster in NifH proteins as indicated by the conserved cysteine residues, which serve as ligands of the cluster (see above), and by characteristic EPR spectra. In the dithionite-reduced state the cluster is present in the 1e$^-$-reduced form ($[Fe_4S_4]^+$). In that state the *A. vinelandii* VnfH protein ($Av2^V$) [7, 103], the *R. capsulatus* AnfH protein ($Rc2^{Fe}$) [104], and the *R. rubrum* AnfH protein ($Rr2^{Fe}$) [10] show, like all NifH proteins, a rhombic S = 1/2 signal (central g-value: ~ 1.94) which changes to an axial signal upon addition of MgATP [7,104]. In addition to the S = 1/2 signal, the $[Fe_4S_4]^+$ cluster may exhibit a broad S = 3/2 signal, the detectable part of which is in the g = 4-6 region [105]. This "spin mixture" has also been demonstrated in $Rc2^{Fe}$ and $Av2^V$ samples [7,103,104]. The proportion of the S = 3/2 signal within such a spin mixture depends on the protein preparation, the presence or absence of MgATP and particularly on the solvent [103,105]. It should also be mentioned that in the case of NifH, the protein can further be reduced to an "all-ferrous" $[Fe^4Fe^4]^0$ form by flavodoxin, methyl viologen or titanium citrate [106]. Whether this "all-ferrous" cluster is of mechanistic relevance, potentially serving as two-electron donor remains to be established.

5.2. The P cluster

The structures of both the P cluster and the FeMoco have been elucidated on the basis of crystallographic analyses of the *A. vinelandii* and *C. pasteurianum* MoFe proteins refined up to 2.0 Å resolution [100,101]. The P cluster is a large FeS cluster (Fe_8S_7) which consists of two Fe_4S_3 sub-clusters linked by two bridging cysteine thiolate groups and an additional hexacoordinate sulfur atom bonded to three Fe atoms from each sub-cluster. The P cluster is located at the α/β subunit interface ligated by three cysteines from both subunits. Since the MoFe protein exhibits an $\alpha_2\beta_2$ structure, the holo-protein contains two P clusters (and two FeMo cofactors).

Evidence that the P cluster present in alternative nitrogenases is structurally identical with the P cluster of Mo nitrogenases comes from protein sequence analysis, EPR spectroscopic studies and Mössbauer studies. As mentioned above, the six cysteine residues binding the P cluster to the protein are conserved in all the three nitrogenases. Less convincing evidence is based on EPR spectra. This is due to the fact that P clusters (unlike the $[Fe_4S_4]$ cluster and the FeMoco) do not display a very specific and easily detectable fingerprint signal, which would be more or less identical in all nitrogenases regardless of the origin. P clusters are stronger influenced by slight modifications in the protein environment. The only signal which might serve as a unique and characteristic fingerprint is a weak signal of an S = 3 system near g = 12. This signal has been described for MoFe and VFe proteins but not yet for FeFe proteins [107,108]. It should be remembered that integer spin systems, even if they are definitely present, are often very difficult to detect by EPR. Such systems are preferably measured under parallel mode conditions. EPR measurements on FeFe proteins, however, have so far only been performed under perpendicular conditions.

More conclusive data concerning the structural identity of P clusters are derived from Mössbauer studies on ^{57}Fe-enriched dinitrogenase preparations ($Av1^V$, $Rc1^{Fe}$) although analysis of the P cluster spectra is rendered difficult due to overlapping signals of the respective cofactor spectra. Nevertheless, the three characteristic spectral subcomponents (called Fe^{2+}, S and D) comprising the P cluster spectrum, are clearly recognizable for alternative nitrogenases. The quantitative compositions of these components as well as their parameters (isomer shifts and quadrupole coupling constants) are very similar for both the $Av1^{V}$ and $Rc1^{Fe}$ proteins [109,110], and corresponds well with the data on MoFe proteins [111].

5.3. The structures of FeV and FeFe cofactors

The MoFe protein contains two FeMo cofactors (one per α subunit), with a total composition of each cofactor corresponding to Fe_7MoS_9-homocitrate (also see Chapter 2). The FeMoco formally consists of an Fe_4S_3 and an $MoFe_3S_3$ cluster bridged by three inorganic sulfides, and homocitrate as a simple organic constituent of the cofactor (Figure 5; [100,112]). The Mo site exhibits an octahedral coordination geometry, and Mo is bound to three cluster sulfur atoms, two oxygen atoms of homocitrate, and one nitrogen atom from a histidine residue (His442 in *A. vinelandii* NifD). The second protein ligand of FeMoco is a cysteine residue (Cys275 in *A. vinelandii* NifD) coordinating to the only Fe site with tetrahedral geometry. One of the structural peculiarities of the FeMoco is that the six other iron atoms are trigonally coordinated forming a central prismatic arrangement.

There is strong evidence for the structural similarity of the clusters FeMoco, FeVco and FeFeco resulting from biochemical and genetic data, as well as from recent spectroscopic studies (EXAFS, Mössbauer). (i) Under *in-vivo* conditions the FeFeco of *A. vinelandii* and *R. capsulatus* can be replaced by the FeMoco leading to formation of hybrid enzymes [60,113]. In a comparable way, under *in-vitro* conditions a FeMoco-deficient apo-dinitrogenase (Av1) can be reconstituted and activated with FeFeco from $Rr1^{Fe}$ [10]. Similarly, the FeVco extracted from $Ac1^{V}$ can be incorporated into the *K. pneumoniae* apo-dinitrogenase [114]. These experiments demonstrate that FeMoco, FeVco and FeFeco can functionally substitute for each other. (ii) The FeMoco is covalently bound in the MoFe protein through a cysteine and a histidine residue. As mentioned above, these amino acid residues are highly conserved in V and Fe nitrogenases. (iii) The *nifV* gene product, homocitrate synthase [115], is essential for the biosynthesis of all three cofactors indicating that homocitrate is an essential component not only of FeMoco but also of FeVco and FeFeco [30,116]. (iv) All three nitrogenase systems are dependent on the *nifB* gene [30,36,117], which encodes a protein responsible for synthesis of an FeS cluster called NifBco [118]. NifBco may be a common heterometal-free intermediate in the biosynthetic pathway of FeMoco, FeVco and FeFeco [111]. (v) Mössbauer spectra (subspectra "M") measured with conventional and alternative dinitrogenase proteins from *A. vinelandii* and *R. capsulatus* yield very similar isomer shifts and quadrupole splittings of FeMoco, FeVco and FeFeco [109-111]. Spectra of $Rc1^{Fe}$ suggest that FeFeco, in contrast to FeMoco and FeVco, is diamagnetic in the dithionite-reduced state, and that it contains eight iron atoms, whereas FeMoco and FeVco each contain only seven [109]. (vi) Vanadium-EXAFS measurements with $Av1^{V}$ and $Ac1^{V}$ showed that the types of atom (Fe, S, N/O) and their distances from the V atom are very similar to those of the Mo system, except for a longer V-Fe distance (2.76 Å in Av1^{V}) compared to the Mo-Fe distance (2.67 Å in Av1^{Mo}) [111,119,120]. Fe-EXAFS spectra for Av1^{V}, as well as for isolated FeVco from Ac1^{V} and from Rc1^{Fe}, exhibit a marked similarity in key features compared to the respective MoFe protein data [121-123]. The ~2.6 Å and the ~3.7 Å Fe-Fe interactions, unique to FeMoco, are retained in the isolated FeVco as well as in the native VFe and FeFe proteins (Figure 5).

Due to the variation or absence of heterometal atom EPR spectra and redox behaviour of the protein-bound FeV cofactor and the FeFe cofactor differ distinctly from the respective FeMo cofactor data. Sodium dithionite ($Na_2S_2O_4$)-reduced MoFe proteins show the characteristic rhombic S = 3/2 EPR signal of the FeMo cofactor with g values at about 4.3, 3.7 and 2.01 [123]. The redox state of the FeMoco giving rise to this signal (midpoint potential E_m = −50 mV) has been designated as "semi-reduced" state or "resting state". Upon one-electron-oxidation as well as upon one-electron-reduction (under turnover conditions in the presence of ATP, a substrate and low amounts of dinitrogenase reductase) the FeMoco becomes diamagnetic and the S = 3/2 signal disappears ([124]; S. Siemann, K. Schneider, M. Dröttboom, and A. Müller, unpublished results).

In the dithionite-reduced state the best-investigated FeV protein (Av1^{V}) exhibits an S = 3/2 EPR signal as well, but with quite different g values (5.68, 5.45) [108]. This signal disappears upon oxidation, but not under turnover conditions.

In none of the EPR spectra of all three known FeFe proteins (Rc1^{Fe}, Rr1^{Fe}, Av1^{Fe}) was an S = 3/2 signal present [10,98,125] supporting the conclusion from the Mössbauer study that the dithionite-reduced FeFeco is diamagnetic (see above) [109]. In contrast to the protein-bound FeMo and FeV cofactors, the dithionite-reduced state of the FeFe cofactor (studied in detail with Rc1^{Fe}) can be converted to paramagnetic states by oxidation (resulting in an S = 1/2 signal at g = 2.27, 2.07) as well as by further reduction (S = 1/2 signal at g = 1.96, 1.92, 1.77) ([98]; S. Siemann, K. Schneider, M. Dröttboom, and A. Müller, unpublished results).

In summary, the only fundamental structural difference between the three cofactors seems to be that the Mo atom of FeMoco is replaced by either V or Fe in the alternative cofactors (Figure 5). With regard to Fe nitrogenases, it must be emphasized that binding and reduction of N_2 can be catalyzed by a cofactor, which is actually based on Fe as sole metal. Furthermore, this heterometal-free FeFeco (Fe_8S_9) is currently the largest member of the family of naturally occuring FeS clusters.

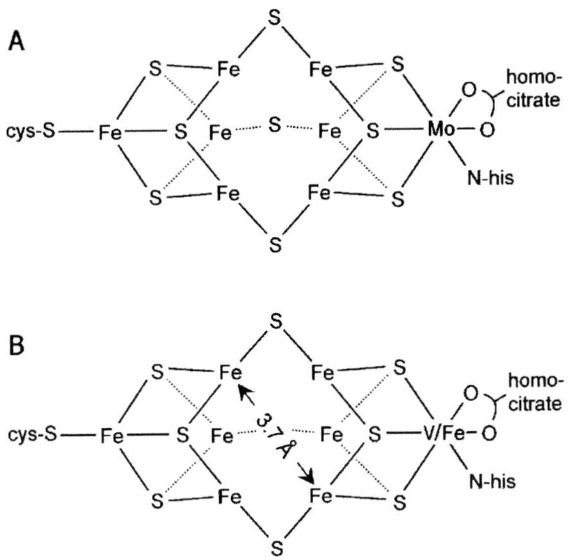

Figure 5. Proposed structure of dinitrogenase cofactors. (A) Schematic representation of the FeMo cofactor as deduced from crystal structures. (B) In the proposed structure of the FeV and FeFe cofactors of alternative nitrogenases the Mo atom is replaced by either V or Fe. Similar to FeMoco, it is assumed that the central Fe atoms of FeVco and FeFeco form a trigonal prismatic arrangement as implied by the characteristic 3.7 Å FeFe interaction.

6. BIOSYNTHESIS OF FeVco AND FeFeco

The products of more than one third of the *nif* genes of *K. pneumoniae* and *A. vinelandii* are somehow involved in FeMo cofactor biosynthesis and/or insertion of FeMoco into apo-dinitrogenase [19]. To analyze the pathway of FeMoco biosynthesis, first an *in-vitro* system has been developed, which involves mixing extracts from two different mutant strains defective in the synthesis of the cofactor. The *in-vitro* assays revealed that at least the products of *nifQ*, *nifB*, *nifE*, *nifN*, *nifV*, and *nifH* are required for FeMoco biosynthesis [126]. Later on, an *in-vitro* system based on purified components identified previously unknown factors (like NifX) that play a role in cofactor biosynthesis [127]. To date no similar *in-vitro* system for FeVco or FeFeco biosynthesis has been established. However, VnfH can replace NifH in both the *in-vitro* biosynthesis of FeMoco and in the maturation of apo-dinitrogenase, suggesting that the dinitrogenase reductase proteins do not specify the heterometal incorporated into the cofactors of the corresponding nitrogenases [99].

Since biosynthesis of the FeMoco and the FeVco has been reviewed recently [19], we will only summarize briefly the current model of cofactor biosynthesis. In this model, the NifU and NifS proteins are involved in the mobilization of Fe and S to NifB, which in turn is responsible for synthesis of NifBco (Fe_7-S_x). Since *nifB* is essential for all the three nitrogenases of *A. vinelandii* [36], NifBco seems to resemble the FeS precursor of FeMoco, FeVco and FeFeco. NifBco might be donated either to NifEN (in the FeMoco biosynthetic pathway) or to VnfEN (in the FeVco biosynthetic pathway). As mentioned above, NifEN and VnfEN are somehow shared by all three nitrogenases [25] suggesting that either NifEN or VnfEN are involved in biosynthesis of the FeFe cofactor of the Fe nitrogenase. In the next step, NifH or VnfH may interact with the NifEN-NifBco or the VnfEN-NifBco complex thereby inducing a conformational change within the complex, which might be essential for incorporation of the heterometal (Mo or V) *via* NifX or VnfX, respectively. Despite this requirement of NifX for FeMoco biosynthesis, genetic studies demonstrated that *nifX* is not essential for diazotrophic growth in *K. pneumoniae*, *A. vinelandii* and *R. capsulatus*. NifX and VnfX also seem to be involved in incorporation of homocitrate, resulting in the finished cofactors [128]. Incorporation of FeMoco into apo-NifDK involves the γ protein (in *A. vinelandii*) or NifY (in *K. pneumoniae*). In a NifH-dependent manner, the γ protein or NifY deliver FeMoco to apo-dinitrogenase and then dissociate from the holo-enzyme. Similarly, the VnfG protein might be involved in insertion of FeVco into apo-VnfDK. In contrast to the γ protein or NifY, VnfG finally remains (as regular subunit δ) bound tightly to the holo-enzyme. The presence of a VnfG-like δ subunit (AnfG) in the Fe dinitrogenase suggests that incorporation of FeFeco into apo-AnfDK might proceed similarly.

7. CATALYTIC FEATURES OF ALTERNATIVE NITROGENASES

7.1. Oxygen sensitivity and general protein stability

Although Mo-containing and Mo-independent nitrogenases are very similar with respect to structure and function, they exhibit different stability and catalytic properties (Table 2). The Mo dinitrogenase component is very oxygen sensitive, and the dinitrogenase reductase is extremely sensitive [129]. Compared to the conventional nitrogenase, the V and Fe nitrogenases are even more unstable in the presence of O_2. Exposure of isolated VFe protein ($Ac1^V$) to air results in a rapid loss of activity (half life $t_{1/2}$ = 40 s at 30 °C), whereas the MoFe protein ($Ac1^{Mo}$) is considerably more stable ($t_{1/2}$ = 8 min). Similarly, the Fe nitrogenase of *R. capsulatus* is extremely O_2 sensitive as demonstrated by *in-vivo* studies showing 50 % inhibition of activity of the alternative nitrogenase at ~ 0.5 % oxygen in the gas phase [130]. In contrast, 50 % inhibition of activity of the Mo nitrogenase of *R. capsulatus* are achieved at ~ 4.5 % oxygen.

In addition to O_2 sensitivity, purified Fe nitrogenase component proteins are generally unstable. $Rc1^{Fe}$ is particularly labile towards dilution and freezing-thawing procedures [98,104]. This might be the main reason for the very low specific activities determined for most of the Fe nitrogenase preparations described in the literature (for example, see the values of the *A. vinelandii* Fe nitrogenase in Table 2). In contrast, relatively high specific activities were obtained for the *R. capsulatus* Fe nitrogenase after optimization of Mo starvation and other factors affecting derepression, combined with a gentle cell disruption procedure and elaboration of a rapid one-column purification step [98].

7.2. Substrate reduction rates and particular catalytic features

The Mo nitrogenase is able to react with a large number of small, reducible substrates, most of which contain triple bonds. Studies on the reactivity of alternative nitrogenases have so far been limited to the natural substrate N_2, to protons (H^+) and to the artificial substrate acetylene, which is used in the standard activity assay.

Comparison of the N_2 reduction rates (measured as NH_3 formation) shown in Table 2 suggests that activities decrease in the order Mo nitrogenase > V nitrogenase > Fe nitrogenase. However, there are significant discrepancies between the representatives within one enzyme system. This may (partly) be due to real catalytic differences or to differences in the assay conditions, but the main reason might be the different quality of the preparations (see above). If one considers that the higher activities appear to be the more reliable ones, the maximal specific activities of Mo nitrogenases range around 1,000 U (nmol product formed per min and mg of protein), whereas V and Fe nitrogenases reach about 2/3 and 1/3 of that value, respectively.

Specific activity [U/mg protein] [a]						
Enzyme [b]	$N_2 \rightarrow NH_3$	$C_2H_2 \rightarrow C_2H_4$ [c]	$C_2H_2 \rightarrow C_2H_6$ [c]	$H^+ \rightarrow H_2$ (Ar)	$H^+ \rightarrow H_2$ (Ar/C_2H_2) [c]	$H^+ \rightarrow H_2$ (N_2)
Kp-Mo	990	1690	0	2100	304	648
Av-Mo	1040	2200	n.d. [d]	2220	303	n.d. [d]
Rc-Mo	470	1200	< 0.1 %	1300	140	210
Av-V	660	220	n.d. [d]	1400	n.d. [d]	n.d. [d]
Ac-V	350	608	~ 3 %	1350	1000	930
Av-Fe	38	28	n.d. [d]	253	202	213
Rc-Fe	350	260	~ 2 %	2400	1300	1700

[a] N_2 reduction was measured as NH_3 formation, acetylene reduction was measured as ethylene and ethane formation, and proton reduction was assayed as H_2 evolution.
[b] Enzyme abbreviations: Kp-Mo (*Klebsiella pneumoniae* Mo nitrogenase [129]), Av-Mo (*Azotobacter vinelandii* Mo nitrogenase [7]), Rc-Mo (*Rhodobacter capsulatus* Mo nitrogenase [98]), Av-V (*A. vinelandii* V nitrogenase [7]), Ac-V (*Azotobacter chroococcum* V nitrogenase [131]), Av-Fe (*A. vinelandii* Fe nitrogenase [8]), Rc-Fe (*R. capsulatus* Fe nitrogenase [98]).
[c] Assays were performed under an atmosphere of 90 % argon and 10 % acetylene.
[d] n.d. (not determined).

Table 2. Comparison of the specific activities of Mo, V and Fe nitrogenases

N_2H_4 has been postulated to be an intermediate during the reduction of N_2 to ammonia. However, in studies with Mo nitrogenase, N_2H_4 can only be detected when the reaction mixture is quenched with acid or alkali [132]. In contrast, N_2 reduction by the *A. chroococcum* V nitrogenase results in the formation of a clearly detectable amount of N_2H_4 as a by-product [133], which is a direct evidence for an enzyme intermediate at the four-electron-reduced level during the reduction of N_2. Similar studies on Fe nitrogenases have not yet been carried out.

The decrease of C_2H_2 reduction rates (measured as C_2H_4 formation) in the sequence Mo nitrogenase > V nitrogenase > Fe nitrogenase is consistent with the N_2 reduction rates (Table 2). However, the differences between the acetylene reduction rates displayed by the three nitrogenases seem to be greater compared to the rates for reduction of the natural substrate N_2. Probably these discrepancies are due to the fact that the standard assay conditions for acetylene reduction (with an amount of acetylene corresponding to a partial pressure of 10 kPa) are optimal for the Mo nitrogenase but not for the alternative nitrogenases. Consistently, acetylene has repeatedly been described as a poor substrate for V and Fe nitrogenases [7,8,131]. K_m values of ~ 1 kPa have been reported for Mo nitrogenases, whereas K_m values determined for the *A. chroococcum* V nitrogenase and the *R. capsulatus* Fe nitrogenase were 6 kPa and 12.5 kPa, respectively [98,134]. Therefore, the use of 10 kPa C_2H_2 does not yield saturating substrate concentrations for alternative nitrogenases thus leading to a clear underestimation of C_2H_2 reduction activities, particularly for Fe nitrogenases. For example, the acetylene-reduction activity of the *R. capsulatus* Fe nitrogenase is about two times higher when measured under a pure C_2H_2 atmosphere (100 kPa) compared to the activity obtained at 10 kPa [98].

A general catalytic feature of Mo-independent nitrogenases is the capability to reduce acetylene (C_2H_2), not only to ethylene (C_2H_4) but also partially to ethane (C_2H_6) [135]. The proportions of ethane and ethylene may differ between *in-vivo* and *in-vitro* acetylene reduction assays, and for *in-vitro* assays this ratio of products strongly depends on the assay conditions (pH, temperature, molar ratio of the component proteins). Under standard *in-vitro* conditions (pH 7.4, 30 °C, 10 kPa C_2H_2, presence of saturating amounts of dinitrogenase reductase) both the V and the Fe nitrogenase produce predominantly C_2H_4 and a relative amount of 2-3 % C_2H_6 [98,134]. However, the mechanism by which C_2H_6 is formed remains to be elucidated. Controversial reports on the effect of the pH value on ethane production complicate elucidation even more. For example, ethane production mediated by *R. capsulatus* Fe nitrogenase decreases from 5.5 % at pH 6.6 to about 1.6 % at pH 7.8 and almost zero at pH 9.0 [98], whereas ethane production mediated by *A. chroococcum* V nitrogenase remains almost constant over a wide pH range (6.5-7.5) and decreases only slightly above pH 7.5 [134].

Surprisingly, comparison of proton reduction rates (measured as H_2 evolution in the absence of substrates like N_2 or acetylene) reveals that specific activities of all nitrogenases (except the *A. vinelandii* Fe nitrogenase) are of the same order of magnitude (Table 2). Remarkably, the *R. capsulatus* Fe nitrogenase exhibits an even higher H_2-evolving activity than the Mo nitrogenase, although the Fe nitrogenase exhibits relatively low N_2 reduction activity, suggesting that the reduction of substrates other than protons is rate-limiting. However, fundamental differences occur when H_2 production is measured in the presence of N_2 or C_2H_2 (Table 2). H_2 production catalyzed by Mo nitrogenase is drastically inhibited by N_2 and acetylene, whereas the inhibitory effect of these substrates on H_2 evolution catalyzed by alternative nitrogenases is relatively weak. Molar ratios of H_2 produced per N_2 reduced are approximately 1:1 for Mo nitrogenase [98,111,136], 3-5:1 for V nitrogenase [131,136], and 7-10:1 for Fe nitrogenase [8,98]. Mo nitrogenase allocates 75 % of the total electron flux to N_2 reduction and the remaining 25 % to proton reduction. In contrast, Fe nitrogenase allocates only a small proportion of electrons (25-30 %) to N_2 reduction, whereas the bulk of electrons is utilized for proton reduction. V nitrogenase allocates almost equal amounts of electrons to either of the two substrates [131]. Taken together, electron distributions imply the following equations:

Mo nitrogenase: $N_2 + 8\,e^- + 8\,H^+ \rightarrow 2\,NH_3 + H_2$
V nitrogenase: $N_2 + 12\,e^- + 12\,H^+ \rightarrow 2\,NH_3 + 3\,H_2$
Fe nitrogenase: $N_2 + 24\,e^- + 24\,H^+ \rightarrow 2\,NH_3 + 9\,H_2$

The predominance of the H_2 evolution reaction mediated by alternative nitrogenases indicates that in competing for electrons, protons are clearly favoured as electron acceptors above N_2. It has been postulated that N_2, if bound to the FeFeco, becomes more difficult to reduce than if it were bound to FeMoco or FeVco due to a substantially larger highest-occupied molecular orbital/lowest-unoccupied molecular orbital (HOMO-LUMO) energy separation [137]. In other words, the FeFeco cannot accumulate electrons required for N_2 reduction as easily as FeMoco and FeVco, and therefore, the presence of a heterometal guarantees better and more effective nitrogen fixation.

7.3. Formation and catalytic activity of hybrid nitrogenases

As mentioned above, the FeFe cofactor of the Fe nitrogenase of both *R. capsulatus* and *A. vinelandii* can be replaced *in vivo* by the conventional FeMo cofactor resulting in formation of hybrid enzymes [60,113]. In *R. capsulatus* this can be achieved by addition of Mo to a culture derepressed for the Fe nitrogenase (in a *nifHDK* deletion background) resulting in a dramatic increase of production of C_2H_6 from C_2H_2 (up to ~ 60 % of C_2H_4) [60]. Evidence that this so-called "Mo effect" is due to formation of a hybrid enzyme resulting from exchange of FeFeco for FeMoco comes from EPR spectroscopy. The hybrid enzyme shows the typical S = 3/2 FeMoco signal. Furthermore, cofactor exchange does not require *de novo* protein synthesis after Mo addition, but clearly depends on the presence of NifEN. Similar, growth of an *A. vinelandii* mutant strain deleted for the structural genes of both the Mo and the V nitrogenase carrying an additional mutation allowing synthesis of the Fe nitrogenase in the presence of molybdate also results in an enormously enhanced C_2H_6 formation (26 % of C_2H_4)

[113]. In this case, instead of cofactor replacement, probably a hybrid nitrogenase is formed directly. Metal analysis of such a hybrid enzyme demonstrated the presence of one Mo atom per FeFe protein. Therefore, apparently the two cofactor sites within one FeFe protein seem to be occupied by one FeMoco and one FeFeco. However, high ethane production is an intrinsic feature of hybrid nitrogenases, emphasising the great influence of the protein environment on the catalytic activity. Indeed, some *nif*-encoded dinitrogenases with site-directed changes in the primary sequence also reduce acetylene to ethylene plus ethane [138].

8. RELEVANCE OF ALTERNATIVE NITROGENASES FOR DIAZOTROPHIC BACTERIA UNDER NATURAL CONDITIONS

Many diazotrophic bacteria harbor one or two alternative Mo-independent nitrogenases in addition to the Mo nitrogenase. Since these alternative enzymes exhibit lower specific activities than Mo nitrogenases, paralleled by unfavourable allocation of electrons towards N_2 (at least under standard laboratory conditions, Table 2), the existence of these enzymes poses questions concerning the roles that they play under natural conditions. Both V and Fe nitrogenases are expressed only in the absence of Mo, suggesting that these alternative nitrogenases are (of course) advantageous under conditions of Mo limitation. Such a limitation can be due either to natural low abundance of Mo or to formation of Mo-depleted microzones through Mo accumulation by different microorganisms.

A. vinelandii has a powerful Mo accumulation system (Mo storage protein), which acts as a supply for Mo nitrogenase synthesis when extracellular Mo has been exhausted [139]. However, the Fe nitrogenase is advantageous for diazotrophic growth of *A. vinelandii* even on a solid medium containing molybdenum [140] indicating that Mo becomes limiting within bacterial colonies formed on agar plates. Similarly, Mo may become limiting in microzones in the soil, during water blooms or in cyanobacterial mats.

Recently, nitrogen fixation in the symbiotic microbial community in the gut of the termite *Neotermes koshunensis* was analyzed by culture-independent methods [141]. Nitrogen fixation by the symbiotic microorganisms inhabiting the gut of termites is important, since termites thrive on a nitrogen-poor diet. DNA sequence analysis and expression studies indicate that genes exhibiting strong similarity to *anf* genes are the most important group of genes responsible for nitrogen fixation in *N. koshunensis*.

In addition to Mo deficiency, temperature may affect the requirement of alternative nitrogenases. Low temperature favours N_2 reduction by the V nitrogenase as demonstrated by *in-vitro* studies with purified Mo and V nitrogenases of *A. chroococcum* [47]. In contrast with Mo nitrogenase, N_2 is preferentially reduced by V nitrogenase as the temperature is decreased from 30°C to 5°C. This temperature effect is associated with the Fe protein of V nitrogenase. In addition, temperature indirectly affects Mo-dependent repression of the alternative nitrogenases in *A. vinelandii* [48]. Mo is taken up efficiently at 30°C but is poorly transported into cells grown at lower temperatures, and consistently Mo-mediated repression of the *vnfH* and *anfH* operons is relatively weak at 20°C and completely absent at 14°C.

9. REFERENCES

1. H. Bortels, Zentralbl. Bakteriol. Parasitenkd. Infektionskr. II. Abt. 95 (1936) 193.
2. J. R. Benemann, C. E. McKenna, R. F. Lie, T. G. Traylor, and M. D. Kamen, Biochim. Biophys. Acta 264 (1972) 25.
3. P. E. Bishop, D. M. L. Jarlenski, and D. R. Hetherington, Proc. Natl. Acad. Sci. USA 77 (1980) 7342.
4. R. Premakumar, E. M. Lemos, and P. E. Bishop, Biochim. Biophys. Acta 797 (1984) 64.
5. P. E. Bishop, R. Premakumar, D. R. Dean, M. R. Jacobson, J. R. Chisnell, T. M. Rizzo, and J. Kopczynski, Science 232 (1986) 92.
6. R. L. Robson, R. R. Eady, T. H. Richardson, R. W. Miller, M. Hawkins, and J. R. Postgate, Nature 322 (1986) 388.
7. B. J. Hales, E. E. Case, J. E. Morningstar, M. F. Dzeda, and L. A. Mauterer, Biochemistry 25 (1986) 7251.
8. J. R. Chisnell, R. Premakumar, and P. E. Bishop, J. Bacteriol. 170 (1988) 27.
9. K. Schneider, A. Müller, U. Schramm, and W. Klipp, Eur. J. Biochem. 195 (1991) 653.
10. R. Davis, L. Lehman, R. Petrovich, V. K. Shah, G. P. Roberts, and P. W. Ludden, J. Bacteriol. 178 (1996) 1445.
11. Y.-T. Chien, V. Auerbuch, A. D. Brabban, and S. H. Zinder, J. Bacteriol. 182 (2000) 3247.
12. T. M. Loveless and P. E. Bishop, Can. J. Microbiol. 45 (1999) 312.
13. R. L. Robson, P. R. Woodley, R. N. Pau, and R. R. Eady, EMBO J. 8 (1989) 1217.
14. R. D. Joerger, M. R. Jacobson, R. Premakumar, E. D. Wolfinger, and P. E. Bishop, J. Bacteriol. 171 (1989) 1075.
15. M. Ribbe, D. Gadkari, and O. Meyer, J. Biol. Chem. 272 (1997) 26627.
16. W. Arnold, A. Rump, W. Klipp, U. B. Priefer, and A. Pühler, J. Mol. Biol. 203 (1988) 715.
17. T. R. Hoover, in: E. W. Triplett (ed.), Prokaryotic Nitrogen Fixation, Horizon Scientific Press, Norfolk, U. K., 2000, p. 131.
18. P. E. Bishop and R. Premakumar, in: G. Stacey, R. H. Burris, and H. J. Evans (eds.), Biological Nitrogen Fixation, Chapman & Hall, New York, London, 1992, p. 736.
19. P. Rangaraj, C. Rüttimann-Johnson, V. K. Shah, and P. W. Ludden, in: E. W. Triplett (ed.), Prokaryotic Nitrogen Fixation, Horizon Scientific Press, Norfolk, U. K., 2000, p. 55.
20. G. Blanco, M. Drummond, P. Woodley, and C. Kennedy, Mol. Microbiol. 9 (1993) 869.

21. M. R. Jacobson, K. E. Brigle, L. T. Bennett, R. A. Setterquist, M. S. Wilson, V. L. Cash, J. Beynon, W. E. Newton, and D. R. Dean, J. Bacteriol. 171 (1989) 1017.
22. R. D. Joerger and P. E. Bishop, J. Bacteriol. 170 (1988) 1475.
23. S. I. Waugh, D. M. Paulsen, P. V. Mylona, R. H. Maynard, R. Premakumar, and P. E. Bishop, J. Bacteriol. 177 (1995) 1505.
24. R. Raina, U. K. Bageshwar, and H. K. Das, Mol. Gen. Genet. 236 (1993) 459.
25. E. D. Wolfinger and P. E. Bishop, J. Bacteriol. 173 (1991) 7565.
26. R. D. Joerger, M. R. Jacobson, and P. E. Bishop, J. Bacteriol. 171 (1989) 3258.
27. P. V. Mylona, R. Premakumar, R. N. Pau, and P. E. Bishop, J. Bacteriol. 178 (1996) 204
28. J. Liang, M. Madden, V. K. Shah, and R. H. Burris, Biochemistry 29 (1990) 8577.
29. R. Overbeek, N. Larsen, G. D. Pusch, M. D´Souza, E. Selkov Jr., N. Kyrpides, M. Fonstein, N. Maltsev, and E. Selkov, Nucleic Acids Res. 28 (2000) 123.
30. K. Schüddekopf, S. Hennecke, U. Liese, M. Kutsche, and W. Klipp, Mol. Microbiol. 8 (1993) 673.
31. T. Thiel, J. Bacteriol. 175 (1993) 6276.
32. T. Thiel, E. M. Lyons, and J. C. Erker, J. Bacteriol. 179 (1997) 5222.
33. T. Thiel, E. M. Lyons, J. C. Erker, and A. Ernst, Proc. Natl. Acad. Sci. USA 92 (1995) 9358.
34. J. A. Leigh, in: E. W. Triplett (ed.), Prokaryotic Nitrogen Fixation, Horizon Scientific Press, Norfolk, U. K., 2000, p. 657.
35. P. S. Kessler, C. Daniel, and J. A. Leigh, J. Bacteriol. 183 (2001) 882.
36. M. Drummond, J. Walmsley, and C. Kennedy, J. Bacteriol. 178 (1996) 788.
37. C. Kennedy and D. Dean, Mol. Gen. Genet. 231 (1992) 494.
38. J. Walmsley, A. Toukdarian, and C. Kennedy, Arch. Microbiol. 162 (1994) 422.
39. P. Yuvaniyama, J. N. Agar, V. L. Cash, M. K. Johnson, and D. R. Dean, Proc. Natl. Acad. Sci. USA 97 (2000) 599.
40. B. Masepohl and W. Klipp, Arch. Microbiol. 165 (1996) 80.
41. H.-S. Jeong and Y. Jouanneau, J. Bacteriol. 182 (2000) 1208.
42. H. Kumagai, T. Fujiwara, H. Matsubara, and K. Saeki, Biochemistry 36 (1997) 5509.
43. M. Schmehl, A. Jahn, A. Meyer zu Vilsendorf, S. Hennecke, B. Masepohl, M. Schuppler, M. Marxer, J. Oelze, and W. Klipp, Mol. Gen. Genet. 241 (1993) 602.
44. G. Gennaro, P. Hübner, U. Sandmeier, A. F. Yakunin, and P. C. Hallenbeck, J. Bacteriol. 178 (1996) 3949.
45. S. Siemann, K. Schneider, K. Behrens, A. Knöchel, W. Klipp, and A. Müller, Eur. J. Biochem. 268 (2001) 1940.
46. M. J. Dilworth, M. E. Eldridge, and R. R. Eady, Biochem. J. 289 (1993) 395.
47. R. W. Miller and R. R. Eady, Biochem. J. 256 (1988) 429.
48. J. Walmsley and C. Kennedy, Appl. Environmental Microbiol. 57 (1991) 622.
49. F. Arsene, P. A. Kaminski, and C. Elmerich, J. Bacteriol. 178 (1996) 4830.
50. R. A. Monteiro, E. M. Souza, S. Funayama, M. G. Yates, F. O. Pedrosa, and L. S. Chubatsu, FEBS Lett. 447 (1999) 283.

51. A. Paschen, T. Drepper, B. Masepohl, and W. Klipp, FEMS Microbiol. Lett. 200 (2001) 207.
52. A. Herrero, A. M. Muro-Pastor, and E. Flores, J. Bacteriol. 183 (2001) 411.
53. G. Scherings, H. Haaker, H. Wassink, and C. Veeger, Eur. J. Biochem. 135 (1983) 591.
54. H. Haaker and C. Veeger, Eur. J. Biochem. 63 (1976) 499.
55. J. Oelze, FEMS Microbiol. Rev. 24 (2000) 321.
56. R. Dixon, R. R. Eady, G. Espin, S. Hill, M. Iaccarino, D. Kahn, and M. Merrick, Nature 286 (1980) 128.
57. F. Luque and R. N. Pau, Mol. Gen. Genet. 227 (1991) 481.
58. R. Premakumar, R. N. Pau, L. A. Mitchenall, M. Easo, and P. E. Bishop, FEMS Microbiol. Lett. 164 (1998) 63.
59. R. D. Joerger, E. D. Wolfinger, and P. E. Bishop, J. Bacteriol. 173 (1991) 4440.
60. U. Gollan, K. Schneider, A. Müller, K. Schüddekopf, and W. Klipp, Eur. J. Biochem. 215 (1993) 25.
61. G. Wang, S. Angermüller, and W. Klipp, J. Bacteriol. 175 (1993) 3031.
62. L. J. Lehman and G. P. Roberts, J. Bacteriol. 173 (1991) 5705.
63. H.-M. Fischer, Microbiol. Rev. 58 (1994) 352.
64. M. J. Merrick and R. A. Edwards, Microbiol. Rev. 59 (1995) 604.
65. A. J. Ninfa and M. R. Atkinson, Trends Microbiol. 8 (2000) 172.
66. P. Hübner, J. C. Willison, P. M. Vignais, and T. A. Bickle, J. Bacteriol. 173 (1991) 2993.
67. R. G. Kranz and D. Foster-Hartnett, Mol. Microbiol. 4 (1990) 1793.
68. B. Masepohl, T. Drepper, A. Paschen, S. Groß, A. Pawlowski, K. Raabe, K.-U. Riedel, and W. Klipp, J. Mol. Microbiol. Biotechnol. (2001), in press.
69. P. J. Cullen, W. C. Bowman, and R. G. Kranz, J. Biol. Chem. 271 (1996) 6530.
70. W. C. Bowman and R. G. Kranz, Genes Development 12 (1998) 1884.
71. D. Foster-Hartnett, P. J. Cullen, E. M. Monika, and R. G. Kranz, J. Bacteriol. 176 (1994) 6175.
72. S. Elsen, W. Dischert, A. Colbeau, and C. E. Bauer, J. Bacteriol. 182 (2000) 2831.
73. J. J. Buggy, M. W. Sganga, and C. E. Bauer, J. Bacteriol. 176 (1994) 6936.
74. M. Kern, P.-B. Kamp, A. Paschen, B. Masepohl, and W. Klipp, J. Bacteriol. 180 (1998) 1965.
75. W. T. Self, A. M. Grunden, A. Hasona, K. T. Shanmugam, Res. Microbiol. 152 (2001) 311.
76. A. M. Grunden, R. M. Ray, J. K. Rosentel, F. G. Healy, and K. T. Shanmugam, J. Bacteriol. 178 (1996) 735.
77. M. Kutsche, S. Leimkühler, S. Angermüller, and W. Klipp, J. Bacteriol. 178 (1996) 2010.
78. P. S. Solomon, A. L. Shaw, M. D. Young, S. Leimkühler, G. R. Hanson, W. Klipp, and A. G. McEwan, FEMS Microbiol. Lett. 190 (2000) 203.
79. R. N. Pau, W. Klipp, and S. Leimkühler, in: G. Winkelmann and C. J. Carrano (eds.), Transition Metals in Microbial Metabolism, Harwood Academic Publishers, Amsterdam, The Netherlands, 1997, p. 217.

80. T. Drepper, S. Groß, B. Masepohl, and W. Klipp, in: F. O. Pedrosa, M. Hungria, G. Yates, and W. E. Newton (eds.), Nitrogen Fixation: From Molecules to Crop Productivity, Kluwer Academic Publishers, Dordrecht, Boston, London, 2000, p. 129.
81. B. Masepohl, R. Krey, and W. Klipp, J. Gen. Microbiol. 139 (1993) 2667.
82. C. M. Halbleib and P. W. Ludden, J. Nutr. 130 (2000) 1081.
83. B. Förster, K. Maner, F. Fassbinder, and J. Oelze, FEMS Microbiol. Lett. 170 (1999) 167.
84. J. Pierrard, P. W. Ludden, and G. P. Roberts, J. Bacteriol. 175 (1993) 1358.
85. A. F. Yakunin, T. V. Laurinavichene, A. A. Tsygankov, and P. C. Hallenbeck, J. Bacteriol. 181, 1994.
86. S. Austin and J. Lambert, J. Biol. Chem. 269 (1994) 18141.
87. R. Premakumar, T. M. Loveless, and P. E. Bishop, J. Bacteriol. 176 (1994) 6139.
88. E. Frise, A. Green, and M. Drummond, J. Bacteriol. 176 (1994) 6545.
89. E. Morett and M. Buck, Proc. Natl. Acad. Sci. USA 85 (1988) 9401.
90. P. Woodley, M. Buck, and C. Kennedy, FEMS Microbiol. Lett. 135 (1996) 213.
91. T. R. Hoover, E. Santero, S. Porter, and S. Kustu, Cell 63 (1990) 11.
92. E. Santero, T. R. Hoover, A. K. North, D. K. Berger, S. C. Porter, and S. Kustu, J. Mol. Biol. 227 (1992) 602.
93. R. Wassem, E. M. de Souza, M. G. Yates, F. de Oliveira Pedrosa, and M. Buck, Mol. Microbiol. 35 (2000) 756.
94. R. D. Joerger, T. M. Loveless, R. N. Pau, L. A. Mitchenall, B. H. Simon, and P. E. Bishop, J. Bacteriol. 172 (1990) 3400.
95. M. M. Georgiadis, H. Komiya, P. Chakrabarti, D. Woo, J. J. Kornuc, and D. C. Rees, Science 257 (1992) 1653.
96. H. Schindelin, C. Kisker, J. L. Schlessman, J. B. Howard, and D. C. Rees, Nature 387 (1997) 370.
97. W. N. Lanzilotta and L. C. Seefeldt, Biochemistry 35 (1996) 16770.
98. K. Schneider, U. Gollan, M. Dröttboom, S. Selsemeier-Voigt, and A. Müller, Eur. J. Biochem. 244 (1997) 789.
99. R. Chatterjee, R. M. Allen, P. W. Ludden, and V. K. Shah, J. Biol. Chem. 272 (1997) 21604.
100. J. T. Bolin, A. E. Ronco, T. V. Morgan, L. E. Mortenson, and N. H. Xuong, Proc. Natl. Acad. Sci. USA 90 (1993) 1078.
101. J. W. Peters, M. H. Stowell, S. M. Soltis, M. G. Finnegan, M. K. Johnson, and D. C. Rees, Biochemistry 36 (1997) 1181.
102. C. Z. Blanchard and B. J. Hales, Biochemistry 35 (1996) 472.
103. Y. A. Onate, M. G. Finnegan, B. J. Hales, and M. K. Johnson, Biochim. Biophys. Acta 1164 (1993) 113.
104. K. Schneider, U. Gollan, S. Selsemeier-Voigt, W. Plass, and A. Müller, Naturwissenschaften 81 (1994) 405.
105. W. R. Hagen, R. R. Eady, W. R. Dunham, and H. Haaker, FEBS Lett. 189 (1985) 250.

106. S. J. Yoo, H. C. Angove, B. K. Burgess, M. P. Hendrich, and E. Münck, J. Am. Chem. Soc. 121 (1999) 2534.
107. A. J. Pierik, H. Wassink, H. Haaker, and W. R. Hagen, Eur. J. Biochem. 212 (1993) 51.
108. R. C. Tittsworth and B. J. Hales, Biochemistry 35 (1996) 479.
109. E. Krahn, B. J. R. Weiss, M. Kröckel, J. Groppe, G. Henkel, S. P. Cramer, A. K. Trautwein, K. Schneider, and A. Müller, JBIC (2001), in press.
110. N. Ravi, V. Moore, S. G. Lloyd, B. J. Hales, and B. H. Huynh, J. Biol. Chem. 269 (1994) 20920.
111. R. R. Eady, Chem. Rev. 96 (1996) 3013.
112. J. Kim and D. C. Rees, Science 257 (1992) 1677.
113. R. N. Pau, M. E. Eldridge, D. J. Lowe, L. A. Mitchenall, and R. R. Eady, Biochem. J. 293 (1993) 101.
114. B. E. Smith, R. R. Eady, D. J. Lowe, and C. Gormal, Biochem. J. 250 (1988) 299.
115. T. R. Hoover, J. Imperial, P. W. Ludden, and V. K. Shah, Biochemistry 28 (1989) 2768.
116. B. Masepohl, S. Angermüller, S. Hennecke, P. Hübner, C. Moreno-Vivian, and W. Klipp, Mol. Gen. Genet. 238 (1993) 369.
117. E. M. Lyons and T. Thiel, J. Bacteriol. 177 (1995) 1570.
118. V. K. Shah, J. R. Allen, N. J. Spangler, and P. W. Ludden, J. Biol. Chem. 269 (1994) 1154.
119. J. M. Arber, B. R. Dobson, R. R. Eady, P. Stevens, S. S. Hasnain, C. D. Garner, and B. E. Smith, Nature 325 (1987) 327.
120. G. N. George, C. L. Coyle, B. J. Hales, and S. P. Cramer, J. Am. Chem. Soc. 110 (1988) 4057.
121. J. Chen, J. Christiansen, R. C. Tittsworth, B. J. Hales, S. J. George, D. Coucovanis, and S. P. Cramer, J. Am. Chem. Soc. 115 (1993) 5509.
122. I. Harvey, J. M. Arber, R. R. Eady, B. E. Smith, C. D. Garner, and S. S. Hasnain, Biochem. J. 266 (1990) 929.
123. D. J. Lowe and B. E. Smith, Biochem. J. 209 (1985) 207.
124. D. J. Lowe, K. Fisher, and R. N. Thorneley, Biochem. J. 292 (1993) 93.
125. B. J. Hales, Adv. Inorg. Biochem. 8 (1990) 165.
126. V. K. Shah, J. Imperial, R. A. Ugalde, P. W. Ludden, and W. J. Brill, Proc. Natl. Acad. Sci. USA 88 (1986) 1636.
127. V. K. Shah, P. Rangaraj, R. Chatterjee, R. M. Allen, J. T. Roll, G. P. Roberts, and P. W. Ludden, J. Bacteriol. 181 (1999) 2797.
128. C. Ruttimann-Johnson, P. Rangaraj, V. K. Shah, and P. W. Ludden, J. Biol. Chem. 276 (2001) 4522.
129. R. R. Eady, B. E. Smith, K. A. Cook, and J. R. Postgate, Biochem. J. 128 (1972) 655.
130. E. Krahn, K. Schneider, and A. Müller, Appl. Microbiol. Biotechnol. 46 (1996) 285.
131. R. R. Eady, R. L. Robson, T. H. Richardson, R. W. Miller, and M. Hawkins, Biochem. J. 244 (1987) 197.

132.D. J. Lowe and R. N. F. Thorneley, Biochem. J. 224 (1984) 877.
133.M. J. Dilworth and R. R. Eady, Biochem. J. 277 (1991) 465.
134.M. J. Dilworth, R. R. Eady, and M. E. Eldridge, Biochem. J. 249 (1988) 745.
135.M. J. Dilworth, R. R. Eady, R. L. Robson, and R. W. Miller, Nature 327 (1987) 167.
136.M. J. Dilworth, M. E. Eldridge, and R. R. Eady, Biochem.J. 289 (1993) 395.
137.W. Plass, J. Mol. Struct. (Theochem.) 315 (1994) 53.
138.D. J. Scott, D. R. Dean, and W. E. Newton, J. Biol. Chem. 267 (1992) 20002.
139.P. T. Pienkos and W. J. Brill, J. Bacteriol. 145 (1981) 743.
140.R. H. Maynard, P. Premakumar, and P. E. Bishop, J. Bacteriol. 176 (1994) 5583.
141.S. Noda, M. Ohkuma, R. Usami, K. Horikoshi, and T. Kudo, Appl. Environment. Microbiol. 65 (1999) 4935.

Nitrogen Fixation at the Millennium.
G. Jeffery Leigh, Editor.

CHAPTER 9

Advances Towards the Mechanism of Nitrogenases

Richard A. Henderson
The Department of Chemistry, University of Newcastle, Newcastle-upon-Tyne, NE1 7RU, UK

1. INTRODUCTION TO BIOLOGICAL NITROGEN FIXATION

In this chapter the mechanism of biological nitrogen fixation will be described. However, despite the appreciable advances that have been made particularly over the last decade, our understanding of how nitrogenases transform dinitrogen into ammonia at the molecular level is still rudimentary.

Since the 1960's research has established: (i) a detailed structure of the enzyme; (ii) a kinetic model of the enzyme action and (iii) an in-depth understanding of how dinitrogen is bound and transformed into ammonia at simple abiological metal sites. So, what is the problem? Surely, you might think, we can piece together the pertinent features from these three areas and the mechanism of biological nitrogen fixation will emerge. But two key parts of the jigsaw are missing: how is dinitrogen bound and transformed at Fe-S clusters and how does the polypeptide modulate the reactivity. Only in the last couple of years have these aspects been addressed.

The chapter will start with a brief description of the basic structural features of the nitrogenase proteins, paying particular attention to the established functions of the various Fe-S-based clusters within the proteins. Subsequently, the mechanism of nitrogenase will be described starting at the level of the protein interactions and finishing with studies on the protonation of coordinated dinitrogen at simple metal sites. Discussion of the chemistry of Fe-S-based clusters including the extracted FeMo-cofactor will be presented throughout as appropriate.

The mechanism of nitrogen fixation has always been of interest to theoreticians and this involvement has escalated in the last few years. Although the major conclusions from these calculations will be presented in this chapter the focus will be on experimental work.

2. THE MECHANISM OF BIOLOGICAL NITROGEN FIXATION

In essence, the "nitrogen fixation problem" has always been one of kinetics and mechanism. Thermodynamics are not a problem as has been been discussed several times in the past [1]. I am not going to labour these arguements again here, but the important point is that the conversion of dinitrogen plus dihydrogen into ammonia is a thermodynamically favourable reaction (Equation 1).

$$N_2 + 3H_2 \leftrightarrows 2NH_3 \quad \Delta H^o_{f(g)} = -92.6 \text{ kJ mol}^{-1} \tag{1}$$

$$N_2 + H_2 \leftrightarrows N_2H_2 \quad \Delta H^o_{f(g)} = +203 \text{ kJ mol}^{-1} \tag{2}$$

It is only the formation of the diazene intermediate that is a thermodynamic problem as shown in Equation (2). After forming diazene everything else is thermodynamically downhill. An important qualification is that the thermodynamics present in Equations (1) and (2) refer to gaseous molecules. One of the main tenets of the action of nitrogenase is that dinitrogen is bound to, and transformed at, a metal site.

Up to 1986 the consensus view was there was only one type of nitrogenase. Since then it has become clear that there are at least four types [2,3]. These nitrogenases are distinct at the genetic level and the principle chemical difference between them is the metal content, and more specifically the metal content of the cofactor. The "usual" nitrogenase contains molybdenum, whilst another contains vanadium (V-based nitrogenase) in the place of molybdenum, and another contains only iron (Fe-only nitrogenase). In addition, an apparently entirely new type of nitrogenase has recently been isolated from *Streptomyces thermoautropicus* which contains molybdenum, but requires dioxygen and consumes carbon monoxide [4]. The lion's share of this chapter will be devoted to studies on the "usual" Mo-based nitrogenase. I make no apologies for this. The Mo-based nitrogenase is the one that has been studied most extensively.

2.1 The electron-transfer pathway of nitrogenase

At its crudest level the mechanism of nitrogenase is the electron-transfer pathway shown in Figure 1, which is common to all nitrogenases. Electrons are supplied by a ferredoxin or flavodoxin *in vivo* and transferred initially to the Fe-protein of nitrogenase, which contains a single $\{Fe_4S_4\}$ cuboidal cluster [5,6]. The electrons are subsequently transferred to the MoFe-protein and thence to reduction of dinitrogen. The electron transfer to dinitrogen is coupled to proton transfer and the hydrolysis of MgATP.

The MoFe-protein contains two structurally unique Fe-S-based clusters. The P-clusters almost certainly act as capacitors, storing electrons until they are required by the active site, for the conversion of dinitrogen into ammonia. The active site is called FeMo-cofactor.

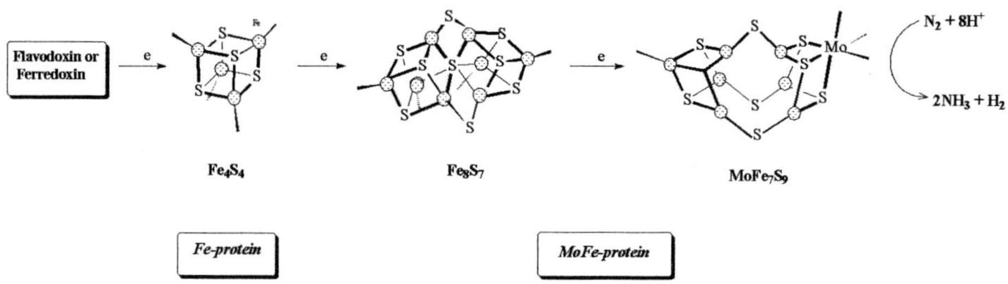

Figure 1. Electron transfer pathway for Mo-based nitrogenase.

The biological role of nitrogenase is to convert dinitrogen into ammonia. However, in the laboratory, the enzyme is capable of transforming a variety of other (mostly unsaturated) molecules and ions [3] such as H^+, RNC, N_3^-, CN^-, cyanamide alkynes, cyclopropene and N_2O. All these so-called "alternative" substrates are transformed by sequences of coupled electron and proton transfer reactions. Although there have been several mechanistic studies on the reactions of these alternative substrates with nitrogenase, this chapter will restrict its attention to dinitrogen.

2.2 The Fe-protein

The Fe-protein plays three roles in nitrogen fixation: (i) transfer of electrons; (ii) biosynthesis of FeMo-cofactor and (iii) regulation of the biosynthesis of the alternative nitrogenases. The Fe-protein has been isolated and purified from a variety of different organisms. Irrespective of their origin, they are all much the same size (M_r ~ 65 kDa) and have an α_2 dimer structure with a single cuboidal $\{Fe_4S_4\}$ cluster bound between the subunits *via* cysteinate amino acid residues. The cluster is bound near the surface of the protein and is easily accessible to solvent [2]. The amino acid sequence of Fe proteins from more than 20 types of bacteria have been shown to have an identity of *ca.* 45-90%.

The addition of MgATP and MgADP to the Fe-protein causes a shift in the mid-point potential from –350 mV to –450 mV [7] and a change in EPR and CD spectra, attributed to conformational changes in the reduced state with S = ½ spin states. For many years it was believed that the Fe-protein acted as a one-electron donor to the MoFe-protein. However, it has now been shown that the two-electron transfer from the Fe-protein occurs at a potential which is accessible under physiological conditions.

Thus, the true role of the protein could be the supply of electrons in a pairwise fashion [8,9].

2.3 The MoFe-protein

The MoFe-protein isolated from a number of different bacterial sources are all similar with $M_r \sim 220$ kDa and an $\alpha_2\beta_2$ tetramer structure. The two $\alpha\beta$ dimers interact predominantly between the helices of the β-subunits at the tetramer interface. A rather open channel (diameter *ca.* 8 Å) goes through the centre of the tetramer. Each $\alpha\beta$ subunit contains one P-cluster and one FeMo-cofactor [10-12].

The VFe- and FeFe-proteins are $\alpha_2\beta_2\gamma_2$ hexamers, with the $\alpha\beta$ subunits being very similar to those of the MoFe-protein. There is considerable similarity in the primary sequences of MoFe-protein with FeFe-protein and VFe-protein of the alternative nitrogenases (32%), and the similarity between the Fe-only protein and VFe-protein primary structures is even greater (55%).

2.3.1 P-Clusters and the Storage of Electrons

The P-clusters are buried within the MoFe-protein at the interface of the α and β subunits with each subunit supplying three cysteinate ligands: one cysteinate bridges two iron atoms and the other two cysteinates each bind single iron atoms. The P-clusters are close to the interface between the Fe-protein and the MoFe-protein and comprise $\{Fe_8S_7\}$ clusters. Electrons are stored in the P-clusters until they are required by the FeMo-cofactor. The two P-clusters, in the different sub-units, are *ca.* 70 Å apart whilst the distance between the P-cluster and the FeMo-cofactor in the same subunit is *ca.* 20 Å.

The structure of the P-clusters was the cause of some dispute. However, the situation has now been clarified and discussed in detail elsewhere [3]. It turns out that the P-cluster core structure changes upon oxidation as shown in Figure 2.

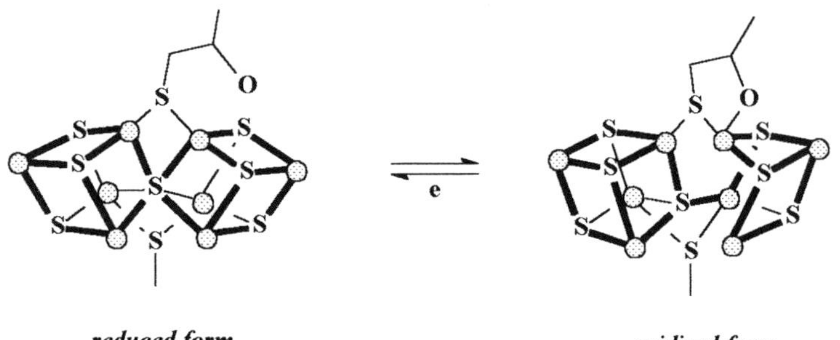

Figure 2. Structural changes in P-clusters with redox level.

2.3.2 FeMo-cofactor: The Active Site

The FeMo-cofactor is enclosed within the α-subunit of the MoFe-protein, *ca.* 10 Å from the surface. The cluster has a $\{MoFe_7S_9\}$ core and is bound to the polypeptide *via* Cys-α275 (to the unique tetrahedral iron) and His-α442 (to molybdenum). In contrast to the P-clusters, there is no evidence that FeMo-cofactor undergoes major structural changes on electron transfer. *R*-homocitrate is also bound to the molybdenum as a bidentate ligand, coordinated through alkoxy and carboxylate groups, with the carboxylate groups of the $-CH_2CH_2CO_2^-$ and $-CH_2CO_2^-$ arms free. The molybdenum end of the cofactor is surrounded by water, whilst the tetrahedral Fe end is effectively anhydrous.

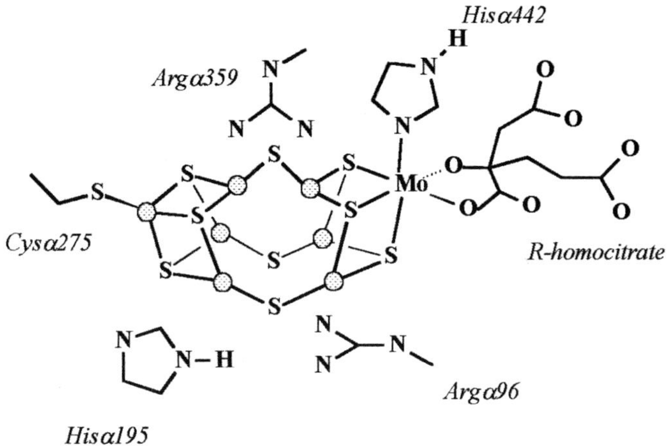

Figure 3. The main components of the active site in Mo-based nitrogenase.

A number of amino acid residues in the active-site cavity hydrogen bond to the FeMo-cofactor as shown in Figure 3, and are essential to catalysis. The key residues are Arg-α96, Arg-α359, Phe-α381 which span, and hydrogen bond to, the centre of the cluster. Arg-α277 is close to Cys-α275, which is the ligand to iron. When Arg-α277 is mutated to His the resulting enzyme does not reduce dinitrogen but it does reduce acetylene, cyanide, azide, and protons. Gln-α191 is hydrogen-bonded to the homocitrate and when changed to Lys the mutant enzyme is unable to reduce dinitrogen but does reduce acetylene to ethane and ethylene [13,14]. Finally, His-α195 is hydrogen bonded to the cofactor. Mutating His-α195 to Gln produces an enzyme that binds dinitrogen (as demonstrated by its inhibition of acetylene and proton reduction) but does not reduce dinitrogen [15]. Clearly, the amino acid residues surrounding cofactor play crucial roles in modulating the cluster's reactivity.

The most compelling evidence that the FeMo-cofactor is the substrate-binding site comes from studies on the *nifV* mutant nitrogenase [16]. The *nifV* mutant nitrogenase has citrate bound in the place of R-homocitrate, and has an altered substrate specificity. Most notably the *nifV* mutant nitrogenase is a poor dinitrogen fixer and dihydrogen evolution is inhibited by CO. When the FeMo-cofactor from the *nifV* mutant is removed from the MoFe-protein and then added to apo-MoFe-protein from the wild type nitrogenase, the product has the same substrate specificities as the *nifV* mutant.

There have been few studies on the alternative V-based and Fe-only enzymes but it is clear that they have slightly different product specificities to those of the Mo-based nitrogenase. Thus the Mo-based nitrogenase reduces dinitrogen exclusively to ammonia and acetylene to ethylene. In contrast, the V-based nitrogenase produces some hydrazine in the reduction of dinitrogen [17], and some ethane in the reduction of acetylene [18]. The mechanistic origin of these product specificities is unclear at the moment.

2.4 The Fe-Protein: MoFe-Protein Interaction

Both the Fe-protein and the MoFe-proteins contain redox-active sites. Consequently it is a non-trivial task to map out the electron-transfer pathway, both between the Fe- and MoFe-proteins and within the MoFe-protein. It is initially simplest to consider the Fe-protein. The Fe-protein redox cycle is shown in Figure 4 and indicates both the electron-transfer and MgATP-hydrolysis [19] reactions that occur on the Fe-protein during catalysis. Although this cycle was developed for the Mo-based nitrogenase of *Klebsiella pneumoniae*, it appears to be equally applicable to nitrogenases from other organisms and to the V-based and Fe-only nitrogenases.

Having first been reduced by the external ferredoxin or flavodoxin, the Fe-protein binds to the oxidised MoFe-protein. The binding of the two proteins occurs at the diffusion-controlled rate, or close to it ($k > 5 \times 10^7$ $dm^3\ mol^{-1}\ s^{-1}$). The adduct thus formed undergoes electron transfer and finally the two proteins dissociate.

It is clear from Figure 4 that the structure of the adduct between the Fe- and MoFe-proteins is central to the proposed mechanism for nitrogenase. Recently, the X-ray crystal structure of the adduct was reported [20]. By using $[AlF_4]^-$ to bind the Fe- and MoFe-proteins together the dissociation of the two proteins has a half life of *ca.* 20 h, rather than the *ca.* 100 ms observed during enzyme turnover. The structure of the adduct reveals that the Fe-protein has undergone significant conformational changes on binding to the MoFe-protein. The $\{Fe_4S_4\}$ cluster is 14 Å from the P-cluster, which is mid-way between the $\{Fe_4S_4\}$ cluster and FeMo-cofactor. In addition, ADP is bound parallel to the Fe-protein axis rather than across the dimer interface as it is in the isolated protein.

Further studies [21] on the Fe-protein:MoFe-protein adduct show that the mid-point potential of the Fe-protein is decreased to –620 mV (200 mV lower than the free protein), and the redox potential of the P-cluster is lowered to –380 mV (80 mV lower than the free protein). There is no change in the redox potential of the FeMo-cofactor but ENDOR studies show that non-exchangable protons in the vicinity of the cofactor are altered upon binding of the Fe- and MoFe-proteins [22].

After electron transfer within the adduct, the dissociated, oxidised Fe-protein is reduced by the external reductant (Flavodoxin or Ferredoxin, or $S_2O_4^{2-}$ *in vitro*), in a MgATP-dependent process. That the Fe-protein must dissociate before it can be reduced is contentious. Dissociation does not seem to be necessary with reductants other than dithionite [23].

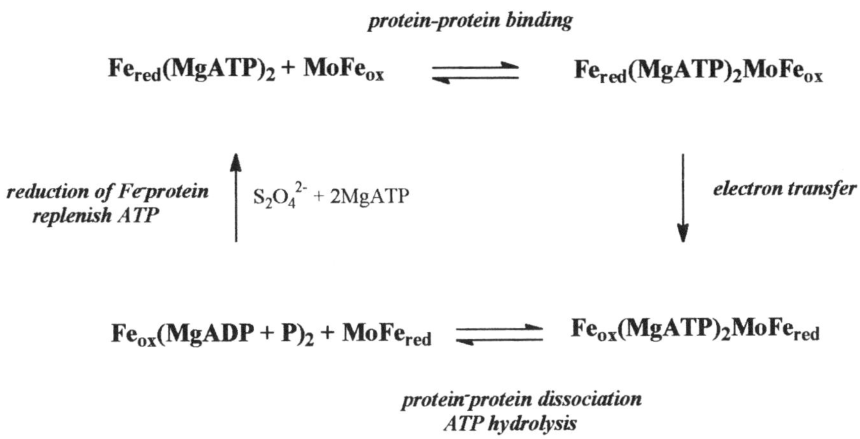

Figure 4. The Fe-protein cycle.

2.5 The MoFe-protein cycle

It is now time to consider the processes that result in the binding and transformation of dinitrogen on the protein. The sequence of reactions which changes the redox state of the MoFe enzyme is shown in Figure 5. The four elementary steps of the Fe-protein cycle (Figure 4) are required every time an electron is transferred to the MoFe-protein. Thus, the single arrows between each state E_n in Figure 5 correspond to the sequence shown in Figure 4. This same sequence of reactions is performed a total of eight times to complete one cycle of the MoFe-protein. The rate-limiting step of nitrogenase is dissociation of the component This same sequence of reactions is performed a total of eight times to complete one cycle of the MoFe-protein. The rate-limiting step of nitrogenase is dissociation of the component proteins ($k = 6.4\ s^{-1}$) [19].

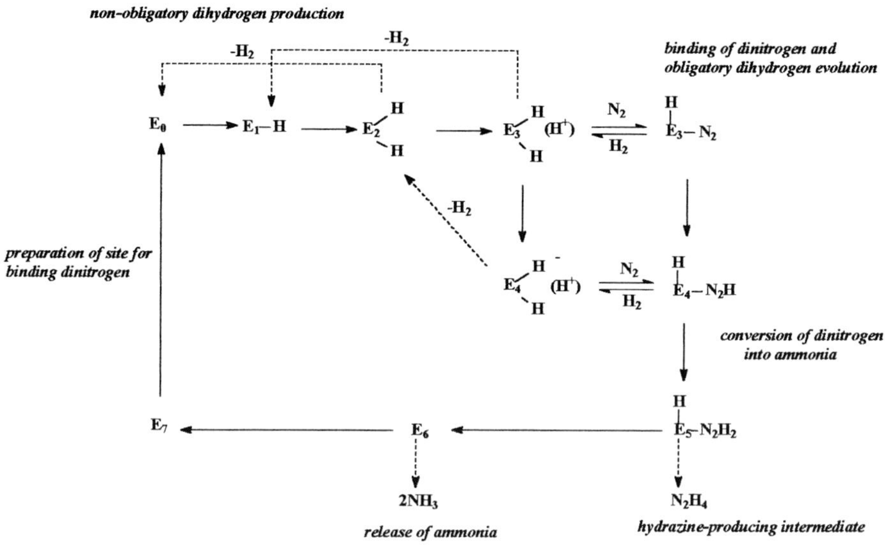

Figure 5. The MoFe-protein cycle.

In Figure 5 the various redox states of the MoFe-protein are represented by E_n. Specifically, the subscript n corresponds to the total number of electrons transferred from the Fe-protein to the MoFe-protein. Eight states of the MoFe-protein are required because the limiting stoichiometry of Mo-based nitrogenase is that shown in Equation (3), in which one dihydrogen is produced for every dinitrogen transformed into ammonia. As well as the six electrons necessary for the reduction of dinitrogen, two electrons go into formation of dihydrogen

$$H_2 + 8H^+ + 8e \rightarrow 2NH_3 + H_2 \qquad (3)$$

The main features of the mechanism have been derived from studies and simulations of the pre-steady-state time-courses of ammonia and hydrazine production and dihydrogen release from acid quench experiments [19]. A particularly crucial factor in analysing the system has been understanding the lag phases observed with each substrate. These lags correspond to the enzyme in its resting state (E_0) preparing itself for the binding of the substrate. For dinitrogen this binding occurs at E_3.

E_0 represents the state of the MoFe-protein as isolated. E_1H and E_2H_2 have one and two electrons transferred, respectively. It is assumed that the negative charge introduced on transfer of the electron is neutralised by the addition of a proton It has been proposed that the protons form metal-hydrides that can produce dihydrogen (Section 3.3). E_2H_2 is the first state that can evolve dihydrogen. Addition of the third electron gives $E_3H_2(H^+)$. The identity of this species needs some further explanation. $E_3H_2(H^+)$ represents a reduced dihydride site with a proton bound to an adjacent amino acid side chain. $E_3H_2(H^+)$ and its successor, $E_4H_2(H^+)$, are the first species at which dinitrogen can bind.

There have been few kinetic studies on the V-based nitrogenase. However, it has been shown that the rate of electron-transfer from the Fe-protein to the VFe-protein (within the adduct) is appreciably slower ($k = 46\ s^{-1}$) than that to the MoFe-protein ($k = 220\ s^{-1}$) [24].

2.5.1 Electron-Transfer Sequence from {Fe_4S_4} to FeMo-cofactor

The electron-transfer from the {Fe_4S_4} cluster of the Fe-protein to the FeMo-cofactor is very rapid. Intermediate stages in the electron-transfer have been detected only under two circumstances. (i) Changes in the electronic spectrum have been observed in the first 600 ms of enzyme turnover, and these have been correlated with EPR spectral features g = 5.4 and 5.7, but only under argon. The EPR signals have been attributed to one-electron oxidation of the P-cluster [25]. (ii) At high salt concentrations there is a lag period after the {Fe_4S_4} is oxidised, but before the FeMo-cofactor is reduced. It is proposed that this lag corresponds to the electron residing on the P-cluster for an appreciable length of time [26].

The role of MgATP hydrolysis remains ambiguous. Certainly this hydrolysis is essential for nitrogen fixation and it is generally accepted that two MgATP's are required for each electron transferred, under optimal conditions. Two roles have been proposed for MgATP hydrolysis. (a) Conformational change in the Fe-protein to allow dissociation. Certainly, the Fe-protein contracts by *ca.* 2 Å upon binding of MgATP [27]. (b) Conformational changes in the Fe- and MoFe-proteins that gates the electron transfer from P-cluster to FeMo-cofactor [3,28].

2.6 Detection of intermediates on the enzyme

Nitrogenase is "designed" to bind dinitrogen rapidly, convert dinitrogen into ammonia and release the product and so it is difficult to detect intermediates during turnover. The problems are compounded by the variety of different states that the MoFe-protein can attain, as shown in Figure 5.

Only one enzyme-bound dinitrogen intermediate has been detected. During enzyme turnover under dinitrogen, acid or alkali quench produces hydrazine [29]. This does not correspond to the formation of hydrazine as an intermediate since no free hydrazine is produced during the conversion of dinitrogen into ammonia. Rather this hydrazine is released after the acid or alkali quench, from a "hydrazine-level" intermediate. Based on comparison of the known chemistry of dinitrogen at mononuclear metal complexes (see Section 3.4.4) it was proposed that the most likely intermediate is a coordinated hydrazide, $M{=}NNH_2$ (M = metal). Realistically, the enzyme-bound intermediate could be anything from $M{-}N_2$ through to $M{-}NHNH_3$ (or M-

NH_2NH_2). It seems reasonable that only after N-N cleavage does the reaction become incapable of producing hydrazine.

Recently, the binding of CO to nitrogenase has been monitored using stopped-flow FTIR spectrometry [30]. Although CO is not a substrate, it is an inhibitor for nitrogenase and the isoelectronic relationship between CO and dinitrogen may mean that indirect information concerning dinitrogen-binding can be obtained from information on CO. Mixing the Fe- and MoFe-proteins with MgATP and CO produced a complex series of transient IR absorptions, of which the most notable were, one relatively intense peak at 1936 cm^{-1} and two smaller bands at 1958 and 1906 cm^{-1}. The 1906 cm^{-1} band is short-lived, reaching maximum intensity after *ca.* 7 s. In contrast, the other two peaks are relatively long-lived with the 1936 cm^{-1} band peaking at *ca.* 60 s and the 1958 cm^{-1} band peaking at *ca.* 100 s. The data indicate that three distinct terminally bound CO species are formed. One problem is that the enzyme contains three different Fe-S-based clusters, all of which are, in principle, capable of binding CO. FTIR spectroelectrochemical studies on FeMo-cofactor *extracted* from nitrogenase have shown that reduced FeMo-cofactor binds CO and gives rise to bands at 1870, 1910 and 1929 cm^{-1}. These bands are all within the spectral domain of those observed with the enzyme [31].

3. THE BINDING OF DINITROGEN

Little is known about how dinitrogen binds to FeMo-cofactor, or how the dinitrogen is subsequently transformed into ammonia at this site. Consequently, our understanding of this chemistry is currently based on theoretical calculations and studies of simple mononuclear and binuclear complexes.

3.1 Theoretical studies on the binding of dinitrogen to FeMo-cofactor

It was long assumed that once the structure of the FeMo-cofactor was determined the nitrogen fixation problem would be effectively solved: the structure of the active site would reveal how dinitrogen is activated. Nothing could be further from the truth. If anything, knowing the structure of the cofactor has opened-up a plethora of possible binding modes for dinitrogen. Ever since the structure of FeMo-cofactor was reported, theoreticians have been calculating where and how dinitrogen binds to the cluster. The results of these studies are summarised in Figure 6.

Thus, Extended Huckel calculations [32] indicate thatdinitrogen binds in an end-on fashion to four iron atoms in the centre of the cluster. In contrast, *ab initio* Density Functional calculations [33] indicate side-on binding of dinitrogen to the same iron atoms. Intermediate Neglect of Differential Overlap (INDO) calculations [34] favour binding of dinitrogen to the inside of the trigonal cavity in the centre of the cofactor, with the nitrogen atom bound tighter to the molybdenum end than the iron end. However, the cavity is 0.5Å too small to accommodate dinitrogen. Clearly some expansion of the cavity has to occur before incorporation of dinitrogen. EXAFS studies indicate that the volume of FeMo-cofactor actually *decreases* upon one-electron reduction [35].

Semi-empirical Complete Neglect of Differential Overlap (CNDO) calculations indicate that dinitrogen binds with one nitrogen atom in the centre of the trigonal cavity of cofactor whilst the other nitrogen atom is on the outside [36]. Recently Density Functional calculations have been used to investigate further the binding of CO, dinitrogen or dihydrogen to a FeMo-cofactor model [37]. The key findings of these calculations are as follows. (i) CO binds more tightly than dinitrogen. (ii) Dihydrogen absorption is facile and can take place at the same position as dinitrogen. The dihydrogen can dissociate and when it does so the most stable position of the hydrogen atoms is the μ_2-S atoms. (iii) Dinitrogen binds in an end-on fashion to the cluster but in order to spend an appreciable time bound the cluster has to be at least one-electron reduced (an associated proton is bound to a nearby amino acid residue). The formation of the NNH residue is rate-limiting in ammonia formation, provided electrons and protons are in a plentiful excess. In the absence of hydrogen atoms, NNH decays to dinitrogen and μ_2-SH. If three μ_2-SH's are present then irreversible formation of hydrazine ensues. (iv) Finally, it is proposed that the reasons why FeMo-cofactor is such a large cluster is because it is necessary to load the site with electrons and protons for nitrogen fixation. In addition, to bind the unreactive dinitrogen requires an unsaturated transition metal site. However, this site needs to be unreactive to other potential ligands. In order to accomplish this the iron atoms are surrounded by a sulfur shell. Effectively, the cluster looks like a passivated metallic iron. In order to bind substrates a distortion occurs which results in one of the iron atoms sticking out. This distortion can only occur if the cluster is large.

Finally, B3LYP Hybrid Density Functional calculations indicate that dinitrogen binds to the molybdenum atom after dissociation of *R*-homocitrate. We will return to this model in Section 5.2.2.

3.2 Biomimetic studies and their relationship to the enzyme action

Studies on simple chemical systems have developed a sophisticated understanding of how dinitrogen binds to metal sites and the subsequent protonation and electron-transfer chemistry that results in the formation of ammonia [38,39]. This chemistry is the basis for the enzyme's action.

At this stage it is worth refocusing our attention on the MoFe-protein cycle in Figure 5. This cycle can be divided into three main areas: (i) binding of dinitrogen and evolution of dihydrogen (E_0 to E_3); (ii) the conversion of dinitrogen into hydrazide species (E_3 to E_5) and (iii) the conversion of hydrazide into ammonia (E_5 to E_6). The remainder of the cycle corresponds to the oxidised MoFe-protein being reduced back to the E_0 state. In the remainder of this article we will outline the chemical studies that aim to model these stages.

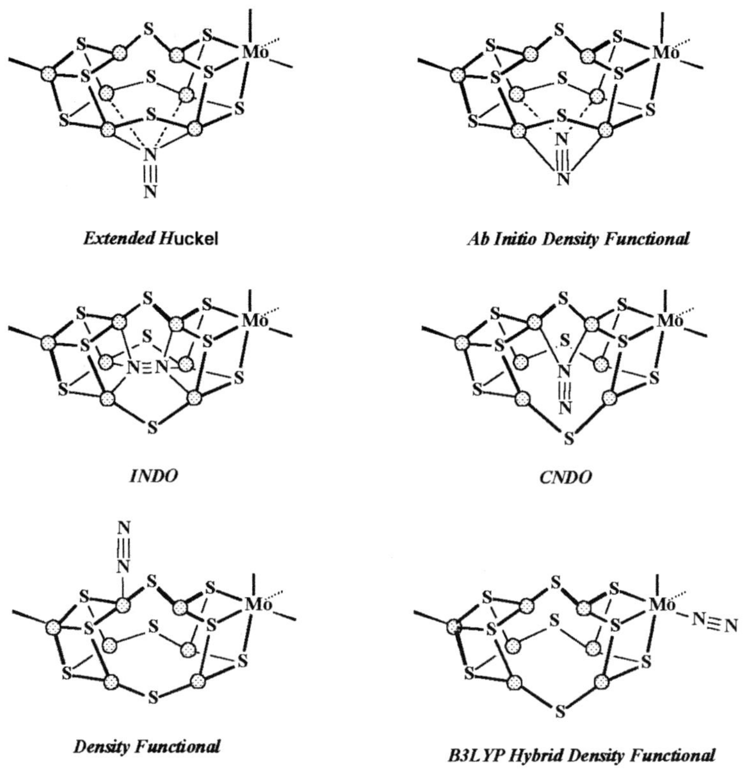

Figure 6. Proposed binding of dinitrogen to FeMo-cofactor based on calculations.

3.3 Dinitrogen displacing dihydrogen

In the absence of any substrate, nitrogenase reduces protons to dihydrogen. Indeed, even under 50 atmospheres of dinitrogen [40] dihydrogen evolution cannot be entirely suppressed. As we saw in Equation (3), for the Mo-based nitrogenase at least one mole of dihydrogen is produced for every mole of dinitrogen reduced. An early proposal [41] suggested that this so-called obligatory evolution of dihydrogen came about by decomposition of a coordinated diazene intermediate as shown in Equation (4). Although the decomposition of free diazene into dinitrogen and dihydrogen [42] is well established in the gas phase, there is no evidence that coordinated diazene has a similar reactivity.

$$M\text{-}N_2H_2 \longrightarrow M\text{-}N_2 + H_2 \qquad (4)$$

It has been known for some time that molecules and ions can bind to metal sites by displacing dihydrogen. The usual pathway for this generic reaction is shown in Equation (5), in which the reductive-elimination of dihydrogen generates a vacant site on M at which L binds.

$$M\begin{matrix}H\\H\end{matrix} \xrightarrow{-H_2} M \xrightarrow{L} M{-}L \qquad (5)$$

The similarities of Equations (3) and (5) led to the suggestion that dinitrogen binds to the FeMo-cofactor by displacing dihydrogen. The obligatory evolution of dihydrogen is implicitly built into the MoFe-protein cycle shown in Figure 5. The species E_1H, E_2H_2 and $E_3H_2(H^+)$ correspond to a proposed monohydride, dihydride and reduced dihydride, respectively.

Three observations have been instrumental in suggesting a metal-hydride binding site in nitrogenase: (i) the obligatory evolution of dihydrogen as implicated in the limiting stoichiometry of the Mo-based enzyme; (ii) dihydrogen is a competitive inhibitor of dinitrogen reduction and (iii) dinitrogen accelerates the formation of HD in the reaction of the enzyme under an atmosphere of D_2.

It has been proposed that the species $E_3H_2(H^+)$ in Figure 5 is a metal dihydride which is activated by protonation from the adjacent protic amino acid side chain. There are two features of this mechanism that need attention. First, it is implied that protonation of a hydride site facilitates the release of dihydrogen from the site. Secondly, proton attack does not occur at the metal since this would produce a trihydride, which is inconsistent with studies in the presence of D_2 (or T_2). Specifically, the following experimental observations need to be rationalised in any proposition concerning the hydridic binding site. (i) During turnover under D_2, no HD is formed. (ii) In experiments under T_2, no T^+ is incorporated into the solution. Chemists have been particularly imaginative in developing metal-hydride models for the substrate-binding site. These models will now be described.

3.3.1 Protonation of Metal Hydrides

It has been known for a long time that dinitrogen will displace dihydrogen at a metal site, as illustrated in Equation (6).

$$[CoH_3(PPh_3)_3] + N_2 \rightarrow [CoH(N_2)(PPh_3)_3] + H_2 \qquad (6)$$

Similarly, the ability of protons to facilitate the release of dihydrogen frommetal hydrides has been known for some time. The combination of these two elementary reactions to bind dinitrogen by acid-catalysed displacement of dihydrogen was realised [43,44] in the (idealised) reactions shown in Figure 7.

The reactions in Figure 7 are complete within a few seconds, under ambient conditions. In contrast, in the absence of acid, dissociation of dihydrogen from $[MoH_4(Ph_2PCH_2CH_2PPh_2)_2]$ has a half-life of *ca.* 2 h. It is evident that with some substrates (such as alkynes and CO), no further transformation of the substrate occurs, upon binding to the molybdenum, whereas other substrates (*e.g.*, N_3^-, MeNC and N_2) become protonated. When bound, dinitrogen is protonated to form the hydrazide.

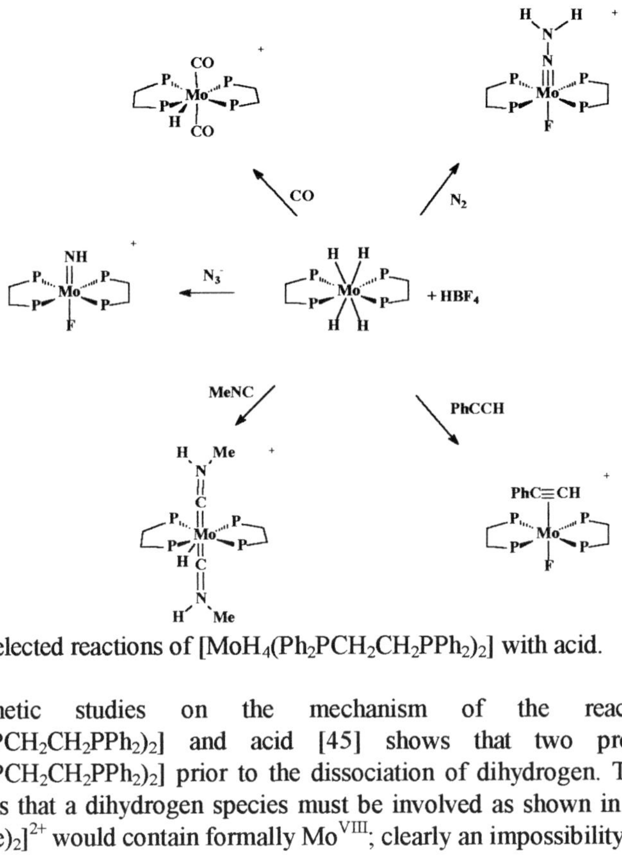

Figure 7. Selected reactions of $[MoH_4(Ph_2PCH_2CH_2PPh_2)_2]$ with acid.

Kinetic studies on the mechanism of the reaction between $[MoH_4(Ph_2PCH_2CH_2PPh_2)_2]$ and acid [45] shows that two protons bind to $[MoH_4(Ph_2PCH_2CH_2PPh_2)_2]$ prior to the dissociation of dihydrogen. This observation demonstrates that a dihydrogen species must be involved as shown in Figure 8, since $[MoH_6(dppe)_2]^{2+}$ would contain formally Mo^{VIII}; clearly an impossibility!

Initial protonation of molybdenum, produces $[MoH_5(Ph_2PCH_2CH_2PPh_2)_2]^+$. Further protonation of $[MoH_5(Ph_2PCH_2CH_2PPh_2)_2]^+$ at molybdenum is not possible since the metal is already in its maximum oxidation state (+VI). Further protonation can only occur under two circumstances. (i) Intramolecular coupling of hydride ligands generates $[MoH_3(\eta^2\text{-}H_2)(Ph_2PCH_2CH_2PPh_2)_2]^+$. Hydride-hydride coupling effectively releases two electrons to molybdenum to produce a Mo^{IV} species that can be protonated further at the metal. (ii) Alternatively, protonation can occur directly at one of the hydride ligands. This work highlights the possible role of dihydrogen ligands in the binding of dinitrogen to metal sites.

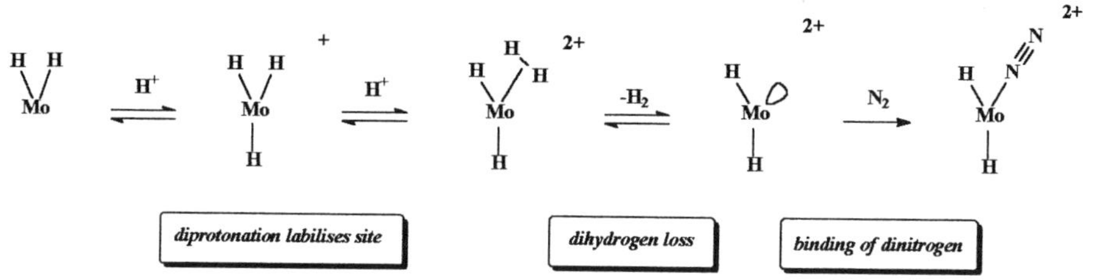

Figure 8. Binding of dinitrogen at a hydridic site and catalysed by protons.

Evidence for direct protonation of hydride ligands has been obtained from studies [46] on the reaction shown in Equation (7). Subsequently, other investigations have shown the generality of the mechanism developed in this study [47].

$$[WH_4(PMe_2Ph)_4] + HCl \rightarrow [WH_5(PMe_2Ph)_4]^+ + Cl^- \quad (7)$$

Although Equation (7) is stoichiometrically simple, the kinetics of this reaction are complicated. The kinetic data are consistent with a mechanism in which initial protonation at a hydride ligand produces the transient $[WH_3(\eta^2\text{-}H_2)(PMe_2Ph)_4]^+$, which subsequently undergoes intramolecular H-H cleavage to form the classical pentahydride product $[WH_5(PMe_2Ph)_4]^+$. The mechanism by which nitrogenase is proposed to bind dinitrogen is shown in Figure 9. Many of the features observed in the model systems described above are incorporated into the enzyme mechanism.

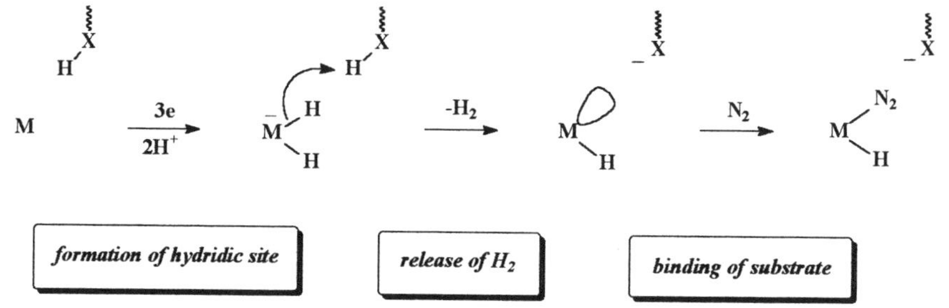

Figure 9. Mechanism proposed for the binding of dinitrogen to a hydridic site.

The model systems discussed so far have been addressing the details of protonation of hydride sites and binding of dinitrogen. Such models are incomplete since they do not include the non-obligatory evolution of dihydrogen which occurs with the enzyme in the absence of substrate. There is one model which combines the obligatory and non-obligatory evolution of dihydrogen. In common with the systems discussed so far in this Section, this model proposes that dinitrogen binds to a hydridic site, but in this case by initial displacement of a carboxylate ligand. Dihydrogen evolution occurs subsequently.

$[MoH_2(\eta^2\text{-}O_2CMe)(Ph_2PCH_2CH_2PPh_2)_2]^+$ catalyses the reduction of protons to dihydrogen [48] by the mechanism shown in Figure 10. Two-electron reduction of $[MoH_2(\eta^2\text{-}O_2CMe)(Ph_2PCH_2CH_2PPh_2)_2]^+$, under an atmosphere of dinitrogen, results in the dissociation of the carboxylate ligand and generation of a vacant site on the molybdenum at which dinitrogen can bind. Transient formation of the detected $[MoH_2(N_2)(Ph_2PCH_2CH_2PPh_2)_2]$ is followed by the ultimate production of *trans*-$[Mo(N_2)_2(Ph_2PCH_2CH_2PPh_2)_2]$.

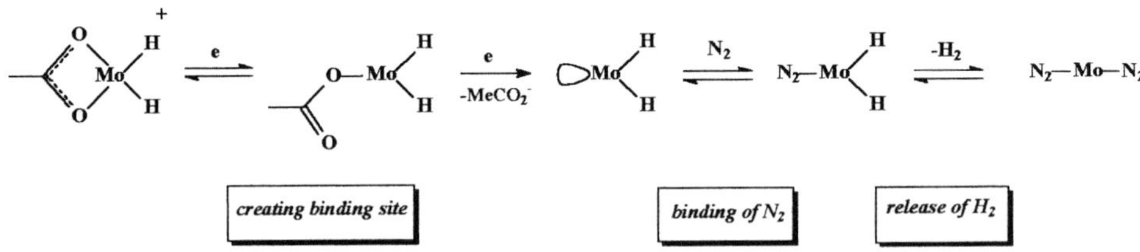

Figure 10. Dihydrogen-evolving pathway facilitated by dinitrogen.

In the absence of dinitrogen (or any other substrate) $[MoH_2(\eta^2\text{-}O_2CMe)(Ph_2PCH_2CH_2PPh_2)_2]^+$ reduces protons to dihydrogen. Reduction of $[MoH_2(\eta^2\text{-}O_2CMe)(Ph_2PCH_2CH_2PPh_2)_2]^+$ to $[MoH_2(OCMeO)(Ph_2PCH_2CH_2PPh_2)_2]$ by one electron produces a compound in which the acetate is monodentate [49]. Protonation can now occur at the more open molybdenum site and dihydrogen is released when the carboxylate reforms the chelate.

Figure 11 shows how the chemistry observed with $[MoH_2(\eta^2\text{-}O_2CMe)(Ph_2PCH_2CH_2PPh_2)_2]^+$ could operate in nitrogenase and proposes some more specific identities for the various E_n states [50]. In this mechanism the obligatory evolution of dihydrogen is not a pre-requisite for dinitrogen binding, but rather is a consequence of dinitrogen binding to a hydridic site. It has been proposed that the active site of the enzyme is naturally hydridic because protonation of molybdenum facilitates reduction of the site and labilises carboxylate at a higher potential than could otherwise be achieved. The R-homocitrate bound to the FeMo-cofactor could provide the carboxylate leaving group to allow dinitrogen to bind.

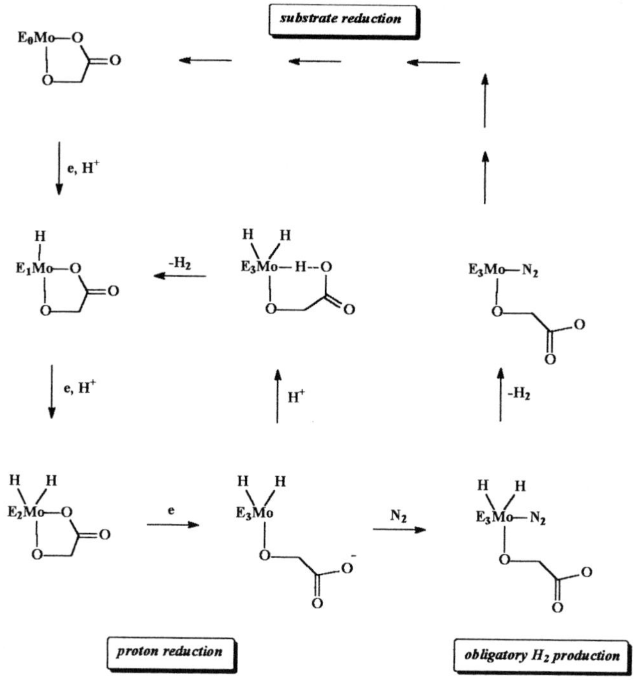

Figure 11. Dihydrogen-evolving and dinitrogen-binding scheme.

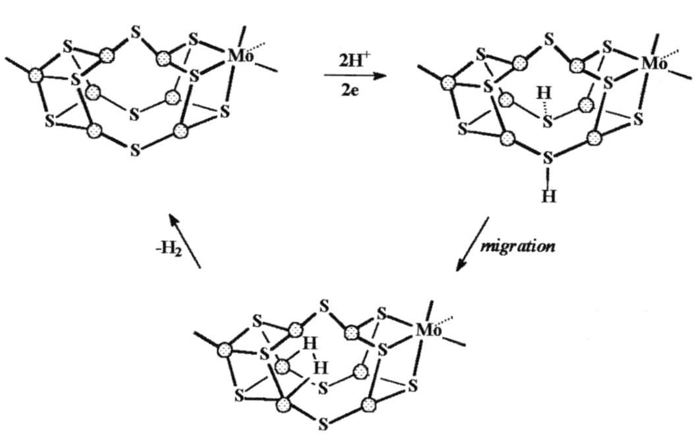

Figure 12. Proposed dihydrogen-evolving mechanism on FeMo-cofactor.

The intimate mechanism by which FeMo-cofactor reduces protons and evolves dihydrogen remains unknown. One pathway proposed from DFT calculations is illustrated in Figure 12 [51]. In this mechanism protons bind initially to the μ_2-S atoms, which weakens the μ_2-S-Fe bonds, lengthening the Fe-S bonds and shortening Fe-Fe separation. The hydrogens then move together making Fe-H bonds, and forming dihydrogen that separates from the cluster. Another picture of proton reduction by the FeMo-cofactor has been presented, also derived from DFT calculations. In this proposal the first three hydrogen atoms bind to μ_2-S and the fourth hydrogen binds to an iron atom, producing a state that is metastable compared to the state where dihydrogen is bound to iron [37].

The kinetics of the reduction of protons by $[Fe_4S_4(SPh)_4]^{3-}$ have been investigated [52]. The study shows that protonation of one thiolate and two μ_3-S ligands is essential before dihydrogen is produced. However, the details of the H-H coupling steps are unknown.

3.3.2 Why Bind Dinitrogen by Displacing Dihydrogen?

The model systems discussed above are consistent with the proposal that metal hydrides are involved in the binding of dinitrogen to nitrogenase. But further considerations reveal some concerns about the evidence for dinitrogen binding in this fashion.

As noted above, in order to get to a state of the enzyme which will bind dinitrogen {$E_3H_2(H^+)$, Figure 5} the enzyme "resting state" (E_0) has to bind three electrons. For each electron transferred it is generally accepted that a minimum of two ATP molecules have to be hydrolysed. Having consumed these three electrons (and hence six ATP molecules), it is surprising that the enzyme now wastes two electrons (and four ATP molecules), by evolving dihydrogen. Why would the enzyme be prepared to waste these electrons and energy? The reason is far from clear, particularly when studies on simple chemical compounds show that dinitrogen can bind to metal sites by displacing a variety of other ligands including water, ammonia and carboxylate groups.

For several years the limiting stoichiometry of the Mo-based nitrogenase and the inhibitory effect of dihydrogen on ammonia-formation was accepted as evidence for the involvement of a hydridic site for dinitrogen binding. The discovery and purification of the V-based and Fe-only nitrogenases, and determination of the limiting stoichiometries of these enzymes, have cast doubt on the role of metal hydride sites in binding dinitrogen. The V-based and Fe-only nitrogenases have a limiting stoichiometry in which more than one molecule of dihydrogen is produced for every molecule of dinitrogen transformed, as shown in Equations (8) and (9), respectively [2]. It is clearly evident why more than one dihydrogen must be released to bind only one dinitrogen.

$$N_2 + 12H^+ + 12e \longrightarrow 2NH_3 + 3H_2 \quad (8)$$

$$N_2 + 21H^+ + 21e \longrightarrow 2NH_3 + 7.5H_2 \tag{9}$$

There are problems in nitrogenase in measuring accurately the ratio of dihydrogen released to dinitrogen reduced. Although there is no doubt that for the Mo-based enzyme the ratio of dihydrogen evolved to dinitrogen reduced is about 1, it cannot be claimed with any certainty that it is exactly 1.0. Indeed, as early as 1975 the value of this ratio was reported [53] to be in the range 0.6-0.9.

A recent study on the reactions of $[Fe_4S_4(SPh)_4]^{3-}$ shows a behaviour which may be more pertinent in rationalising the limiting stoichiometries of the nitrogenases [52]. Studies on the protonation of $[Fe_4S_4(SPh)_4]^{3-}$ in MeCN have shown that at high concentrations of $[lutH]^+$ (lut = 2,6-dimethylpyridine) the cluster binds protons to one thiolate and two μ_3-S ligands. It is only this triprotonated cluster which is capable of evolving dihydrogen or converting acetylene into ethylene. In the absence of any substrate the triprotonated cluster evolves stoichiometric amounts of dihydrogen, according to Equation (10).

$$[Fe_4S_2(SH)_2(SPh)_3(SHPh)] \longrightarrow [Fe_4S_3(SH)(SPh)_3(SHPh)] + 0.5H_2 \tag{10}$$

$$2C_2H_2 + 6H^+ + 6e \longrightarrow 2C_2H_4 + H_2 \tag{11}$$

In the presence of acetylene the triprotonated cluster produces ethylene and evolves less dihydrogen. An important point is that if electrons are not going into the formation of dihydrogen they are going into the formation of ethylene, as shown in Figure 13. As the concentration of acetylene is increased, progressively more ethylene, and less dihydrogen, is produced. However, the production of dihydrogen can never be entirely suppressed, and the limiting product distribution comprises *ca*70% ethylene and *ca*30% dihydrogen, as described by Equation (11). The important point is that the $[C_2H_4]:[H_2]$ ratio is only reflecting the inability of the bound acetylene to route all the electrons towards its reduction. It is conceivable that the limiting stoichiometries of the various nitrogenases {Equations (3), (8) and (9)} merely reflect the ability of dinitrogen in each enzyme to divert the electron-flux away from dihydrogen production [6].

FeMo-cofactor extracted from the MoFe-protein is capable of reducing protons to dihydrogen. The electrochemical behaviour of FeMo-cofactor has been investigated and two redox waves have been observed at –0.32 V and *ca* –1.0 V (relative to NHE). These potentials are modified slightly by binding of thiol [54]. If a relatively acidic thiol such as C_6F_5SH is used then electrochemical studies reveal a catalytic process in which dihydrogen is being produced, and this occurs at the relatively high potential of –0.28 V.

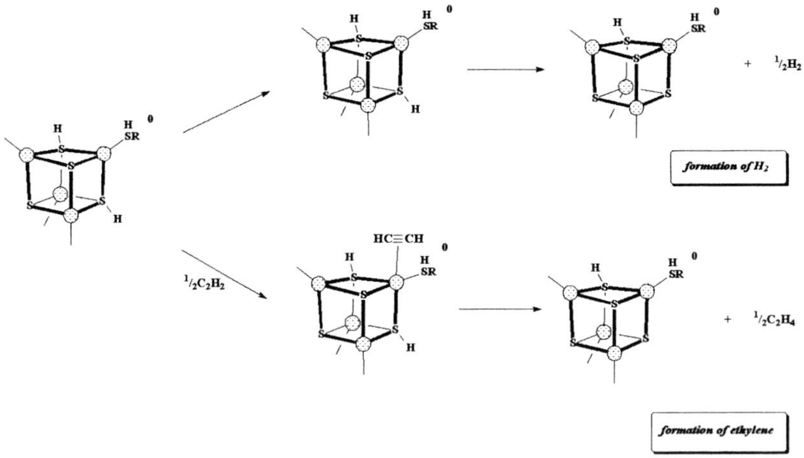

Figure 13. Dihydrogen-evolving and acetylene-reducing pathway at $[Fe_4S_4(SPh)_4]^{3-}$.

3.3.3 Formation of HD

Further evidence for the involvement of hydrido-species in the action of nitrogenase is the formation of HD from D_2 in the presence of dinitrogen as the enzyme turns over. Selective formation of HD is observed when a solution of $[RhH(CO)(buS_4)]$ ($buS_4 = SC_6H_2(^tBu)_2SCH_2CH_2SC_6H_2(^tBu)_2S$) is stirred under an atmosphere of D_2 in a solution containing a protic acid, HD is selectively produced [55]. The mechanism of this reaction is shown in Figure 14.

The D_2/H^+ exchange is catalysed by a pathway involving heterolytic cleavage of D_2 on the metal site with transfer of D^+ to the thiolate sulfur. Rapid exchange of S-D with free H^+ in solution produces S-H. In a subsequent re-coupling reaction HD is produced selectively. The selective formation of HD is merely a consequence of proton exchange of a metal-hydride being much slower than for S-H.

4. THE PROTONATION OF DINITROGEN

4.1 The protonation of dinitrogen coordinated end-on

We now turn our attention to the most obvious role for protons in nitrogen fixation: the protonation of dinitrogen and its conversion into ammonia. In the 1970's and 1980's simple consideration of the most common ways dinitrogen coordinates to metal sites led to three proposals for the coordination of dinitrogen to nitrogenase: end-on to one metal; side-on to one metal and bridging between two metals. Obviously these proposals were suggested before the structure of FeMo-cofactor was known. However, we have seen in Section 3.1 that these proposals have been adapted to FeMo-cofactor.

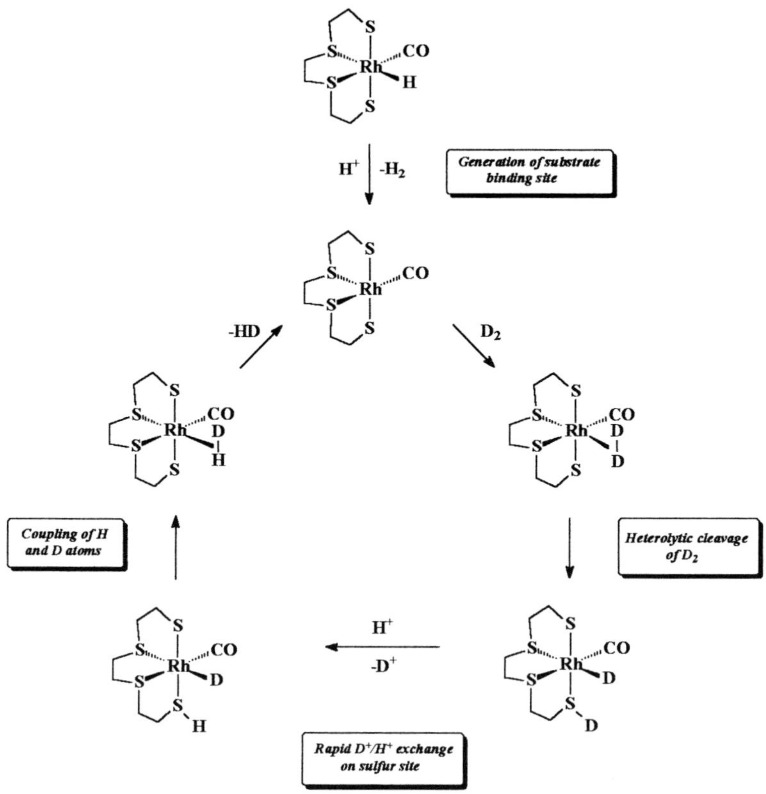

Figure 14. Mechanism for the formation of HD from D_2.

In Section 4 we will discuss the protonation chemistry of dinitrogen in structurally well-defined mononuclear and binuclear complexes. Whilst these complexes bear little resemblance to the active site in the enzyme, the important feature is that they allow definition of the protonation chemistry. Nowhere is this more evident than for the conversion of end-on coordinated dinitrogen to ammonia, as shown in Figure 15.

The first demonstration [56] that a structurally well-defined coordinated dinitrogen could be transformed into ammonia under ambient conditions was from the reaction of acid with $[Mo(N_2)_2(PR_3)_4]$ as shown in Equation (12).

$$[M(N_2)_2(PR_3)_4] + 6HX \longrightarrow M^{VI} + 2NH_3 + N_2 + 4PR_3 + 6X^- \qquad (12)$$

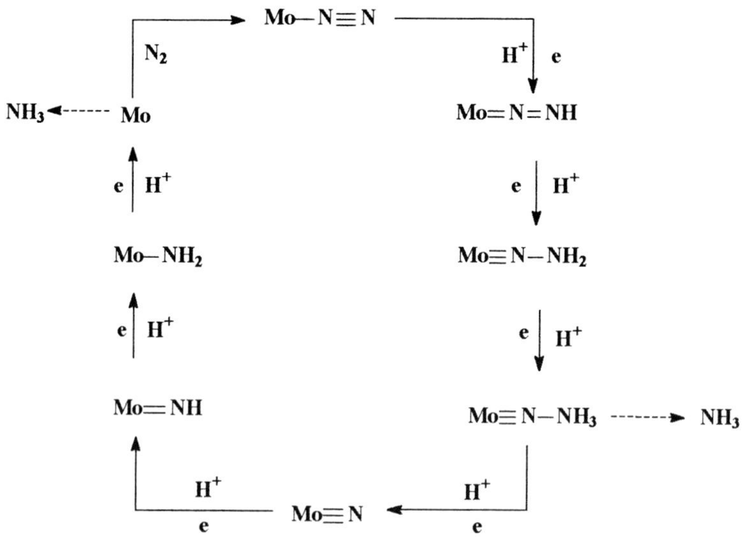

Figure 15. The conversion of dinitrogen into ammonia at a single metal site

Further studies on analogous systems used a variety of different phosphine coligands which modulate the protonation chemistry and facilitate the isolation of the various intermediates [38,39,56] shown in Figure 15. A major reactivity difference between the action of the enzyme and the chemistry of Equation (12) is the source of the electrons. In the enzyme, the sequential addition of electrons comes from the external reductant, with the metal complexes the six electrons necessary for the conversion of dinitrogen into ammonia come entirely from the metal.

The important features of the species shown in Figure 15 are as follows. (i) Complexes containing all the nitrogenous species have been isolated. (ii) The structure of the nitrogenous residues have all been established by X-ray crystallography. (iii) The kinetics and mechanisms of the interconversions have been studied and the factors controlling the sites and rates of protonation of the nitrogenous residues delineated.

4.1.1 Diprotonation of Dinitrogen

The addition of the first two protons to end-on coordinated dinitrogen occurs at the nitrogen atom remote from the metal and produces hydrazide species ($M{=}N{-}NH_2$).

Complexes containing bidentate phosphine ligands, such as that shown in Equation (13), have been studied extensively since they react with acid in an aprotic solvent to form the corresponding hydrazide, and the reaction proceeds no further.

$$[M(N_2)_2(Ph_2PCH_2CH_2PPh_2)_2] + 2HX \longrightarrow [M(NNH_2)X(Ph_2PCH_2CH_2PPh_2)_2]X + N_2 \quad (13)$$

Kinetic studies [57,58] on Equation (13) demonstrate the following general features. (i) Protonation of the dinitrogen is rapid. We can estimate the rate constant for protonation, $k \geq 1 \times 10^6$ dm^3 mol^{-1} s^{-1}, and this probably represents a diffusion-controlled, or close to diffusion-controlled process. (ii) Protonation of a dinitrogen ligand effectively withdraws electron density from the metal site, thus decreasing π-backbonding to the remaining dinitrogen, and increasing its lability. The dinitrogen in *trans*-$[Mo(NNH)(N_2)(Ph_2PCH_2CH_2PPh_2)_2]^+$ is at least 1×10^6 times more labile than in the parent complex. (iii) Both single and double protonation of a dinitrogen ligand can occur before dissociation of the other *trans*-dinitrogen. The extent of protonation of the dinitrogen affects the lability of the *trans*-dinitrogen. This depends on the strength of the acid, the identity of the metal and the nature of the phosphine ligand. (iv) Dissociation of the *trans*-dinitrogen is invariably the rate-limiting step in the reaction. (v) Changes in the coordination sphere affect the basicity of the dinitrogen ligands. Thus, replacement of the electron-withdrawing dinitrogen ligand in *trans*-$[M(NNH)N_2(Ph_2PCH_2CH_2PPh_2)_2]^+$ by an electron-releasing halide (X) makes the diazenido-ligand sufficiently basic to bind a further proton and form *trans*-$[M(NNH_2)X(Ph_2PCH_2CH_2PPh_2)_2]^+$.

The question of which site is protonated is complex and theoretical studies have not resolved the problem. If nothing else, there is the problem of whether the protonation site is decided by electron density or frontier orbital considerations. The earliest studies [59] were those of DuBois and Hoffman who considered structural aspects of the nitrogenous species shown in Figure 15, including bending of diazenido-ligands as shown in Figure 16. Singly bent diazenido-ligands protonate at the nitrogen atom remote from the metal, but doubly diazenido-ligands protonate at the donor nitrogen atom.

Ab initio methods have shown that there is a remarkable redistribution of electron-density upon protonation. Most of the electron density in d_{xz} orbitals is transferred to the N-H bond. As a consequence of the protonation the N-N bond becomes longer and the M-N bond becomes shorter [60].

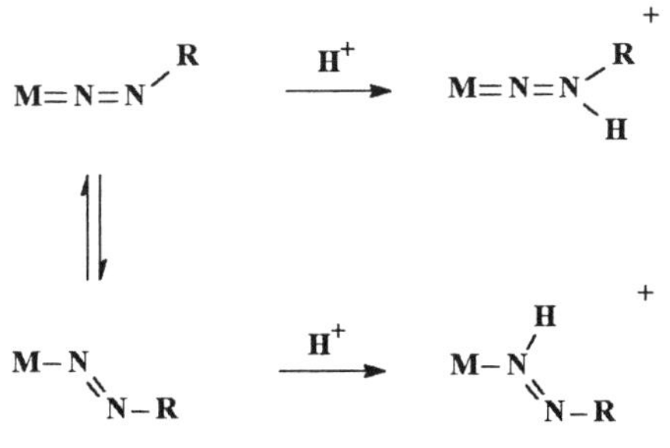

Figure 16. Protonation sites of diazenide ligands.

4.1.2 Basicity of Coordinated Dinitrogen

There is still little quantitative information on the basicity of coordinated dinitrogen. It has not been possible to quantify the basicity of coordinated dinitrogen because many of the protonation studies were performed in aprotic solvents for which we have no information on the strengths of acids. An alternative approach has been more successful: the deprotonation of *trans*-$[Mo(NNH_2)X(Ph_2PCH_2CH_2PPh_2)_2]^+$ (X = F or Br) by MeO^- or NEt_3 in dinitrogen-saturated methanol as shown in Equation (14) [61,62].

$$[M(NNH_2)X(Ph_2PCH_2CH_2PPh_2)_2]^+ + 2\text{base} \xrightarrow[N_2]{} [M(N_2)_2(Ph_2PCH_2CH_2PPh_2)_2] + X^- + 2\text{baseH}^+ \quad (14)$$

TABLE

Summary of pK_a's of *trans*-$[Mo(L)X(Ph_2PCH_2CH_2PPh_2)_2]^+$ (L = NNH_2, NNH or NH; X = Br or F)

X	L	pK_a
Br	NNH_2	9.8
	NNH	12.1
	NH	<10
F	NNH_2	<11
	NNH	14.2
	NH	12.7

The kinetics of Equation (14) are consistent with rapid deprotonation of the hydrazide ligand, followed by rate-limiting dissociation of the *trans*-halide and finally rapid binding of dinitrogen to yield the product. Analysis of the kinetics yields the equilibrium constant for the deprotonation of the hydrazide and hence the corresponding pK_a can be calculated. The Table shows the effect that the *trans*-halide has on the pK_a. Similar studies have been performed on the analogous *trans*-$[Mo(NH)X(Ph_2PCH_2CH_2PPh_2)_2]^+$ [63] and comparison of the analogous hydrazide and imide complexes allows us to probe the effect that distance from the metal site and halide has on the pK_a of the N-H moiety.

4.1.3 Protonation of Dinitrogen versus Protonation of Metal

In order to bind and protonate dinitrogen it is necessary to coordinate the substrate to an electron-rich site. Consequently, protonation at the metal can occur under certain conditions. Protonation of the metal deactivates coordinated dinitrogen as shown by the study illustrated in Figure 17. The reaction of *trans*-$[Mo(N_2)_2(Et_2PCH_2CH_2PEt_2)_2]$ with an excess of anhydrous HCl in tetrahydrofuran initially produces $[MoH(N_2)_2(Et_2PCH_2CH_2PEt_2)_2]^+$ [64]. At low concentrations of HCl,

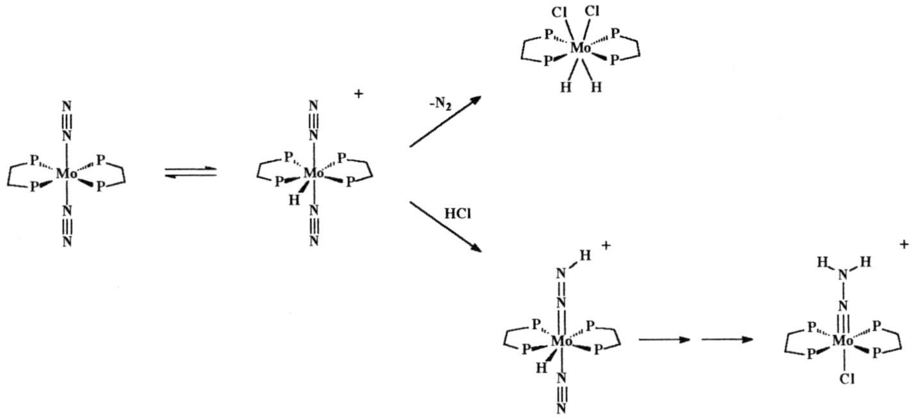

Figure 17. Protonation at the metal and the subsequent formation of hydrazide complex.

dinitrogen dissociates from $[MoH(N_2)_2(Et_2PCH_2CH_2PEt_2)_2]^+$ and $[MoH_2Cl_2(Et_2PCH_2CH_2PEt_2)_2]$ is produced. Thus, initial protonation of the metal produces a hydride and labilises dinitrogen towards dissociation. However, increasing the concentration of HCl leads to a change in the product. Progressively less $[MoH_2Cl_2(Et_2PCH_2CH_2PEt_2)_2]$ and more *trans*-$[Mo(NNH_2)Cl(Et_2PCH_2CH_2PEt_2)_2]^+$ is formed. At high concentrations of HCl, protonation of dinitrogen can occur before dinitrogen dissociation.

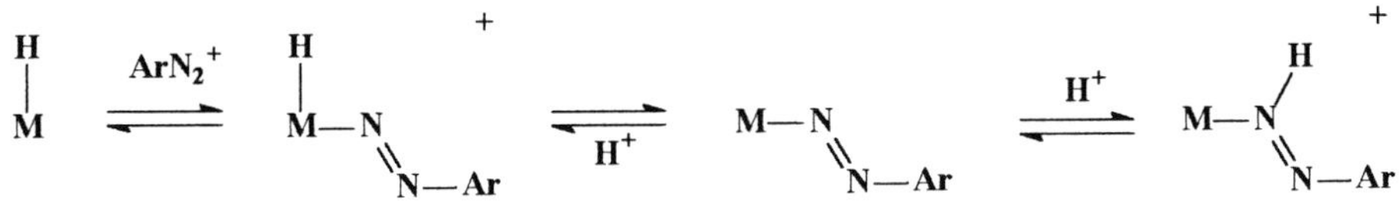

Figure18. The mechanism of "insertion" of diazonium cation into metal-hydride bond

Intramolecular proton migration between metal and nitrogenous ligand is rare. The transfer of a proton from metal to dinitrogen is usually accomplished by a sequence of acid-base catalysed reactions. This is exemplified in the pathway shown in Figure 18. Kinetic studies [65] on the reaction of diazonium cations (ArN_2^+) with $[RhHCl_2(PEtPh_2)_3]$ and $[PtHCl(PEt_3)_2]$ to form aryldiazene complexes indicate that the reaction involves initial binding of the diazonium cation to the metal. This facilitates the release of the hydride as a proton, then recapture of H^+ by the nitrogen atom adjacent to the metal, to give the aryldiazene complex.

Intramolecular proton transfer between metal and ligand has been observed [66]. As shown in Equation (15), $[W(\eta^5\text{-}C_5Me_5)_2H(NNHAr)]^+$ contains a bent hydrazide ligand. Upon warming, the molecule rearranges to form a side-on coordinated hydrazide(1-) ligand.

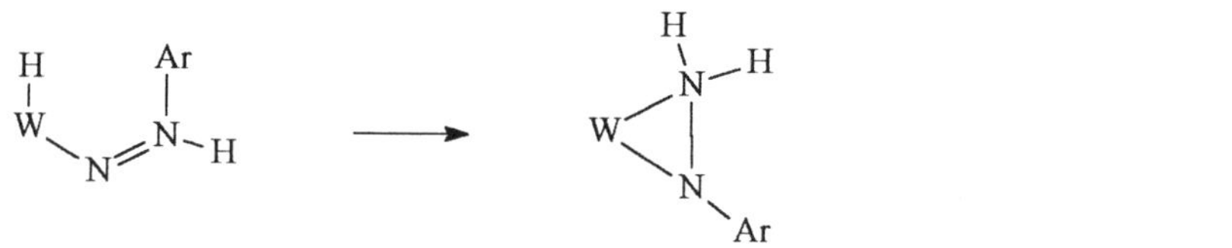

(15)

4.1.4 Reactivity of Coordinated Hydrazide: Forming Hydrazine or Ammonia

In order to understand how hydrazide ligands are converted into amines or the corresponding hydrazines it is necessary to identify where protons bind on hydrazide complexes, at what stage the N-N bond breaks and what factors discriminate between amine and hydrazine formation?

There have been only a few mechanistic studies on the reactions of acid with hydrazide complexes. The major problem is finding analytically clean systems. For example, although the reactions of acids with $[Mo(NNH_2)X_2(PMe_2Ph)_3]$ give ammonia the yields are not very reproducible with some hydrazine always produced, the metal products are poorly defined, and the reactions take several hours to complete. These problems have meant that systematic studies on amine and hydrazine formation have so far been restricted to alkyl- and aryl-substituted hydrazide complexes.

In principle, protonation of a hydrazide complex can occur at three possible sites: the metal {to form $M(H)NNH_2$}; the remote nitrogen atom (to form $M{=}NNH_3$); and the nitrogen atom adjacent to the metal (to form $M\text{-}NHNH_2$). Complexes of all

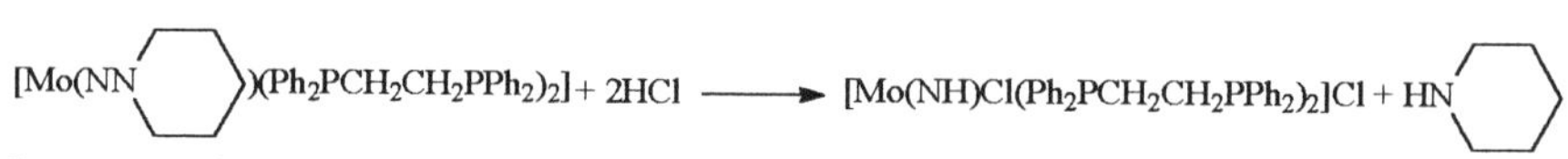

(16)

three types have been isolated [67]. Early theoretical studies [68] implied that the site of protonation is a major factor controlling whether hydrazide complexes produce amines or hydrazines, and specifically, that $NHNH_2$ is on the hydrazine-forming pathway whilst NNH_3 is on the ammonia-forming route. However, the only kinetic study on the mechanism of N-N cleavage is of the reaction shown in Equation (16) and indicates that N-N cleavage can occur from either $Mo{=}NNH_3$ and $Mo\text{-}NHNH_3$ species [69].

$$[Mo(NNRR')_2(S_2CNEt_2)_2] + 2HCl \longrightarrow [MoCl_2(NNRR')(S_2CNEt_2)_2] + H_2NNRR' \quad (17)$$

Similarly, kinetic studies [70] on the reaction of Equation (17) indicate that hydrazines are produced from both M=NNHRR' and M-NHNHRR': the equivalent degrees of protonation that result in N-N cleavage in (16). It is clear that the factor deciding whether hydrazine or amines are produced is not the position or extent of protonation but rather the metal site. It seems likely that if the metal site favours the formation of strong M-N multiple bonds then N-N cleavage will ensue.

Ab initio calculations on $LiNHNH_2$ and LiNNH indicate that coulombic forces are at least as significant as π-bonding in defining structures. More important, such calculations show that the energy difference between various configurations of the same ligand may be rather small [71,72]. This is particularly marked for NNH_2, NNH and $NHNH_2$ ligands. Although complexes containing side-on hydrazide(1-) ligands have been isolated, the question remains, "Are they intermediates on the ammonia- or hydrazine-forming pathways?"

$[Mo(\eta^2\text{-NHNMePh})(NNMePh)(S_2CNEt_2)_2]^+$ can be prepared by the reaction of one mole-equivalent of acid with $[Mo(NNMePh)_2(S_2CNEt_2)_2]$. Kinetic studies [72] show that $H_2NNMePh$ is formed rapidly in the reaction of $[Mo(NNMePh)_2(S_2CNEt_2)_2]$ with acid. Under identical conditions $[Mo(\eta^2\text{-NHNMePh})(NNMePh)(S_2CNEt_2)_2]^+$ forms $H_2NNMePh$ only slowly. Thus, Equation (17) does not involve the intermediacy of $[Mo(\eta^2\text{-NHNMePh})(NNMePh)(S_2CNEt_2)_2]^+$. It appears that the conversion of the NNMePh or NHNMePh ligands from end-on to side-on coordination is slow, and so the fastest pathway for hydrazine formation involves the end-on hydrazide(1-) species.

It has been possible to study the acid-base chemistry of side-on coordinated hydrazide(1-). Two complementary studies [70,73] demonstrate that the deprotonation of hydrazide(1-) ligands can occur by the two routes shown in Figure 19.

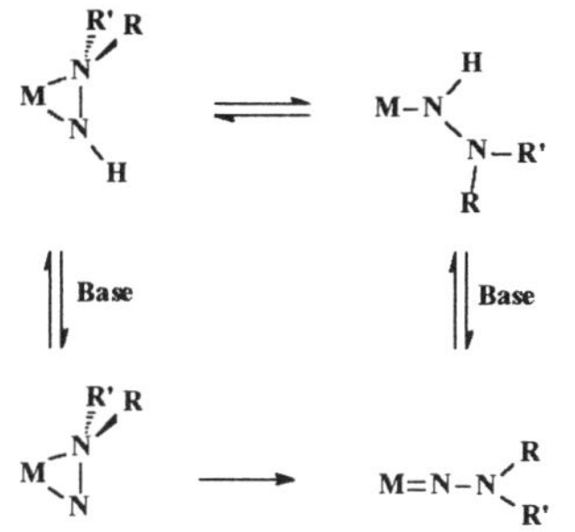

Figure 19. Pathways for interconversion of hydrazide(2-) and hydrazide(1-) ligands.

Initial deprotonation (requiring strong bases) generates a side-on coordinated hydrazide(2-) species which then rapidly rearrange to the end-on form. Alternatively, rearrangement of the side-on hydrazide(1-) to the end-on coordinated hydrazide(1-) occurs initially and subsequent rapid deprotonation of the end-on form produces the hydrazide(2-).

4.1.5. Protonation of Coordinated Imide

Following N-N cleavage from the protonated hydrazide complexes, and after the release of the first ammonia, the MN or MNH species then reacts further to form ammonia. This is not a conceptually very difficult pathway and has been realised experimentally from both a chemical and an electrochemical standpoint [63,74]. In the electrochemical route shown in Figure 20, the integrity of the $\{Mo(dppe)_2\}$ core is retained and M-NH_2 species has been detected electrochemically.

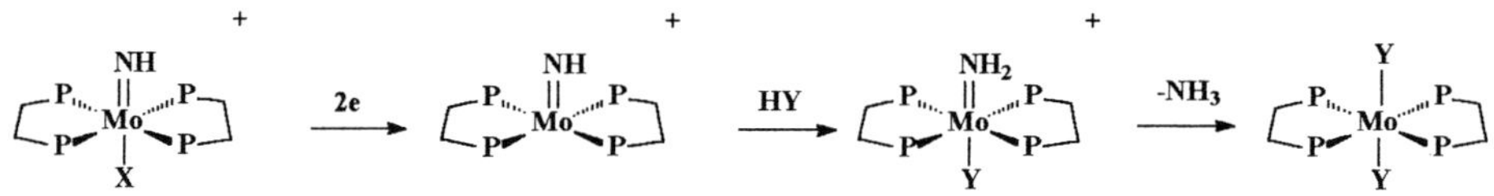

Figure 20. Conversion of imide ligand to ammonia.

4.2 The protonation of bridged coordinated dinitrogen

In simple transition metal complexes, after end-on coordination, the most common type of coordination is dinitrogen bridged between two metal sites. Some of the best artificial nitrogen-fixing systems have been proposed to operate *via* bridged dinitrogen complexes [1]. It has been argued [1] that bridged dinitrogen is easier to convert into ammonia than mononuclear dinitrogen formation for two reasons. First, two metals can mediate the reduction of dinitrogen better than one, since less dramatic changes in oxidation state of each metal is involved. Secondly, two metals are able to increase the basicity of dinitrogen better than one, hence facilitating protonation.

Typical of the artificial nitrogen fixing systems is that comprising mixed vanadium(II)-magnesium(II) hydroxide. The transformation of dinitrogen occurs in aqueous or alcohol suspensions of freshly prepared mixtures of VCl_2 and $MgCl_2$. High rates of formation of hydrazine and ammonia are produced at room temperature and 1 atmosphere pressure, even under air [75-77]. The binding site and proposed initial activation of the dinitrogen is shown in Figure 21.

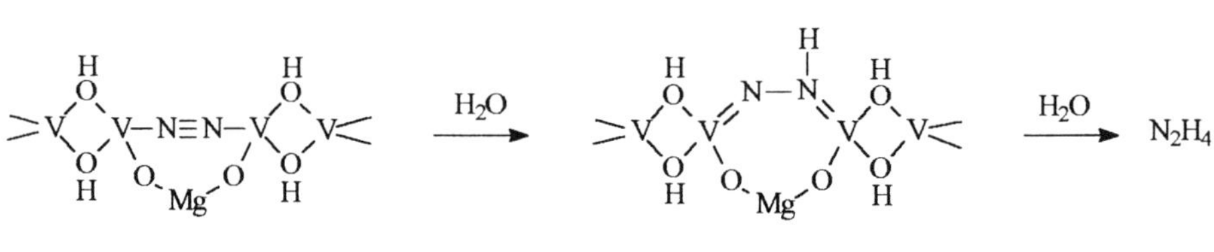

Figure 21. Proposed binding and activation of dinitrogen on $V(OH)_2$.

The problem with the vanadium systems is that they are made up from a large number of components, making it difficult to establish the identity and structure of the "active species" and the mechanisms by which they operate. The evidence that these systems operate *via* bridged dinitrogen complexes is thus circumstantial.

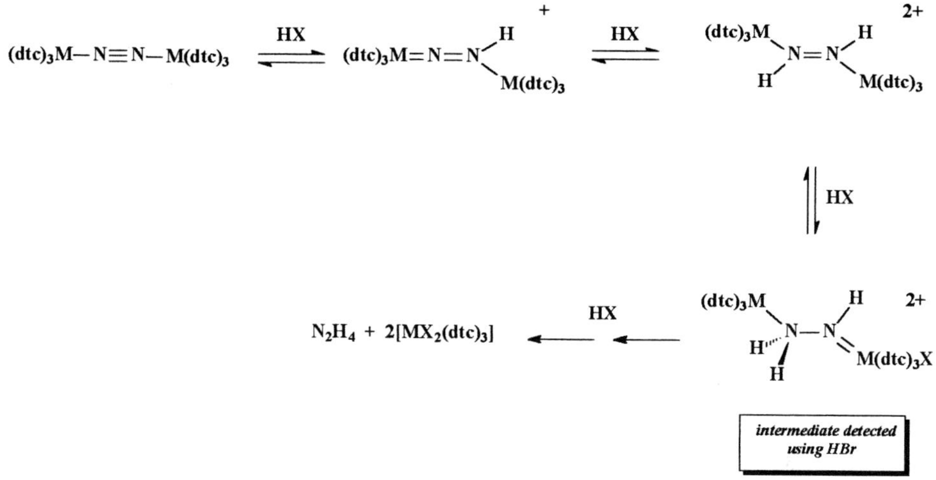

Figure 22. Mechanism of Hydrazine Formation in [$\{M(S_2CNEt_2)_3\}_2(\mu\text{-}N_2)$].

The detailed mechanism by which bridged dinitrogen is converted into hydrazine has been defined in studies on [$\{M(S_2CNEt_2)_3\}_2(\mu\text{-}N_2)$] (M = Ta or Nb) [78,79]. The mechanism of the reaction is shown in outline in Figure 22. The key features of this mechanism are as follows. (i) The rate of the first protonation is slow. This stands in stark contrast to that of end-on coordinated dinitrogen where protonation is diffusion-controlled or close to it. Protonation of the bridged dinitrogen is slow because there are no lone pairs of electrons to which the proton can bind. Thus, protonation must involve disruption of the triple bond between the nitrogen atoms: an energetically more demanding process. (ii) Addition of the second proton is also slow and is believed to produce a bridged diazene species. (iii) With strong acids, tri-protonation of the dinitrogen ligand is observed before the bridge cleaves. It is conceivable (if not likely) that weakly bound bridged-dinitrogen complexes react by a pathway in which cleavage of the bridge is the first step, and the ensuing protonation reactions are those of mononuclear, end-on coordinated dinitrogen. (iv) The Nb and Ta systems produce stoichiometric amounts of hydrazine. Since each Nb or Ta site is formally M^{III} only a total of four electrons can be supplied; insufficient to produce ammonia.

A key intermediate in the reaction of bridged dinitrogen is the diazene species. Complexes containing bridged diazene have been isolated [39], including [$\{Cr(CO)_5\}_2(\mu\text{-}N_2H_2)$] and [$\{Mn(\eta^5\text{-}C_5H_5)(CO)_2\}_2(\mu\text{-}N_2H_2)$], and most recently on the metal-sulfur site [80] shown in Figure 23.

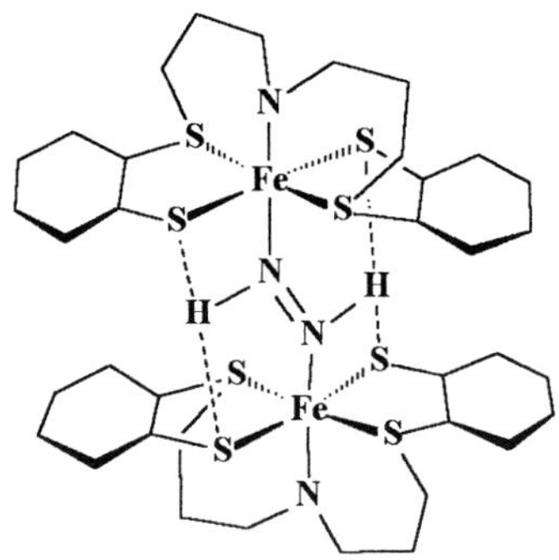

Figure 23. Structure of bridged diazene ligand showing hydrogen bonds to sulfur atoms.

Before leaving bridged dinitrogen species, it is important to mention the cleavage of bridged-dinitrogen complexes [81] typified by the system shown in Equation (18).

$$[(PhBu^tN)_3Mo—N{\equiv}N—Mo(NBu^tPh)_3] \rightarrow 2[N{\equiv}Mo(NBu^tPh)_3] \qquad (18)$$

Although the biological relevance of Equation (18) is dubious, the reaction is unusual since the N-N bond is effectively wrenched apart. Theoretical studies are consistent with a so-called "zig-zag" transition state. The calculations indicate activation parameters which are close to those observed experimentally ($\Delta H^{\ddagger}$ = 5.6 kJ mol^{-1}; $\Delta S^{\ddagger}$ = 0.6 J K^{-1} mol^{-1}.

In the 1960's it was proposed that dinitrogen bound side-on to the active site [82]. No study on the mechanism of protonation of a structurally-authenticated side-on dinitrogen has been reported.

5. DINITROGEN REACTIVITY ON Fe-S-CLUSTERS

A major contemporary challenge is to establish the binding and protonation chemistry of dinitrogen (and other substrates) bound to Fe-S-based clusters in general, and FeMo-cofactor, in particular.

5.1 Reduction of nitrogenous substrates by synthetic Fe-S-based clusters

A variety of different systems involving synthetic Fe-S based clusters have been shown to reduce dinitrogen and/or hydrazine, either stoichiometrically or catalytically. Multi-component systems containing molybdenum complexes and Fe-S clusters (acting as electron donors), reduce dinitrogen to ammonia under mild conditions [83], using ancillary reductants such as BH_4^-. Similarly, hydrazine reduction to ammonia occurs [84] in mixtures of $Na_2[MoO_4]$, *L*-cysteine and $NaBH_4$. Catalytic reduction of both dinitrogen and hydrazine to ammonia has been reported by the electrochemically reduced clusters $[Fe_4S_4(SPh)_4]^{2-}$, $[\{MoFe_3S_4(SPh)_3\}_2(\mu\text{-}SPh)_3]$ in protic solvents, at a Hg electrode [85]. The reduction of hydrazine can be performed in water or MeOH/thf mixtures containing $[Fe_4S_4(SR)_4]^{2-}$ or $[\{MoFe_3S_4(SR)_3\}_2(\mu\text{-}SPh)_3]^{3-}$ (R = Ph or CH_2CH_2OH) as suspensions [86]. However, it seems likely that in many of these experiments the clusters undergo decomposition during these reactions.

Single cubane clusters such as $[MoFe_3S_4Cl_3(\text{polycarboxylate})(NCMe)]^{2-}$ and $[VFe_3S_4Cl_3(dmf)_3]^-$ (dmf = *N,N*-dimethylformamide) reduce hydrazine to ammonia or acetylene to ethylene [87,88] using 2,6-dimethylpyridinium salts as acid and $[Co(\eta^5\text{-}C_5H_5)_2]$ as reductant. Precipitation of the ammonium salts prohibited a detailed kinetic study, but there are strong indications that the heterometal (Mo or V) is the hydrazine binding site. Thus, $[VFe_3S_4Cl_3(HBpz_3)]^{2-}$ ($HBpz_3$ = hydrotris(pyrazolyl)borate) does not transform the hydrazine or acetylene since it contains the non-labile $HBpz_3$ ligand which occupies all potential substrate binding sites on the vanadium atom. However, clusters where the heterometal has no vacant site (or labile ligand) still transform hydrazine or acetylene, albeit slowly. It has been suggested that the heterometal is the preferred binding site, substrates can also bind at the Fe sites of the cluster core.

There are a few examples of isolable clusters containing substrates or products bound to the cluster. Thus, $[VFe_3S_4Cl_3(bipy)(PhNHNH_2)]^-$ has been isolated and characterised by X-ray crystallography, demonstrating that the hydrazine is bound to the vanadium in an end-on fashion. In addition, $[MoFe_3S_4Cl_3(tccat)(NCMe)]^{2-}$ reacts with *cis*-MeN=NMe to give exclusively $MeNH_2$. $[MoFe_3S_4Cl_3(tccat)(NH_2Me)]^{2-}$ (tccat = tetrachlorocatecholate) has been isolated, with $MeNH_2$ bound to molybdenum [89].

5.2 Studies on extracted FeMo-cofactor

To date FeMo-cofactor has not been prepared in the laboratory. Its unique structural features make this synthesis one of the most challenging problems in contemporary inorganic chemistry. However, it is possible study this cluster since FeMo-cofactor can be extracted from the MoFe-protein and into *N*-methylformamide [90]. This extraction must involve the cleavage of the Fe-cysteinate and Mo-histidine bonds (see Figure 3), and it is reasonably assumed that the extracted cofactor has solvent bound at these two positions. Free from the polypeptide and the hydrogen bonds which modulate its reactivity, we are in a position to investigate FeMo-cofactor's reactivity. Outside the protein the catalytic capabilities of the cluster are greatly curtailed, in part because of the limited redox chemistry. Thus, the extracted FeMo-cofactor is capable of reducing protons to dihydrogen and acetylene to ethylene but this occurs (rather suspiciously) only in the presence of borohydride as reductant [91].

5.2.1 Substrate Binding Sites on Extracted FeMo-cofactor

For some time there has been spectroscopic evidence that substrates such as CN^-, N_3^-, H^+, and Bu^tNC can bind to the extracted FeMo-cofactor. The major problem is that the spectroscopic responses to the binding of these substrates are poor. It is clearly necessary to develop new, more sensitive methods of detecting this binding. Using an extension of a kinetic approach developed with the synthetic Fe-S-based clusters [92] it has been possible to map-out effectively where a variety of different substrates bind to FeMo-cofactor. A summary of where various species appear to bind in FeMo-cofactor is shown in Figure 24.

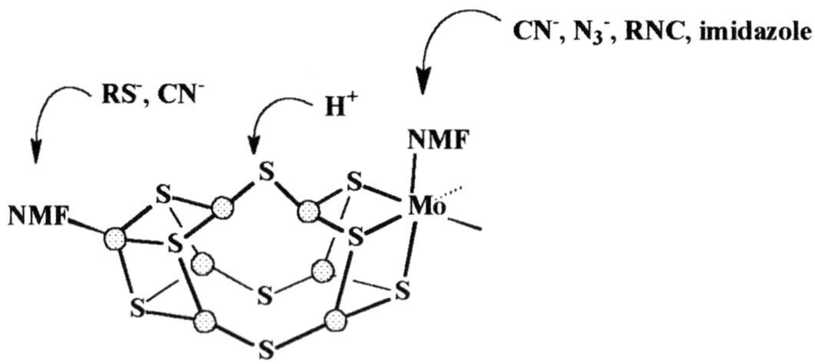

Figure 24. Binding sites of various molecules and ions on extracted FeMo-cofactor.

5.2.2 The Role of R*-Homocitrate*

It has been known for some time that *R*-homocitrate is essential for the enzyme to be a good nitrogen fixer (see Section 2.3.2), and that the enzyme is very specific in its requirement for this polycarboxylate. Thus, the *nifV* mutant nitrogenase has citrate bound to FeMo-cofactor in place of *R*-homocitrate. The only difference between citrate and homocitrate is that the latter has an extra $-CH_2-$ group, which is in the long arm of the homocitrate not coordinated to the cofactor. Yet, this extra $-CH_2-$ group makes all the difference. The mutant nitrogenase is only 7% as good a nitrogen fixer as wild-type nitrogenase.

Recently, kinetic studies on the reactivity of extracted FeMo-cofactor from the wild type enzyme and the *nifV* mutant nitrogenase showed that when N_3^-, CN^-, Bu^tNC or H^+ bind to either cofactor, the reactivities of the corresponding cofactors were indistinguishable. Only when imidazole is bound is there a difference in reactivity of the two cofactors [94]. Molecular mechanics calculations indicate that the difference in reactivity of the imidazole-derivatised cofactors from wild type and mutant nitrogenases is a consequence of hydrogen bonding between the long arm of the *R*-homocitrate and the imidazole N-H group in wild type FeMo-cofactor. The carboxylate-to-imidazole hydrogen bond imposes imidazolate-like character thus affecting the reactivity of the FeMo-cofactor. For the *nifV* cofactor the citrate ligand does not have a sufficiently long pendant arm and consequently cannot hydrogen bond to the imidazole N-H.

Consideration of the crystal structure of the MoFe-protein, together with molecular modelling calculations, suggest that a similar hydrogen bond can occur in the protein but only if the imidazole of Hisα442 rotates appreciably, or if the *R*-homocitrate becomes a monodentate ligand by dissociation of the carboxylate group, as shown in Figure 25.

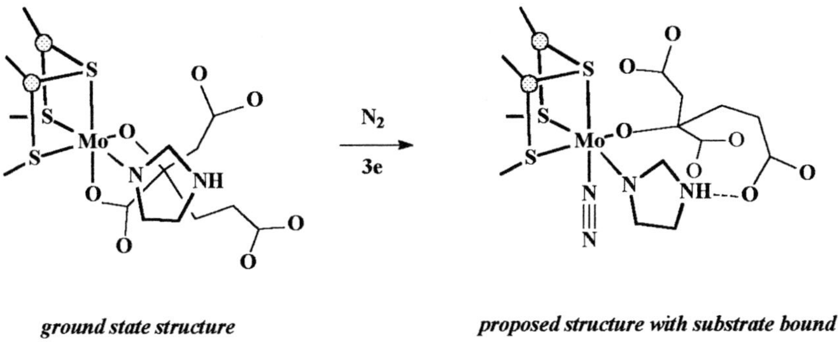

Figure 25. Proposed hydrogen bond between *R*-homocitrate and His-α442.

Again, for the *nifV* mutant nitrogenase, the citrate ligand cannot form a hydrogen bond with the imidazole residue of Hisα442 because it lacks the long $CH_2CH_2CO_2^-$ arm. It has been proposed that the hydrogen bond between monodentate R-homocitrate and Hisα442 facilitates binding of dinitrogen to cofactor. Specifically, dissociation of the carboxylate group of *R*-homocitrate from molybdenum generates a vacant site on the metal at which dinitrogen can bind. Simultaneously, formation of the hydrogen bond effectively releases electron density to the cluster thus making FeMo-cofactor more electron-rich, facilitating the binding of dinitrogen and its subsequent protonation to ammonia. This is, to date, the only proposal which rationalises the unique structural features of *R*-homocitrate with the observed reactivity.

Theoretical calculations [95] using hybrid density functional method B3LYP have explored the structure and stability of the polycarboxylate ligand bound to the molybdenum, and in particular protonation of the R-homocitrate. The major conclusions from the theoretical studies are as follows. (i) Protonation of the coordinated hydroxyl-residue is more favourable than the coordinated carboxylate for the reduced species. In the oxidised form the proton affinities are reversed. (ii) Protonation of the homocitrate elongates the Mo-O distance by 0.2-0.3Å, reduces the Mo-polycarboxylate binding energy but only slightly changes the Mo-N imidazole bond. (iii) The proposed spontaneous bidentate-to-monodentate rearrangement of the homocitrate does not occur upon the addition of up to two protons, or one-electron reduction. It only occurs upon single protonation, single-electron reduction and coordination of the substrate to the cluster. (iv) The coordinated imidazole is not protonated and it is the carboxylate on the long arm which is preventing the protonation.

5.2.3 Other Models for the Action of FeMo-cofactor

In the discussion so far all of the models for the reactivity of FeMo-cofactor have considered only the structure established in the X-ray crystal structure of the MoFe-protein. However, it has recently been proposed that FeMo-cofactor opens one of

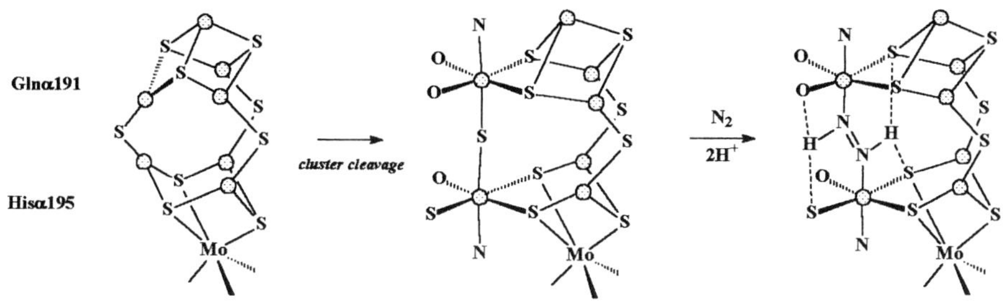

Figure 26. Proposed rupture of FeMo-cofactor to bind dinitrogen.

the Fe-S-Fe bonds [55] as shown in Figure 26. The amino acid side chains of Glu-α191 and His −α195 are sufficiently close to coordinate to the iron sites making these metals six coordinate. The more open structure allows dinitrogen to bind.

In another suggestion it has been proposed that dinitrogen binds initially to the prismatic centre of the cofactor [96], as shown in Figure 27. The initial four-electron reduction of dinitrogen to form bound hydrazine occurs on the central Fe_6 unit.

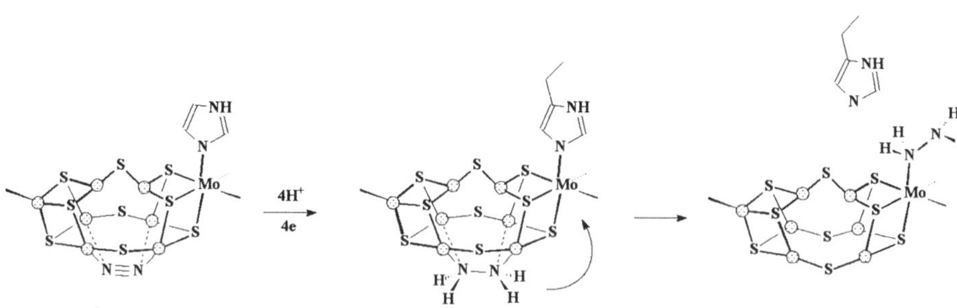

Figure 27. Proposed mechanism for dinitrogen reduction involving migration

The hydrazine is then transferred to the heterometal where reduction to ammonia occurs.

Finally, it has been suggested that the role of the heterometal is merely to modulate the reactivity of the Fe sites [97]. Certainly, the heterometals have been shown to modulate the substitution mechanisms of Fe sites in synthetic Fe-S-based clusters. It could be that the heterometal makes the cofactor a poorer proton reducer (hydrogenase), which is consistent with both the V-based and Fe-only nitrogenases being poorer nitrogen fixers.

6. FUTURE CHALLENGES

No right-minded person would consider predicting the future discoveries in the area of nitrogen fixation, and I for one am not even going to attempt to second guess the mechanism of nitrogenase. The question paramount to a chemist always used to be, "Which metal is the active site: Mo or Fe?" Today the questions are much more subtle. "What is the role of the metals: both those that bind substrate and the others? How is the polypeptide modulating the reactivity at the metals?"

The future challenges in mechanistic aspects of nitrogen fixation are simple to state, but demanding to accomplish. To complete our understanding of the mechanism of nitrogenase the following goals have to be achieved. (i) Establish how dinitrogen binds to FeMo-cofactor. (ii) Define how dinitrogen is transformed into ammonia at the active site. This requires, at one level, establishing the protonation chemistry of dinitrogen bound to FeMo-cofactor, the kinetics of the transformation and identification of the intermediates involved. At another level we need to understand how the active site operates including the sequence of electron- and proton-transfer reactions, and the roles for various amino acid side chains which hydrogen bond to the cofactor. Hidden amongst these challenges are questions about how FeMo-cofactor change shape during catalysis. How does the protein modulate the reactivity? What is the role of MgATP?

One thing is abundantly clear about the mechanism of nitrogenase, and has been so for many years. No one individual is going to solve the problem. The collaboration between chemists, enzymologists and geneticists is essential to establish the mechanism by which nitrogenase converts dinitrogen into ammonia.

7. REFERENCES

1. T. A. Bazhenova and A E Shilov, Coord. Chem. Rev. 144 (1995) 69, and refs therein.
2. R. R. Eady, Chem. Rev. 96 (1996) 3013, and refs therein.
3. B. E. Smith, Adv. Inorg. Chem. 47 (1999) 159, and refs therein.
4. M. Ribbe, D .Gadkari and O. Myer, J. Biol. Chem. 272 (1997) 26627.
5. D. J. Evans, R. A. Henderson and B. E. Smith, Bioinorganic Catalysis, J Reedijk and E Bouwman ed, Marcel Dekker, New York, 1999, Chapter 7, and references therein.
6. B. Burgess and D. J. Lowe, Chem. Rev. 96 (1996) 2983 and refs therein.
7. L. C. Seefeldt and D. R. Dean, Acc. Chem. Res. 30 (1997) 260, and references therein.
8. G. D. Watt and K. R. N. Reddy, J. Inorg. Biochem. 53 (1994) 281.
9. H. C. Angove, S. J. Yoo, B. K. Burgess and E. Munck, J. Am. Chem. Soc. 119 (1997) 8730.
10. J. Kim and D. C. Rees, Science 257 (1992) 1677.

11. J. Kim and D. C. Rees, Nature 360 (1992) 553.
12. J. W. Peters, M. H. B. Stowell, S. M. Soltis, M. C. Finnegan, M. K. Johnson and D. C. Rees, Biochemistry 36 (1997) 118.
13. J. Shen, D. R. Dean and W. E. Newton, Biochemistry 36 (1997) 4884.
14. D. J. Scott, H. D. May, W. E. Newton, K. E. Brigle and D. R. Dean, Nature 343 (1990) 188.
15. C. H. Kim, W. E. Newton and D. R. Dean, Biochemistry 34 (1995) 2798.
16. T. R. Hawkes, P. A. McLean and B. E. Smith, Biochem J. 217 (1984) 317.
17. M. J. Dilworth and R. R. Eady, Biochem J., 277 (1991) 465.
18. M. J. Dilworth, R. R. Eady, R. L. Robson and R. W. Miller, Nature 327 (1987) 167.
19. R. N. F. Thorneley and D. J. Lowe, Molybdenum Enzymes, T J Spiro ed, J Wiley and Son, New York, 1985, p. 221, and refs therein.
20. H. Schindelin, C. Kisker, J. L. Schlessman, J. B. Howard and D. C. Rees, Nature 387 (1997) 370.
21. W. N. Lazilotta and L. C. Seefeldt, Biochemistry 36 (1997) 12976.
22. D. J. Lowe and C. Mitchell, abstract of contribution to conference Nitrogen Assimilation: Molecular and Genetic Aspects, University of Southern California, 1997.
23. S. Y. Druzhinin, L. A. Syrtsova, A. M. Uzenskaja and G. I. Likhtenstein. Biochem J. 290 (1993) 627.
24. R. N. F. Thorneley, J. Bergstrom, R. R. Eady and D. J. Lowe, Biochem. J. 271 (1989) 789.
25. D. J. Lowe, K. Fisher and R. N. F. Thorneley, Biochem. J. 292 (1993) 93.
26. M. G. Duyvis, R. E. Mensink, H. Wassink and H. Haaker, Biochem. Biophys. Acta, Bioenergetics 1320 (1997) 34.
27. D. J. Lowe, G. A. Ashby, M. Brune, H. Knights, M. R. Webb and R. N. F. Thorneley, Nitrogen Fixation: Fundamentals and Applications, I. A. Tikhonovich, N. A. Provorov, V. I. Romanov and W. E. Newton (eds.), Kluwer Academic Press, Dordrecht, 1995, p. 103.
28. M. J. Ryle and L. C. Seefeldt, Biochemistry 35 (1996) 4766.
29. R. N. F. Thorneley, D. J. Lowe and R. R. Eady, Nature 272 (1978) 557.
30. S. J. George, G. A. Ashby, C. N. Wharton and R. N. F. Thorneley, J. Am. Chem. Soc. 119 (1997) 6450.
31. S. K. Ibrahim, K. Vincent, C. A. Gormal, B. E. Smith, S. P. Best and C. J. Pickett, Chem. Commun.. (1999) 1019.
32. M. Deng and R. Hoffman, Angew. Chem., Int. Ed. Engl. 32 (1993) 1061.
33. I. G. Dance, Aust. J. Chem. 47 (1994) 979.
34. K. K. Stavner and M. C .Zerner, Chem. Eur. J. 2 (1996) 83.
35. J. Christiansen, R. C. Tittsworth, B. J. Hales and S. P. Cramer, J. Am. Chem. Soc. 117 (1995) 10017.
36. S. J. Zhong and C. W. Liu, Polyhedron 16 (1997) 653.
37. T. H. Rod and J. K. Nørskov, J. Am. Chem. Soc. 122 (2000) 12751.

38. J. Chatt, J. R. Dilworth and R. L. Richards, Chem. Rev. 78 (1977) 589 and references therein.
39. R. A. Henderson, G. J. Leigh and C. J. Pickett, Adv. Inorg. Radiochem. 27 (1983) 198 and references therein.
40. F. F. Simpson and R. H. Burris, Science 224 (1984) 1095.
41. W. E. Newton, J. L. Corbin and J. W. McDonald, Proceedings First International Symposium on Nitrogen Fixation, Washington State University Press, 1976, p. 53.
42. C. Willis, R. A. Back, J. G. Purdon and J. M. Parsons, J. Am. Chem. Soc. 99 (1977) 4451, and references therein.
43. R. Ellis, R. A. Henderson, A. Hills and D. L. Hughes, J. Organomet. Chem. 333 (1987) C6.
44. D. Dawson, R. A. Henderson, A. Hills and D. L. Hughes, Polyhedron 8 (1989) 1870.
45. R. A. Henderson, J. Chem. Soc., Chem. Comm. (1987) 1670.
46. R. A. Henderson and K. E. Oglieve, J. Chem. Soc., Chem. Commun. (1991) 584
47. M. G. Basallote, J. Durain, J. Fernandez-Trujillo and A. Manez, J. Organomet. Chem. 609 (2000) 29.
48. D. L. Hughes, S. K. Ibrahim, G. Querne, A. Laouenan, J. Talarmin, A. Queiros, A. Fouseca and C. J. Pickett, Polyhedron 13 (1994) 3341.
49. S. A. Fairhurst, R. A. Henderson, D. L. Hughes, S. K. Ibrahim and C. J. Pickett, J. Chem. Soc., Chem. Commun. (1995) 1569.
50. C. J. Pickett, J. Biol. Inorg. Chem. 1 (1996) 601, and refs therein.
51. I. G. Dance, J. Chem. Soc., Chem. Comm. (1998) 523, and refs therein.
52. K. L. C. Gronberg, R. A. Henderson and K. E. Oglieve, J. Chem. Soc., Dalton Trans. (1998) 3093.
53. J. M. Rivera-Ortiz and R. H. Burris, J. Bacteriol. 123 (1975) 537.
54. B. E. Smith, M. C. Durrant, S. A. Fairhurst, C. A. Gormal, K. L. C. Gronberg, R. A. Henderson, S. K. Ibrahim, T. LeGall and C. J. Pickett, Coord. Chem. Rev. 185 (1999) 669.
55. D. Sellmann and J. Sutter, Acc. Chem. Res. 30 (1997) 460, and refs therein.
56. J. R. Dilworth and R. L. Richards, Comprehensive Organometallic Chemistry, Volume 8, G Wilkinson, F G A Stone and E W Abel (eds.), Pergamon, Oxford, 1982, p. 1073, and refs therein.
57. R. A. Henderson, J. Organomet. Chem. 208 (1981) C51.
58. R. A. Henderson, J. Chem. Soc., Dalton Trans. (1982) 917.
59. D. L. DuBois and R. Hoffman, Nouv. J. Chim. 1 (1977) 479.
60. T. Yamabe, K. Hori and K Fukui, Inorg. Chem. 21 (1982) 2046.
61. J. D. Lane and R. A. Henderson, J. Chem. Soc., Dalton Trans. (1986) 2155.
62. J. D. Lane and R. A. Henderson, J. Chem. Soc., Dalton Trans. (1987) 197.
63. R. A. Henderson, G. Davies, J. R. Dilworth and R. N. F. Thorneley, J. Chem. Soc., Dalton Trans. (1981) 40.
64. R. A. Henderson, J. Chem. Soc., Dalton Trans. (1984) 2259.

65. V. L. Frost and R. A. Henderson, J. Chem. Soc., Dalton Trans. (1985) 2059.
66. J. A. Carroll and D. Sutton, Inorg. Chem. 19 (1980) 3137.
67. J. E. Barclay, A. Hills, D. L. Hughes, G. J. Leigh, C. J. Macdonald, M. Abu-Bakar and H. Mohd-Ali, J. Chem. Soc., Dalton Trans. (1990) 2503, and references therein.
68. T. Yamabe, K. Hori and K. Fukui, Inorg. Chem. 21 (1982) 2816.
69. R. A. Henderson, G. J. Leigh and C. J. Pickett, J. Chem. Soc., Dalton Trans. (1989) 425.
70. J. R. Dilworth, R. A. Henderson, P. Dahlstrom, T. Nicholson and J. A. Zubieta, J. Chem. Soc., Dalton Trans. (1987) 529.
71. J. R. Dilworth, A. Garcia-Rodriguez, G. J. Leigh and J. N. Murrell, J. Chem. Soc., Dalton Trans. (1983) 455.
72. R. J. Deeth and S. A. Langford, J. Chem. Soc., Dalton Trans. (1995) 1.
73. P. S. Wagenknecht and J. R. Norton, J. Am. Chem. Soc. 117 (1995) 1841.
74. M. Y. Mohammed and C. J. Pickett, J. Chem. Soc., Chem. Commun (1988) 1119.
75. A. E. Shilov, New Trends in the Chemistry of Nitrogen Fixation, J Chatt, L. M. da Camara Pina and R. L. Richards (eds.), Academic Press, London, 1980, p. 130 and refs therein.
76. A. E. Shilov, Pure and Applied Chem 64 (1992) 1409, and referencestherein.
77. N. P. Luneva, S. A. Mironova, A. E. Shilov, M. Yu Antipin and Y. T. Struchkov, Angew. Chem., Int. Ed. Engl.32 (1993) 1178.
78. R. A. Henderson, S. H. Morgan and A. N. Stephens, J. Chem. Soc., Dalton Trans. (1990) 1101.
79. R. A. Henderson and S. H. Morgan, J. Chem. Soc., Dalton Trans. (1990) 1107.
80. D. Sellmann, W. Soglowek, F. Knoch and M. Moll, Angew. Chem., Int. Ed. 28, (1989) 1271.
81. C. E. Laplaza, M. J. A. Johnson, J. C. Peters, A. L. Odom, E. Kim, C. C. Cummins, G. N. George and I. J. Pickering, J. Am. Chem. Soc. 118 (1996) 8623.
82. G. N. Schrauzer in reference 75, Chapter 4.
83. G. N. Schrauzer, G. W. Kiefer, K. Tano and P. A. Doemeny, J. Am. Chem. Soc. 96 (1974) 641.
84. G. N. Schrauzer, P. R. Robinson, E. L. Moorehead and T. M. Vickrey, J. Am. Chem. Soc. 98 (1976) 2815.
85. K. Taanaka, Y. Hozumi and T. Tanaka, Chem. Lett. (1982) 1203.
86. Y. Hozumi, Y.Imasaka, K. Tanaka and T. Tanaka, Chem. Lett. (1983) 897.
87. S. M. Malinak, K. D. Demadis and D. Coucouvanis, J. Am. Chem. Soc. 117 (1995) 3126.
88. K. D. Demadis, S. M. Malinak and D. Coucouvanis, Inorg. Chem. 35 (1996) 4038.
89. S. M. Malinak, A. M. Simeonov, P. E. Mosier, C. E. McKenna and D. Coucouvanis, J. Am. Chem. Soc. 119 (1997) 1662.
90. B. K. Burgess, Chem. Rev. 90 (1990) 1377, and references therein.

91. V. K. Shah, J. R. Chisnell and W. J. Brill, Biochem. Biophys. Res. Comm. 81 (1978) 232.
92. V. R. Almeida, C. A. Gormal, K. L. C. Gronberg, R. A. Henderson, K. E. Oglieve and B. E. Smith, Inorg. Chim. Acta 291 (1999) 212, and references therein.
93. K. L. C. Gronberg, C. A. Gormal, B. E. Smith and R. A. Henderson, Chem. Commun. (1997) 913.
94. K. L. C. Gronberg, C. A. Gormal, M. C. Durrant, B. E. Smith and R. A. Henderson, J. Am. Chem. Soc. 120 (1998) 10613.
95. R. K. Szilagyi, D. G. Musaev and K. Morokuma, Inorg. Chem. 40 (2001) 766.
96. D. Coucouvanis, J. Biol. Inorg. Chem. 1 (1996) 594, and references therein.
97. R. A. Henderson and K. E. Oglieve, J. Chem. Soc., Dalton Trans. (1993) 1473.

Nitrogen Fixation at the Millennium.
G. Jeffery Leigh, Editor.

CHAPTER 10

A Novel Nitrogenase: Superoxide-Dependent Nitrogen Fixation

D. Gadkari

Chair of Microbiology, University of Bayreuth, 95440 Bayreuth, Germany

1. INTRODUCTION

Streptomyces thermoautotrophicus is able to fix dinitrogen and it harbours a completely unusual nitrogen-fixing system, which differs from "conventional" nitrogenases. *S. thermoautotrophicus* has been isolated from the covering soil of a burning charcoal pile (Figure 1). During the burning process, the surface temperature at different locations of the charcoal pile ranges from 17 to 90 °C (Figure 2). Combustion of wood liberates various components, such as the gases CO_2 (49%), CO (34%), CH_4 (13%), C_2H_4 (2%) and H_2 (2%) [1], and, in addition, compounds such as methanol, acetone, methyl acetate and acetic acid [1,2,3]. Gases and other compounds escape with smoke through the vents, which are provided for access of the limited amount of oxygen necessary for the combustion of wood. This establishes favourable natural conditions around the vents for the enrichment and growth of various bacteria. In general the organisms proliferating there are resistant to toxic components of smoke such as CO.

1.1 Description of *Streptomyces thermoautotrophicus*

Streptomyces thermoautotrophicus forms relatively stable, short, branching vegetative hyphae with a diameter of 0.2 to 0.5 µm. A chain of 2 to 8 oval spores occurs in a sheath residing on a scanty whitish mycelium. On a solid surface it produces whitish rough colonies. In an unshaken liquid medium *S. thermoautotrophicus* tends to produce a layer of floating mycelium on the surface of liquid. No endospores, synematta, sporangia or sclerotia are produced. It is an aerobic, non-motile, gram- and catalase-positive bacterium [4].

Figure 1. Burning charcoal piles.
The charcoal pile in front was set up one week earlier than the pile behind it. A student is measuring a temperature profile of the charcoal pile. The white flags indicate the points where the temperature profile has been measured. The smoke exhausting from the vents is a mixture of gases.

The cell wall of *S. thermoautotrophicus* contains *LL*-diaminopimelic acid and ribose, which indicates that it has a type I cell wall. Mycolic acids are absent. The predominant menaquinone is MK-9(H_4), and a small amount of MK-9(H_6) also occurs. The phospholipid pattern is composed of phosphatidylethanolamine, phosphatidylglycerol, diphosphatidylglycerol, phosphatidylinositol and phosphatidylinositol mannosides (PL-type 2). *Iso* and *anteiso* branched fatty acids are synthesized, with *iso*-C_{17}, *anteiso*-C_{17}, and *iso*-C_{16} being predominant. Small amounts of cyclopropane fatty acids and traces of 2-hydroxy fatty acids are also present. The G+C content of DNA is 70.6 +/- 0.19 mol % [4].

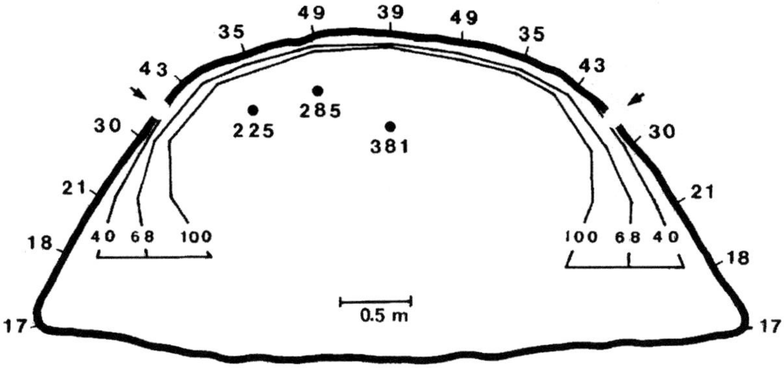

Figure 2. Temperature profiles of a 7 day-old burning charcoal pile. If not indicated, values refer to temperatures in degrees Celsius. Isotherms at 40, 68 and 100 °C are calculated values. Arrows show the positions of the holes for ventilation [4].

S. thermoautotrophicus utilizes the gases as sources of energy, carbon and nitrogen (gasotrophy). It grows in an inorganic medium [5] with either carbon monoxide as the sole source of carbon and energy (*t*d = 8 h) or with hydrogen (H_2) and carbon dioxide (CO_2) as energy and carbon source, respectively (*t*d = 6 h) [4]. *S. thermoautotrophicus* can utilize car exhaust fumes or gas produced by incomplete combustion of wood [4]. Complex media or heterotrophic substrates such as sugars, organic acids, amino acids and alcohols were not utilized for growth. Though CO and H_2 are known inhibitors of "conventional" nitrogenases, *S. thermoautotrophicus* is able to fix dinitrogen [6] under conditions specified in Table 1. Cultures grown under static systems with CO plus CO_2 plus a mixture of $^{14}N_2/^{15}N_2$ incorporate a remarkably high atom % of ^{15}N. A similar situation is found when the growth gas mixture was H_2 plus CO_2 plus a mixture of $^{14}N_2/^{15}N_2$ (Table 1) [6]. A comparatively high incorporation of $^{15}N_2$ has been also reported for *Rhizobium* strains [7,8]. A pure culture of *S. thermoautotrophicus* grows in a desiccator under an atmosphere of 5% CO_2, 45% CO and 50% air by volume. Growth of *S. thermoautotrophicus* occurs between 40 and 65 °C, optimal growth is at 65 °C and pH optimum is 7.5. This bacterium belongs to the group of carboxidotrophic bacteria and it is an obligate chemolithoautotrophic thermophilic bacterium.

Growth condition	$^{15}N_2$ excess atom %	$^{14}N_2/^{15}N_2$
Substrate: CO plus CO_2		
N_2	12.210	7.173
N_2 plus NH_4Cl	0.0002	272.75
N_2 (autoclaved culture)	0.0003	272.67
Substrate : H_2 plus CO_2		
N_2	13.980	6.185
N_2 (autoclaved culture)	0.0005	272.52

[a] The isotope ratio $^{14}N_2/^{15}N_2$ in natural air is 272.90 [6].

Table 1. Incorporation of $^{15}N_2$ into intact cell material of *S. thermoautotrophicus* [a].

2. SUPEROXIDE-DEPENDENT NITROGEN FIXATION

The nitrogen-fixing system of *S. thermoautotrophicus* is comprised of three proteins, a heterotrimeric molybdenum-containing dinitrogenase (St1), a homodimeric manganese-containing superoxide oxidoreductase (St2) and a heterotrimeric molybdenum-containing carbon monoxide dehydrogenase (CODH or St3). These proteins differ entirely from the known nitrogenase protein components. All three proteins of the dinitrogen-fixing system of *S. thermoautotrophicus* are located in the cytoplasmic fraction (at some growth phases CODH is associated with membrane [9]) and are stable towards oxygen. Moreover, for the reduction of N_2 to NH_4^+ superoxide radicals are an essential requirement [10]. The fixation of nitrogen is coupled to the oxidation of CO. CODH oxidizes CO and transfers the electrons released to O_2, thereby producing superoxide ($O_2^{\cdot-}$). The Mn-superoxide oxidoreductase reoxidizes $O_2^{\cdot-}$ to O_2 and transfers the electrons to a Mo-dinitrogenase which in turn reduces dinitrogen to ammonium. In contrast to the electronic coupling in "conventional" nitrogenases, *via* ferredoxin/flavodoxin, superoxide oxidoreductase and dinitrogenase establish a molecular coupling *via* $O_2^{\cdot-}$. As the reduction of N_2 depends obligately on superoxide, the N_2-fixing system of *S. thermoautotrophicus* has been named **superoxide-dependent nitrogenase (sdn)**.

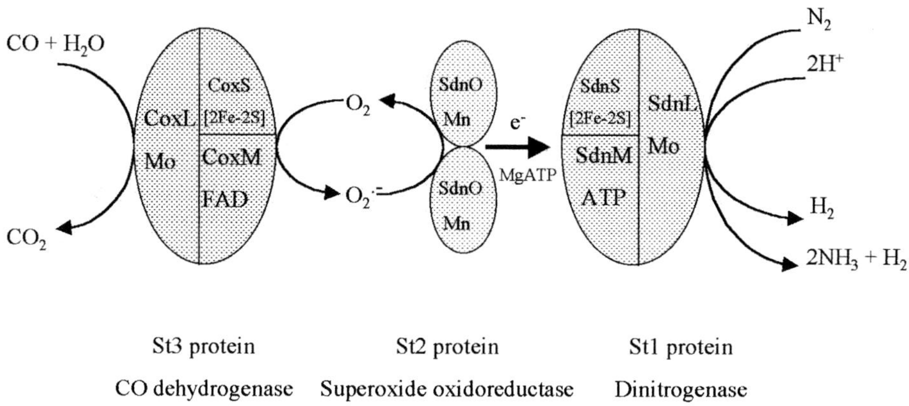

Scheme 1. Schematic representation of N_2 fixation in *S. thermoautotrophicus*. Superoxide is produced by CO dehydrogenase (St3 protein) through the oxidation of CO and the transfer of the electrons to O_2. Subsequently, the superoxide is reoxidized by a superoxide oxidoreductase (St2 protein) that delivers the electrons to a dinitrogenase (St1 protein). The dinitrogenase is capable of reducing N_2 and H^+. CODH is represented as a single heterotrimer. The stoichiometries given in the scheme are arbitrary [10].

3. COMPONENTS OF SUPEROXIDE-DEPENDENT NITROGENASE

3.1 Dinitrogenase (St1 protein)

The polypeptide dinitrogenase (St1 protein) is a 140.85 kDa protein and is composed of three non-covalently bound subunits, which are designated as sdnL (90.98 kDa), sdnM (31.54 kDa) and sdnS (18.33 kDa) [10,11]. The isoelectric point of St1 protein is 6.4. It contains molybdenum, iron and acid-labile sulfide in a molar ratio of 1:17:11. There is also a significant amount of zinc (Table 2), but the role of zinc is unknown.

The St1 protein contains molybdopterin as a cofactor. The nucleotide present in the molybdopterin is cytosine dinucleotide [di(cam)]. The spectrum of [di(cam)] of St1 is similar to that of CODH from *Hydrogenophaga pseudoflava* [12,13].

The N-termini of the St1 polypeptide sequences are as follows.

MALPQTELRPMGKPILRKEDP (sdnL)
MFPNAFKYEAPASVDEAVRLLAEYGYDGKV (sdnM)
MKIRVKVNGTLYEADVEP (sdnS).

The N-termini of sdnL and the dinitrogenase β-subunits of *Anabaena* 7120 or *Klebsiella pneumoniae* show 42.1 or 44.4% sequence similarity and 26.3 or 16.7% identity, respectively. The N-termini of sdnM and the dinitrogenase α-subunits of *Anabaena* 7120 or *Klebsiella pneumoniae* show 55.2 or 52.4% sequence similarity and 38.1 or 20.7% identity, respectively. The N-terminus of sdnS reveals no significant sequence similarity to the α- or β-subunits of other nitrogenases or to the δ-subunit of alternative dinitrogenases. The complete primary sequences of sdnL, sdnM and sdnS polypeptides show a high degree of similarity to the corresponding polypeptides or domains of the molybdenum-hydroxylase family (see genetics section).

3.2 Superoxide oxidoreductase (St2 protein)

The polypeptide superoxide oxidoreductase (St2 protein) is a 48 kDa protein and is composed of a single non-covalently bound 24 kDa subunit, designated as sdnO [11]. The St2 protein has a homodimeric subunit structure. The isoelectric point of St2 is 6.4. It is a manganese-containing enzyme (Table 2). The N-terminus of St2 polypeptide is as shown.

MFELPPLPYPYDALEPYFDAKKMEIHYYGGHGA

This has no significant sequence similarity to sequenced nitrogenase polypeptides. Instead, sdnO shows 96.0% sequence similarity and 72.0% identity to the N-termini of manganese-containing superoxide dismutases (SODs) of *Bacillus stearothermophilus* [14] or *Bacillus caldotenax* [15]. This is also true for the entire sequence of sdnO (see genetics section). In assays containing xanthine oxidase, xanthine and O_2 for the production of superoxide anion radicals ($O_2^{\cdot-}$), the St2 protein revealed $O_2^{\cdot-}$-oxidizing activities. This reaction is insensitive to 3.5 mM potassium cyanide, 60 mM sodium azide, or 10 mM H_2O_2 [10]. Although the St2 protein displays many properties similar to a manganese SOD [16] it actually functions as a superoxide oxidoreductase, transferring electrons to the St1 protein. St2 is unable to convert $O_2^{\cdot-}$ to H_2O_2 [10]. Dithionite-reduced St2 protein can transfer 2.3 $\pm$ 0.7 electrons to oxidized phenazine methosulfate [10]. Another suitable electron acceptor is 2,6-dichlorophenolindophenol

(DCPIP), whereas viologen dyes, methylene blue, ferricyanide, INT, nitro blue tetrazolium, NAD(P), FAD, FMN and riboflavin are ineffective [10].

Protein	Metals					Acid-labile sulfide	Sulfhydryls	
	Mo	Fe	Zn	Mn	V		Available[a]	Total[b]
St1	0.7 ± 0.1	12 ± 1	3.4	ND[c]	<0.06	8 ± 1	4 ± 0.1	12 ± 1
St2	<0.04	<0.07	0.1	0.24 ± 0.03	<0.02	<0.02	0	0

[a] Number of sulfhydryl groups available for 5,5′-dithio-bis-nitrobenzoic acid at the surface of native protein.
[b] Total number of sulfhydryl groups of denaturated protein.
[c] Not determined [10].

Table 2. Contents of metals, acid-labile sulfide and sulfhydryls of St1 and St2 proteins from *S. thermoautotrophicus*. Molecular masses of 144 kDa (St1) and 48 kDa (St2) were used for calculations. The values of acid-labile sulfide, molybdenum, and iron are the mean of six, manganese the mean of two, and the sulfhydryls the mean of four independent determinations.

3.3 Carbon monoxide dehydrogenase (CODH or St3 protein)

Carbon monoxide dehydrogenase (St3 protein) is a 272.6 kDa protein. CODH is a prototype of the molybdenum-hydroxylases sequence family [17]. This protein is generally a dimer of two heterotrimers $(LMS)_2$, and thus it contains six polypeptides. However, it can occur as a single heterotrimer [10,18,19] or as a dimer of heterotrimers [19,20]. Both forms are active. Two of our co-workers [10,18] found only a single heterotrimer on native PAGE. Another worker [19], who reported that these forms are equally active, detected both simultaneously. In contrast, in another report only a dimer of heterotrimers was detected on native PAGE [20]. It appears that this phenomenon depends upon the growth conditions and the time of the harvesting. It is also possible that during the purification process the two monomers, each composed of a heterotrimer, sometimes remain together and sometimes occur as two separate functional monomers. The molecular weight of the dimeric heterotrimer of *S. thermoautotrophicus* corresponds to CODHs from various aerobic carboxidotrophic bacteria [21,22,23]. Each heterotrimer is composed of three non-covalently bound subunits, a molybdo-copper protein, coxL (large subunit, 86.4 kDa), a flavoprotein, coxM (the medium sized subunit, 32.2 kDa) and an iron-sulfur protein, coxS (the small subunit, 17.7 kDa). CODH of *S. thermoautotrophicus* carries a molybdopterin cytosine dinucleotide (MCD)-type of molybdenum cofactor [20]. The metal content of St3 is given in Table 3. The isoelectric point ranges from 4.5 to 5.1. In this range 8 bands of protein are detected. The CODH activity test performed on the gel revealed that all bands showed distinct CODH activities but with different intensities, suggesting that these bands are isoforms of CODH of *S. thermoautotrophicus* [20]. This type of phenomenon of purified protein is known as microheterogeneity. There are many reasons for the occurrence of microheterogeneity. Frequently, the post-transcriptional or the post-translational modifications or uncontrolled fragmentation are the main causes, among many others, for the appearance of several isoforms [24].

Protein	Metals						Acid-labile sulfide	Sulfhydryls	
	Mo	Fe	Zn	Mn	Cu	Se		Available	total
St3	0.77	7.54	0.39	<0.001	0.8	<0.001	8.4	16.6	25

Table 3. Contents of metals, acid-labile sulfide and sulfhydryls of St3 protein from *S. thermoautotrophicus* [18].

The N-termini of the St3 polypeptide (CODH) sequences are as follows.

ALPQTELRPMGKPILRKEDPRLI (coxL)
MFPNAFKYEAPASVDEAVRL (coxM)
MKIRVKVNGTLYEADVEPRTL (coxS) [20].

The amino acid sequences of N-termini of coxL, coxM and coxS have 100% identitiy to sdnL, sdnM and sdnS, respectively. Not only that, but the entire sequences of coxL, coxM and coxS of *Oligotropha carboxidovorans* [21] show high similarity to the corresponding sdnL, sdnM and sdnS, respectively [11]. Similar to other known CODHs [25], St3 transfers electrons to O_2 during oxidation of CO, thereby producing hydrogen peroxide (H_2O_2) and superoxide (O_2^-) [10]. The CODH of the carboxidotrophic bacteria *O. carboxidovorans* does not reduce physiological electron acceptors such as NAD^+ and FAD^+ [26,27]. Similarly the CODH of *S. thermoautotrophicus* also does not reduce NAD^+ and FAD^+. However, it can reduce many artificial electron acceptors such as thionine, methylene blue, DCPIP, INT+ PMS, PMS, NBT, and cytochrome c. Viologen dyes such as benzyl viologen and methyl viologen are reduced to a lesser degree [4,20].

CODH is a key enzyme for the utilization of CO as a growth substrate in aerobic CO-utilizing bacteria. It catalyzes the oxidation of CO with H_2O yielding CO_2, two electrons, and two H^+ [9,28]. In its membrane-associated state, CODH is involved in energy transduction. The electrons formed through CO oxidation are channelled *via* cytochrome b_{561} into a CO-insensitive respiratory chain that generates a proton motive force [9,29]. In its soluble state, CODH transfers the electrons to O_2, thereby producing hydrogen peroxide (H_2O_2) and superoxide ($O_2^{\cdot-}$) [10,25].

4. REDUCTION OF N_2 AND OTHER REACTIONS CATALYZED

4.1 Reduction of N_2 by cytoplasmic fraction

In cytoplasmic fractions where all the required proteins, St1, St2 and St3 are available, the reduction of N_2 occurs under various conditions. The results are summarized in Table 4, which shows that when all three proteins, St1, St2 and St3 are present N_2 reduction can occur with CO plus air or H_2 plus air. With any other system (*e.g.*, xanthine plus xanthine oxidase or riboflavin plus light) that is able to provide electrons, reduction of N_2 also occurs in the presence of St1, St2 and St3 proteins.

4.2 Reduction of N_2 with purified St1 and St2 proteins

Purified St1 and St2 proteins of *S. thermoautotrophicus* reduce dinitrogen following equation (1) [10]:

$$N_2 + 4\text{-}12ATP + 8H^+ + 8e^- \rightarrow 2NH_3 + H_2 + 4\text{-}12ADP + 4\text{-}12P_i \quad (1)$$

In vitro, the components required for the reduction of N_2 to NH_4^+ are St1 protein, St2 protein, ATP (2.5 mM), $MgCl_2$ (5 mM), dithionite (25 mM) and N_2. ATP-regenerating components (creatine phosphate plus creatine phosphokinase) which are generally necessary for the N_2 reduction test by "conventional" nitrogenases are not required. From Table 5 it is evident that only intact St1 plus St2 can reduce N_2, provided that Mg^{2+}, ATP and dithionite are present in the assay in appropriate concentrations. ADP does not support the reduction of N_2 to NH_4^+, however, it also does not inhibit the reduction of N_2.

Conditions	Nitrogenase activity[a]
CO plus air[b]	22.6
CO plus air plus SOD (*E. coli* or *B. stearothermophilus*)[c]	0.0
H_2 plus air[d]	25.2
Xanthine oxidase plus xanthine plus air[e]	35.3
Xanthine oxidase plus hypoxanthine plus air[e]	32.1
Riboflavin plus light plus air[f]	20.6
CO plus N_2[g]	0.0
CO plus N_2, O_2 was added after 3 h[h]	33.2

Assay conditions: 100 mL Wheaton vials containing 15 ml of phosphate buffer (50 mM, pH 7.5), 2.5 mM ATP, 5 mM $MgCl_2$, and 30 mg of cytoplasmic fraction.

[a] Specific activities are in nmol NH_4^+/mg of cytoplasmic protein/h.

[b] Assays were sparged with (% vol/vol) 50 CO plus 50 air.

[c] Assays were supplied with SOD from *E. coli* (0.45 mg) or *B. stearothermophilus* (0.15 mg) and sparged with (% vol/vol) 50 CO plus 50 air.

[d] Assays were sparged with (% vol/vol) 50 H_2 plus 50 air.

[e] Assays were supplied with bovine milk xanthine oxidase (6.4 mg), xanthine (0.7 mM), or hypoxanthine (0.7 mM).

[f] Assays containing 10 mM riboflavin were sparged with air and illuminated with light (50 W/m^2).

[g] Assays were sparged with (% vol/vol) 50 CO plus 50 N_2 under anoxic conditions.

[h] Assays were sparged for 3 h with (% vol/vol) 50 CO plus 50 N_2 under anoxic conditions, and after 3 h N_2 was replaced by air [10].

Table 4. Nitrogenase activities in cytoplasmic fractions of *S. thermoautotrophicus* under different conditions.

Assay components							Ammonium formation[a]
St1	St2	St3	Mg^{2+}	ATP	ADP	Dithionite	
+	+	-	+	+	-	+	296
+[b]	+[b]	-	+	+	-	+	0
+[c]	+[c]	-	+	+	-	+	0
+[d]	+[d]	-	+	+	-	+	0
+	+	-	-	-	-	-	0
+	-	-	+	+	-	+	0
-	+	-	+	+	-	+	0
+	+	-	+	-	-	-	0
+	+	-	-	+	-	-	0
+	+	-	-	-	-	+	0
+	+	-	+	+	-	-	0
+	+	-	+	-	-	+	0
+	+	-	-	+	-	+	0
+	+	-	+	-	+[e]	+	0
+	-	+	+	+	-	+	0
-	+	+	+	+	-	+	0
-	-	+	+	+	-	+	0

[a] N_2 reduction is in nmol NH_4^+/mg protein/h.
[b] St1 and St2 were boiled for 10 min.
[c] St1 and St2 (1 mg each) were pre-incubated for 16 h at 20 °C with protease K (12.5μg) and then assayed for activity.
[d] same as[c], except protease K was replaced by trypsin (50 μg).
[e] same results were obtained with 3 or 30 mg of ADP per assay [10].

Table 5. Requirements of nitrogenase for the reduction of N_2 to NH_4^+.

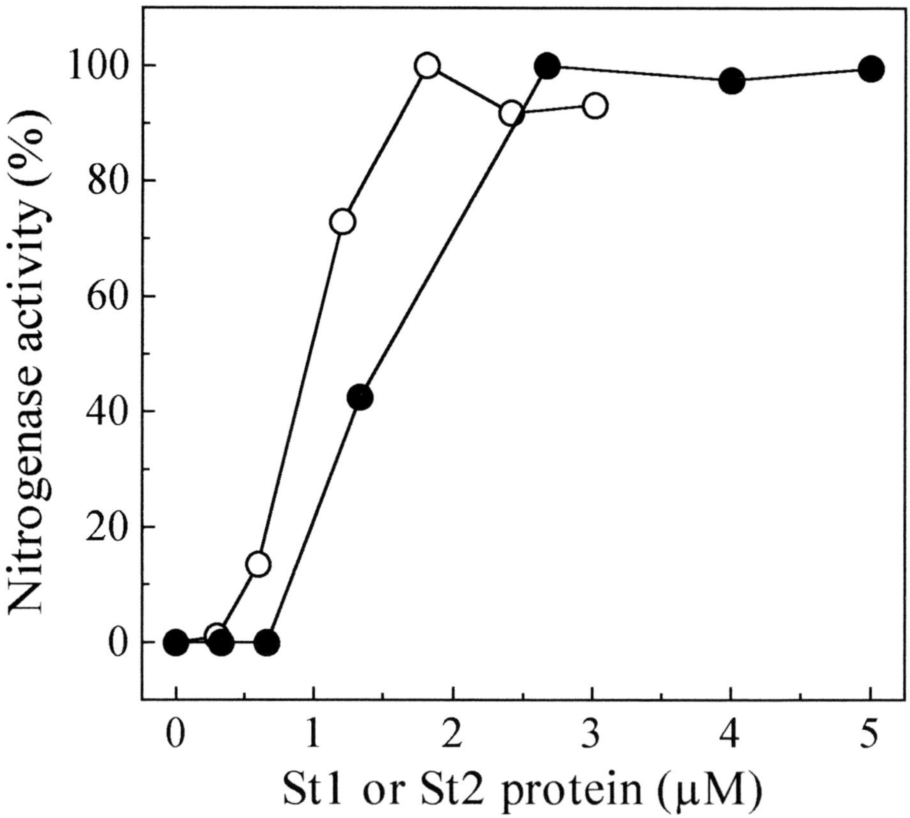

Figure 3. Dependence of nitrogenase activity on the molar ratio of St1 and St2 protein at pH 7.5 and 65 °C. The electron donor was dithionite. Symbols: ○, 2.7 μM St1 protein; ●, 1.2 μM St2 protein. The maximum activities, set 100%, were 74 nmol of NH_4^+/h (○) and 80 nmol NH_4^+/h (●) [10].

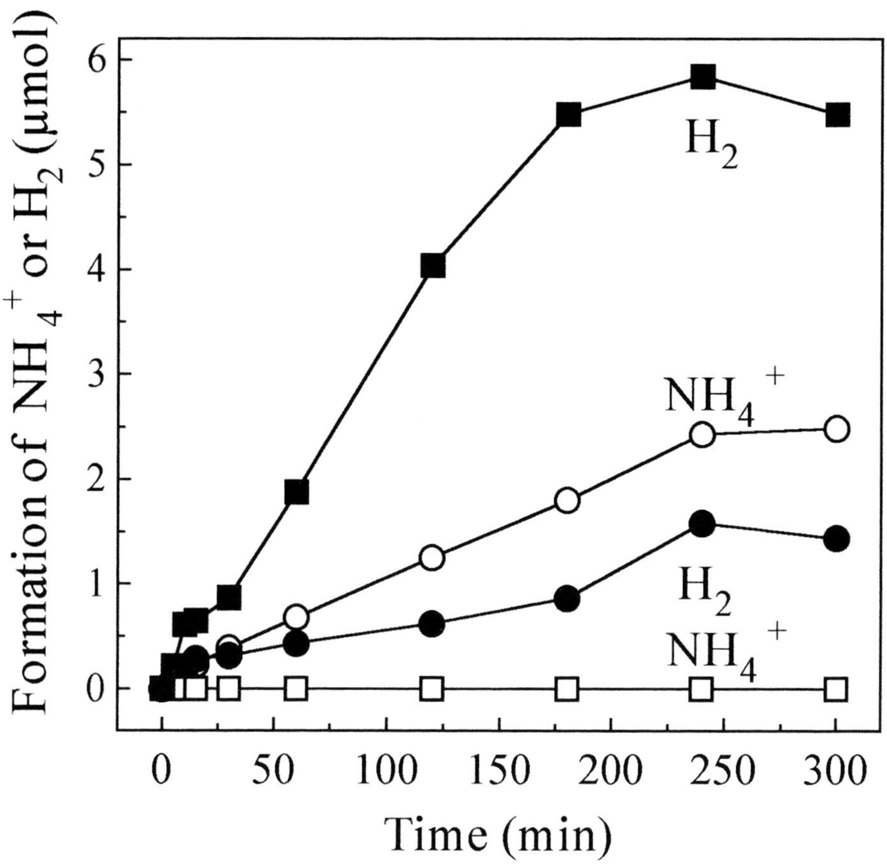

Figure 4. Formation of ammonium or H_2 by the proteins St1 and St2. The electron donor was dithionite. Symbols: ● and ■, formation of H_2 ; O and □, formation of ammonium; O and ●, gas atmosphere of pure N_2; □ and ■, gas atmosphere of pure helium. Assays contained 41.7 nmol of St1 plus 15 nmol of St2 [10].

Figure 3 reveals that in the nitrogenase system of *S. thermoautotrophicus* the optimal reduction of N_2 occurs at a molar ratio of St1:1St2 of 2.2. Theoretically, in "conventional" nitrogenases at least two moles of Fe protein to each mole of MoFe protein are necessary to provide sufficient electrons for the reduction of N_2. In practice a much higher ratio of Fe protein to MoFe protein has been used. The optimal temperature for the reduction of N_2 to NH_4^+ is 65 °C. During reduction of N_2, hydrogen is produced and in the assay where N_2 is replaced by helium only protons are reduced, to H_2. Figure 4 shows clearly that in the presence of helium there is no production of NH_4^+ and only H_2 has been evolved. "Conventional" nitrogenases behave in a similar way.

4.3 Reduction of alternative substrates

In spite of crucial differences (*e.g.*, different proteins with entirely different cofactors, different genes) between the nitrogenase system of *S. thermoautotrophicus* and the "conventional" nitrogenases, both systems have one feature in common: these proteins are able to reduce N_2 to the NH_4^+. This indicates that though the origins, carriers and paths of electrons required for the reduction of N_2 vary (Scheme 1) these systems may have at least one similar step that allows the reduction of N_2. It is known that "conventional" Mo-containing nitrogenases are able to reduce a variety (more than 25) of physiological and non-physiological low-molecular-weight compounds. All such reducible compounds contain N-N, N-O, C-N and C-C double or triple bonds, with exception of hydrazine that contains a single N-N bond [30,31]. It is also known that carbonyl sulfide (COS) and carbon dioxide (CO_2), compounds with C=S and C=O bonds, can also serve as substrates for Mo-nitrogenase [32]. In addition, thiocyanate (SCN^-), cyanate (OCN^-) and carbon disulfide (CS_2) can be reduced by this type of nitrogenase [33]. Based on this knowledge of the reduction of these substrates, it was possible to develop schemes [34,35,36] that give insights into their binding site/s, and also the state of reduction of the enzyme and mechanism of N_2 reduction. The nitrogenase system of *S. thermoautotrophicus* cannot reduce ethyne [10], a widely used substrate in the study of nitrogen fixation, to ethene or to ethane. Besides ethyne, azide and cyanide have been extensively investigated substrates for many "conventional" nitrogenases. Therefore, in our study attention was focused on substrates such as azide and cyanide and, in addition, nitrite.

4.3.1 Reduction of azide by St1 and St2 protein

Under appropriate assay conditions with dithionite as electron donor, the purified *S. thermoautotrophicus* nitrogenase components St1 and St2 are capable of catalyzing the reduction of azide to ammonium (Figure 5). Maximum NH_4^+ formation activity of 1.29 μmole NH_4^+ h^{-1} x mg^{-1} occur at 25 mM of N_3^-. Concentrations above 25 mM neither increase reduction activity nor are they inhibitory.

Figure 5 shows the time-course of ammonium formation by purified *S. thermoautotrophicus* nitrogenase in the presence or absence of N_2 or NaN_3. After 1 h of incubation at 65 °C ammonium in the control was 0.10 mM, and the corresponding amounts of ammonium with N_3^- under dinitrogen and under helium were 0.46 mM and 0.51 mM, respectively. After 6 h of incubation the activities with N_3^- under dinitrogen and under helium increased 3.9 fold and 4.6 fold compared to that of the control. Apparently, N_3^- is a preferred substrate of the nitrogenase of *S. thermoautotrophicus*. The products of N_3^- reduction are N_2 and NH_4^+ and, in the presence of 20 mM N_3^-, 2.4 μmole N_2 and 1.63 μmole NH_4^+ were produced. The molar ratio of N_2 to NH_4^+ was 1.4. Azide can be reduced by one of the following three equations [37].

$$N_3^- + 3H^+ + 2e^- \rightarrow NH_3 + N_2 \qquad (2)$$

$$N_3^- + 7H^+ + 6e^- \rightarrow NH_3 + N_2H_4 \qquad (3)$$

$$N_3^- + 9H^+ + 8e^- \rightarrow 3NH_3 \qquad (4)$$

It appears that the nitrogenase system of *S. thermoautotrophicus* for the reduction of N_3^- follows Equation (2) that requires 2 electrons. Hydrazine, as indicated in Equation (3), was not detected, and the high ammonium formation shown in Equation (4) was also not found.

Kinetic properties of nitrogenase of *S. thermoautotrophicus*

Substrate: **NaN_3 (HN_3)**

Values were determined by using Lineweaver-Burk and Hanes Plots.

A K_m of 2.38 mM for the reduction to NH_4^+ was determined.

V_{max} was 26.6 nmole of NH_4^+ formed (min^{-1} mg^{-1}).

These values are within the range (K_m 7.1 μmole to 4.5 mM and V_{max} 40 to 113 nmole min^{-1} mg^{-1}) reported for "conventional" nitrogenases [38,39].

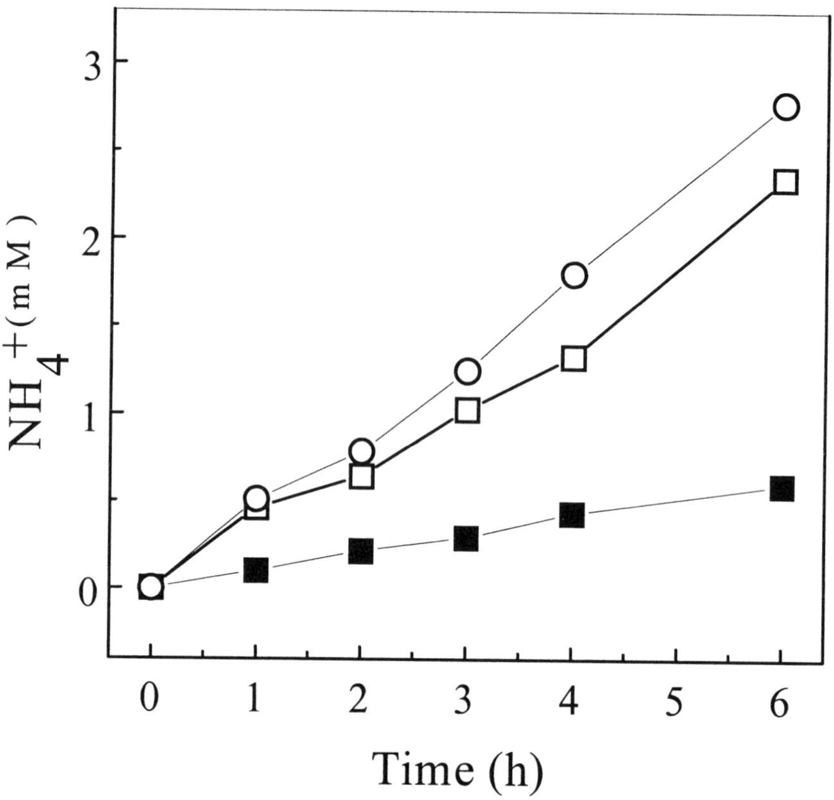

Figure 5. Ammonium formation from azide catalyzed by purified St1 and St2 components of *S. thermoautotrophicus*. Reaction mixtures contained 2.3 mL KH_2PO_4/NaOH buffer (50 mM, pH 7.5), 2.5 mM ATP, 5.0 mM $MgCl_2$, 10.0 mM dithionite, 0.82 mg St1, 0.13 mg St2 and 25 mM NaN_3 where present. Assays were performed at 65 °C under pure N_2 or pure helium. Symbols: ■ without azide under pure N_2, □ with azide under pure N_2, O with azide under pure helium [43].

4.3.2 Influence of carbon monoxide on azide reduction

Carbon monoxide is a powerful inhibitor of "conventional" nitrogenases [40,41,42]. It binds at the E_0 redox level of FeMoco. CO pressures as low as 0.0008 atm. reduced nitrogenase activity by 66%, and at 0.005 atm. complete inhibition occurred [42]. CO also inhibits the general catalytic ability to reduce substrates other than protons. It specifically influences the substrate-reduction sites without interfering in the enzyme-catalyzed hydrolysis of MgATP and it is a non-competitive inhibitor of all nitrogenase substrates except protons.

Figure 6 shows that in the presence of increasing amounts of CO the reduction of N_3^- (25 mM) correspondingly decreases. In both assays (with and without azide) with 15% CO the NH_4^+ formation rates were lowered by about 80%. In contrast, in assays with and without N_3^-, in the presence of CO there was an increase in H_2 evolution corresponding to the CO concentration. After 4 h of incubation the ratio of NH_4^+ to H_2 in the control assay without CO was 1:0.18 which changed with increasing concentrations of CO in the assay and at the CO concentration of 15% it was 1:1.6 (Figure 6). It appears that, in the presence of 15% CO, most of the electrons have been transferred to protons. A similar change in the ratio was observed in assays with N_3^-, 1:0.015 and 1:0.45, without and with CO (15%), respectively. Though these results show a similar pattern of CO inhibition to that known for the "conventional" nitrogenases, the amount of CO required for inhibition is distinctly higher. It is obvious that CO acts as a specific inhibitor of substrate reduction or of thebinding site for the nitrogen-fixing system of *S. thermoautotrophicus*. However proton reduction is not affected. The required higher concentration of CO suggests that it acts only as a competitive inhibitor for superoxide-dependent nitrogenase.

4.3.3 Reduction of cyanide by St1 and St2 protein

Cyanide can be also used as a substrate by St1 and St2 components of *S. thermoautotrophicus* (Figure 7). The maximum concentration of cyanide that is tolerated by St1 and St2 is 20 mM. Above this concentration, the activity of cyanide reduction decreases and at 25 mM the activity entirely ceases. The maximum activity, 0.42 μmole $NH_4^+ h^{-1} \times mg^{-1}$ occurs at 10 mM KCN.

The identified products of cyanide reduction are methyl amine (CH_3NH_2) or NH_3 plus CH_4 [44].

$$HCN + 4H^+ + 4e^- \rightarrow CH_3NH_2 \quad (5)$$

$$HCN + 6H^+ + 6e^- \rightarrow NH_3 + CH_4 \quad (6)$$

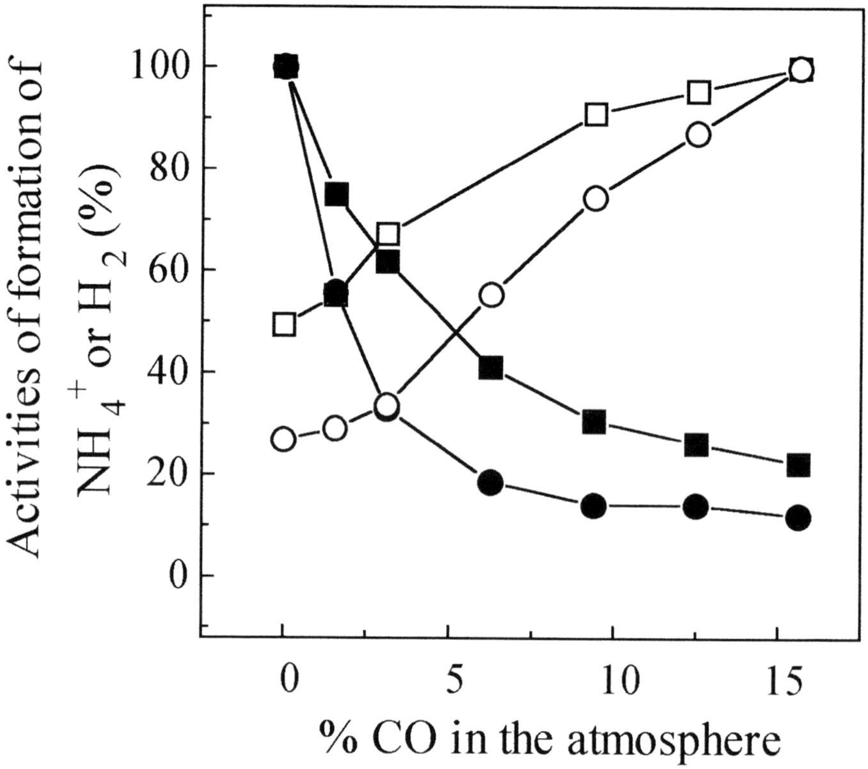

Figure 6. Influence of CO in the presence of azide on NH_4^+ and H_2 formation by St1 and St2 of *S. thermoautotrophicus*. Reaction mixtures contained 2.3 mL KH_2PO_4/NaOH buffer (50 mM, pH 7.5), 2.5 mM ATP, 5.0 mM $MgCl_2$, 10.0 mM dithionite and 25 mM NaN_3 where present. NH_4^+ was analyzed after 4 h of incubation at 65 °C. In all assays gas atmospheres were pure N_2 containing the indicated concentrations of CO. Symbols: ■ NH_4^+ formation in the absence of azide, □ H_2 formation in the absence of azide, ● NH_4^+ formation in the presence of azide, O H_2 formation in the presence of azide. For NH_4^+ activities of 1.6 μmole NH_4^+ h^{-1} x mg^{-1} and 0.28 NH_4^+ μmole h^{-1} x mg^{-1} were set to 100% for assays with and without azide, respectively. For H_2, activities of 0.13 μmole H_2 h^{-1} x mg^{-1} and 0. 43 μmole H_2 h^{-1} x mg^{-1} were set to 100% for assays with and without azide, respectively [43].

Figure 7 shows ammonium formation in the presence and absence of KCN or N_2. The reaction occurs with similar efficiency in the presence and in the absence of N_2 (Figure 7). After 6 h of incubation at 65 °C the total NH_4^+ formation with KCN was 0.88 mM and with N_2 (control) it was 0.45 mM. Since NH_4^+ was detected in the reduction of HCN, the second expected product was CH_4, but it was not detected. Nevertheless, production of NH_4^+ indicates that the reaction follows Equation (6). It appears that cyanide is a substrate for the nitrogenase of *S. thermoautotrophicus*.

Kinetic properties of nitrogenase of *S. thermoautotrophicus*

Substrate: **KCN (HCN)**

Values were determined by using Lineweaver-Burk and Hanes Plots.

A K_m of 5.21 mM for the reduction to NH_4^+ was determined.

V_{max} was 66.7 nmole of NH_4^+ formed (min^{-1} x mg^{-1}).

These values are within the range (K_m 1.1 μM to 4.5 mM and V_{max} 113 to 158 nmole min^{-1} x mg^{-1}) reported for "conventional" nitrogenases [39,44].

4.3.4 Influence of carbon monoxide on cyanide reduction

Increasing amounts of CO (up to 15%) in the assays have hardly any adverse effect on the reduction of KCN (Figure 8). The presence or absence of CO did not affect NH_4^+ formation from HCN, 0.57 and 0.67 μmoles h^{-1} x mg^{-1} were produced, respectively. Even 47% CO in the assay do not affect the NH_4^+ formation. Though, there was no distinct inhibitory effect on formation of NH_4^+ an increase in H_2 formation was observed. It was 4-fold more in the assay with 15% CO than in the assay without CO. This indicates that in the presence of CO, though NH_4^+ formation was hardly affected, there are surplus electrons available which are directed to proton reduction, causing an increase in H_2 evolution.

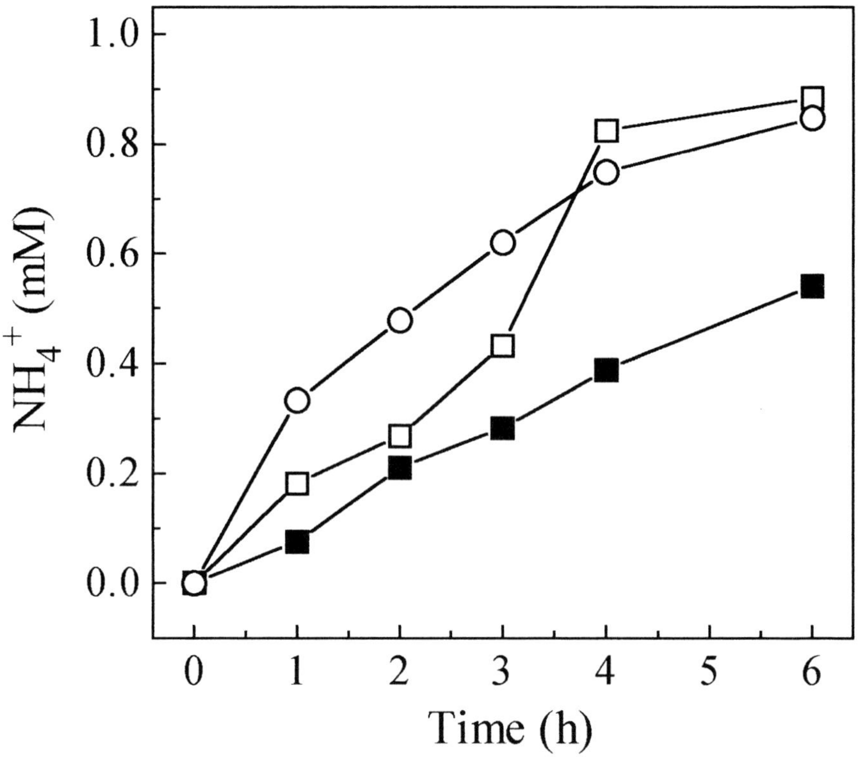

Figure 7. Ammonium formation from cyanide catalyzed by purified St1 and St2 components of *S. thermoautotrophicus.* Reaction mixtures contained 2.3 mL KH_2PO_4/NaOH buffer (50 mM, pH 7.5), 2.5 mM ATP, 5.0 mM $MgCl_2$, 10.0 mM dithionite, 0.82 mg St1, 0.13 mg St2 and 10 mM KCN where present. Assays were performed at 65 °C under pure N_2 or pure helium. Symbols: ■ without cyanide, under pure N_2, □ with cyanide, under pure N_2, O with cyanide under pure helium [43].

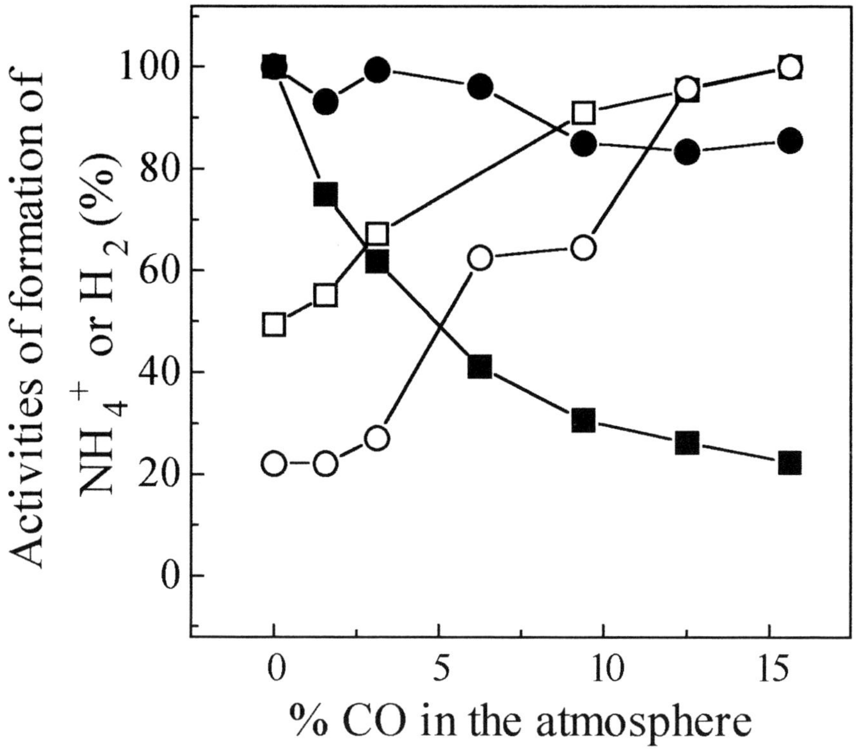

Figure 8. Influence of CO in the presence of cyanide on NH_4^+ and H_2 formation by St1 and St2 components of *S. thermoautotrophicus*. Reaction mixtures contained 2.3 mL KH_2PO_4/NaOH buffer (50 mM, pH 7.5), 2.5 mM ATP, 5.0 mM $MgCl_2$, 10.0 mM dithionite and 10 mM KCN where present. NH_4^+ was analyzed after 4 h of incubation at 65 °C. In all assays gas atmospheres were pure N_2 containing the indicated concentrations of CO. Symbols: ■ NH_4^+ formation in the absence of cyanide, □ H_2 formation in the absence of cyanide, ● NH_4^+ formation in the presence of cyanide, ○ H_2 formation in the presence of cyanide. For NH_4^+ the activities 0.67 µmole NH_4^+ h^{-1} x mg^{-1} and 0.28 NH_4^+ µmole h^{-1} x mg^{-1} were set 100% for assays with and without cyanide, respectively. For H_2, activities of 0.11 µmole H_2 h^{-1} x mg^{-1} and 0.43 µmole H_2 h^{-1} x mg^{-1} were set 100% for assays with and without cyanide, respectively [43].

4.3.5 Reduction of nitrite by St1 and St2 protein

Under appropriate assay conditions with dithionite as electron donor, the nitrogenase components St1 and St2 of *S. thermoautotrophicus* are able to catalyze the reduction of NO_2^- to NH_4^+. There was a linear increase in NH_4^+ formation up to 9 mM NO_2^-. In an assay with NO_2^- at 65 °C the specific activity of a control without NO_2^- was 0.82 µmole NH_4^+ h^{-1} x mg^{-1} and with NO_2^- (9 mM) it was 1.60 µmole NH_4^+ h^{-1} x mg^{-1}. Concentrations higher than 9 mM inhibited ammonium formation. At 35 mM NO_2^- the ammonium formation was 27% less than in the control without NO_2^-. For the reduction of nitrite 6 electrons are required, as in Equation (7) [45].

$$NO_2^- + 10H^+ + 6e^- \rightarrow NH_3 + 2H_2O \qquad (7)$$

Kinetic properties of nitrogenase of *S. thermoautotrophicus*

Substrate: $\mathbf{NO_2^-}$

Values were determined by using Lineweaver-Burk and Hanes Plots.

A K_m of 2.79 mM for the reduction to NH_4^+ was determined.

V_{max} was 27.9 nmoles of NH_4^+ formed (min^{-1} x mg^{-1}).

Vaughn and Burgess [45] suggested a K_m value of 10-20 mM.

It is obvious that, in addition to N_2 and H^+, the nitrogenase system of *S. thermoautotrophicus* is capable of reducing other substrates that are known as alternative substrates for the "conventional" nitrogenases. It is likely that many other alternative substrates that are reduced by "conventional" nitrogenases can also be reduced by the oxygen-insensitive nitrogenase system of *S. thermoautotrophicus*.

4.3.6 Reduction of azide by carbon monoxide dehydrogenase (CODH)

Considering the structural similarities between St1 and St3, it was very interesting to discover whether CODH (St3 protein), which functions as electron donor of nitrogenase in intact cells, can utilize any of the alternative substrates used by nitrogenase. Using appropriate conditions with dithionite as electron donor, the purified CODH of *S. thermoautotrophicus* was found capable of catalyzing the reduction of azide to NH_4^+ (Figure 9), where, after 25 h of incubation at 65 °C from 10 mM azide 10.8 mM NH_4^+ was liberated. In control vials, where only CODH, or CODH plus dithionite, or CODH plus N_3^-, are present, neither NH_4^+ nor N_2 were formed. CODH requires only dithionite for the reduction of azide to NH_4^+, and ATP or St2 were not necessary. During necessary. During N_3^- reduction NH_4^+ and N_2 were formed in stoichiometric proportions. From 5 mM of azide, similar amounts of NH_4^+ and of N_2 were produced, which indicates that the reduction of azide follows Equation (2) as suggested by Dilworth and Thorneley [37]. It is remarkable that CODH, which is believed to be an oxidizing enzyme, is able to reduce a substrate such as azide, thereby producing NH_4^+ and N_2.

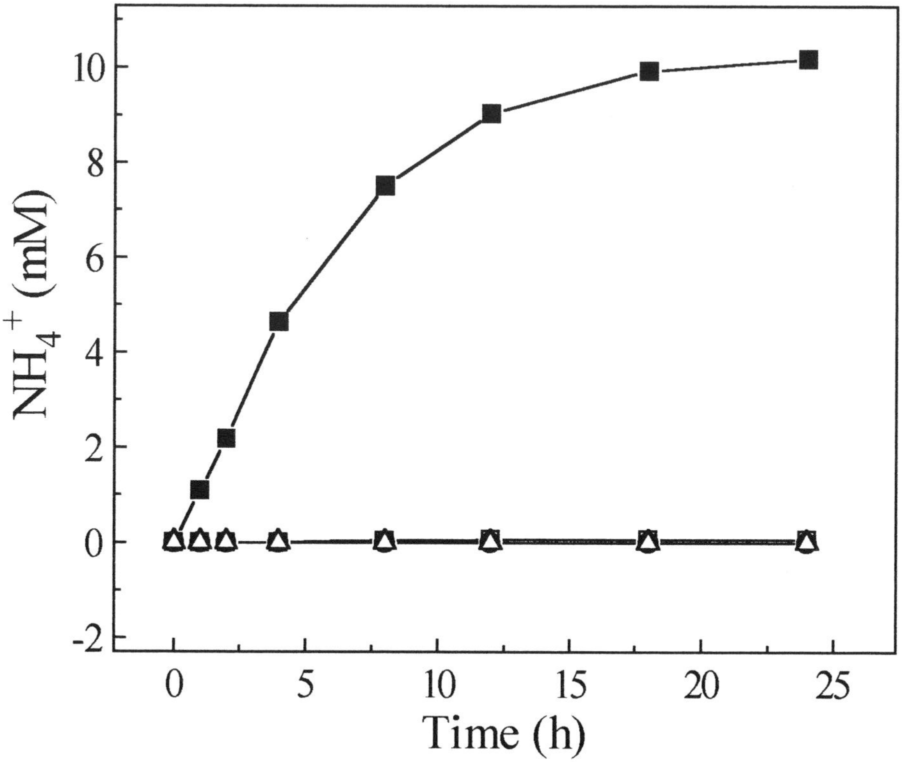

Figure 9. Ammonium formation from azide (10 mM) catalyzed by CODH *of S. thermoautotrophicus*. Reaction mixtures contained 2 mL of KH_2PO_4/NaOH buffer (50 mM, pH 7.5), 10 mM dithionite and 2.8 mg CODH. Assays were performed under pure N_2 at 65 °C. symbols: ■ CODH + dithionite + N_3^-, □ CODH + N_3^-, ● CODH + dithionite, Δ dithionite + N_3^- [43].

5. GENETICS

Nif structural genes are highly conserved in all known diazotrophic bacteria. *NifHDK* probes of Mo-nitrogenase, V-nitrogenase and Fe-nitrogenase are significantly homologous to each other. Southern hybridisation of genomic DNA of *S. thermoautotrophicus* with *nifHDK*-probes derived from *Frankia alnus* and *Klebsiella pneumoniae* reveals no signals [6]. It is obvious that the genes envolved in the nitrogenase system of *S. thermoautotrophicus* are distinctly different from the *nif, vnf* or *anf* enes of "conventional" nitrogenases.

DIG-labelled oligonucleotides probes corresponding to the N-termini of the large (L), medium (M) and small (S) subunits of St1 protein were used for screening total digested DNA of *S. thermoautotrophicus*. Three *sdn*-genes coding for St1 were identified. Their nucleotide sequences showed three open riding frames, designated as *sdnL, sdnM* and *sdnS*. The genes are arranged in the transcriptional order (5′) *sdnM* (885 bp) ➔ *sdnS* (516 bp) ➔ *sdnL* (2493 bp) (3′). The N-termini of the deduced amino acid sequences and of the sdn-polypeptides are identical. The amino acid sequences of the St1 polypeptides sdnM, sdnS and sdnL show high similarity to molybdenum-containing hydroxylases (Figure 10) and only negligible similarity to dinitrogenases. Digested genomic DNA of *S. thermoautotrophicus* was used in southern hybridisations with labelled oligonucleotide probes derived from the N-terminus of the St2 protein. On a sub-cloned fragment, an open reading frame of 624 nucleotides designated as *sdno* was identified. The N-termini of the deduced amino acid sequence and the sdnO-polypeptide are identical in the first 21 positions. *sdnO* is not a part of the *sdnMSL* cluster and is surrounded by genes which have no known function in nitrogen fixation.

5.1 SdnM

SdnM reveals significant similarity to the matching sequences in other Mo-hydroxylases (Figures 10 and 11). With few exceptions, most molybdenum-containing hydroxylases are flavoproteins [46] and it is generally assumed that FAD is associated with the medium-sized subunit or domain. SdnM does not possess FAD. However, at the N-terminus of sdnM, a motif ^{32}AGGQS^{36} is located and this is known to interact with the pyrophosphate of FAD. The motif ^{111}TIGG^{114}, which interacts with adenine nucleotides of cofactor, is situated at C-termini. These motifs are present in corresponding coxM subunit of *O. carboxidovorans,* though in the former motif histidine occurs instead of glutamin [21]. In addition, the amino acids ^{53}DING^{56} and ^{284}AKELQGKV^{291} represent, with minor modifications, the Walker motif B and A [47] at N- and C-termini, respectively. These motifs are typical for ATP-binding proteins and indicate that sdnM is able to bind ATP.

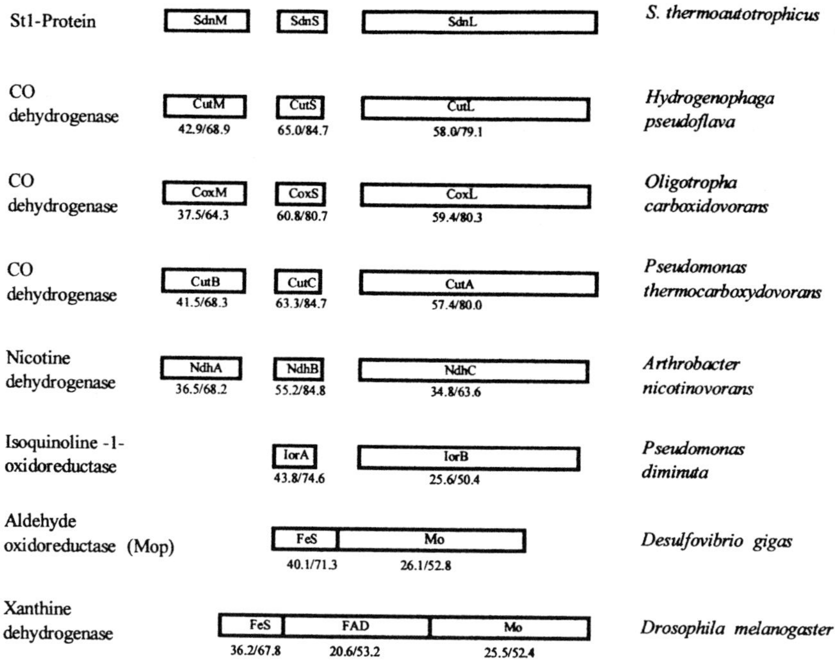

Figure 10. Comparison of Sdn-polypeptides with other molybdenum-hydroxylases. The Sdn-polypeptides are compared with the closely related polypeptides or domains of molybdenum-ydroxylases of different organisms. The bars are proportional to the size of the polypeptides or domains. The numbers refer to identities/similarities relative to the Sdn-polypeptides [11].

5.2 SdnS

SdnS reveals significant sequence similarities to the 2Fe2S-containing subunits or domain of the members of the sequence family of molybdenum-hydroxylases (Figures 10 and 11). The C^{39}-X_4-C^{44}-X_2-C^{47}-X_{11}-C^{59} at the N-terminus of sdnS is homologous to the [2Fe2S] type II binding site of bacterial and chloroplast-type ferredoxins. The C-terminal motif H^{94}-X_3-C^{98} and C^{133}-R-C^{135} is indicative of the unusual type I [2Fe2S] center of Mo-hydroxylases. Obviously, sdnS is the iron-sulfur protein of St1.

5.3 SdnL

The large subunit sdnL reveals significant sequence similarities to the large subunits or domain of the members of the sequence family of molybdenum-hydroxylases harbouring the Molybdenum cofactor (Figures 10 and 11). The sequence ^{248}GGGFGNKV^{255} of sdnL [47] corresponds closely to the conserved sequence GGGFGG of xanthine dehydrogenase [48], which has been shown by mutational analysis to be involved in binding the molybdenum cofactor in *D. melanogaster*. This motif, and further 4 segments which interact with Moco, are structurally identified in CODH from *O. carboxidovorans* [49] and in Mop from *Desulfovibrio gigas* [50] as amino acids contacting the molybdopterin-cytosine-dinucleotide cofactor. These segments are highly conserved, in terms of sequence, relative location within the protein, and amino acid environment, in molybdenum-containing hydroxylases.

These assignments are consistent with the Mo-MCD (molybdopterin cytosine dinucleotide) properties of St1. The fluorescence spectrum revealed that St1 contains at least molybdopterin as a cofactor. The spectrum showed excitation maxima at 281, 313 and 397 nm and an emission maximum at 463 nm, and is almost identical with the typical fluorescence spectrum of form A from the CODH of *H. pseudoflava* [13]. The preceding analysis revealed that the nature of the nucleotide present in the molybdopterin cofactor is molybdopterin cytosine dinucleotide [di(cam)], since the spectrum of [di(cam)] derivate was similar to the [di(cam)] derivate of CODH of *O. carboxidovorans* and *H. pseudoflava* [12,13]. The absorbance maxima of [di(cam)] derivate of St1 were at 279 and 371 nm and the A_{279}/A_{371} ratio was 3.27, which is similar to the absorbance ratio of MCD-containing xanthine dehydrogenase of *Veillonella atypica* [51].

5.4 SdnO

The homology of sdnO to Mn-SOD's is about 80% (Figure 12). The similarity to dinitrogenase reductases is not significant. The amino acid sequence of sdnO carries the motif ^{164}DVWEHAY^{170}, which is the Mn- and Fe-superoxide dismutase signature D-X-WEH-(S,T,A)-(F,Y) [52] as well as the other ligands involved in the coordination of Mn (His^{26}, His^{81}, Figure 11).

Ribbe *et al.* [10] reported that it is not possible to replace St2 by Mn- or Fe-SODs in activity assays. Moreover, these proteins inhibit the formation of ammonium under assay conditions. SODs catalyze the disproportion of superoxide anion radicals to oxygen and peroxide and thus scavenge the radicals, which are the driving force of N_2 reduction.

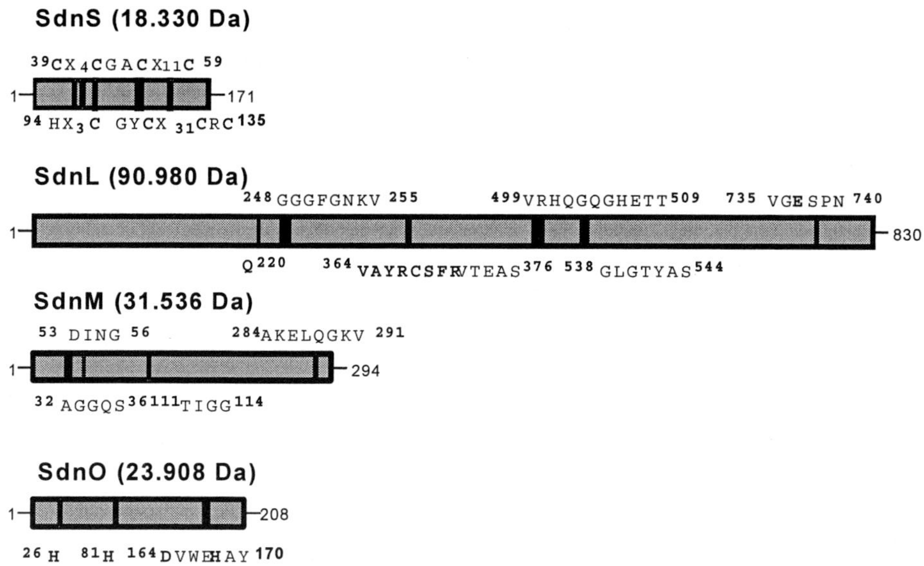

Figure 11. Signatures on the Sdn-polypeptides indicative of the molybdopterin cofactor, 2Fe2S-centers and SOD-type Mn. SdnS: The bold letters represent cysteines, ligands of the 2Fe2S-centers. SdnL: The motif ^{248}GGGFGNKV^{255} refers to the binding site of the molybdenum cofactor. The other segments bind MCD. The motif ^{364}VAYRCSFR^{371} is the active site loop of CO dehydrogenases. SdnM: The amino acids responsible for ATP binding are ^{53}DING^{56} and ^{284}AKELQGKV^{291}, Walker motif B and A, respectively. SdnO: ^{26}H, ^{81}H, ^{168}H and ^{164}D are the presumed ligands of Mn [47].

5.5 Heterologous expression of sdnO

The amino acid sequence of sdnO has 77.1% similarity and 58.5 % identity to the Mn-SOD of *Escherichia coli*. This fact suggested investigation whether SdnO protein can be over-expressed in *E. coli*. In addition, in sdnO there are no cysteines to interfere in the assembly of disulfide bridges leading to an inactive recombinant protein or to the formation of inclusion bodies. For heterologe expression the vector pET11a and expression strain *Escherichia coli* BL21(DE3) were used. With pET11a and using PCR and other standard genetic techniques the *sdnO* gene was transformed into *E. coli* BL21(DE3). The clone selected for the over-expression study was designated as *E. coli* BL21(DE3) pETSDNO (47). For expression, cultures were grown at 30 °C in 2 L Ehrlenmeyer flasks containing 1000 mL Luria broth medium. With the help of 0.1 mM IPTG (isopropyl-β-D-thiogalactopyranoside), expression of *sdnO* gene was induced. The N terminus of recombinant sdnO was identical to the deduced amino acid sequence

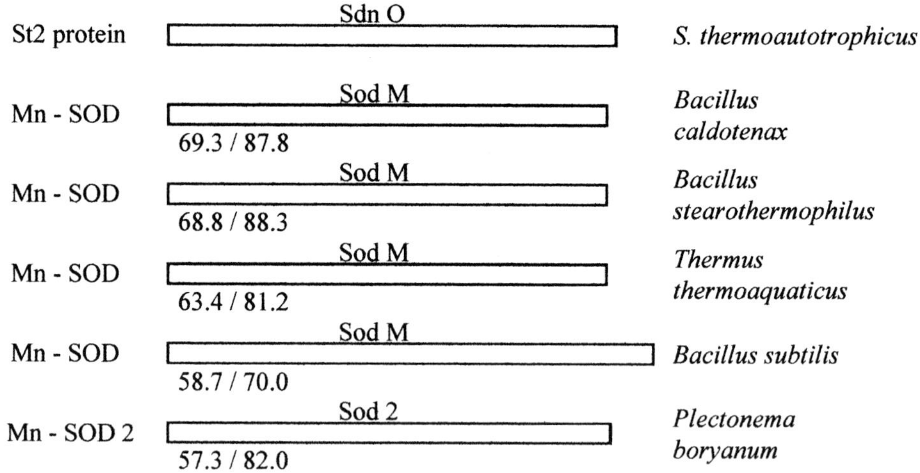

Figure 12. Comparison of the SdnO-polypeptide with Mn-superoxide dismutases. The SdnO-polypeptide is compared with the closely related Mn-SODs from different organisms. The bars are proportional to the sizes of the polypeptides. The numbers given are identity and similarity, respectively [11].

of sdnO polypeptide. *E. coli* itself possess 3 different (Mn, Fe and Cu/Zn) SODs. However, the amino acid sequence of the Mn-containing sdnO polypeptide differs from those of Mn- and Fe-containing SODs of *E. coli* (Figure 13).

The yield of recombinant sdnO in *E. coli* was 30-40% of the total cytoplasmic protein. Recombinant sdnO was entirely soluble and fully assembled as a Mn-containing homodimer. It was purified to homogeneity employing a heat step at 65 °C, since most of the proteins of the host *E. coli* are denatured at this temperature. Purified sdnO was catalytically active, and was able to use superoxide anion radicals as substrate and to transfer the electrons to St1, driving the formation of ammonia (Table 6).

	Formation of NH_4^+ (nmole NH_4^+ h^{-1} mg^{-1})	Oxidation of superoxide (units mg^{-1})[a]
recombinant sdnO + St1	266	-
recombinant sdnO	0	114
St1	0	0
St2	0	104
St1 + St2	109[b]	-
St1 + control[c]	0	-
Control[c]	0	0

[a] PMS/NADH as source of superoxide, as described [53]
[b] Reference [10]
[c] Cytoplasmic fraction of *E. coli* BL21(DE3) pET11a without sdnO after heatstep

Table 6. The activities of purified recombinant SdnO [11].

	1				5					10					15					20
PCR Amplificate								P	Y	P	Y	D	A	L	E	P	Y	F	D	A
SdnO	M	F	E	L	P	P	L	P	Y	P	Y	D	A	L	E	P	Y	F	D	A
Mn-SOD	S	Y	T	L	P	S	L	P	Y	A	Y	D	A	L	E	P	H	F	D	K
Fe-SOD	S	F	E	L	P	A	L	P	Y	A	K	D	A	L	A	P	H	I	S	A

Figure 13. Letters represent amino acids. The amino acids which differ from sdnO and Mn- or Fe-containing SOD of *E. coli* are marked [47].

6. CONCLUSIONS AND PERSPECTIVES

The dinitrogen molecule is chemically very stable and the triple bonds are one of the strongest bonds known. The triple bonds in carbon monoxide (C≡O) are even stronger and are also very difficult to break. *S. thermoautotrophicus* is a carboxidotrophic bacterium and is able to break the strong triple bonds of CO and uses it for growth as the sole source of carbon and energy. It is conceivable solely on grounds of energy consideration that this bacterium might also capable of breaking the triple bonds of N_2. Apart from that, the N_2 molecule is isosteric and isoelectronic with CO. These molecules have many similar physical properties (Table 7). Considering all these facts, it is not surprising that this bacterium is able to break the triple bonds of N_2 and consequently to use it as a nitrogen source.

	N_2(|N≡N|)	CO (|C≡O|)
Melting point (°C)	-210	-205.1
Boiling point (°C)	-195.8	-191.5
Vapor density)	0.796	0.793
Solubility in water at 0° C, L gas/ L H_2O	0.023	0.033

Table 7. Some common physical properties of N_2 and CO.

The superoxide-dependent nitrogen-fixing system of *S. thermoautotrophicus* is unusual. In the absence of a source of combined nitrogen, intact cells of *S. thermoautotrophicus* are able to incorporate a substantial amount of labelled dinitrogen ($^{15}N_2$) during growth with CO plus CO_2 plus air and $^{15}N_2$, or H_2 plus CO_2 plus air and $^{15}N_2$ (Table 1). It has been also reported that in the presence of helium the St1 and St2 proteins produce only H_2, rather than NH_4^+, indicating that, because of the lack of substrate, the electrons have been transferred to the protons [10]. Similarly, it was reported that under dinitrogen (85%) and CO (15%) the formation of NH_4^+ is significantly decreased (Figures 6 and 8). There is a corresponding increase in H_2 formation [43]. This is a typical reaction of "conventional" nitrogenases, since CO is a non-competitive inhibitor of all their known substrates except protons.

CO acts only as competitive inhibitor for the nitrogenase system of *S. thermoautotrophicus*. However, in the presence of CO the production of H_2 increase significantly (Figures 6 and 8). It is to be expected that the nitrogenase system of *S. thermoautotrophicus* would not be as sensitive as the "conventional" nitrogenases, since

CO serves as a carbon source for this carboxidotrophic bacterium. The St1 and St2 proteins are able to reduce azide to N_2 and NH_4^+ in a stoichiometric ratio of 1.4 [43]. Similarly, cyanide and nitrite can be reduced to NH_4^+ by these proteins [43]. It appears that the superoxide-dependent nitrogen-fixing system and the "conventional" nitrogenases have some properties in common. The nitrogenase system of *S. thermoautotrophicus* can reduce low-molecular-weight compounds such as azide, cyanide and nitrite, incorporation of $^{15}N_2$ into intact cells, production of H_2 in the absence of substrate, the production of N_2 *in vitro* during the reduction of azide, and the reduction of cyanide and nitrite to NH_4^+ are the typical features of this nitrogenase. This evidence confirms the existence of the superoxide-dependent nitrogen-fixing system in *S. thermoautotrophicus*. However, until now it has not been possible to show the conversion of $^{15}N_2$ into $^{15}NH_4^+$ using *in vitro* reactions.

The superoxide-dependent nitrogenase has some extraordinary properties. (1) All components, St1, St2 and St3, are oxygen-insensitive. Not only that but, *in vivo,* oxygen is an essential requirement. (2) It is resistant to CO or H_2. (3) Under stringent conditions, *nifH* and *nifdk* gene probes derived from *Klebsiella pneumoniae* and *Frankia alnus* did not hybridize with total DNA from *S. thermoautotrophicus*. (4) Amino acid sequences of the nitrogenase components of *S. thermoautotrophicus* have hardly any homology with nitrogenase components of "conventional" nitrogenases. (5) The St1 protein (dinitrogenase) belongs to the sequence family of a molybdenum-hydroxylase. (6) The St1 protein contains molybdopterin cytosine dinucleotide (MCD) as a cofactor. (7) The St2 protein shows a high similarity to the Mn-SODs. (8) CODH, which delivers electrons to St2 protein, has absolutely no similarity with ferredoxin or flavodoxin. (9) The requirement of ATP for the reduction N_2 is very low and an ATP-regenerating system is not essential.

The superoxide-dependent nitrogenase system, though differing entirely in its protein components and cofactors from the "conventional" nitrogenases, performs at least one similar chemical reaction. It is able to reduce N_2 to NH_4^+ and at the same time produce hydrogen. Compared to the "conventional" nitrogenases, the nitrogen-fixing system of *S. thermoautotrophicus* produces a small amount of hydrogen. However, during the inhibition by CO the hydrogen production increases remarkably (Figures 6 and 8). It is obvious that CO inhibits electron transport to N_2, however, the flow of electrons to protons is not affected. Although the nitrogen-fixing system of *S. thermoautotrophicus* is not able to reduce ethyne, it can reduce alternative substrates such as azide, cyanide and nitrite. In addition to being able to reduce a variety of low-molecular-weight substrates, like conventional "nitrogenases", it may also be able to reduce some additional compounds.

It appears that *S. thermoautotrophicus* has used a very effective way to develop its nitrogenase system employing a principal of building blocks. The primary sequences of the St1 polypeptides are very similar to those of the polypeptides of St3, which is an essential requirement for all carboxidotrophic bacteria. The St2 protein, which is very similar to manganese-containing SODs is a second building block. It is conceivable that this bacterium has used this principal efficiently and with few variations. For example, in primary sequences or some post-translational modifications it has used the same genes and perhaps the same precursors of proteins already present in the bacterium. The acquired additional property may be beneficial for survival in adverse environments such as in a charcoal pile. This may be also one of the reasons why the St3 protein (CODH) of *S. thermoautotrophicus* is able to reduce azide to NH_4^+, thus converting the toxic substance into a usable product.

Bacteria possessing a superoxide-dependent nitogenase system have a few additional advantages compared to other dinitrogen-fixing bacteria. In nature other bacteria can fix atmospheric nitrogen only when anaerobic conditions are available or when bacteria have developed systems (*e.g.*, a branched respiratory system, high oxygen consumption, development of slime or special cells like heterocysts) to protect the highly oxygen-sensitive components of nitrogenase from O_2. These extra requirements need additional energy. Only to fix dinitrogen, the demand of energy is already very high for "conventional" nitrogenases. Theoretically, for the reduction of 1 mole dinitrogen to 2 moles ammonia, these nitrogenases require 16 moles of ATP. *S. thermoautotrophicus* in general needs only 4 moles of ATP to reduce 1 mole dinitrogen to 2 moles of ammonia. Because of its oxygen-insensitivity, no special protective mechanism is essential. Since *S. thermoautotrophicus* is also a chemolithoautrophic bacterium it can occur in habitats where chemolithotrophs proliferate preferentially, that is, in environments where the carbon source is very limited. In addition, its ability to grow between 50 and 65 °C allows this bacterium a relatively selective ecological niche where, because of the lack of carbon and nitrogen sources, the higher temperature and the presence of toxic gases, many organotrophic bacteria cannot grow.

Superoxide-dependent nitrogenase could be a useful tool to produce transgenic plants which, equipped with this nitrogen-fixing system, might entirely or partially fulfil their nitrogen requirement. This system has a great advantage over the "conventional" nitrogenases because all components of this system are stable towards oxygen. Not only that, but oxygen is essential for the reduction of dinitrogen to ammonia. During photosynthesis plants produce oxygen that is absolutely lethal for "conventional" nitrogenase components. It is also not evident that the plants are in a position to spare sufficient ATP for the dinitrogen reduction by "conventional" nitrogenases. Superoxide-dependent nitrogenase requires considerably less ATP for the same reaction. The assimilating nitrate reductase, which almost all plants possess, contains the molybdenum cofactor, and therefore the necessary genes for the synthesis of the molybdenum cofactor are already located on plant chromosomes. The superoxide radicals necessary for the reaction are also part of the photochemical reactions occurring in the chloroplasts.

Very few bacteria from diverse phylogenic origins are able to utilize dinitrogen as a source of nitrogen. Those that can make dinitrogen accessible to the remaining living beings (bacteria, fungi, plants and animals). The oxygen-insensitive nitrogen-fixing system of *S. thermoautotrophicus* seems to be an additional system. It works in the presence of O_2, it is carbon monoxide- and hydrogen-insensitive and it can also use these gases together with CO_2 for growth. Bacteria such as *S. thermoautotrophicus*) that possess such a nitrogen-fixing system could be helpful in aerobic habitats where abundant amounts of CO or H_2 are present, since both of these are inhibitory to the "conventional" nitrogenases. In addition, CO is harmful to most living things. Such bacteria can also fulfil this requirement in moderate thermophilic (45-65 °C) regions, where nearly all mesophilic dinitrogen-fixing bacteria cannot. Few organisms are known which are able to fix dinitrogen at temperatures in the range of 50-65 °C. A cyanobacterium *Mastigocladus laminosus* is able to fix dinitrogen at 45-50 °C [54]. A sulfur bacterium, *Chlorobium tepidum,* fixes dinitrogen at 48-50 °C [55]. The anaerobic bacterium *Clostridium thermosacchrolyticum* can fix dinitrogen up to 60 °C [56,57]. Archaebacteria such as *Methanococcus thermolithotrophicus, Methanosarcina barkeri* 227 and *Methanobacterium thermoautotrophicum* are able to fix dinitrogen at about 60-65 °C [58, 59]. However, all these thermophilic bacteria possess only the "conventional" nitrogenases and their structural genes for nitrogen fixation are highly similar to the structural genes of "conventional" nitrogenases. It is possible that superoxide-dependent nitrogenase systems, such as in *S. thermoautotrophicus,* may exist in many other bacteria, even in some as yet unknown to science.

7. REFERENCES

1. K. Winnacker and L. Küchler, Holzverkohlung in Chemische Technologie, Volume 3, K. Winnacker and L. Küchler ed, Carl Hanser Verlag, Munich, 1971, p. 417.
2. M. Schmitt, O. Meyer, W. Meyer, D. Gadkari, H. Zellmann and K. Schricker, DGMK Congress Report 9206 (1992) 125.
3. O. Meyer, Köhlerei im Fichtelgebirge, Frankenwald und Bayrischen Wald, Verlag Erich Goltze, Göttingen.1997.
4. D. Gadkari, K. Schricker, G. Acker, R. M. Kroppenstedt and O. Meyer, Appl. Environ. Microbiol. 56 (1990) 3727.
5. O. Meyer and H. G. Schlegel, Ann. Rev. Microbiol. 37 (1983) 277.
6. D. Gadkari, G. Mörsdorf and O. Meyer, J. Bacteriol. 174, (1992) 6840.
7. L. J. Rasmussen, G. K. Peters and R. H. Burris, Phykos. 28 (1989) 64.
8. D. H. Kohl, B. A. Bryan and G. Shearer, Plant Physiol. 73 (1983) 514.

9. O. Meyer, K. Frunzke and G. Mörsdorf, Microbial Growth on C_1 compounds, J. C. Murrell and D. P. Kelly (eds.), Itercept Ltd, Andover, United Kingdom, 1993, p. 433.
10. M. Ribbe, D. Gadkari and O. Meyer, J. Biol. Chem. 272 (1997) 26627.
11. C. Hofmann-Findeklee, D. Gadkari and O. Meyer, Nitrogen Fixation: From Molecules to Crop Productivity, F. O. Pedrosa, M. Hungria, M. G. Yates and W. E. Newton (eds.), Kluwer Academic Publishers, Dordrecht, 2000, p. 23.
12. J. L. Johnson, K. V. Rajagopalan and O. Meyer, Arch. Biochem. Biophys. 283 (1990) 542.
13. J. Tachil and O. Meyer, FEMS Microbiol. Lett. 148 (1997) 203.
14. C. Bowler, L. Van Kaer, W. Van Camp, M. Van Montagu, D. Inze and P. Dhaese, J. Bacteriol. 172 (1990) 1539.
15. S. P. Chambers, J. K. Brehm, N. P. Michael, T. Atkinson and N. P. Minton, FEMS Microbiol. Lett. 91 (1992) 277.
16. H. P. Misra, and I. Fridovich, Arch. Biochem. Biophys. 189 (1971) 317.
17. R. Hille, J. Biol. Chem. Rev. 96 (1996) 2757.
18. D. Willbold, Diplom thesis, University of Bayreuth, 1991.
19. K.-P. Fuchs, Diplom thesis, University of Bayreuth, 1992
20. E. Heck, Diplom thesis, University of Bayreuth, 2000.
21. U. Schübel, M. Kraut, G. Mörsdorf and O. Meyer, J. Bacteriol, 177 (1995) 2197.
22. B. S. Kang and Y. M. Kim, J. Bacteriol. 181 (1999) 5581.
23. D. M. Pearson, C. O'Reilly, J. Colby and G. W. Black, Biochim. Biophys. Acta 1188 (1994) 432.
24. B. Lorber and R. Giege, Crystallization of Nucleic Acids and Proteins, A. Ducruix and R. Giege (eds.), Oxford University Press, New York, NY, 1992, p.19.
25. O. Meyer, L. Gremer, R. Ferner, M. Ferner, H. Dobbek, M. Gnida, W. Meyer-Klaucke and R. Huber, Biol. Chem. 381 (2000) 865.
26. O. Meyer and H. G. Schlegel, J. Bacteriol. 141 (1980) 74.
27. B. Santiago, Diplom thesis, University of Bayreuth,1990.
28. O. Meyer, J. Biol. Chem. 257 (1982) 1333.
29. H. Cypionka and O. Meyer, J. Bacteriol. 156 (1983) 1178.
30. M. G. Yates, Biological Nitrogen Fixation, G. Stacey, R. H. Burris, H. J. Evans (eds.), Chapman and Hall, London. 1992, p. 685.
31. B. Burgess, Molybdenum Cofactors, and Model system, E. I. Stiefel, D. Coucouvanis and W. E. Newton (eds.), American Chemical Society, Washington, DC, 1993, p. 144.
32. L. C. Seefeldt, M. E. Rasche and S. A. Ensign, Biochemistry 34 (1995) 5382.
33. M. E. Rasche and L. C. Seefeldt, Biochemistry 36 (1997) 8574.
34. D. J. Lowe and R. N. F. Thorneley, Biochem. J. 224 (1984) 877.
35. R. N. F. Thorneley and D. J. Lowe, Biochem. J. 224 (1984) 887.
36. C. J. Pickett, JBIC 1 (1996) 601.
37. M. J. Dilworth and R. N. F. Thorneley, Biochem. J. 193 (1981) 971.

38. J. F. Rubinson, B. K. Burgess, J. L. Corbin and M. J. Dilworth, Biochemistry 24 (1985) 273.
39. J. Shen, D. R. Dean and W. E. Newton, Biochemistry 36 (1997) 4884.
40. R. W. F. Hardy, E. Knight and A. J. D`Eustachio, Biochem. Biophys. Res. Commun. 20 (1965) 539.
41. W. A. Bulen, J. R. LeComte, R. C. Burns and J. Hinkson, Non-heme Iron Protein: Role in Energy Conversion, A. San Pietro (eds.), Antioch Press, Yellow Springs, Ohio, 1965, p. 261.
42. J. C. Hwang, C. H. Chen and R. H. Burris, Biochem. Biophys. Acta 292 (1973) 256.
43. D. Gadkari, A. Sarjiya and O. Meyer, Proceedings 13th International Conference on Nitrogen Fixation, Hamilton, Canada, 2001.
44. J. G. Li, B. K. Burgess and J. L. Corbin, Biochemistry 21 (1982) 4393.
45. S. A. Vaughn and B. K. Burgess, Biochemistry 28 (1989) 419.
46. J. C. Wootton, R. E. Nicholson, J. M. Cock, D. E. Walters, J. F. Burke, W. A. Doyle and R. C. Bray, Biochem. Biophys. Acta 1057 (1991) 157.
47. C. Hofmann-Findeklee, Ph.D thesis, University of Bayreuth, 2000.
48. K. Hughes, W. A. Doyle, A. Chovnick, R. S. Whittle, J. F. Burke, and R. C. Bray, Biochem. J. 285 (1992) 507.
49. H. Dobbek, L. Gremer, O. Meyer and R. Huber, Proc. Natl. Acad. Sci. U.S.A. 96 (1999) 8884.
50. M. J. Romao, M. Archer, I. Moura, J. J. G. Moura, J. LeGall, R. Engh, M. Schneider, P. Hof and R. Huber, Science 270 (1995) 1170.
51. L. Gremer and O. Meyer, Eur. J. Biochem. 238 (1996) 862.
52. M. W. Parker and C. C. F. Blake, FEBS Lett. 229 (1988) 377.
53. F. E. Ewing and J. R. Janero, Annal. Biochem 232 (1995) 243.
54. E. S. Stevens, V. B. Mehta and L. S. Lane, Nitrogen Fixation and CO_2 Metabolism P.W. Ludden and J. E. Burns (eds.), Elsevier Science Publishing Co, New York, NY, 1985, p.235.
55. T. M. Wahlund and M. T. Madigan, J. Bacteriol. 175 (1993) 474.
56. M. Bogdahn and D. Kleiner, Arch. Microbiol. 144 (1986) 102.
57. M. Bogdahn and D. Kleiner, Arch. Microbiol. 145 (1986) 159.
58. N. Belay, R. Sparling and L. Daniels, Nature 312 (1984) 286.
59. A. L. Lobo and S. H. Zinder, J. Bacteriol. 172 (1990) 6789.

Nitrogen Fixation at the Millennium.
G. Jeffery Leigh, Editor.

CHAPTER 11

Dinitrogen Chemistry

G.J. Leigh
School of Chemistry, Physics and Environmental Science, University of Sussex, Brighton, BN1 9QJ, UK

1. INTRODUCTION

That dinitrogen is a gas and constitutes about 80% of the earth's atmosphere is a commonplace. That nitrogen is not really one of the most abundant elements on earth is also well known. Nitrogen in the earth's crust constitutes about 20 parts per million (ppm), and in seawater its concentration is generally higher [1]. Dinitrogen is generally regarded as an unreactive molecule, but the nitrogen cycle is a concept that is well established and widely recognised [2], even if detailed definition can be very subjective. One can design a nitrogen cycle to suit one's particular requirements, for example, within the sea, within a forest, or within an agricultural system. A typical and very simple nitrogen cycle is shown in Figure 1.

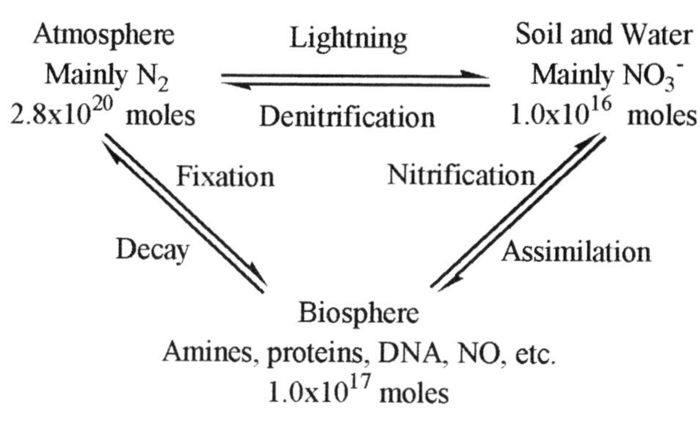

Figure 1. A typical simple nitrogen cycle.

It is evident that amount of nitrogen in the biosphere constitutes only a small proportion of the total nitrogen formally available, but that Nature seems to be able to concentrate within the biosphere that amount that is outside the atmosphere. The purpose of this Chapter is to describe what it is that makes dinitrogen so difficult for Nature to assimilate from dinitrogen, and to describe what chemists have been able to do to overcome this difficulty. It is not intended to be a comprehensive account of the literature, but rather a survey of the current development of the science. This will be presented in a historical context, so that the reader can begin to appreciate the extraordinary achievements of the industrial chemists of the first part of the twentieth century, and also to understand the immense conceptual problems that still exist in understanding how Nature fixes nitrogen.

2. PROPERTIES OF DINITROGEN

2.1. Physical and chemical properties

Dinitrogen is a gas under normal conditions and it melts at –209.86 °C (*ca.* 63 K) and boils at –195.806 °C (77 K) under atmospheric pressure [3]. The liquid range is therefore quite short. One can obtain a measure of the likely reactivity from consideration of properties of dinitrogen such as the energy (ionisation potential of dinitrogen) for the process shown in Equation 1.

$$N_2 \rightarrow N_2^+ + \text{electron} \qquad (1)$$

This is 15.6 eV, and compares to 14.0 eV for the analogous ionisation of carbon monoxide. The value is very similar to that of the "inert gas" argon, which is about 15.8 eV [3]. Clearly a simple ionisation of dinitrogen is an unlikely chemical process, though the ion N_2^+ has been detected in the gas phase. The alternative to oxidation, reduction to give an ion N_2^-, is also a possibility. To understand what this might involve requires some consideration of the electron affinity of the dinitrogen molecule, and its electronic structure.

Chemists have traditionally written the structure of dinitrogen using valence-bond formalism as N≡N. This triple bond is very hard to break homolytically, that is, to break the molecule apart into single uncharged nitrogen atoms. The bond strength is reckoned to be about 9.8 eV, which makes it one of the strongest bonds known [4]. Only the corresponding bond in carbon monoxide, which may be written as a triple bond, is stronger, about 11.1 eV, and that bond is also difficult to break homolytically. Interestingly, a nitrogen-nitrogen double bond has a dissociation energy of about 4.3 eV and a corresponding single bond has a dissociation energy of only about 1.7 eV [5]. Clearly the major problem is to break the first bond of the triple bond, but quite how this might be achieved in practice is not obvious.

Molecular orbital theory gives a much better insight into the problem. Figure 2 shows the calculated electronic structure of the valence-shell electrons of dinitrogen, together with a pictorial representation of the shapes of the corresponding molecular orbitals. The energies quoted are all relative to that of a free electron at infinity, set at zero. It is evident that all the ten valence electrons, five from each atom, are in bonding orbitals, so that they all contribute to strengthening the N-N bond [6]. The addition of a further electron, the energy cost of which is described by the electron affinity, would mean occupancy of orbitals of relatively high energy, and so this can be achieved only by very strong reductants, such as calcium and lithium metals. The ion N_2^- is very

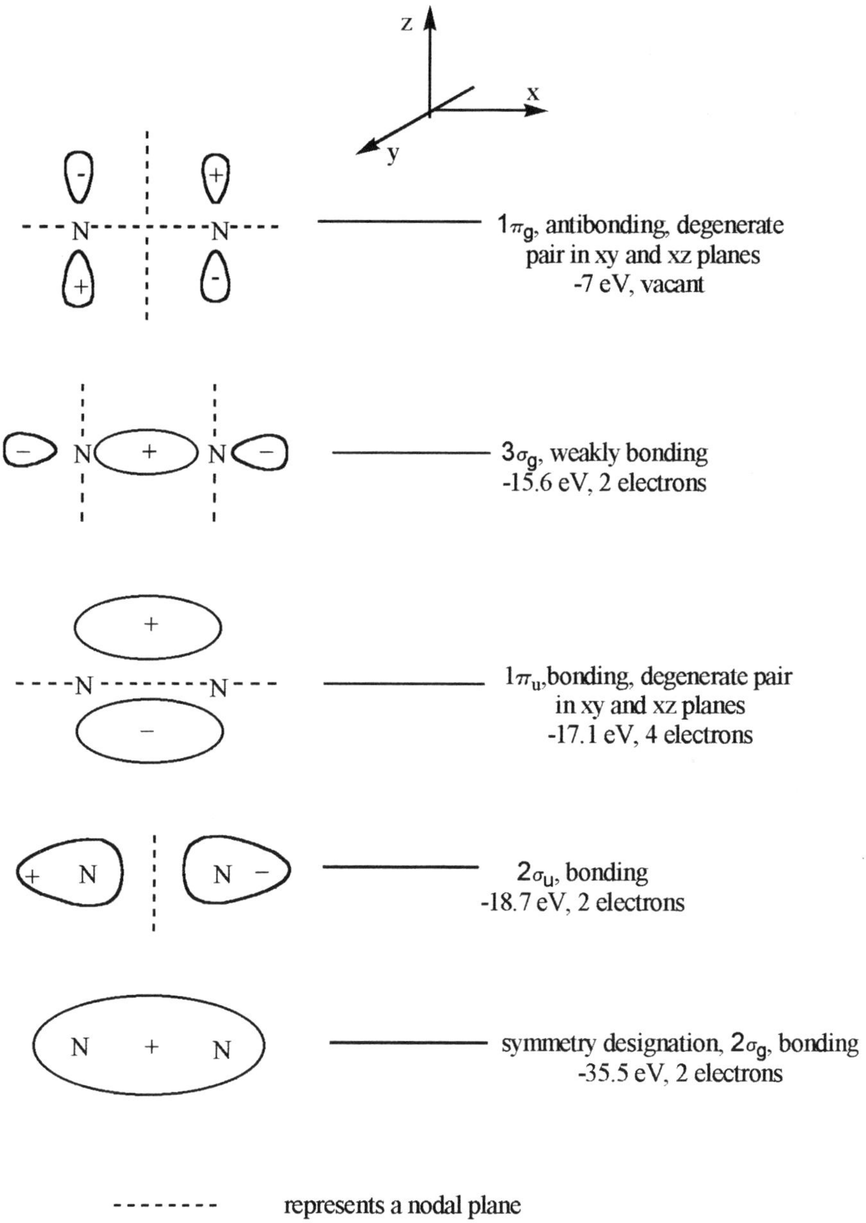

Figure 2. The electronic structure of dinitrogen. The zero of energy is a static electron at infinity.

reactive and does not form stable salts, though it can be stabilised under rather special circumstances. All this confirms that dinitrogen is unlikely to be fixed under mild conditions by simple reagents, and that rather complicated strategies such as those adopted by Nature, will have to be used. However, even if we exclude simple reactions, we cannot exclude a more complex type of reaction, perhaps one in which bonds are made as well as broken, or in which electrons are transferred both to and from the dinitrogen in a synergistic fashion that tends to maintain electro-neutrality.

The IR/Raman spectrum of dinitrogen should contain a single absorption, assigned to the N≡N triple bond stretching vibration, ν(N≡N), but for symmetry reasons in free dinitrogen it is only Raman-active. In complexes it is often IR active and has been used to detect bound dinitrogen and also to assess how much the triple bond is weakened by interaction with metals. It is found at 2331 cm^{-1} in dinitrogen gas, and at somewhat lower wave numbers, down to *ca.* 1900 cm^{-1}, in complexes (see below) [7].

Several common molecules also contain triple bonds, and many of them have been investigated as alternative substrates for nitrogenases. They all tend to have similar strong bonds, but they are often polarised because the two atoms involved in the triple bond are different. This sometimes permits a reactivity that is not available to dinitrogen. The most important such molecule is acetylene (in current nomenclature now called ethyne), HC≡CH, and its substituted derivatives. The reduction of ethyne is often used as a test for nitrogenase activity. Others molecules include carbon monoxide, CO, and organic cyanides RC≡N and the cyanide ion, $C{\equiv}N^-$, and even organic isocyanides, RN≡C. It will be evident that many of these species must be polarised. For example, if the formula of CO is written in the form C≡O, then carbon must be negative and oxygen positive if the eight-electron rule for each atom is to be may be satisfied. This is because carbon brings only four valence electrons to the compound whereas oxygen brings six.

2.2. Reactivity of dinitrogen

Dinitrogen was not recognised as an element in ancient times. Indeed, this could not have happened until the concept of gases had been established, and it became evident that air is a mixture, some of which could support life, and most of which could not.

A lack of knowledge of the development of ideas can cause some confusion if one is trying to read earlier literature. At the beginning of modern chemistry, it was recognised that there was some spirit that supported fire and also supported life, and which somehow was responsible for the fertility of soils. Once a soil became exhausted, it could be allowed to regenerate naturally by allowing it to rest, when it collected this spirit from the air. Manures were known to be rich in this spirit. Adopting concepts that were really first enunciated clearly by Paracelsus, this spirit was termed “nitre”. Some of the older literature can be easily misunderstood if it is read with an unwitting assumption of a modern understanding of nitrogen chemistry. For example, Houghton, in 1692 [8], wrote that: "some in Essex have their Fallow after Turneps, which feeds their Sheep in Winter, by which means their Turneps are scooped, and so made capable of holding Dews and Rain water, which by corrupting imbibes the Nitre of the Air, and when the Shell breaks, it runs about and fertilizes." This was not an early report of nitrogen fixation. The fields were and undoubtedly fertilised by the whole process, but by the sheep dung and not by anything else.

That air is a mixture of gases was not realised until the second half of the eighteenth century. The famous experiment of burning a candle in a closed system until all the dioxygen is exhausted was probably undertaken first by Hero of Alexandria. Several other researchers repeated this experiment during the centuries, including Fludd, Mayow, Cavendish and Priestley. The modern accepted interpretation of the overall process was probably due to Rutherford, a student of Black in Glasgow, who, in 1772, burnt a candle in a closed system until the gas would no longer support combustion, and showed that what was left in the container was a mixture of a gas that was absorbed by sodium hydroxide (actually carbon dioxide) with a gas that neither supported a flame nor reacted with alkali [9]. This was dinitrogen, although the name nitrogen was not suggested until about 1790, by Chaptal [10]. It had taken some time to elucidate the relationship between the various nitrogen species, dinitrogen, nitrogen(II) oxide or nitric oxide, NO, nitrogen(IV) oxide, NO_2, nitric acid, HNO_3, nitrate salts containing the ion NO_3^- and ammonia, NH_3. However, by 1800 it was already clear that dinitrogen is a very unreactive species. That it was not completely unreactive was also known [11].

Cavendish in about 1780 had demonstrated that dinitrogen and dioxygen react together when subjected to an electric spark to form oxides of nitrogen and nitric acid [11]. In 1785, C. L. Berthollet finally established [12] the composition of ammonia, by subjecting it to an electric spark, when dihydrogen and dinitrogen were formed. He was not the first to observe this decomposition. Davy in 1807 probably observed the synthesis of ammonia by what was effectively the reverse of this reaction [13], but it was to take another 100 years for this to be developed into an industrial process.

The science of agricultural chemistry was developed first by Davy, and, above all, by Liebig. Both chemists realised that elements other than carbon, hydrogen and nitrogen were necessary for plant growth, but neither was convinced that biology could actually fix as unreactive material as dinitrogen. In fact, Liebig conclusively demonstrated [14] to his own satisfaction that ammonia circulated from decaying animal and vegetable matter through the atmosphere and rain into the soil, where plants again absorbed it. Such was Liebig's authority that few questioned his opinion. The unequivocal proof of the reality of biological nitrogen fixation came about by the work of Wilfarth and Hellriegel only by *ca.* 1886 [15]. By the end of the nineteenth century it was evident that nitrogen could be fixed in plant systems under mild conditions, but also that the industrialised world could not obtain sufficient quantities of fertiliser of whatever kind to guarantee that it could feed its population.

Thus it was in 1898 that Sir William Crookes made his famous address to the British Association [16] in which he asked chemists to meet the challenge of fixing nitrogen in an industrially efficient and commercially viable fashion. This address was actually couched in terms that would be regarded today as racist and offensive. Civilisation, he claimed, was developed by the eaters of wheat rather than by the eaters of rice and other inferior foods, and it was threatened by the lack of fixed nitrogen to support the necessary agriculture to produce wheat. Without sufficient wheat to sustain it, civilisation would suffer. This analysis is not acceptable today, but the threatening shortage of fixed nitrogen to feed a hungry population, whatever it chose to eat, was indeed a fact. Nevertheless, chemists were soon able to do industrially what biology apparently does so effortlessly, and Crookes' plea was answered.

3. INDUSTRIAL NITROGEN FIXATION

Attempts to develop industrial processes for nitrogen fixation were reported and patented throughout the second half of the nineteenth century. It was evident to many, and not just to Sir William Crookes, that a reliable and plentiful supply of fixed nitrogen was needed for the intensive agriculture that was then developing. The Chilean sources of nitrate were being exhausted, and countries such as Germany, in particular, could not guarantee supplies if the sea were controlled by maritime powers, especially Great Britain. It is therefore not surprising that much of the work on nitrogen fixation was supported in Germany, and by BASF in particular.

It was already evident that any fixation process would require a large input of energy. In Norway, Eyde of Norsk Hydro [17] was buying up stretches of river that would be exploitable for hydro-electricity, because he realised that Norway might then be ideally placed to exploit new industrial processes. In 1903 he attended a lecture by Birkeland who was describing his plans for an electric gun. These never came to fruition, but what he described was essentially a furnace capable of generating a high temperature very rapidly in an enclosed volume of gas. The resultant pressure could then be used to drive a projectile from a gun. Eyde realised that this was fundamentally what was required to generate a temperature high enough to cause dinitrogen to burn in air, since this phenomenon was already recognised to occur when an electric spark was passed through air. This was the genesis of the Norwegian Arc process, the basis of which is given in the equations below.

$$N_2 + O_2 \rightarrow 2NO \quad (2)$$

$$2NO + O_2 \rightarrow 2NO_2 \quad (3)$$

$$H_2O + 2NO_2 \rightarrow HNO_2 + HNO_3 \quad (4)$$

The obstacles to producing a process were severe, and required the development of furnaces capable of withstanding temperatures of 2000 °C or more, and of a size to allow production on a scale of many tonnes per day and not of a few grammes. In addition, the reaction products were highly corrosive. Furthermore, this was likely to be a process that was very wasteful in energy, and only organisations as ideally placed as Norsk Hydro were likely to be able to operate it successfully. The work required a large input of capital, and it was not until ten years later that the process was finally established. In the later stages of development the German company BASF was also involved. BASF were already been working on a variant of the Arc process, the Schönherr furnace, and were interested in what Hydro might be able to do. They were also pursuing a completely different approach that ultimately became the Haber process. This process eventually drove the Arc process to extinction.

Haber was an ardent German nationalist of Jewish origins who seems to have been a very combative individual. He had been working on the reaction of dinitrogen and dihydrogen at the beginning of the century because it was already known that this reaction is reversible.

$$N_2 + 3H_2 \leftrightarrows 2NH_3 \tag{5}$$

The feasibility of exploiting this reaction industrially depended upon where the position of equilibrium might actually lie, and one of the early luminaries of physical chemistry, Nernst, had published data that were purported to show that this was unlikely. On the basis of his own early work, Haber disagreed, and he set out to show that Nernst was wrong [18]. Some intensive studies with Rossignol allowed him to develop a sophisticated bench-top apparatus to synthesise ammonia directly form its elements. This necessitated finding a suitable catalyst, making the gases circulate over it at high temperatures and pressures (dihydrogen was expensive and the unused gas economically could not be released after a single pass over a rather inefficient catalyst), and developing a method of removing the ammonia without losing the unreacted gases. This Haber achieved, and by 1909 he was in a position to approach BASF with his process.

BASF appointed Bosch, who belatedly received the Nobel prize in 1931, some 13 years after Haber, to turn Haber's laboratory process into something commercially viable. Bosch effectively invented high-pressure industrial chemistry, and by 1913 BASF were operating "Haber" plants capable of producing thousands of tonnes of ammonia per year. During the First World War, Britain and France soon became aware that Germany had successfully exploited atmospheric nitrogen as a source of ammonia and that this had enabled Germany to continue to fight despite the sea blockade on imported fixed nitrogen. When the War ended, both the British and French did their best to obtain the secrets of how it was done. The reaction itself was no mystery, but the technology certainly was. Despite the reluctance of the Germans to supply their former enemies with the details, by the middle of the 1920s Bosch's technology was being applied in several countries. The energy efficiency of the Bosch method spelled the death of the Arc process, and by about 1930 the Haber process was the only industrial route used to prepare ammonia.

Some of this ammonia produced by the Haber process was used to prepare nitric acid, by catalytic oxidation in air, a method that was first observed more than 150 years ago and that is still employed today. The primary product of the Arc process was thus obtained by an efficient transformation of ammonia produced by the Haber process, and this was probably a further reason why the Norwegian technology was no longer economically viable.

$$4NH_3 + 5O_2 \rightarrow 4NO + 6H_2O \qquad (6)$$

$$4NO + 3O_2 + 2H_2O \rightarrow 4HNO_3 \qquad (7)$$

The largest modern Haber plants produce as much as 2000 tonnes of ammonia per day. They are essentially the same as those developed by Bosch, though the dihydrogen is produced from natural gas rather than from coal and steam. The catalysts are generally of metallic iron with various promoters, although there are now signs that a new generation of catalysts may finally displace iron [19]. It is believed that the iron catalyst acts to split the dinitrogen to produce nitrogen atoms on the metal surface. It is perhaps a matter of taste as to whether one prefers to consider these as atoms or nitride ions. However, it seems that one particular surface of the iron crystals, the (111) face, is especially effective in splitting the dinitrogen and that the energy input required to do this is balanced by the energy output obtained by forming three single iron-nitrogen bonds. The catalyst also splits the dihydrogen to give surface dihydrogen atoms, and ammonia is formed by the combination of the nitrogen and hydrogen atoms through intermediates such as NH and NH_2.

The Haber process triumphed ultimately because it was much more energy-efficient than the alternative. However, there was a further process for fixing nitrogen that was also energy-intensive and that, in addition, produced an intractable by-product, yet which survived much longer [20]. It also required high temperatures, and, like the Arc process, it was based upon a cheap supply of electrical energy. This was the Cyanamide Process, which was developed to use hydroelectric power from places such as the Niagara Falls. The essential chemistry is shown below.

$$CaCO_3 \rightarrow CaO + CO_2 \qquad (8)$$

$$CaO + 3C \rightarrow CaC_2 + CO \qquad (9)$$

$$CaC_2 + N_2 \rightarrow CaNCN + C \qquad (10)$$

This requires temperatures of the order of 1200 °C, and was developed as a method for making cyanides. These had a use in the dyeing industry. However, calcium cyanamide, CaNCN, can also be used as a fertiliser because it hydrolyses upon exposure to moisture to produce ammonia and lime. The process has survived because of its value in producing a material with many industrial uses.

$$CaNCN + 4H_2O \rightarrow Ca(OH)_2 + CO_2 + 2NH_3 \quad (11)$$

Its value as a source of fertiliser nitrogen is now only marginal and the quantity of nitrogen fixed this way is relatively small. In contrast, the Haber process now fixes about 100×10^6 tonnes of nitrogen annually, and is unlikely to be displaced within the foreseeable future. The ammonia it produces is too important for Western food systems, and it is also costs relatively little compared to the other expenses of intensive, industrialised agriculture. Its cost judged by the budgets of subsistence farmers in the developing world is another matter.

4. METALS IN BIOLOGICAL NITROGEN FIXATION

After the development of the Haber process and its victory over rival technologies, little significant happened in the way of chemical nitrogen fixation for perhaps thirty years. Interestingly, studies of biological nitrogen fixation also proceeded along conventional lines, and both areas of research, chemical and biological, required some novel stimulus to revitalise them. Studies on whole organisms are not appropriate to discover the detailed chemistry and biochemistry of any life function, let alone one as complex as biological nitrogen fixation. However, there was one important piece of research that did pose interesting questions, even though part of it was ignored for thirty years.

In 1930 Bortels [21] published a study that showed that the growth of nitrogen-fixing cultures of *Azotobacter chroococcum* is strongly promoted by a dilute mixture of salts containing molybdenum, tungsten and uranium. Further research showed that the increased growth is due to the presence of molybdenum rather than of the other two metals. Growth rates were compared to rates in appropriate standard media, either sterile or based on soil extracts. This molybdenum-dependence became an article of faith amongst most researchers, and is generally found to hold true.

In 1933 and 1936 [22] Bortels extended these observations to show that molybdenum (and sometimes vanadium) strongly promoted the growth of some nitrogen-fixing legumes, but the report concerning vanadium was generally ignored, though there were some notable exceptions [23]. These were the only metals of some forty investigated that were found to stimulate fixation. Nevertheless, the findings encouraged chemists to think about how dinitrogen might interact with metal ions such as of molybdenum, with the implicit assumption that such ions were indeed involved in the initial uptake of dinitrogen in the biological fixation systems.

5. DINITROGEN COMPLEXES

5.1. The initial discoveries

Since dinitrogen is isoelectronic with carbon monoxide and with ethyne, there seemed to be no reason in 1960 why dinitrogen complexes analogous to the known carbonyls and ethyne complexes should not exist. None were known at that time. Inspection of Figure 2 shows that the dinitrogen molecule possesses a symmetry property common to homonuclear diatomic molecules and, moreover, a property that is not possessed either by ethyne or to carbon monoxide. The orbitals that might be involved in bonding to a metal ion can be donor orbitals (in the conventional Lewis base sense) or acceptor orbitals (in the Lewis acid sense). The orbital of lowest energy, $2\sigma_g$ is probably too low to release its electrons easily to a metal ion. The next, $2\sigma_u$, may be involved in donation, but it will form a σ-type bond along the molecular axis, but a π-type bond at right angles to it. In that case the acceptor orbital on the metal ion should have π-type symmetry, as found for d-orbitals, p-orbitals, and for suitable hybrid orbitals, such as ds-hybrids. Similar considerations apply to all the other full and empty orbitals. This means that dinitrogen might be expected *a priori*, to bond to metal ions either end-on or side-on. However, there is also no symmetry reason why bonding to metal ions cannot occur in both senses at the same time, or indeed at both ends and at both sides at the same time. The decisive factor would appear to be not symmetry but energy. In other words, is energy given out if a metal ion bonds to dinitrogen?

The option to bind end-on or side-on does not arise with ethyne, in which the end-on mode is blocked by hydrogen atoms, nor is carbon monoxide exactly the same as dinitrogen because it is not symmetrical enough. Indeed, carbonyl complexes such as the famous $[Ni(CO)_4]$ have been known for more than a century. The first dinitrogen complex appeared in 1965.

The first dinitrogen complex was isolated by Allen and Senoff, [24] who were attempting to make $[Ru(NH_3)_6]^{2+}$ by a reaction of hydrazine hydrate with hydrated ruthenium(III) chloride. They were clever enough to realise that they were not obtaining the material they wanted, and by applying the relatively new (to inorganic chemists) technique of IR spectroscopy, they observed a new strong band that they assigned to the N-N stretching vibration, ν(N≡N). The material they had isolated contained the ion $[Ru(NH_3)_5(N_2)]^{2+}$, and this is a normal octahedral complex with dinitrogen bound singly at one end. The value of ν(N≡N) falls in the range 2100-2170 cm^{-1}, depending on the counter anion. This discovery was serendipitous. Thermodynamically unstable hydrazine decomposes violently in the presence of many metal ions, giving ammonia and dinitrogen, and the isolation of this particular material was due to the appropriate but accidental suitability of solvent, concentration and temperature. The nitrogen-nitrogen bond length was later shown to be 1.12.Å. The lowering of the stretching frequency upon complexation and the lengthening of the bond length are typical of all dinitrogen complexes, though the amount of change varies from complex to complex [25].

Despite an early report claiming that the dinitrogen in this complex is convertible to ammonia, this is certainly not the case under mild protonation or reductive conditions Shortly after this, Taube and his collaborators showed [26] that dinitrogen could be made to bridge two ruthenium(II) centres in $[Ru(NH_3)_5(N_2)Ru(NH_3)_5]^{4+}$ and that this ion could accessed by displacing water from $[Ru(NH_3)_5(H_2O)]^{2+}$. Despite that fact that ruthenium is not a "biological metal, these complexes appeared to be good models for the supposed interaction of molybdenum with dinitrogen in nitrogenase enzymes. The question of the reactivity of coordinated dinitrogen was left in abeyance.

5.2. The requirements for binding dinitrogen to metal ions

In subsequent years a flood of dinitrogen complexes was reported. Note that, with one possible exception, no stable complex is yet known with dinitrogen bonded sideways to a single metal ion. A very large number of dinitrogen complexes are now known, and it is not intended to produce a complete list here. The reader is referred to appropriate references [27] and to the current literature. Only the most significant compounds and systems will be referred to in the rest of this article.

Complexes of dinitrogen with osmium, rhenium, and the Group 6 metals molybdenum and tungsten soon followed. Now dinitrogen complexes are known with most of the transition metals, including uranium. Exceptions seem to be platinum, palladium, and silver and gold. The major interest here is in their structural variety and in the chemical transformations available to the coordinated dinitrogen.

Early and novel complexes of titanium and zirconium are interesting from both structural and reaction points of view. Bercaw *et al.* [28] showed that it is possible to prepare complexes that contain both bridging and terminal dinitrogen.

$$[Ti(Cp^*)_2Cl_2] + N_2 + Na/Hg \rightarrow [(N_2)Ti(Cp^*)_2(\mu\text{-}N_2)Ti(Cp^*)_2(N_2)] \qquad (12)$$
$$(Cp^* = \eta^5\text{-}C_5Me_5)$$

This remarkable compound showed different N-N bond lengths for the two kinds of bound dinitrogen, and the terminal dinitrogens, but not the bridging dinitrogen, were found to be exchangeable. The compound reacts with hydrochloric acid to generate hydrazine and to reform the initial dichloride. The four electrons that reduce the dinitrogen seem to come from the metal ions, two from each. Remarkably, when the complex with $^{14}N_2$ in the bridge and $^{15}N_2$ in the terminal positions is treated with the acid, the resulting hydrazine was found to be a 1:1 mixture of $^{14}N_2H_4$ and $^{15}N_2H_4$. This observation has yet to be explained.

Further early observations that stimulated the development of our ideas about the possibilities of dinitrogen chemistry included the discovery that dinitrogen could displace both dihydrogen and ammonia from complexes of a suitable kind, and that alkali metal ions could somehow help to stabilise dinitrogen binding. These are demonstrated by the Reaction 13 [29] and the typical structure [30] shown below, Figure 3. The ability of an $Mg(thf)_4$ unit to form a bridge between two cobalt dinitrogen residues in $[\{Co(PMe_3)_3(N_2)\}_2\{\mu\text{-}Mg(thf)_4\}]$, with the structural motif Co-N-N-Mg-N-N-Co, was established at about the same time [31]. These observations lent some credence to the suggestions that in nitrogenases dinitrogen displaces either ammonia or dihydrogen (or both, even if not water) when it binds to the metal, and that electrons pass to the dinitrogen from the metal ion, presumably molybdenum, to which it first binds.

$$[CoH_3(PPh_3)_3] + N_2 \rightarrow [CoH(N_2)(PPh_3)_3] + H_2 \qquad (13)$$

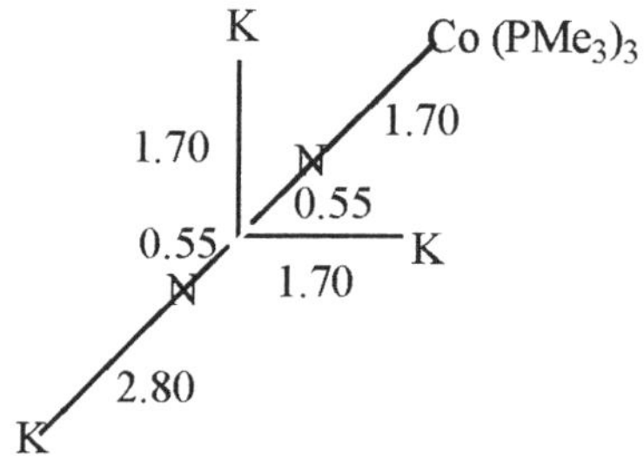

Figure 3. Essential motif of the structure of $K[Co(PMe_3)_3(N_2)]$, distances in Å [30].

A very significant advance was the preparation of dinitrogen complexes of molybdenum [32] and tungsten [33]. These were of the form *trans*-$[M(N_2)_2(\text{diphosphine})_2]$ (M = Mo or W, diphosphine = compounds such as $Me_2PCH_2CH_2PMe_2$ and $Ph_2PCH_2CH_2PPh_2$). Subsequently related materials with triphosphines and with monophosphines were described. One stable compound, $[\{W(N_2)_2(PEt_2Ph)_3\}_2(N_2)]$ [34] actually contains three dinitrogen molecules bound to each metal, though one of the dinitrogen molecules is bridging. Isoelectronic vanadium complexes, $[V(N_2)_2(\text{diphosphine})_2]^-$, were also announced [35]. Thus, about fifteen years after the discovery of the first dinitrogen complex a whole range of them were known and the following generalisations could be made. (1) The binding of dinitrogen could occur both end-on and side-on, and this resulted in the lengthening of the N-N bond, though generally not as far as to justify describing it as a double, let alone a single, bond. The IR stretching frequency, $\nu(N{\equiv}N)$, was also lowered, sometimes by as much as about 400 cm^{-1}. (2) On the basis of a range of spectroscopic studies, this was ascribed to the weak donation of electrons from bonding orbitals (generally of σ-type) of the dinitrogen to the metal ion and stronger acceptance of electrons from the metal into the π-type antibonding orbitals of the dinitrogen. The overall consequence is to make the dinitrogen more basic then if it were free. (3) Because dinitrogen was clearly a poor base, electron acceptance rather than electron donation from dinitrogen would be likely to be the more significant factor in determining whether dinitrogen could bind to a metal ion. When it did bind it should do so predominantly to metal ions in low oxidation states, and to metals that would release electrons most easily. These would be to the left-hand side of the transition series.

Since then these conclusions have not changed to any degree, but it was also evident that the preparation of dinitrogen complexes such as those of molybdenum(0) and vanadium(-I) cited above required much stronger reducing agents (sodium amalgam is a typical example) than were likely to be available to a biological system. The simple complexes described so far cannot be typical of the kind of chemistry that occurs in biological systems, because it is highly improbable that molybdenum or vanadium, if they are at the active sites, would be able to access such low oxidation states within a living and functioning cell.

6. REACTIONS OF DINITROGEN

6.1. Nitriding reactions

Dinitrogen has been known for many years to react with metals, such as lithium amongst the alkali metals and calcium, under very mild conditions. Such reactions are recognised to be initially surface tarnishing reactions, and the ultimate bulk product is the metal nitride, such as Li_3N. The initial step in such reactions is the transfer of an electron from the metal, which is highly reducing, to the dinitrogen to form N_2^-. It is this kind of mechanistic information that has led to the idea that the systems first described by Volpin and his collaborators in 1964 [36] involve the formation of metal nitrides.

Volpin showed that a mixture of one of many transition metal salts, including those of titanium, molybdenum and iron, with a strong reducing agent, such as an organoaluminium compound, a metal and a Grignard reagent, and exposed to dinitrogen under mild conditions, formed a material that yielded ammonia upon hydrolysis with acids. The initial reaction mixture recalls the formulations of Ziegler-Natta catalysts, and Some researchers have claimed that they knew of some kind of reaction of dinitrogen with Ziegler-Natta catalysts even in the 1950s, because such catalysts seemed to deteriorate when exposed to the gas. By using a non-protic Lewis acid, aluminium tribromide, rather than an aqueous acid, Volpin and his group [37] were even able to make the system truly catalytic in titanium compound. For example, as much as 200 moles of ammonia were obtained from a mole of $TiCl_4$ in the presence of aluminium and its tribromide under dinitrogen at 50 °C. All of these systems ultimately require hydrolysis to release the ammonia. The use of H_2 as a reductant was also described [37]. This method was adapted to produce organo-nitrogen compounds; for example, exposure to dinitrogen at room temperature followed by hydrolysis of a mixture of $[TiCp_2Cl_2]$

and LiPh yields aniline and aminobiphenyl [38]. If the reductant is magnesium metal and ketones are added to the system after reaction with dinitrogen, then secondary amines are formed upon hydrolysis. Although the mechanism of these reactions is still not elucidated, it seems highly likely that nitrides are involved. Likely intermediates are complexes such as $[TiCp_2Ph(N_2)TiCp_2Ph]$ which were prepared and characterised [39] independently. The reaction of this complex with LiPh and Li in ether produces, after hydrolysis, some aniline and also ammonia. The mechanism of this reaction has not been definitively determined [40].

Another system that may also involve nitrides at some stage is due to Schrauzer and his collaborators [41]. He has described how iron-doped rutile can catalyse the reaction of dinitrogen and water to give ammonia and dioxygen upon UV irradiation. This is, in essence, the reaction that is performed by biological systems. It is uphill thermodynamically, and light is an appropriate energy source. Much more recently, this kind of approach has been taken up again. Perchlorate-doped electro-polymerised 3-methylthiophene composited with titanium dioxide gradually forms ammonium perchlorate under 1 atmosphere of dinitrogen at 20 °C and irradiation. [42].

Several systems show some hope of eventually challenging the Haber process. Yields are often small and hardly commercial in most cases. For example, potassium-doped lamellar graphite can catalyse the reaction of methane and dinitrogen to yield ammonia at ambient temperatures [43]. A process not requiring the generation of dihydrogen would be a great advance. It is claimed that dinitrogen and dihydrogen combine in a proton-conducting cell reactor at 570 °C [44]. Another electrochemical reaction of dinitrogen and water is claimed to occur in a solid polymer electrolyte cell at atmospheric pressure and at temperatures below 100 °C [45]. Even new Haber-type catalysts are on the horizon. A barium-promoted ruthenium catalyst supported on boron nitride (many newer ruthenium catalysts are supported on the structurally related graphite) works well at pressures down to 50 bar and at temperatures as low as 160 °C. Even more interesting is the report that caesium-promoted cobalt molybdenum nitride [C. J. H. Jacobsen in reference 19] is a more active catalyst for the Haber reaction than commercial iron catalysts. This suggests a new class of catalyst.

The reaction mechanisms associated with none of these systems have been described. It is, however, likely that most if not all operate by splitting dinitrogen to yield covalently-bound nitride ions.

6.2. Nitrogen fixation in aqueous media

Quite apart from these reactions, evidence began to accumulate that fixation could occur in another way. Schrauzer and his group published [46] a series of papers over many years entitled: The chemical evolution of a nitrogenase model. Initially, the essence of their work was that an aqueous solution of molybdate, thiol and iron, suitably buffered and in the presence of a reducing agent, reduces dinitrogen to ammonia, and also reduces several of the alternative nitrogenase substrates. Amounts of ammonia produced were always rather low, but the system is apparently reproducible.

Kinetic analysis of such a complicated system is clearly difficult, but Schrauzer interpreted his data in terms of a key intermediate in which dinitrogen is bound sideways to a single molybdenum atom in the oxidation state IV. It is fair to say that this interpretation of the data was not accepted as accurate in some circles. Like most of reactions discussed in this Section, the absence of dioxygen seems to be obligatory, as the systems are very air-sensitive.

At about the same time, Shilov's group in the Soviet Union developed another set of fixation reactions. [47] Typically, strongly alkaline solutions or suspensions of molybdenum(III) together with a suitable reducing agent (Ti^{III}, Cr^{II} or Ta^{III}) or vanadium(II) with magnesium(II) produce both ammonia and hydrazine. Such systems are active over a considerable range of temperatures and pressures of dinitrogen. They apparently consist of suspensions of metal hydroxides, and the precise reaction pathway has not been clarified to the satisfaction of everyone, Nevertheless, these systems have given rise to genuine catalytic systems for fixing nitrogen under aqueous conditions, as described below, and some interesting observations have been made as to the requirements of a metal ion likely to bind dinitrogen, perhaps the most significant being that d^3 ions are the best suited.

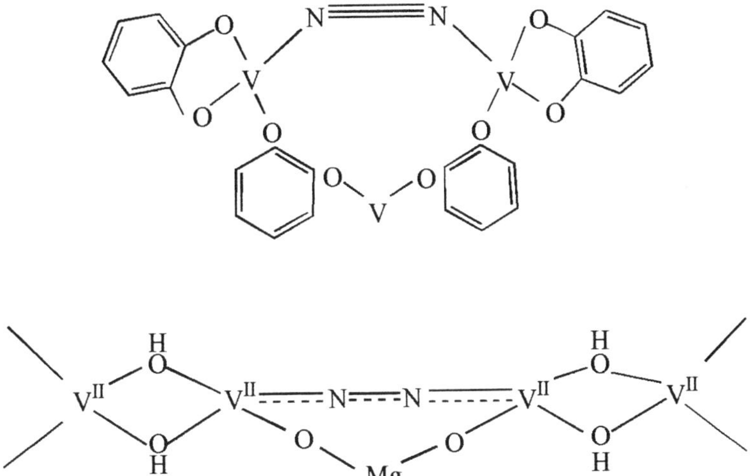

Figure 4. Intermediates proposed in vanadium nitrogen-fixing systems.

The vanadium systems were elaborated to show that vanadium(II) in the presence of catechol and in a narrow range of pH, from about 9 to 11, can reduce dinitrogen to ammonia and hydrazine [48]. At the same time considerable amounts of dihydrogen are generate. This work has been reproduced in other laboratories, but the reaction route is still to be clarified to everyone's satisfaction. On the basis of spectroscopic studies, Shilov and his collaborators have proposed a key intermediate in the vanadium-catechol and vanadium-magnesium systems with dinitrogen bound end-on between two vanadium ions, Figure 4. Indeed, this is the kind of dinitrogen-bridging intermediate that is proposed for all the Shilov systems, whether based upon vanadium or otherwise.

Despite that, Schrauzer [49] has interpreted his own results obtained on Shilov's vanadium(II) systems as involving side-on dinitrogen in the key intermediate, rather than a dinuclear dinitrogen bridge. The situation remains unresolved, though the likelihood that polynuclear intermediates are involved in these reductions is strong.

The molybdenum systems have undergone even more interesting developments. It had been noted that alkaline mixtures of Mo^{III} and Ti^{III} reduce dinitrogen. Replacing the Ti by another, but reversible, reducing agent should promote catalytic reduction if the function of molybdenum is simply to bind the dinitrogen. This has been achieved.

Eventually a catalytic system was developed, consisting of sodium amalgam coated with phosphatidyl choline which reacts with a methanolic solution of molybdenum(III) in the presence of magnesium(II) and, for preference, a tertiary phosphine [50,51]. The mechanism of the reduction is not clear, though polynuclear molybdenum-magnesium compounds have been isolated from such systems. In addition, electrocatalytic reduction at a mercury cathode has been achieved using a methanolic solution of molybdenum(III) and titanium(III) buffered with guanidine [51,52].

These systems remain unique in their ability to reduce dinitrogen catalytically and in protic media. The authors favour a key intermediate with dinitrogen bound end-on between two molybdenum ions, which are themselves held in proximity by bridges consisting of magnesium and oxygen atoms. This is still a matter of some conjecture. What the observations do demonstrate unequivocally, however, is that low-oxidation states and elements to the left side of the transition series are not necessary to effect binding and reduction of dinitrogen. Ligating atoms such as oxygen seem efficacious. Since this occurs, the deduction from early work that molybdenum(0) complexes are somehow involved in biological nitrogen fixation seems not to be tenable. This kind of insight was confirmed by the discovery [53] of compounds such as $[\{M(C_5Me_5)Me_3\}_2(N_2)]$ ($M = Mo^{IV}$ or W^{IV}). These complexes yield ammonia but little hydrazine after protonation and base distillation.

7. THE NEW GENERATIONS OF DINITROGEN COMPLEXES

7.1. New kinds of dinitrogen complex

It has already been explained that, in a formal senses, dinitrogen can bind to a metal in both end-on and side-on configurations. In fact, it gradually became clear that both modes of binding may operate at the same time. A most complex, early example [54] is sketched in Figure 5. This was formed by the uptake of dinitrogen by an uncharacterised reduction product of $[Ti(C_5H_5)_2Cl_2]$ and potassium naphthalene. The N-N separation is 1.301(12) Å, and the IR stretching frequency is 1282 cm^{-1}. This indicates considerable reduction of the dinitrogen triple bond, but it is difficult to assign it a precise bond structure. Hydrolysis of this complex yields ammonia. Compounds such as this also show that coordinated dinitrogen can show considerable Lewis base character, and Lewis acids that are metal ions can form adducts.

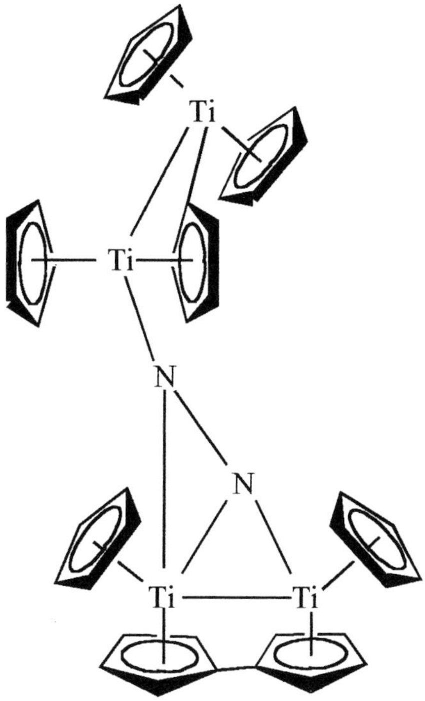

Figure 5. The structure of $[\{Ti_2(C_{10}H_8)(C_5H_5)_2\}(N_2)\{Ti_2(C_5H_5)_3(C_5H_4)\}]$.

7.2. Reactions of coordinated dinitrogen with metal ions

Starting with an end-on singly bound dinitrogen it is possible in favourable circumstances to form bridging dinitrogen complexes, and this was noted relatively early with rhenium [55] and cobalt [31] complexes. The former case involved rather ill-defined materials such as [adducts of $[ReCl(PMe_2Ph)_4(N_2)]$ and $[TiCl_3(thf)_3]$. In the latter case, the formation of, say, $[\{Co(N_2)(PMe)_3\}_2\{Mg(thf)_4\}]$ from $[CoH(N_2)(PMe_3)_3]$ and $MgEt_2$ in thf yields a compound containing a Co-N-N-Mg-N-N-Co grouping and dinitrogen susceptible to hydrolysis to ammonia and hydrazine. This kind of adduct turns out to be relatively common.

In 1991, Schrock demonstrated the stability of a high-oxidation-state dinitrogen compound with double Mg(tmen) (tmen = *N, N, N', N'*-tetramethylethane-1,2-diamine) bridges spanning two end-on W-N-N units [56]. Later he showed [57] that molybdenum and tungsten may form high–oxidation-state dinitrogen complexes with ligand sets other than Cp*Me_3. Examples are [$\{ML\}_2(N_2)$] (M= Mo or W, L = a triamidoamidine(3-) ligand such as $\{N(CH_2CH_2NC_5F_5)\}^{3-}$. Produced by one-electron reduction of [MLCl], they can be further reduced by sodium amalgam to give a product, possibly with a linear Mo-N-N-Na arrangement and this, in turn, reacts with $SiPr^i_3$ and with $SnBu_3Cl$ to give compounds containing Mo-N-N-Si and Mo-N-N-Sn systems, respectively. These last reactions were not without precedent (see below), but the use of the trianionic nitrogen donor suggested that a new approach to dinitrogen activation might be fruitful. Subsequently Schrock showed [58,59] that the compound [$\{MoL\}_2(N_2)$] reacts with magnesium powder in thf under N_2 to yield [$\{MoL\}(N_2)\{Mg(thf)_2\}(N_2)\{MoL\}$]. This adduct contains an Mo-N-N-Mg-N-N-Mo chain bent at the magnesium at 134°. With $ZnCl_2$ this last compound produced not an adduct but mononuclear [$\{MoL\}(N_2)$], which itself easily reverts to the bridging dinitrogen species. With $FeCl_2$ it generates [$\{MoL(N_2)\}_3Fe$]. All these compounds show the expected lengthening of the N-N bond compared to that in free dinitrogen, but it is generally difficult to write simple valence-bond structures for them. An extensive series of heterometal bridging dinitrogen adducts featuring Mo and W on the one hand and Ti, Zr, Hf, Nb and Ta on the other has been described recently [60].

7.3. Bridging dinitrogen complexes

By far the most interesting recent developments in dinitrogen chemistry have concerned bridging complexes. These seem to be more accessible now than are singly-bonded end-on dinitrogen complexes. For example, though the first vanadium dinitrogen complexes were of V^{-I}, and of the form [$V(N_2)_2$(tertiary phosphine)$_4$]$^-$, isoelectronic with the Mo^0 and W^0 complexes. Later vanadium complexes, of vanadium(II) and of vanadium(III), generally contain bridging dinitrogen. Treatment of both vanadium(-I) [35] and vanadium(II) [61] compounds with acids causes protonation of the dinitrogen to

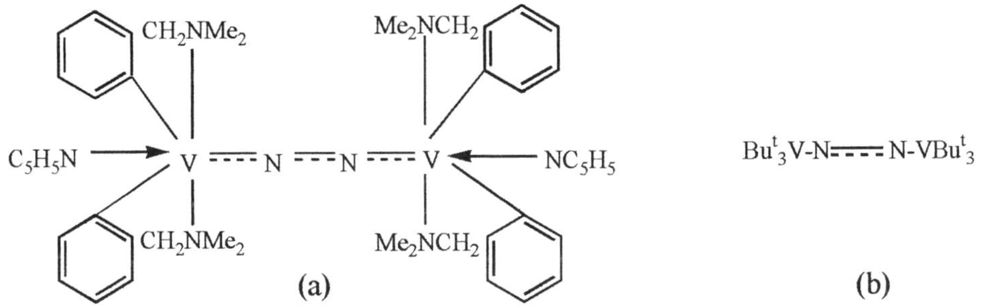

Figure 6. A vanadium(II) (a) [63] and a vanadium(III) (b) [62] dinitrogen complex; bonds to dinitrogen nitrogens are only schematic.

ammonia and/or hydrazine, but vanadium(III) compounds merely release dinitrogen upon such treatment [62]. Examples of these complexes are shown in Figure 6.

The vanadium(III) compound has an N-N separation of 1.250(3) Å, and the vanadium(II) of 1.228(4) Å. Thus N-N bond length is clearly no simple guide to reactivity with acids. A further example of a bridging vanadium(III) dinitrogen complex with a trianionic nitrogen donor is $[(VL)_2(N_2)]$ (L = $\{MeC(CH_2NPr^i)_3\}^{3-}$), N-N = 1.257(6) Å [64]. Vanadium(II) analogues based upon $[(CyN)CH(Ncy)]^{2-}$ (Cy = cyclohexyl) are also known [65]. The N-N bond lengths are very similar.

However, more complex structures have also been uncovered. The first rare-earth dinitrogen complex [66] was also bridging, in $[\{Sm(C_5Me_5)_2\}_2(N_2)]$. The N-N separation is very short, initially suggested to be 1.088(12) Å, but the most interesting aspect is that the dinitrogen is sideways-on to the two samarium atoms, and all four atoms are coplanar. The dinitrogen is not strongly bound, and in the related thulium compounds, $[\{(R_2Me_3C_5)_2Tm\}_2(N_2)]$ [R_2 = Me_2, $Me(SiMe_3)$, or $(SiMe_3)_2$] [67] of similar structure, the stability increases with the number of silicon substituents in the cyclopentadienide rings. The first actinide dinitrogen complex [68], $[(UL)_2(N_2)]$ (L = the trianionic donor $N(CH_2CH_2NSiBu^tMe_2)_3{}^{3-}$), has a similar $\{U_2N_2\}$ structure. The N-N separation is 1.109(7) Å. Probably none of these complexes reacts with acids to protonate the dinitrogen, though as yet no information has been provided. A more complex example is a compound of samarium with octaetylporphyrinogen (oepg) as a ligand. It was obtained by the reduction of $[SmCl_3(thf)_3]$ with $[Li_4(oepg)(thf)_4]$ (oepg = octaethyl porphyrinogen anion) under dinitrogen. The product is an extended cage containing dinitrogen coordinated by 4 lithium ions in a plane, and on either side of the plane is a $Sm(oepg)Li(thf)_2$ unit [69]. There are clearly both Li-N and Sm-N interactions, and the N-N separation is 1.525(4) Å, longer than in hydrazine itself. This implies a six-electron reduction of the dinitrogen if these are really nitrides, and but the samarium is evidently still Sm^{II}. It would again be unwise to assign a valence-bond structure to this compound. Oepg forms similar dinitrogen complexes based upon praseodymium and neodymium [70]. These also contain side-on dinitrogen and the N-N separations are about 1.24 Å, though in the case of Nd there seem to be additional quasi-end-on interactions with two sodium ions. Probably such interactions provide additional stabilisation, but are not absolutely necessary for the formation of rare-earth dinitrogen complexes.

These alkali-metal interactions are consistent with the fact that lithium metal can react with dinitrogen. A dinitrogen complex containing only lithium ions, $[\{Li(thf)_3\}_2(N_2)]^{2+}$ [71] has been described. Each lithium ion is coordinated to the dinitrogen, but the Li_2N_2 arrangement is quasi planar with the lithium bound sideways to the dinitrogen. The N-N separation is effectively the same as in free dinitrogen. In fact, related "clusters" have been known for some years. Examples include

$[\{(PhLi)_3Ni\}_2(N_2)]$ [72] and a similar material containing both sodium and lithium ions [73]. These have N-N separations of about 1.35 Å.

Evidently dinitrogen can be bound by what amounts to a metal cluster, but how it might achieve that situation within a preformed cluster is not evident. For example, the reaction of $[(LAu)_3O]^+$ (L = tertiary phosphine such as PPh_2Pr^i) with hydrazine produces a dinitrogen complex $[(LAu)_6(N_2)]^{2+}$, in which each nitrogen is bound to three gold atoms [74]. The N-N separation is 1.475(14) Å, and although the Au-Au distances within each group of 3 gold atoms indicates some metal-metal interaction, there is none between the two groups of gold atoms. Acidolysis in the presence of a suitable reducing agent yields hydrazine and ammonia. This complex cannot be obtained from dinitrogen gas, only *via* hydrazine.

In the last ten years it has become evident that the sideways dinitrogen-bridging structure to two metal ions, as exemplified in the rather unstable dilithium and the disamarium species described above, is no longer unique. For example, titanium(II) forms end-on bridging complexes as in $[\{TiCl(NR_2)(C_5H_5N)_2\}_2(N_2)]$ (R = $SiMe_3$), with an N-N separation of 1.263(7) Å [75]. As in all such compounds, the metal-N-N-metal system is more or less linear. The analogue with (tmen) replacing the two pyridine ligands, obtained from $[TiCl_2(tmen)_2]$ and $LiN(SiMe_3)_2$ under dinitrogen, has a similar structure, though the donor nitrogens of the diamine are, of course, *cis* rather than *trans*, as in the pyridine complex [76]. Use of more silicon reagent than is required to make the end-on complex produces a novel complex with two side-on dinitrogens, $[\{Ti(NR_2)_2\}_2(N_2)_2]$ (Figure 7).

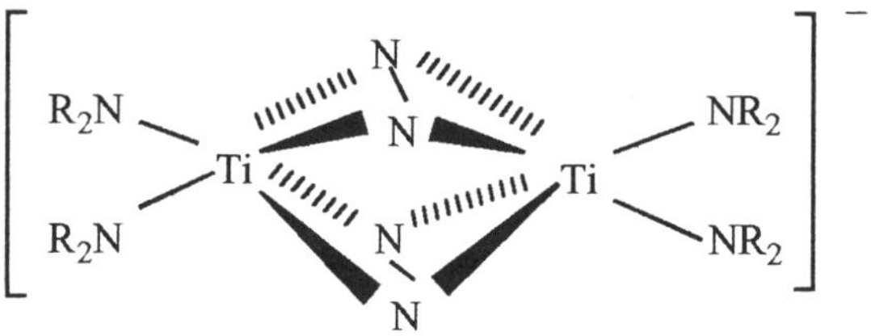

Figure 7. The structure of $[\{Ti(NR_2)_2\}_2(N_2)_2]^-$ (R = $SiMe_3$).

The N-N separations are greater than before, 1.379(4) Å, and some reduction has occurred, the complex formally now containing both Ti^{II} and Ti^{I}, but it is not evident what favours this side-on structure over the end-on structure. Is it simply the addition of a single electron to one metal ion?

This question has, to some extent, been answered by the study of compounds where the conversion of one form of bound dinitrogen to the other has been observed [77]. The zirconium(IV) complex $[ZrCl_3\{(Pr^i_2PCH_2SiMe_2)_2N\}]$ can be reduced with sodium amalgam under dinitrogen to give the side-on bridging complex,

[{ZrCl{($Pr^i_2PCH_2SiMe_2)_2N$}}(N_2)]. This has an enormous N-N separation of 1.548(7) Å, yet treatment with acid yields hydrazine, not ammonia. This is because two zirconium(II) ions can supply only four electrons to the dinitrogen. The fact that there is apparently no N-N bond in the parent compound is misleading. A detailed analysis of the frontier orbital interactions led to the suggestion that what determines the geometry of the bridge is the availability of orbitals of appropriate symmetry on the metal. The most important interactions in the end-on case involve transfer of electrons from the d_{xz} and d_{yz} orbitals of the metal ions to the $1\pi_g$ (see Figure 1) antibonding orbitals of the dinitrogen. This weakens the N-N bond but strengthens the M-N bond. For the side-on case, the degeneracy of the $1\pi_g$ orbitals is broken and one of them then has δ symmetry with respect to the metal-metal axis. Interactions of the dinitrogen with this orbital would hence be expected to be weakened as compared to the end-on case. However if one of the $1\pi_g$ pair of orbitals is preferentially overlapping with an empty orbital of a ligand other than the dinitrogen, then the side-on configuration may become favoured. This postulate was supported, at least in this kind of system, by the preparation of a complex in which the chloride on each zirconium atom in the side-on complex described above is replaced by a cyclopentadienyl ligand, giving [{Zr(Cp){($Pr^i_2PCH_2SiMe_2)_2N$}}(N_2)] (Cp = η^5-C_5H_5). This complex still contains, zirconium(II) and still hydrolyses to give hydrazine, but the dinitrogen is end-on, and the N-N separation is only 1.301(3) Å. The conclusion is that it is necessary to consider the detailed structure of a bridging complex and especially the geometry around the metal ions if one is to be able to predict with some certainty the arrangement of the bridge. Furthermore, if the side-on interaction to a metal is weaker than the end-on, this is probably why side-on bonding to a single metal ion [78] has yet to be proven unequivocally, though it has been recognised in frozen matrices and in excited states [79]. A mixture of end-on and side-on bonding as seen in the titanium complex cited above and, more recently, in tantalum chemistry [77] is turning out to be not uncommon. Theory does not yet say much about poly(dinitrogen) interactions.

8. REACTIONS OF COORDINATED DINITROGEN

8.1. Reactions with Lewis acids

The empirical evidence seems to show quite clearly that coordinated dinitrogen carries more negative charge than the free dinitrogen. In the case of the dinitrogen complexes discussed above, this is often shown by a lengthening of the N-N separation from the value of 1.0976 Å observed in the gas phase. Singly-bonded end-on dinitrogen complexes can often act as donors to typical metal ion Lewis acids forming bridging compounds, and some of these have been described above. They will not be discussed further here. Of course, the family of Lewis acids includes the proton, and the possibility of protonating coordinated dinitrogen to form hydronitrogen derivatives was a very important aim of the chemists who first investigated dinitrogen complexes. Protonation of coordinated dinitrogen is relatively common now, but the discussion above has indicated that one cannot infer a susceptibility to protonation from an observed N-N separation. This is because the metal ions involved in the complex are required not to share their valence electrons, as in covalent metal-dinitrogen bond formation, but to release them completely when a remote nitrogen-hydrogen bond is formed using a proton which carries no electrons for bond formation. This is a function of the ease of oxidation of the metal ion in the particular environment in which it finds itself, and such properties are generally not clearly defined. Nevertheless, a great deal is now known about the protonation of dinitrogen, especially coordinated to molybdenum and tungsten.

8.2. Reactions of coordinated dinitrogen with protons and hydrogen

The first reported reaction of coordinated dinitrogen in a well-defined dinitrogen complex to give a well-defined complex product was a serendipitous discovery. The reaction of $[W(N_2)_2(dppe)_2]$ (dppe = $Ph_2PCH_2CH_2PPh_2$) with $CH_3C(O)Cl$ gives a product $[WCl(N{=}NCOCH_3)(dppe)_2]$ [80], which was as welcome as it was unexpected. In a formal sense, one can write the reaction as a kind of oxidative addition, as shown (diphosphine ligands are omitted).

$$\{N{\equiv}N\text{-}W\text{-}N{\equiv}N\} + CH_3C(O)Cl \rightarrow \{Cl\text{-}W\text{-}N{=}NC(O)CH_3\} + N_2 \tag{14}$$

This is essentially the reverse of a reaction used to prepare an extensive series of dinitrogen complexes of rhenium [81]. The reaction involves the oxidation of the tungsten from W^0 to W^{II}, with the concomitant reduction of one of the two dinitrogen ligands. The complex can be further protonated at the *exo*-nitrogen, which converts the diazenido-ligand to a hydrazido(2-) ligand.

$$\{Cl\text{-}W\text{-}N{=}NC(O)CH_3\} + HCl \rightarrow \{Cl\text{-}W{=}N\text{-}NHC(O)CH_3\}^+ + Cl^- \tag{15}$$

In a formal sense, the tungsten has now become W^{IV} and the dinitrogen residue has received two more electrons from the metal. Since the acetyl chloride is believed to be polarised in the sense $CH_3CO^+...Cl^-$, it was an easy step to employ simple proton acids in this reaction [82]. In fact, acids as weak as methanol are strong enough to protonate one of the two coordinated dinitrogens on this and many related molybdenum and tungsten complexes. From such reactions compounds containing the ligands $(NNH)^-$ and $(NNH_2)^{2-}$ were fully characterised.

A similar reaction occurs if monodentate tertiary phosphines are employed instead of bidentate ditertiary phospines in the starting dinitrogen complexes, but then not only are diazenido and hydrazido(2-) compounds formed, but the protonation can proceed all the way to ammonia. The yield depends to some degree on the reaction conditions, the solvent, and the acid, but at best it can be represented in the following way [83].

$$\{N{\equiv}N\text{-}W^0\text{-}N{\equiv}N\} + 6HCl \rightarrow 2NH_3 + W^{VI} \tag{16}$$

The nature of the final tungsten(VI) compound has not been defined, and the initial dinitrogen complex cannot be reformed and recycled. Presumably some kind of halo-oxide of tungsten is formed. It is remarkable that the tungsten can give up all six of its valence-shell electrons to dinitrogen, but this cannot be the basis of a usable process unless it can be recycled. This was actually achieved [84] by attaching the dinitrogen complex to an electrode and employing trifluoracetic acid as the proton source. The system can then be cycled several times to generate ammonia, the electrons taken up from the electrode preventing the complete degeneration of the coordination about the metal. Overall the protonation sequence can be represented by the following cycle (Figure 8).

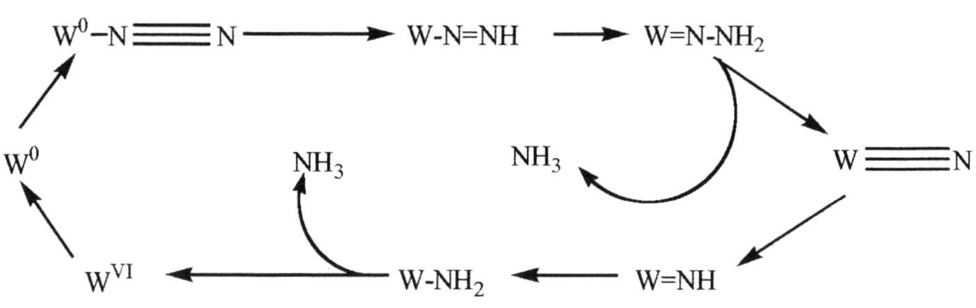

Figure 8. A protonation cycle for tungsten (and molybdenum) dinitrogen complexes.

Compounds containing all the nitrogen species presented in this Figure have been isolated from suitable dinitrogen-protonation reactions. The implications of these observations for biological nitrogen fixation will be discussed elsewhere in this book.

It is not only aqueous acids that can protonate coordinated dinitrogen in these molybdenum and tungsten complexes. In early work it was shown that hydridocarbonyl complexes such as $[FeH_2(CO)_4]$ react with *cis*-$[W(N_2)_3(PMe_2Ph)_4]$ in methanol to yield, after base distillation, ammonia and hydrazine [85]. Such metallohydrides display a wide range of acidities [86], so that this is an imaginative approach. The reaction routes are complex, and in particular cases other reactions, such as dinitrogen/dihydrogen interchange between $[FeH_3(dmpe)_2]$ (dmpe = $Me_ePCH_2CH_2PMe_2$) and $[Mo(N_2)_2(dppe)_2]$ have also been observed [87].

Metallohydrides are sometimes not easy to prepare, but a significant more recent advance has been the use of $[RuCl(Ph_2PCH_2CH_2CH_2PPh_2)_2]$ that can take up one mole of dihydrogen. The complex thus formed can protonate tungsten and molybdenum dinitrogen complexes to yield ammonia, though the precise details of the mechanism are not yet clear. However, the dihydrogen is necessary, and it is possible that a genuine catalytic process might eventually arise from this kind of system. It is clearly an advance over an earlier report on the use of dihydrogen as a reductant in a fixing system consisting initially of a mixture of alkaline Ti^{IV}, Mo^{V}, Mg^{II} and a PtO_2 catalyst [89].

This is not the only kind of hydrogen compound that is capable of protonating coordinated dinitrogen. Some thiols can do this, but the bridging trithiolato-complex

[{$Ir(C_5Me_5)$}$_2$(μ-SH)$_3$] and a similar iron complex can also do so [90]. Ammonia can be generated, though with [$W(N_2)_2$(dppe)$_2$] rather than a monotertiary phosphine complex a hydrazido(2-) product was isolated. Such observations may be relevant to understanding nitrogenase function.

8.3. Reactions with radicals and other electrophiles

At the time that protonation reactions were first being observed, and with the recognition of the ability of $CH_3C(O)Cl$ to attack coordinated dinitrogen, the reactions of alkyl halides were also investigated. Simple alkyl halides do attack coordinated dinitrogen in molybdenum dinitrogen complexes, but the tungsten analogues need irradiation to make the reactions proceed at any reasonable rate [91]. An extensive study of the mechanism revealed that it is overall a radical process, and that it does not, as was originally believed, involve positive alkyl ions analogous to $CH_3C(O)^+$(Figure 9).

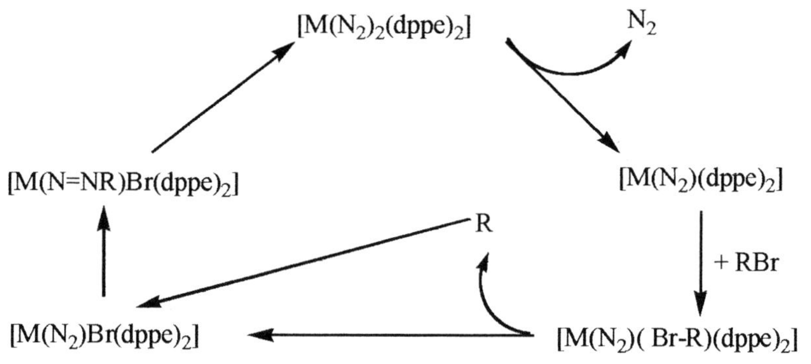

Figure 9. Radical attack on coordinated dinitrogen (M = Mo, or W plus irradiation).

The alkyldiazenido-complexes can be converted to dialkylhydrazido(2-)-complexes by reaction with further alkyl halide. Reduction of the alkylated products with, for example, lithium aluminium hydride, yields amines, and controlled electrolytic reduction can split the N-N bond yielding a nitride and an organonitrogen compound [92].

A large variety of halides may be used. Methylene dibromide forms complexes containing the novel group $M=N-N=CH_2$ [93] and trifluoroacetylation has also been observed [94]. Reactions of this kind are not restricted to carbon compounds, and the formation of silyldiazenido-, silylhydrazido(2-)-, and disilylhydrazido(2-)-compounds from appropriate reactions with silyl halides has been reported [95], and the bridgiing complex [{ML}$_2$(N_2)] [L = $N(CH_2CH_2NSiMe_3)_3$] can be alkylated and silylated with displacement of one molybdenum atom [96]. All the N-Si bonds formed in these reactions are susceptible to hydrolysis. There are even reports of the formation of tin derivatives [57,97] though, in view of the fact that similar iron, magnesium and vanadium derivatives are known [98], whether any or all of these reactions involve radicals is questionable.

Comparable arylations of coordinated dinitrogen with simple aryl halides such as PhBr do not appear to occur. However, aryldiazenido-compounds can be attained form tungsten dinitrogen complexes *via* protonation and reaction with the strongly electrophilic 2,4-dinitrofluorobenzene [99]. Presumably halides such as PhBr are not electrophilic enough to attack strongly the *exo*-nitrogen atom of coordinated dinitrogen. Very strong electrophiles run the risk of oxidising the dinitrogen complex, with release of the gaseous element.

8.4. Reactions with nucleophiles

The reaction of certain cationic manganese carbonyl dinitrogen complexes such as $[Mn(C_5H_5)(CO)_2(N_2)]$ with nucleophiles such as methyl and phenyl carbanions has been reported [100]. Here attack apparently occurs at the *endo*-nitrogen rather than at the *exo*. The product contains a grouping represented formally as Mn-N(Ph)=N. Such reactions appear not to have been widely exploited.

8.5. Other reactions of coordinated dinitrogen

In 1999 came the first definitive report of a new kind of reactivity of coordinated dinitrogen [101]. The macrocyclic 2- ligand Q (Figure 10) binds one atom of Zr^{IV} as the dichloride. Reduction of this under dinitrogen yields the complex $[(ZrQ)_2(N_2)]$ with side-on dinitrogen which has a separation of 1.43(1) Å. This complex reacts, in turn, with dihydrogen to produce a dinuclear material which has the two zirconium atoms bridged by a hydrogen atom and side-on diazenide(1-), $(NNH)^{1-}$ [102]. The mean Zr-NH bond lengths are 2.13 Å whereas as the Zr-N separations are 2.09 Å. This is a completely new kind of dinitrogen reactivity. Even more surprising is the reaction of the dinitrogen complex with silanes $RSiH_3$ (R = Ph or Bu. Again the dinitrogen was observed to react, forming a new dinuclear complex with a bridging hydrogen atom and a new side-on diazenido(1-) ligand, $(NNSiR_3)^{1-}$. Both these reactions are without precedent, and there have been no reports of similar observations. Attempts to complete the splitting of the

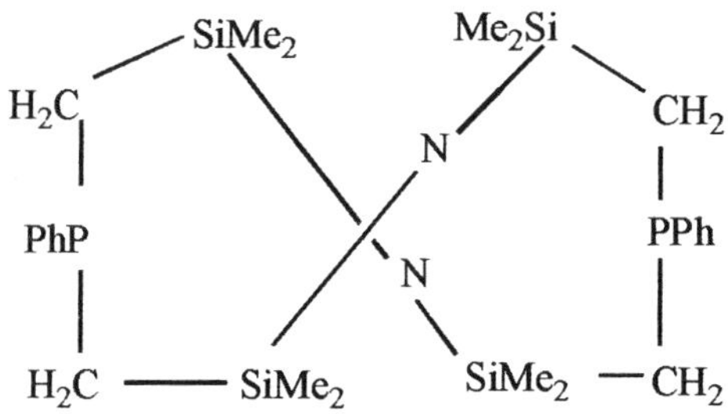

Figure 10. The macrocyclic dianionic ligand Q^{2-}.

dinitrogen N-N link in the dinitrogen complex in a controlled fashion have not been successful to date even in the niobium(II) analogue of the zirconium complex described here [103].

8.6. Splitting the N-N bond in bridging dinitrogen complexes

In the protonation reactions of {M(N_2)$_2$(dppe)$_2$] discussed in Section 7.2, all six electrons for the reduction of the dinitrogen triple bond are supplied by a single metal(0) ion. Clearly, this is likely to be more easily achievable if the six necessary electrons can come from two or more different metal ions. This was first achieved by Cummins [104] who showed how the three-coordinate molybdenum amido-complexes [Mo(NRAr)$_3$] (R = Bu^t, Ar = Ph or substituted phenyl groups) react with dinitrogen at –35 °C to yield an

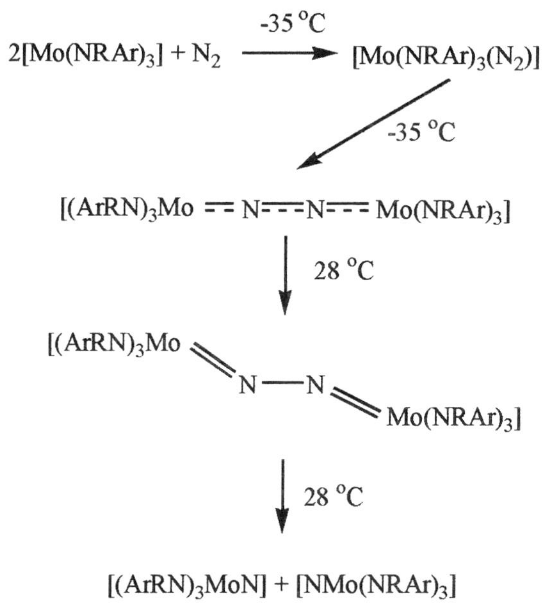

Figure 11. The splitting of dinitrogen mediated by a molybdenum tris(amido)-complex.

unstable bridging dinitrogen complex. This then breaks down to yield two molecules of nitrido-complex, in the sequence of reactions shown in Figure 11.

This unprecedented reaction proceeds spontaneously at room temperature, and represents the limit of N-N bond stretching in which the bond is actually broken. As noted above, this is not necessarily evident from the observed N-N separation that can be longer than a normal N-N single bond yet still retaining a significant degree of N-N bonding. There is no doubt of bond breaking in this case because the individual nitrido complexes can be isolated and completely characterised. The N-N separation in the linear dinitrogen complex stable below – 35 °C is 1.19(2) Å, as determined by EXAFS.

Cummins has since shown [105] that dinitrogen between molybdenum and niobium with similar amido-ligands can also be split to form nitrides.

In a related system, the complex [$MoCl_4(MeOCH_2CH_2OMe)$] reacts with LMgBr (L = 2,4,6-C_6H_2) under dinitrogen at *ca.* 20 °C to give the thermally stable compound [$\{MoL_3\}_2(N_2)$]. [106]. It possesses a linear Mo-N-N-system with an N-N separation of 1.243(13) Å. Upon UV irradiation the dinitrogen splits to give [$(MoL_3)N(MoL_3)$]. This seems to be a mixed-oxidation state compound. Interestingly, the tungsten complex [$MoCl_4(MeOCH_2CH_2OMe)$] can be reduced under dinitrogen to form a family of dinitrogen-containing oligomers [$Cl\{WL_3(N_2)\}_nWL_3$] (n = 1, 2 or 3, with bridging dinitrogens. These structurally characterised materials are currently unique [107].

Since then there have been reports of related reactions. Niobium forms a derivative with calix[4]arene, which can then be reduced by sodium [108]. Under dinitrogen a niobium(III) diniobium end-on bridging dinitrogen complex may be formed. This has an N-N separation 1.390(17) Å. Further reduction with sodium yields a complex with a $Na_2Nb_2N^2$ moiety (Figure 12a) but with an N-N separation of 2.598(8) Å, which must exclude any direct nitrogen-nitrogen interaction.

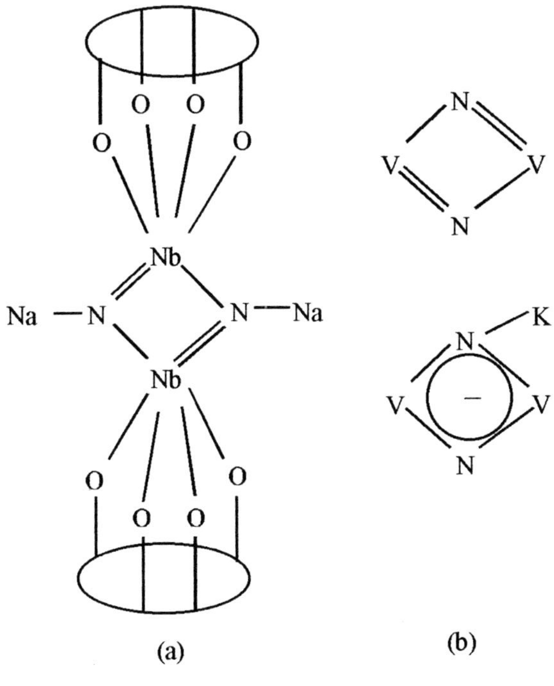

Figure 12. Cores of the nitrido-species produced from dinitrogen in niobium [108] and vanadium [109] systems.

Particularly remarkable in this system is the way that the splitting of the dinitrogen can be controlled by selective use of solvents, and the demonstration that the orientation of the dinitrogen with respect to the two niobium atoms (side-on or end-on bridging)can be controlled in a way that can be convincingly rationalised.

A similar kind of splitting of dinitrogen with formal vanadium (II) has been reported by Cloke *et al.* [109]. The complex $[VCl_3(thf)_3]$ reacts with the dianion $[Me_3SiN(CH_2CH_2NSiMe_3)_2]^{2-}$ to give a dinuclear vanadium(III) monochloride. This can be reduced by potassium under dinitrogen in two distinct steps to yield two complexes with bridging nitrido-groups (Figure 12b). Again, the atom separations exclude direct N-N interactions in these complexes. A related vanadium system also been described earlier [110], but in less detail.

9. CONCLUSIONS

Results such as those described in Section 7 offer the hope that breaking of the dinitrogen triple bond will soon be extended rationally to newer systems. It seems that single metal atoms of low enough oxidation states in suitable ligand environments and with the ability to access oxidation states two or three units higher than that in which they can trap dinitrogen will be able to form kinds of nitrido-complex. This splitting is what seems to occur by a cooperative process on the surface of a Haber-type catalyst.

Nitrido-complexes can usually be protonated to give ammonia, but the control of such reactions is still problematic. The Haber catalysts don't use solvolysis, but seem able also to split dihydrogen to yield dihydrogen atoms that then combine with the surface nitrogen atoms. It is now possible to use dihydrogen to convert dinitrogen to ammonia in selected model systems, though it seems unlikely that the dihydrogen necessarily is split homolytically. In the most promising examples, the dihydrogen is probably heterolysed.

In any case, since the reaction of an end-on bound dinitrogen with a Lewis acid metal ion to form a bridging complex, and with a different Lewis acid, the proton, to form a hydronitrogen species are formally similar, there seems to be at least two general classes of dinitrogen reactivity, one in which the dinitrogen molecule is split before it reacts with a hydrogen species, and one in which it reacts with a hydrogen species as it is split in stepwise fashion. Clearly dinitrogen is a very reactive material with a lot of possibilities for synthetic exploitation. Further discussion of some of the chemistry presented in this Chapter will be found in Chapters 9 and 12. Although empirical [111] and theoretical researches [112] are being pursued vigorously, whether any of this chemistry yet tells us much about biological nitrogen fixation is still an open question.

10. REFERENCES

1. K. B, Kraus and D. K. Bird, Introduction to Geochemistry, Third edition, McGraw Hill, New York, NY, 1995.
2. See, for example, M. Lægrid, A.C. Bøckman and O Kaarstad, Agriculture, Fertilizers and the Environment, CABI Publishing, New York, NY, 1999.
3. A. M. James and M. P. Lord ed., Macmillan's Chemical and Physical Data, Macmillan Press, London, 1992.
4. J. A. Dean ed., Lange's Handbook of Chemistry, Thirteenth edition, McGraw Hill, New York, NY, 1985.
5. D. F. Shriver, P. W. Atkins and C. H. Langford, Inorganic Chemistry, Second edition, Oxford University Press, Oxford, 1994.
6. See, for example, J. Chatt and R. L. Richards in The Chemistry and Biochemistry of Nitrogen Fixation, J. R. Postgate ed.,Plenum Press, London, 1971, p. 76.
7. For an early discussion see J. Chatt and G. J. Leigh, Chem. Soc. Rev. 1 (1972) 121.
8. J.Houghton, Collections of Letters for the Improvement of Husbandry and Trade, quoted by Lord Ernle, English Farming, Past and Present Sixth Edition, Heinemann, London, 1961
9. D. Rutherford, Thesis, Dissertatio inaugralis de aere fixo dicto aut mephitico, Edinburgh, 1772.
10. J. H. C. Chaptal de Canteloup, Elements de Chymie, Vol. 1, Paris, 1790, p. 128.
11. For a comprehensive historical review, see, J. W. Mellor, A comprehensive Treatise on Inorganic and Theoretical Chemistry, Volume 8, Longmans, Green & Co., London, 1962.
12. C. L. Berthollet, Mém. Acad. (1885) 316.
13. H. Davy, Phil. Trans. 97 (1807) 1.
14. J Liebig, Chemistry in its Application to Agriculture and Physiology, L. Playfair ed., Second edition, Taylor and Walton, London, 1842.
15. H. Hellriegel, Tageblatt der 59 Versammlung deutscher Naturforscher und Ärtzte, Berlin , 1886, p. 290;H. Hellriegel and H. Willfarth, Beil. Z. Ver. Dt. Zucklnd. 1888.
16. W. Crookes, Presidential Address to the British Association, 1898. A copy of the original may be obtained from the British Association.
17. La Société Norvégienne de l'Azote, Imprimerie Hemmerlé, Petit & Cie., Paris, 1957.
18. For a detailed discussion, see V. Smil, Enriching the Earth, The MIT Press, Cambridge, Mass., 2001.
19. See, for example, H. Bielawa, O. Hinrichsen, A, Birkner and M. Muhler, Angew. Chem., Int. Ed. Engl. 40 (2001) 1061; C. J. H. Jacobsen, Chem. Commun. (2000) 1057; C. J. H. Jacobsen, J. Cat. 200 (2001) 1; V. B. Shur, and S. M. Yunusov, Russ. Chem. Bull. 47 (1998) 765.
20. L. F. Haber, The Chemical Industry 1900-1930, Oxford University Press, Oxford, 1971.
21. H. Bortels, Arkiv. Mikrobiol. 1 (1930) 333.

22. H. Bortels, Centralbl. F. Bakt. II 87 (1933) 476; H. Bortels, Arkiv. Mikrobiol. 7 (1936) 13.
23. For example, H. L. Jensen and D. Spencer, Proc. Linn. Soc. N. S. Wales 72 (1947) 73.
24. A. D. Allen and C. V. Senoff, J. Chem. Soc., Chem. Commun., (1965) 621.
25. J. Chatt, J. R. Dilworth and R. L. Richards, Chem. Rev. 78 (1978) 589.
26. C. Creutz and H. Taube, Inorg. Chem. 10 (1971) 2664.
27. A reasonable review of this period is R. A. Henderson, G. J. Leigh and C. J. Pickett, Adv. Inorg. Chem. Radiochem. 27 (1983) 197.
28. J. E. Bercaw, E. Rozenberg and J. D. Roberts, J. Am. Chem. Soc. 96 (1974) 612; R. D. Sanner, D. M. Duggan, T. C. McKenzie, R. E. Marsh and J. E. Bercaw, J. Am. Chem. Soc. 98 (1976) 8358; J. Manriquez and J. E. Bercaw, J. Am. Chem. Soc. 96 (1974) 6229; J. M. Manriquez, R. D. Sanner, R. E. Marsh and J. E. Bercaw, J. Am. Chem. Soc. 98 (1976) 8351.
29. A. Sacco and M. Rossi, J. Chem. Soc., Chem. Commun. (1967) 316.
30. R. Hammer, H.-F. Klein, P. Friedrich and G. Huttner, Angew. Chem., Int. Ed. Engl. 16 (1977) 485.
31. R. Hammer, H.-F. Klein, U Schubert, A. Frank and G. Huttner, Angew. Chem., Int. Ed. Engl. 15 (1976) 612; see also A. Yamamoto, Y. Miura, T. Ito, H.-L. Chen, K. Iri, F. Ozawa, K. Miki, T. Sei, N. Tanaka, N. Kasai, Organometallics, 2 (1983) 1429.
32. M.Hidai, K. Tominari, Y. Uchida and A. Misono, J. Chem. Soc., Chem. Commun. (1969) 1239.
33. B. Bell, J. Chatt and G. J. Leigh, J. Chem. Soc., Chem. Commun. (1970) 842.
34. S. N. Anderson, D. L. Hughes and R. L. Richards, J. Chem. Soc., Dalton Trans. (1982) 1291.
35. H. Gailus, C. Woitha and D. Rehder, J. Chem. Soc., Dalton Trans. (1994) 3471, and references therein.
36. M. E. Volpin and V. B. Shur, Doklady Akad. Nauk. S. S. S. R. (1964) 591.
37. M. E. Volpin and V. B. Shur, Organometallic Reactions, Volume 1, John Wiley & Sons, New York, NY, 1970, p. 55; for later developments see K. Ueda, Y. Sato and M. Mori, J. Am. Chem. Soc. 122 (2000) 10722.
38. V. B. Shur, E. G. Berkovich, S. M. Yunusov, Sh. M. Mamedov, L. L. Nizker and M. E. Volpin, Izvestia Akad. Nauk S. S. S. R. (1977) 2841.
39. See, for example, F. W. Van der Weij, H. Scholtens and J. H. Teuben, J. Organomet. Chem. 127 (1977) 299.
40. M. E. Volpin, J. Organomet. Chem. 200 (1980) 319; E. G. Berkovich, V. S. Lenenko, L. I. Vyshinskaya, G. A. Vasileva, V. B. Shur and M. E. Volpin, J. Organomet. Chem. 535 (1997) 169.
41. G. N. Schrauzer and T. D. Guth, J. Am. Chem. Soc. 99 (1977) 7189. Compare O. Rusina, A. Eremenko, G. Frank, H. P. Strunk and H. Kisch, Angew. Chem., Int. Ed. Engl. 40 (2001) 3993.
42. K. Hoshino, M. Inui, T. Kitamura, and H. Kokado, Angew. Chem., Int. Ed. Engl. 39 (2000) 2509; K. Hoshino, Chem. Eur. J. 7 (2001) 2727.

43. M. Ichikawa, T. Kondo, K. Kawase, M. Sudo, T. Onishi and K. Tamaru, J. Chem. Comm., Chem. Commun (1972) 176.
44. G. Marnellos and M. Stoukides, Science 282 (1998) 98.
45. V. Kordalis, G. Kyriacou and Ch. Lambert, Chem. Commun. (2000) 1673.
46. G. N. Schrauzer, G. W. Kiefer, T. Kano and P. R. Robinson, J. Am. Chem. Soc. 97 (1975) 6088; G. N. Schrauzer, P. R. Robinson, E. L. Moorhead and T. M. Vickery, J. Am. Chem. Soc. 97 (1975) 7069; G. N. Schrauzer, P. R. Robinson, E. L. Moorhead and T. M. Vickery, J. Am. Chem. Soc. 98 (1976) 2815; E. L. Moorhead, P. R. Robinson, T. M. Vickery and G. N. Schrauzer, J. Am. Chem. Soc. 98 (1976) 6555; S. I. Zones, T. M. Vickery, J. G. Palmer and G. N. Schrauzer, J. Am. Chem. Soc. 98 (1976) 7289, and many other related papers.
47. For a comprehensive review, see A. E. Shilov, Metal Complexes in Biomimetic Chemical Reactions, CRC Press, Boca Raton, NY, 1997;see also N. T. Denisov, V. F Shuvalov, N. I. Shuvalova, A. K. Shilova and A. E. Shilov, Kinet. Katal11 (1970) 813; A. E. Shilov, N. T. Denisov, O. N. Efimov, V. F. Shuvalov, N. I. Shuvalova and A. K. Shilova, Nature 231 (1971) 469; N. T. Denisov, A. G. Ovcharenko, V. G. Svirin, A. E. Shilov and A. E. Shilov, Nouveau. J. Chim. 3 (1979) 403; T. A. Bazhenova, M. A. Bazhenova, G. N. Petrova, A. K. Shilova and A. E. Shilov, Izvestia Akad. Nauk., Ser. Khim. Nauk. (1998) 890.
48. L.A. Nikonova, A. G. Ovcharenko, O. N. Efimov, V. A. Avilov and A. E. Shilov, Kinet. Katal. 13 (1972) 1602.
49. G. N. Schrauzer, N. Strampach, M. R. Palmer and S. I. Zones, Nouveau J. Chim. 5 (1981) 5.
50. L. P. Didenko, O. K. Gavrilina, E. E. Yablonskaya, A. K. Shilova and A. E. Shilov, Nouveau J. Chim. 7 (1983) 605.
51. L. P. Didenko, A. B. Gavrilov, A. K. Shilova, V. V. Strelets, V. N. Tsarev and A. E. Shilov, Nouveau J. Chim. 10 (1986) 583.
52. S. I. Kulakovskaya, L. P. Vasilieva and O. N. Efimov, React. Kinet. Katal. Lett. 14 (1980) 181.
53. M. B. O'Regan, A. H. Liu, W. C. Finch, R. R. Schrock and W. M. Davis, J. Am. Chem. Soc. 112 (1990) 4331; R. R. Schrock, R. M. Kolodziej, A. H. Liu, W. M. Davis and M. G. Vale, J. Am. Chem. Soc. 112 (1990) 4338.
54. G. P. Pez, P Apgar and R. K. Crissey, J. Am. Chem. Soc. 104 (1982) 482.
55. J. Chatt, J. R. Dilworth, G. J. Leigh and R. L. Richards, J. Chem. Soc., Chem. Commun. (1970) 955; J. Chatt, R. C. Fay and R. L. Richards, J. Chem. Soc. A (1971) 702; M. Mercer, R. H. Crabtree and R. L. Richards, J. Chem. Soc., Chem. Commun, (1973) 808.
56. T. E. Glassman, A. H. Liu and R. R. Schrock, Inorg. Chem. 30 (1991) 4723.
57. M. Kol, R. R. Schrock, R. Kempe and W. M. Davis, J. Am. Chem Soc. 116 (1994) 4382.
58. M. B. O'Donoghue, N. C. Zanetti, W. M. Davis and R. R. Schrock, J. Am. Chem. Soc. 119 (1997) 2753.
59. G. E. Greco and R. R. Schrock, Inorg. Chem. 40 (2001) 3861.

60. H.Oshino, T. Nagano, S. Kuwata, Y. Yokobashi, Y. Ishii and M. Hidai, Organometallics 20 (2001) 188.
61. G. J. Leigh R. Prieto-Alcón and J. R. Sanders, J. Chem. Soc., Chem. Commun. (1991) 922.
62. J.-K. Buijink, A. Meersma and J. H. Teuben, Organometallics, 12 (1993) 2004.
63. J. J. H. Edema, A. Meetsma and S. Gambarotta, J. Am. Chem. Soc. 111 (1989) 6878.
64. N. Desmangles, H. Jenkins, K. B. Ruppa and S. Gambarotta, Inorg. Chim. Acta 250 (1996) 1.
65. P. Berno, S. Hao, R. Minhas and S. Gambarotta, J. Am. Chem. Soc. 116 (1994) 7417.
66. W. J. Evans, T. A. Ulibarri and J. W. Ziller, J. Am. Chem. Soc. 110 (1988) 6877.
67. W. J. Evans, N. T. Allen and J. W. Ziller, J. Am. Chem. Soc. 123 (2001) 7928.
68. P. Roussel and P. Scott, J. Am. Chem. Soc. 120 (1998) 1070.
69. J. Jubb and S. Gambarotta, J. Am. Chem. Soc. 116 (1994) 4477.
70. E. Campazzi, E. Solari, C. Floriani and R. Scopelliti, Chem. Commun. (1998) 2603.
71. J. Ho, R. J. Drake and D. W. Stephan, J. Am. Chem. Soc. 115 (1993) 3792.
72. K. Jonas, Angew. Chem., Int. Ed. Engl. 12 (1973) 997; C. Krüger and Y.-H. Tsay, Angew. Chem., Int. Ed. Engl. 12 (1973) 998.
73. K. Jonas, D. J. Brauer, C. Krüger, P. J. Roberts and Y.-H. Tsay, J. Am. Chem. Soc. 98 (1976) 74.
74. H. Shan, Y. Yang, A. J. James and P. R. Sharp, Science 275 (1997) 1460.
75. N. Beydoun, R. Duchateau and S. Gambarotta, J. Chem. Soc., Chem. Commun. (1992) 244.
76. R. Duchateau, S. Gambarotta, N. Beydoun and C. Bensimon, J. Am. Chem. Soc. 113 (1991) 8986.
77. M. D. Fryzuk, T. S. Haddad and S. J. Rettig, J. Am. Chem. Soc. 112 (1990) 8185; M. D. Fryzuk, T. S. Haddad M. Mylvaganam, D. H. McConville and S. J. Rettig, J. Am. Chem. Soc. 115 (1993) 2782; M. D. Fryzuk, S. E. Johnson and S. J. Rettig, J. Am. Chem. Soc. 120 (1998) 11024.
78. J. Jeffery, M. F. Lappert and P. I. Riley, J. Organomet. Chem. 181 (1979) 25.
79. D. V. Fomitchev, K. A. Bagley and P. Coppens, J. Am. Chem. Soc. 122 (2000) 532.
80. J. Chatt, G. A. Heath and G. J. Leigh, J. Chem. Soc., Chem. Commun. (1972) 444.
81. J. Chatt, J. R. Dilworth and G. J. Leigh, J. Chem. Soc., Chem. Commun. (1969) 687.
82. J. Chatt, A. J. Pearman and R. L. Richards, J. Chem. Soc., Dalton. Trans (1976) 1520; J. Chatt, A. J. Pearman and R. L. Richards, J. Chem. Soc., Dalton. Trans (1977) 1852.
83. J. Chatt, A. J. Pearman and R. L. Richards, Nature 253 (1975) 39.
84. C. J. Pickett, K. S. Ryder and J. Talarmin, J. Chem. Soc., Dalton Trans., (1986) 1453.
85. H. Nishihara, T. Mori, Y. Tsurita, K. Nakano, T. Saito and Y. Sasaki, J. Am. Chem. Soc. 104(1982) 4367; H. Nishihara, T. Mori, T. Saito and Y. Sasaki, Chem. Lett. (1980) 667; M. Hidai, T. Takahashi, I. Yokotake and Y. Uchida, Chem. Lett. (1980) 645.
86. For a discussion see G. Jia, R. H. Morris and C. T. Schweitzer, Inorg. Chem. 30 (1991) 593.
87. G. J. Leigh and M. Jimenez-Tenerio, J. Am. Chem. Soc. 113 (1991) 5862.

88. Y. Nishibayashi, S. Takemoto, S. Iwai and M. Hidai, Inorg. Chem. 39 (2000) 5946.
89. V. V. Abalyaeva, N. T. Denisov, M. L. Khidekel and A. E. Shilov, Izvestia Akad. Nauk., Ser. Khim. Nauk. (1975) 2638.
90. Y. Nishibayashi, S. Iwai and M. Hidai, J. Am. Chem. Soc. 120 (1998) 10559; Y. Nishibayashi, I Wakiji, K. Hirata, M. Rakowski Du Bois and M. Hidai, Inorg. Chem. 40 (2001) 578.
91. J. Chatt, A. A. Diamantis, G. A. Heath, N. E. Hooper and G. J. Leigh, J. Chem. Soc., Dalton Trans. (1977) 688; J. Chatt, R. A. Head, G. J. Leigh and C. J. Pickett, J. Chem. Soc., Dalton Trans. (1979) 1638.
92. P. C. Bevan, J. Chatt, M. Hidai and G. J. Leigh, J. Organomet. Chem. 160 (1978) 165; P. C. Bevan, J. Chatt, G. J. Leigh and E. G. Leelamani, J. Organomet. Chem. 139 (1977) C59.
93. R. Ben-Shoshan, J. Chatt, W. Hussain and G. J. Leigh, J. Organomet. Chem. 112 (1976) C9.
94. H. M. Colquhoun, Trans. Met. Chem. 6 (1981) 57.
95. M. Hidai, K. Komori, T. Kodama, D.-M. Jin, T. Takahashi, |S. Sugiura, Y. Uchida and Y. Mizobe, J. Organomet. Chem. 272 (1984) 155; K. Mori, S. Sugiura, Y. Mizobe, M. Yamada and M. Hidai, Bull. Chem. Soc. Jpn. 62 (1989) 2953; H. Oshita, Y. Mizobe and M. Hidai, Organometallics, 11 (1992) 4116.
96. M. B. O'Donoghue, W. M. Davies and R. R. Schrock, Inorg. Chem. 37 (1998) 5149.
97 G. Greco and R. R. Schrock, Inorg. Chem, 40, (2001) 3861.
98. M.B. O'Donoghue, W. M. Davis, R. R. Schrock and W. M. Reiff, Inorg. Chem. 38 (1999) 243.
99. H. M. Colquhoun, J. Chem. Res. (S) (1979) 325.
100. D. Sellman and W. Weiss, Angew. Chem, Int. Ed. Engl. 17 (1978) 269.
101. H. Basch, D. G. Musaev and K. Morokuma, J. Am. Chem. Soc. 121 (1999) 5754.
102.H. Basch, D. G. Musaev, K. Morokuma, M. D. Fryzuk, J. B. Love, W. W. Seidel, A. Albinati, T. F. Koetzle, W. T. Klooster, S. A. Mason and J. Eckert, J. Am. Chem. Soc. 121 (1999) 523.
103. M. D. Fryzuk in Modern Coordination Chemistry,The Legacy of Joseph Chatt, G. J. Leigh and N. Winterton eds., Royal Society of Chemistry, Cambridge, 2002.
104. C. E. Laplaza, M. J. A. Johnson, J. C. Peters, A. L. Odom, E. Kim, C. C. Cummins, G. N. George and I. J. Pickering, Inorg. Chem. 118 (1996) 8623.
105. D. J. Mindiola, K. Meyer, J.-P. F. Cherry, T. A. Baker and C. C. Cummins, Organometallics, 19 (2000) 1622.
106. E. Solari, C. Da Silva, B, Iacono, J. Hesschenbrouck, C. Rizzoli, R. Scopelliti and C. Floriani, Angew. Chem. Int. Ed. Engl. 40 (2001) 3907.
107. E. Solari, J. Hesschenbrouck, R. Scopelliti, C. Floriani and N. Re, Angew. Chem. Int. Ed. Engl. 40 (2001) 922.
108. A. Zanotti-Gerosa, E. Solari, L. Giannini, C. Floriani, A. Chiesi-Villa and C. Rizzoli, J. Am. Chem. Soc. 120 (1998) 437; A. Caselli, E. Solari, R. Scopelliti, C. Floriani, N. Re, C. Rizzoli and A. Chiesi-Villa, J. Am. Chem. Soc. 122 (2000) 3652.

109. G. K. B. Clentsmith, V. M. E. Bates, P. B. Hitchcock and F. G. N. Cloke, J. Am. Chem. Soc. 121 (1999) 10444.
110. P. Berno and S. Gambarotta, Angew. Chem. Int. Ed. Engl. 34 (1995) 822.
111. J. M. Smith, R. J. Lachicotte, K. A. Pittard, T. R. Cundari, G. Lukat-Rodgers, K. R. Rodgers and P. L. Holland, J. Am. Chem. Soc. 123 (2001) 9222; D. Sellmann. B. Hautsc, A. Rösler and F. W. Heinemann, Angew. Chem. Int. Ed. Engl. 40 (2001) 1505.
112. R. J. Deeth, J. Organomet. Chem. 635 (2001) 165; Y. Chen, M. Hartmann and G. Frenking, Eur. J. Inorg. Chem. (2001) 1441.

Nitrogen Fixation at the Millennium.
G. Jeffery Leigh, Editor.

CHAPTER 12

Chemical Models for Nitrogenase

R. L. Richards
School of Chemistry, Physics and Environmental Science, University of Sussex, Brighton BN1 9QJ, East Sussex, UK.

1. INTRODUCTION

In this Chapter we will be concerned with chemical systems that been designed to reproduce (model) aspects of the structure and function of the nitrogenase enzymes. The X-ray crystal structural data for the active metal centres of molybdenum nitrogenase are summarised below in Section 2, and in more detail in other chapters in this book (see Chapter 2). However, the scope of this Chapter encompasses work that predates the determination of the structure of molybdenum nitrogenase, as well as those studies made subsequently. This early work was based upon knowledge of the biochemistry of nitrogenase, which was accumulated in depth particularly after the preparation of cell-free extracts of the molybdenum enzyme in the 1960's [1]. Some 20 years later, biochemical and genetic studies confirmed early evidence for an alternative nitrogenase with vanadium, rather than molybdenum, at the active centre [2], and also a third system in which iron replaces molybdenum [3]. More recently, a radically different nitrogenase has been isolated from *Streptomyces thermoautrophicus.* This also contains molybdenum, but it requires dioxygen to function and consumes carbon monoxide [4].

The overall aim of chemical modelling is to synthesise a catalyst that reproduces the function of nitrogenase without the necessity for the protein structure of the enzyme. A natural consequence of the pursuit of this aim has been the development of an understanding of the general coordination chemistry of dinitrogen and the development of metal-sulfur coordination chemistry, especially that which involves cluster systems based on iron. More recent has been the use of computational methods to develop theoretical models to aid in understanding the electronic structure and chemical behaviour of the cluster components of the active centre of nitrogenase [5].

1.1. Early biomimetic model systems

Early attempts to reproduce nitrogenase function were based on the biochemcal knowledge of the period. At that time, essentially the decade 1970-1980, it was known that nitrogenase contained an active centre made up from non-haem iron, together with molydenum or vanadium and sulfur ligands, including sulfide and cysteine [6,7]. In order to function, the system required the presence of magnesium and ATP, together with a reducing agent such as sodium dithionite. It was also established that nitrogenase was a versatile reducing agent, able to use electrons (supplied from dithionite in most nitrogenase preparations) to reduce dinitrogen according to equation (1).

$$N_2 + 8H^+ + 8e^- \rightarrow 2NH_3 + H_2 \qquad (1)$$

A number of other unsaturated molecules such as ethyne and methyl isocyanide can also be reduced by nitrogenases, molybenum nitrogenase being the most thoroughly studied enzyme in this respect [8].

Two research groups developed biomimetic systems which were reported to reduce N_2 in aqueous or aqueous/alcoholic conditions, but no intermediate compounds that contained N_2, or one of its reduced derivatives, were isolated. The first group, led by Schrauzer [9,10], used a number of systems, based on molybdenum, which were effective in reducing ethyne to ethene (a classic reaction of molybdenum nitrogenase [11]) but which, in general, gave very low yields of NH_3 or N_2H_4, especially when thiolates were present. The most successful system used the complex $K_2[MoO(CN)_4(H_2O)]$ together with $NaBH_4$, ATP and the cluster $[Fe_4S_4(SPr^i)_4]^{2-}$ in borate buffer. This mixture was reported to give 0.29 mol of NH_3 per Mo, but other workers were unable to reproduce this observation [10,12].

Aqueous or alcoholic systems that reduce N_2 to NH_3 and/or N_2H_4 more successfully were developed by Shilov and co-workers [12,13]. Two types of system were studied. One type is exemplified by an aqueous suspension of V^{II} hydroxide in presence of magnesium hydroxide at pH 14.7, which reduces N_2 to NH_3 and N_2H_4 in 70% yield based on V^{II}. The second type of system that was developed is exemplified by a solution of a V^{II} catecholate in methanol at pH 10, which gives NH_3 in 75% yield [12,13]. Intermediate species containing N_2 bridged between two vanadium centres have been proposed for these systems, but have not been isolated [14].

1.2. Dinitrogen chemistry

Work carried out in parallel with the biomimetic studies was designed to understand the fundamental chemistry of dinitrogen with transition metals such as iron, molybdenum and vanadium, since these are the metals associated with the reactive centres of nitrogenase. The development of dinitrogen chemistry is described in detail in Chapter 11 and has been extensively reviewed [15-21]. This chapter will merely point out the types of reactivity that have been observed, since this knowledge has been an important guide in the design of the synthesis and study of the chemical reactivity of metal complexes which might model the reactions of dinitrogen and alternative substrates with nitrogenase.

There is a variety of ways in which dinitrogen can bind to metals, depending upon the sites involved. These are summarised in Figure 1. The predominant modes of binding are end-on to a metal, 1(a), or linearly bridged between two metal centres, 1(b). The variety of binding therefore allows plenty of scope when constructing models, theoretical of empirical, for the possible binding of dinitrogen to the cofactor clusters. Nevertheless, it has been extremely challenging to provide convincing synthetic models for dinitrogen bound to the metals at the active centre of nitrogenase, as we shall see later in this Chapter.

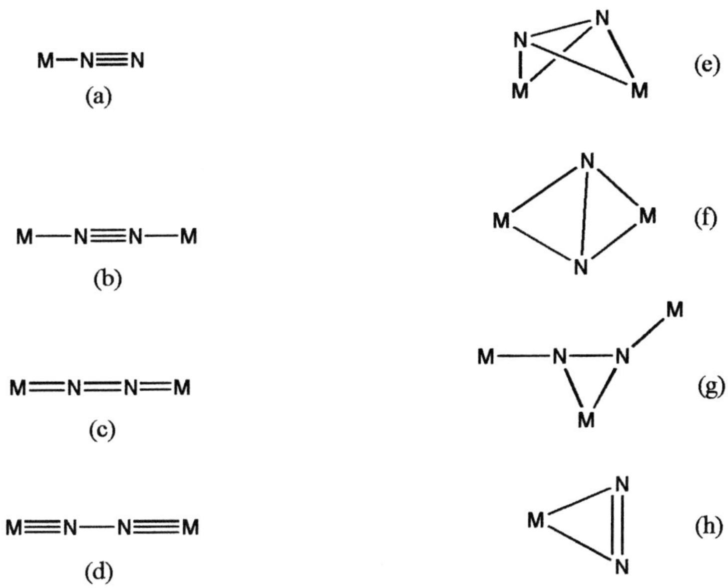

Figure 1. Modes of binding of N_2 to metal centres.

As well as the preparation of dinitrogen complexes, extensive studies of their reactions, particularly to give ammonia and/or hydrazine, pre-dated structural knowledge of nitrogenase centres. In some cases the studies were particularly thorough, in particular a variety of metal complexes that contained species *en route* from dinitrogen to ammonia or hydrazine were isolated, allowing a detailed knowledge of the mechanism of these conversion reactions to be built up, especially when kinetic studies were also undertaken [15-20]. This type of investigation is described in detail in Chapter 9, but a cyclic scheme of N_2 reduction at molybdenum (or tungsten) is shown below in Scheme 1 [20], because the intermediate species shown have been used as target moieties in developing models for the nitrogenase function.

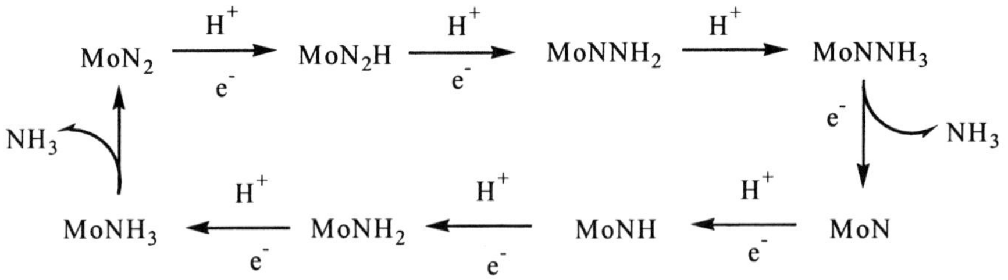

Scheme 1. Cycle of reduction of N_2 at a molybdenum site, typically $\{Mo(Ph_2PCH_2CH_2PPh_2)_2\}$.

A wide variety of ligands have been used as co-ligands with dinitrogen [18,19], these include amines, amides, porphyrins, phosphines, halides, hydride, cyclopentadienyl, arene, alkyl and aryloxide. More pertinent to the nitrogenase environment are sulfur-donor ligands such as dithiocarbamates [22], thioethers [23] and thiolates [24,25,26]. Complexes of the sulfur-donor group of ligands will be discussed in some detail in this Chapter (Section 8).

Although models which involve dinitrogen are most appropriate in terms of understanding nitrogenase function, nevertheless the behaviour at metal sites of the alternative substrates of nitrogenase such as alkynes, and of inhibitors such as CO, continues to provide important information with which to judge the overall pattern of reductions carried out by this enzyme [20].

2. THE STRUCTURES OF THE CLUSTERS OF NITROGENASE

The three known types of nitrogenase consist of two readily separable, oxygen-sensitive proteins. One is a relatively small α_2-type protein which contains a single $\{Fe_4S_4(cysteinyl)_4\}$ and acts as an electron donor to the second, larger $\alpha_2\beta_2$-type protein [27]. The single iron cluster has the well-known cubane structure (Figure 2) [28,29] as shown below. The second protein contains two types of cluster, both of which have structures that are probably unique to nitrogenases. The X-ray structures of these clusters have been obtained for molybdenum nitrogenase [28,30,31]. They are named, essentially for historical reasons, the iron-molybdenum cofactor (FeMoco) and the P-cluster. They are described in detail in Chapter 2 and are summarised very briefly here for convenience.

2.1. The single iron cluster

This cluster links the two subunits of the small α_2-type protein via the cysteinyl groups which are ligands to the irons of the Fe_4S_4(cysteinyl)$_4$ cluster unit. Two cysteines originate from one subunit and two from the other. The structure of the cluster is the distorted cube typically found in this type of cluster, which is widespread in biology [28].

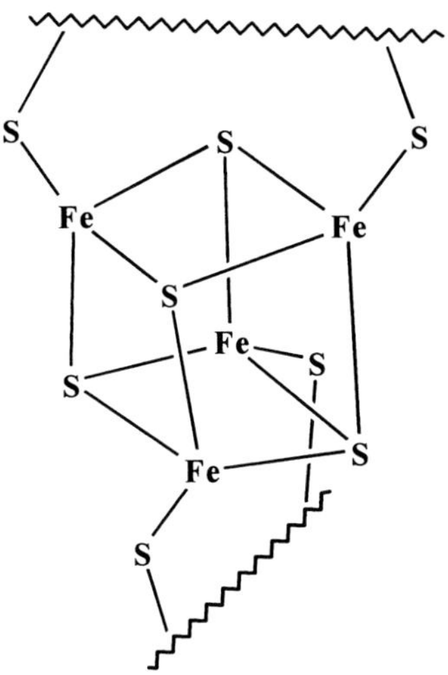

Figure 2. The single iron/sulfur cluster of the small iron protein of nitrogenase.

2.2. The FeMoco cluster structure

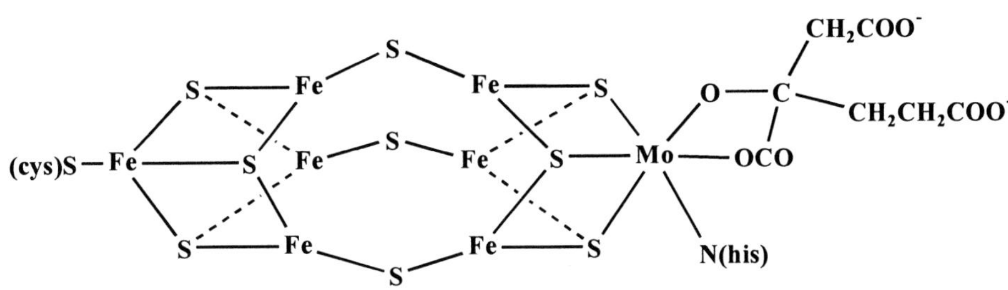

Figure 3. Structural model of FeMoco as deduced from X-ray crystallography.

The structure proposed for this cluster, based on X-ray crystallographic studies, is essentially the same whether it is from *Azotobacter vinelandii* [28,30,31], *Klebsiella pnuemoniae* [32] or *Clostridium pasteurianum* [33], as shown in Figure 3. It consists essentially of two partial cubic clusters, one based on the Fe_4S_3 composition and one based on the $MoFe_3S_3$ composition, bridged by three sulfide ligands. In addition, the terminal Fe carries a cysteinyl ligand and the terminal Mo carries a histidine ligand, together with an *R*-homocitrate bound through the hydroxyl and carboxylate oxygen atoms. Thus the molybdenum is 6-coordinate, the terminal iron 4-coordinate and the central irons appear to be only 3-coordinate. The overall stoichiometry is $MoFe_7S_9$(*R*-homocitrate). It should be pointed out that although the structure of Figure 3 is used as a basis for chemical model studies, it itself is a structural model which has been constructed to fit electron-density maps, using data for synthetic Fe/S and Fe/Mo/S clusters as a basis for the fitting. These synthetic clusters will be described in some detail in later sections of this article, together with the efforts to synthesise models, both for the whole structure of FeMoco outside the protein and also for various components of the structure, such as the apparently three-coordinate iron centres.

2.3. The P-cluster structures

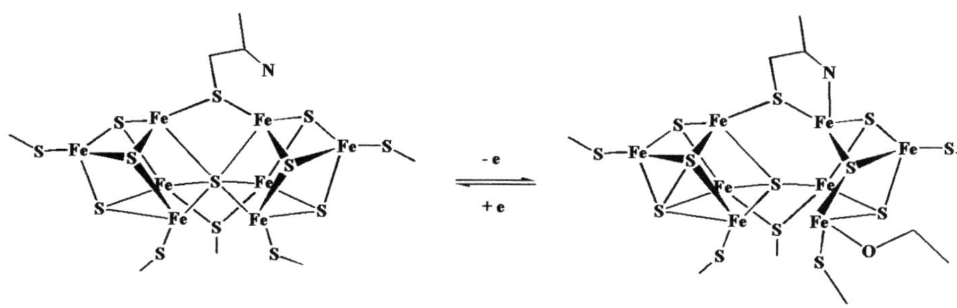

(a) (b)

Figure 4. Structures of the P-clusters deduced from X-ray crystallogaphy: (a) reduced form; (b) oxidised form.

The P-clusters are held via cysteine ligands at the interface of the α and β subunits of the larger protein of nitrogenase and their structure has been the subject of some controversy [30,34]. The resolution of this debate and the details of the structures are dealt with in Chapter 2, it therefore suffices for the purposes of this article to show in Figure 4 the currently accepted picture [32,35,36].

A P-cluster can exist in a dithionite-reduced form [Figure 4(a)] where it consists essentially of two Fe_4S_4-cubanes that are fused at one corner, giving rise to a central sulfur atom that has a distorted octahedral coordination to six iron atoms. Three cysteine side-chains from each subunit bind the assembly to the protein and two of these cysteines act as internal bridges in the cluster. Upon two-electron oxidation, half the P-cluster adopts a more open structure in which two iron atoms move away from the central sulfur atom, cleaving their bonds to it. To replace these bonds, one iron atom binds to the backbone amide nitrogen of a bridging cysteine and the other to a serine hydroxyl, as indicated in Figure 4 (b). It may be that during the function of nitrogenase, electrons (together with protons) are transferred to FeMoco *via* the P-cluster and the above structural changes to the P-cluster are important in this process. This, however, remains to be established [32,36].

3. THEORETICAL STUDIES

The use of mathematical models has an increasing role to play in nitrogen fixation studies. This has arisen, not only because of the growing availability of sophisticated computational techniques that can be applied to model systems of choice, but also because the complex nature of the cluster sites of nitrogenase, seen above, has posed problems which are particularly difficult to address by conventional synthetic modelling techniques. Therefore, in parallel with synthetic studies, theoretical methods have been adopted, particularly to attempt to decide where and how on FeMoco (and the related V and Fe cofactors) N_2 might bind and how it might be reduced to ammonia.

Because the molybdenum atom in FeMoco is six-coordinate and vanadium and iron-only versions of FeMoco exist, the question was raised as to whether the binding of N_2 could be at the iron atoms of FeMoco, rather than at Mo or V. When the structure of FeMoco was first established, it was proposed that N_2 was likely to bind the unsaturated Fe atoms in the centre of the trigonal cavity of FeMoco [30]. This suggestion, although apparently supported by later INDO (Intermediate Neglect of Differential Overlap) calculations [37], did not appear to be likely since the cavity is probably too small to accommodate N_2 and certainly too small to hold a protonated derivative [38]. Moreover, the cavity appears to decrease in volume upon one-electron reduction of FeMoco [39]. Deng and Hoffmann used extended-Hückel molecular orbital (EHMO) calculations to investigate N_2 binding to a FeMoco model framework based on the X-ray structural data and favoured a side-on bridging position between two iron atoms as in Figure 5(a), although the end-on positions to two [Figure 5(b)] or four [Figure 5(c)] of the central Fe atoms were also considered possible [40]. Subsequently the more sophisticated technique of density function theory (DFT) has been applied, generally to sub-sections of the FeMoco structure because the large size of the cofactor would cause data-handling problems.

Early use of the DFT method had laid groundwork in that it had been established that the bonding and reactivity of metal-N_2 compounds could be described by ground-state electron-density distributions [41]. Subsequent studies then concentrated upon the Fe atoms as the potential binding sites of N_2; the binding of H_2 and CO were also considered. Dance [42] concluded that N_2 is bound and activated at four of the central Fe atoms in an asymmetric mode of binding [Figure 5(d)]. Siegbahn *et al.* [43] considered that N_2 could bind a single central Fe, although the binding energy was lower than for N_2 binding four central Fe atoms. In a very thorough DFT study of the central section of FeMoco, Rod *et al.* [44] examined a wide range of possibilities for N_2 binding and concluded that end-on binding to a single iron atom was the only stable mode [Figure 5(e)], varying only slightly with the redox level of FeMoco, although at

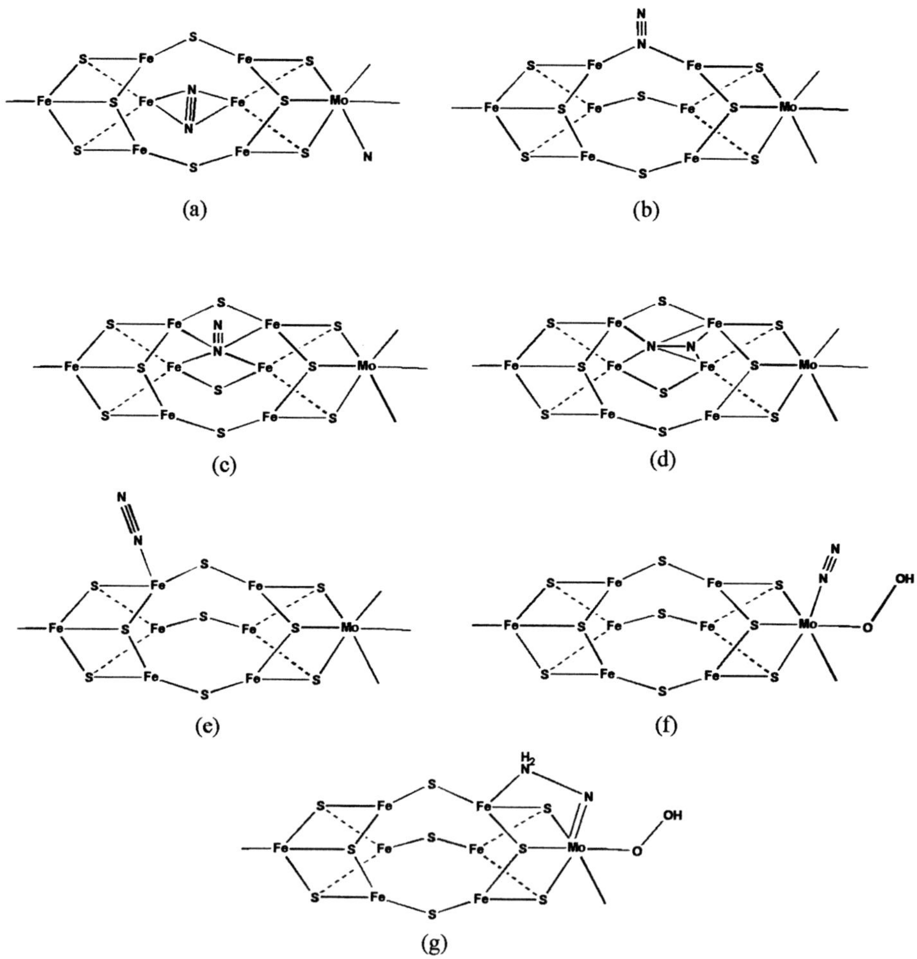

Figure 5. Models for N_2 binding at FeMoco derived from theoretical studies.

least one electron must be added to the cluster before binding can occur. A general conclusion for all studies to this point was that CO binds more strongly than N_2, in keeping with practical observation, and that H_2 binding is facile and can occur at the same site as N_2.

Although the above studies suggest that N_2 is activated at one or more of the central Fe atoms, there is considerable chemical precedent for the reduction of N_2 at molybdenum (Scheme 1) and also that Mo can lose a carboxylate ligand and bind N_2 upon reduction [45]. These observations prompted DFT (hybrid B3LYP) calculations based on the possibility that Mo could be the site of activation. Thus Morokuma *et al.* [46] concluded that decoordination of one of the homocitrate 'legs' (see Figure 3) could occur, but only upon substrate coordination to Mo in the singly-protonated form of the complex. Thus at least one electron (to reduce the system), one proton (to protonate a carboxylate leg), and N_2 are needed for the bidentate → monodentate rearrangement of the homocitrate ligand of FeMoco [see Figure 5(f)]. Durrant [47] also considered the possibility of opening the homocitrate ring at the Mo and concluded that this would give a coordination site that, for both Mo and V, has a higher affinity for N_2 than that of the trigonal iron site(s). Even for the iron-only system, the terminal iron shows a comparable N_2 affinity to that of the trigonal Fe. He suggested further that, following N_2 binding at Mo and its partial reduction, the unique structure and flexibility of FeMoco allows cleavage of a partially-reduced N_2 species across an $MoFeS_2$ face of the cluster [see Figure 5 (g) and Scheme 1].

Thus theoretical models have not given a definitive single possibility for the binding of N_2 to FeMoco. Nevertheless, the conclusions so far act as a stimulus for targeted chemistry to provide new information with which to further refine theoretical models.

4. SYNTHESIS OF MODELS FOR THE SINGLE FOUR-IRON CLUSTER

The $\{Fe_4S_4\}$ cluster core occurs in a wide range of biological systems and many analogues containing it, such as $[Fe_4S_4(SR)_4]^{n-}$ (*n* is commonly 2 or 3) have been obtained by synthetic chemical routes. These are, of course, structural models for the single $\{Fe_4S_4(cysteinyl)_4\}$ cluster of the Fe-protein of nitrogenase. The area has been thoroughly reviewed and it will not be discussed further in this article [48].

5. SYNTHESIS OF MODELS FOR FeMoco AND COFACTORS OF THE ALTERNATIVE NITROGENASES

The structure of FeMoco (Figure 3), has a number of unusual features, as noted above. Therefore devising a synthetic route to the complete cluster structure outside the protein has presented a particularly daunting task, and the major effort has been expended upon the synthesis of fragments of the total structure. Indeed, some models of subunits of the FeMoco structure preceded the X-ray structural determination of FeMoco and in fact were used as input data for the solution of the crystallographic analysis, as pointed out in Section 2.1.This discussion will therefore be subdivided between the various subunit studies.

5.1. Synthesis of models for the Fe_4S_3 unit of FeMoco

The terminal Fe atom of FeMoco may be regarded as the apex of an Fe_4S_3 unit which is then attached to the three bridging sulfur atoms. The terminal Fe has tetrahedral geometry by virtue of its attachment the three bridging sulfides and the cysteine link to the protein [32,33]. A model for this type of Fe is to be found in the cluster $[Fe_4S_4(Cys\text{-}OMe)(L)]^{2-}$ [48], which carries the tridentate thiolate ligand shown in Figure 6 and is designated a 'site-differentiated' cluster. Clusters containing the Fe_4S_3

core occur biologically and can be converted into Fe_4S_3 centres [49,50]. There are few synthetic models for the Fe_4S_3 core although some, such as Roussin's black anion, $[Fe_4S_3(NO)_7]^-$ have been known for many years, together with $[Fe_4S_3(NO)_7]^{2-}$ [50] and $[Fe_4S_3(NO)_4(PPh_3)_3]^{n+}$ (n = 0 or 1) [51]. The presence of the NO ligand in these clusters hampered attempts to use them as synthetic buildingblocks for FeMoco models. The clusters $[MFe_4S_6(PEt_3)_4X]$ (M = Mo or V; X = Cl or SPh) [52] also contain a structurally defined Fe_4S_3 unit and are discussed later in section 5.3.

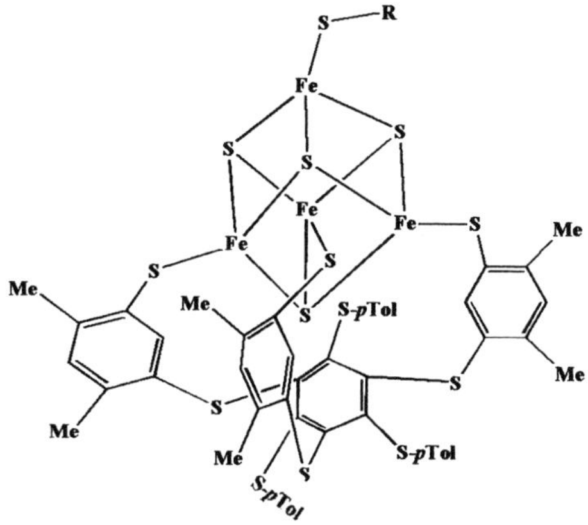

Figure 6. Site-differentiated Fe_4S_4 cluster, SR = cysteine derivative.

5.2. Synthesis of models for the central, trigonal iron atoms of FeMoco

The geometry of the six apparently three-coordinate iron atoms at the centre of the FeMoco cluster might be described as tending towards trigonal pyramidal, rather than trigonal planar, since each of the Fe atoms is displaced below the S_3 plane and towards the centre of the cluster by a mean distance of around 0.051 nm [31,54]. There have not been reports of characterised three-coordinate iron complexes with sulfide ligands, presumably because of the tendency for sulfide readily to act as a bridging ligand to produce polynuclear species.

Synthetic chemists have therefore turned to thiolate ligands to synthesise low-coordinate S-ligated iron, because although thiolate can also readily act as a bridging ligand, this tendency can be attenuated by the use of bulky substituents attached to sulfur. Even then, the high reactivity of low-coordinate iron can result in its attaining a higher coordination number by interaction with substituents on the thiolate ligands. Thus the compounds $[\{Fe(SR)_2\}_2]$ (R = $C_6H_2Bu^t$-2,4,6 or $C_6H_2Ph_3$-2,4,6) were originally thought to contain three-cooordinate iron [55,56], but further examination of the X-ray structural and Mössbauer spectroscopic data for these compounds [57] indicated that the coordination number of their iron is not three but four, so that they have distorted tetrahedral geometry. The fourth coordination position is occupied by hydrogen from the aryl substituents, or the π-electron cloud of phenyl rings on the thiolate, which forms an agostic interaction with the iron. However, in the complex anion $[Fe(SC_6H_2Bu^t_3\text{-}2,4,6)_3]^{3-}$ no agostic interaction exists and this compound is therefore an unambiguous example of a homoleptic three-coordinate iron complex [56] (Figure 7).

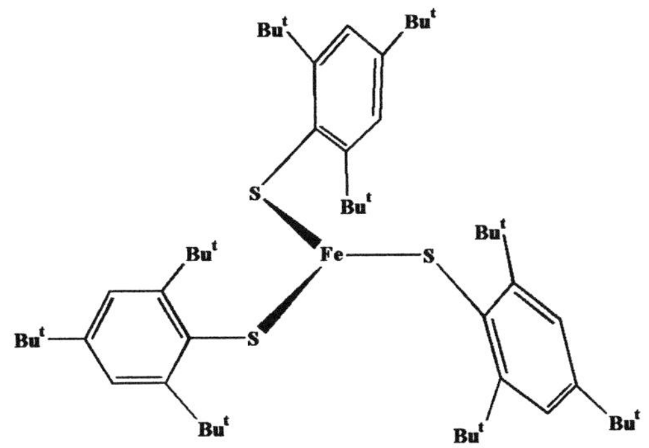

Figure 7. Structure of the anion $[Fe(SC_6H_2Bu^t_3\text{-}2,4,6)_3]^{3-}$.

In an extension of the Fe_3S_4 chemistry described in Section 5.1, Holm has attempted to force the Fe_3S_4 unit to adopt a more planar geometry by the use of bulky trialkylphosphine ligands. This procedure did result in complexes in which an Fe has moved down into the Fe_3 plane, but the coordination number of Fe remained four and it proved difficult to remove ligands, such as NO, to generate lower coordination numbers. Examples of compounds displaying this type of structural feature include $[Fe_4S_3(NO)_4(PR_3)_3]$ (R = Et or Ph) [51,54], $[MFe_4S_6(PEt_3)X]$ (M = Mo or V; X = Cl or Ph) [52,53], $[Fe_6S_6(PBu^n_3)_3Cl_3]$ [58,59], $[Fe_6S_6(PEt_3)_6]^+$ [60] and $[Fe_7S_6(PEt_3)_4Cl_3]$ [61].

5.3. Synthesis of models for the $\{Fe_3S_3Mo(homocitrate)(histidine)\}$ unit

Although the synthetic work described in this section models structural features of the immediate environment of the molybdenum in FeMoco, much of it was in fact carried out a considerable time before the structure of FeMoco was determined. At that time, biochemical studies, including EXAFS data, had shown that the active centre of nitrogenase contained molybdenum (or vanadium), iron and sulfide together with homocitrate, most likely in a cluster array. Cluster complexes had been synthesised which contained these features, typical examples being $[MoFe_3S_4X_3(cat)(solvent)]^{2-}$ (X = Cl or SR; cat = substituted catecholate) and $[VFe_3S_4(dimethylformamide)_3]^-$ [48,62,63,64], as exemplified in Figure 8(a) and Figure 8(b), respectively. As noted in Section 2.1, X-ray parameters obtained for this class of molybdenum compound were subsequently used in the data fitting process to solve the X-ray structure of FeMoco.

Following knowledge of the structure of FeMoco, further refinement of these models was made to modify the Mo coordination sphere so that it resembled more closely the

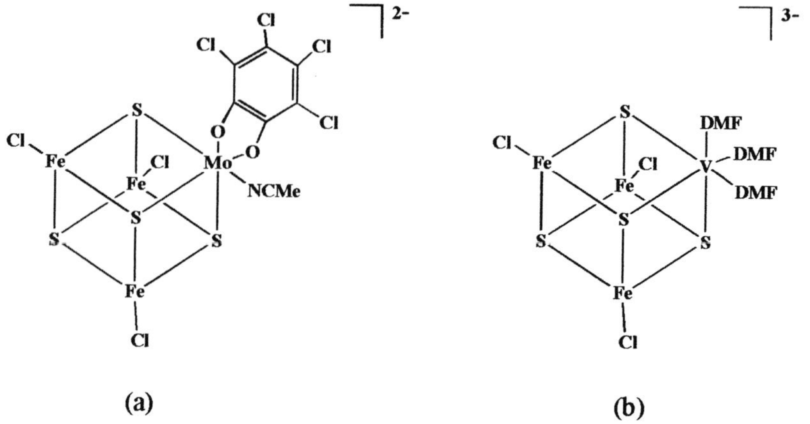

Figure 8. Heteronuclear cubane clusters.

S_3O_2N environment found in FeMoco. These modifications were achieved mainly by treatment of $[NEt_4]_2[MoFe_3S_4Cl_3(Cl_4\text{-cat})(MeCN)]$ (Cl_4-cat = tetrachlorocatecholate-2,3,4,5), Figure 8(a), with various reagents [65]. For example, its reaction with imidazole or histidine dimethyl ester gave products in which the Mo is coordinated by two oxygens of the catecholate and a nitrogen of histidine or histidine [66]. Reaction with $[NEt_4]$(citrate) gave $[NEt_4]_2[MoFe_3S_4Cl_3(\text{citrate})]$ [67] which may be considered as a structural model for the molybdenum site in the nifV modification of FeMoco, in which citrate replaces homocitrate [68]. The product of reaction of (methylimino)diethanoic acid has the Mo bound by two carboxylates and a nitrogen atom, close to the required environment to model FeMoco (Figure 9) [67], but a cluster with homocitrate and histidine (or a close analogue such as imidazole) bound to Mo is still to be attained. The further reactions of this class of cluster, which were designed to test the ability of such complexes to model the functional behaviour of FeMoco, are discussed in Section 7.

Although the above-mentioned clusters model the Mo environment reasonably well, they contain the Fe_3S_4Mo subunit, whereas the Fe_3S_3Mo subunit is required. This unit has been obtained by a further extension of the chemistry of the heteronuclear cubane systems discussed above. In this case, treatment of $[NEt_4]_2[MoFe_3S_4Cl_3(Cl_4\text{-cat})(MeCN)]$ with PR_3 (R = Et or PPr^n_3) in the presence of $NaBPh_4$ or $NaPF_6$, gave the fused, double cubane clusters $[\{MoFe_3S_4(Cl_4\text{-cat})(PR_3)_3\}_2]$ [69,70]. These, when treated with CO, either with or without pyridine (py), gave the clusters $[MoFe_3S_3(Cl_4\text{-cat})(CO)_6(PEt_3)_3]$ [69], $[MoFe_3S_3(Cl_4\text{-cat})(CO)_6(py)(PEt_3)_2]$ [70] and $[MoFe_3S_3(Cl_4\text{-cat})(CO)_4(py)(PPr^n_3)_3]$ [70]. Relative to $[NEt_4]_2[MoFe_3S_4Cl_3(Cl_4\text{-cat})(MeCN)]$, these complexes are reduced by one electron. The lower oxidation level is achieved by the reducing action of the phosphine and is stabilised by the incorporation of CO and PR_3 ligands.

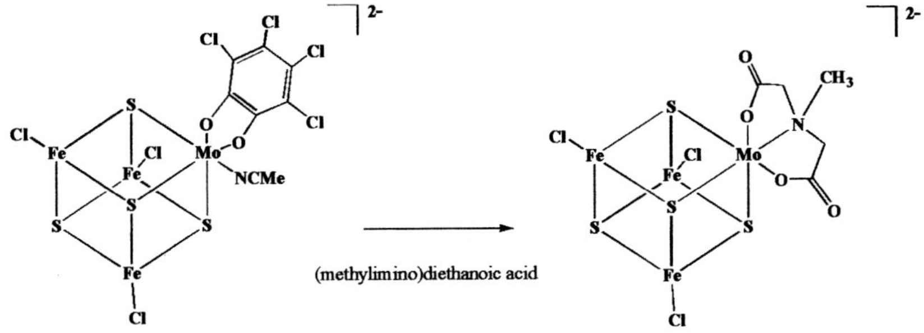

Figure 9. Incorporation of (methylimino)diethanoate into an Fe/Mo cluster.

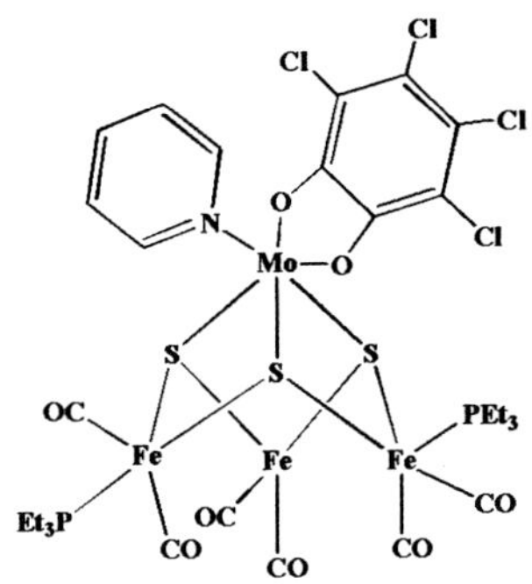

Figure 10. Outline structure of [$MoFe_3S_3(Cl_4$-cat)$(CO)_6$(py)$(PEt_3)_2$].

Figure 10 shows an outline structure of the complex [$MoFe_3S_3(Cl_4$-cat)$(CO)_6$(py)$(PEt_3)_2$] and it can be seen that the structure of the Fe_3S_3Mo unit resembles that of the compounds [$Fe_4S_3(NO)_4(PR_3)_3$] (R = Et or Ph) [51,54] (Section 5.1), which contain a cuboidal core which has one corner sulfide missing. The other compounds of this series have a similar core. Further chemistry of these complexes is awaited.

6. SYNTHESIS OF STRUCTURAL MODELS FOR THE P-CLUSTERS

Early attempts to synthesise models for the P clusters were confused by the uncertainty of the structural determination described in Section 2.2. The original report that the P cluster might contain two Fe_4S_4 units linked by a disulfur bridge led to a number of syntheses of Fe-clusters linked by sulfido-bridges [27], such as the rhomb-bridged double cubane [{$MoFe_3S_2$(μ-S)(Cl_4-cat)$(PEt_3)_3$}$_2$] [69, 71]. The subsequent definitive conclusion, that the reduced P-cluster contains Fe_4S_3-units linked by a μ_6-S atom, then spurred new efforts to find synthetic routes to appropriate structures containing the μ_6-S link. This has proved to be a very difficult synthetic problem and as yet only one such structural unit has been found, and even this is built using Mo-containing sub-units in two complex anions that have complicated structures [72,73].

The first complex anion, $[Mo_6Fe_{20}S_{30}(PEt_3)_6(Cl_4\text{-cat})_6]^{8-}$ is formed by treatment of $[\{MoFe_3S_2(\mu\text{-}S)(Cl_4\text{-cat})(PEt_3)_3\}_2]$ with $[Et_4N]SH$. The structure of this compound can be described as having three fragments, of which the central one, $Mo_2Fe_8S_{14}$, is connected by six μ_3-S-Fe bridges to symmetry-related upper and lower fragments which have the formula $Mo_2Fe_6S_9(Cl_4\text{-cat})_2$ and whose structures are as indicated in Figure 11. This structural unit, with its μ_6-S bridge, can be regarded as having a resemblance to the $Fe_8S_7(SR)_2$ portion of the reduced P-cluster (Figure 4a) where $Fe^{II}(SR)$ is replaced by $Mo(Cl_4\text{-cat})(PEt_3)$ and the terminal and bridging SR units are replaced by S atoms. A second compound which contains this structural fragment, $[Mo_4Fe_{12}S_{20}(PEt_3)_4(Cl_4\text{-cat})_4K_3(DMF)]^{5-}$, has also been prepared [69,70].

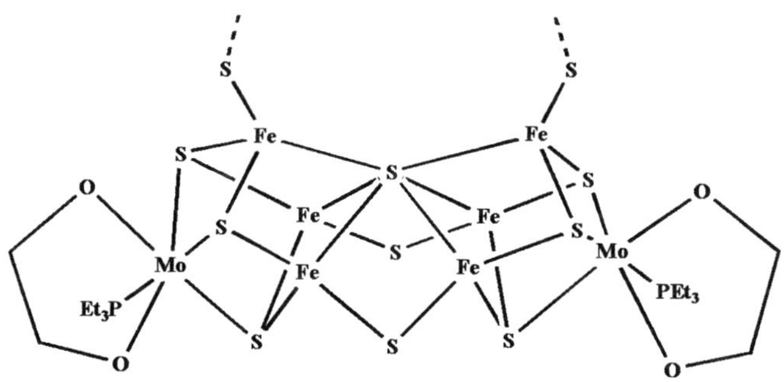

Figure 11: The $Mo_2Fe_6S_9(Cl_4\text{-cat})_2$ structural unit.

7. REACTIVITY OF Fe-S-CLUSTERS INCLUDING FeMoco ITSELF

7.1. Reactions of clusters with dinitrogen

There are no examples of characterised clusters to which N_2 is bound. Nevertheless, reduced metal clusters have been used to reduce N_2 by two techniques, though in neither case are details of the mechanism available. In the first technique, Fe-S clusters were used as electron-transfer agents from a reducing agent, such as BH_4^-, to undefined molybdenum complexes at which the reduction of N_2 occurred (see also Section 1.1) [10,12]. The second technique achieved reduction of N_2 [74] and N_2H_4 [75] by controlled potential electrolysis of the clusters $[Fe_4S_4(SPh)_4]^{2-}$ (by one or two electrons) and $[\{MoFe_3S_4(SPh)_3\}_2(\mu\text{-}SPh)_3]^{3-}$ (by one electron) in protic solvents. Although N_2 reduction was observed, the current efficiency was low and the total yield of NH_3 based on the cluster was 195% over 4 days. It appeared that most of the electrons generated at the electrode were used to produce H_2 from the solvent. No intermediate clusters containing N_2-derived species were isolated from these reaction mixtures.

7.2. Reactions of clusters with hydrazine

N_2H_4 is a substrate of nitrogenase. It has also been observed as a product when the molybdenum enzyme is quenched under turnover conditions [76] and in the reduction of N_2 to NH_3 by the vanadium enzyme [77]. These observations have prompted studies of the generation of N_2H_4 from reactions of reduced clusters with N_2 and the reduction of N_2H_4 by cluster systems. As expected on thermodynamic grounds, it is much easier to reduce N_2H_4 than N_2 [19].

As mentioned in Section 7.1, controlled potential electrolysis of $[Fe_4S_4(SPh)_4]^{2-}$ and of $[\{MoFe_3S_4(SPh)_3\}_2(\mu\text{-}SPh)_3]^{3-}$ in protic solution at pH 12, in presence of N_2H_4, produced NH_3. The molybdenum cluster was the more effective reagent and the maximum current efficiency in MeOH/tetrahydrofuran was 97% [75].

Coucouvanis *et al.* carried out a more extensive investigation of the reduction of N_2H_4. They used single heteronuclear cluster anions of the type $[VFe_3S_4(DMF)_3]^-$ and $[MoFe_3S_4Cl_3(\text{carboxylate})(\text{solvent})]^{2-}$ to reduce N_2H_4 to NH_3 catalytically, using 2,6-dimethylpyridinium salts as acid and $[Co(\eta^5\text{-}C_5H_5)_2]$ as the reductant [78,79]. The most effective reaction system was based on $[MoFe_3S_4Cl_3(\text{H-citrate})]^{3-}$, (see section 5.3), and with a molecular ratio of N_2H_4:cluster of 100:1, the production of NH_3 reached 98% of completion in 12 hours. Other carboxylate analogues produced lower yields under the same conditions, as does the $[VFe_3S_4(DMF)_3]^-$ anion [78,79].

It was not possible to obtain meaningful kinetic data for these systems because of the precipitation of various salts during the reactions. However, use of a variety of clusters allowed some conclusions to be proposed concerning the possible reaction mechanism, as follows. First, control experiments with Fe_4S_4 clusters, which did not reduce N_2H_4, indicated that the Mo atom was the catalytic site and in accordance with this, inhibition studies with ligands which are known to bind to Mo, showed total or pronounced loss of catalytic activity. Second, it appeared that the Mo-bound carboxylate ligands partially dissociate under acid conditions to allow binding of N_2H_4 at the Mo. This led to the suggestion that if the reduction of N_2 on FeMoco occurs at Fe as far as the N_2H_4 stage, then the final reduction steps could be achieved at molybdenum by transfer of the N_2H_4 from Fe to Mo [79,80] (see also Section 3).

7.3. Reactions of clusters and FeMoco with alternative substrates and CO

As would be expected from the early work on reduction of alternative substrates such as ethyne at metal centres (Section 1.1), a variety of cluster systems were shown to reduce C_2H_2 to C_2H_4 [20,27]. These include $[Fe_4S_4(SPh)_4]^{3-}$/ethanoic acid/ethanoic acid anhydride [81], $[Fe_4S_4(SPh)_4]^{3-}/NaBH_4^-$ [82], $[Fe_4S_4(SPh)_4]^{3-}/Mo^{IV}/(\text{Gly-Gly-Cys})_n/NaBH_4$ [83] and $[MoFe_3S_4Cl_3(\text{H-citrate})]^{3-}$/2,6-dimethylpyridinium/$[Co(\eta^5\text{-}C_5H_5)_2]$ [84]. In all these systems, under deuteriation conditions, C_2H_2 is reduced to *cis*-CHDCHD, which mirrors the behaviour of nitrogenase. Controlled potential electrolysis of $[Fe_4S_4(SPh)_4]^{2-}$ and $[\{MoFe_3S_4(SPh)_3\}_2(\mu\text{-}SPh)_3]^{3-}$ in MeOH/thf produces H_2, but, in presence of C_2H_2, H_2

production is greatly reduced and C_2H_4 is obtained. This behaviour is similar to that of nitrogenase, but deuteriation studies showed that, unlike the enzyme, this reduction does not give *cis*-CHDCHD selectively [85].

Although it has been suggested for the system using $[MoFe_3S_4Cl_3(H\text{-citrate})]^{3-}$ and related clusters that the Mo is the site of reduction of C_2H_2 [84], unlike the situation for N_2H_4 there does not appear to be a particular selectivity for metal in these reductions, since Fe-only systems are effective as well as those containing Mo or V. A mechanism for the reduction of C_2H_2 to C_2H_4 by clusters such as $[Fe_4S_4(SPh)_4]^{3-}$ in the presence of a proton source such as 2,6-dimethylpyridinium has been proposed from detailed kinetic studies. It is suggested that three protons must be bound to give $[Fe_4S_2(SH)_2(SPh)_3(SHPh)]$ before substrate can be transformed. Subsequent dissociation of the thiol to generate $[Fe_4S_2(SH)_2(SPh)_3]$ is also essential for C_2H_4 production (together with H_2) [86].

Other substrates of nitrogenase are also reduced by the systems noted above or their analogues, to give the same products as the enzyme; thus H^+ gives H_2, N_3^- gives N_2, N_2H_4 and NH_3, CH_3NC gives CH_4 and NH_3, and NO_2^- gives NH_3 and N_2 [27]. As is generally the case, mechanistic details and knowledge of the substrate-binding site are not available. Carbon monoxide inhibits nitrogenase action and is not reduced. This property of the enzyme is also in general shown by the above systems but the reactions of clusters with CO can be complex, as indicated in Section 5.3.

FeMoco can be isolated from the Mo-Fe protein in *N*-methylformamide solution, when it then has lost the binding of the terminal Fe to cysteine and the histidine to the Mo (see Figure 3) [87]. It is normally isolated in the so-called semi-reduced state and although it has not been shown to interact with N_2 in this condition, its reactions with a number of other nitrogenase substrates have been studied. The binding of CN^- has been studied by a combination of spectroscopic techniques, including X-ray absorption fine structure analysis (EXAFS). Cyanide appears to bind to both Fe and Mo; competitive binding between thiolate and CN^- seems to result in the terminal Fe being bound by thiolate and the Mo by CN^- [88,89,90]. C_2H_2 is catalytically reduced to C_2H_4 by amalgam reducing agents in presence of FeMoco [91]. Reduction of H^+ to H_2 [92] and binding of CO [93] to isolated FeMoco have also been studied by electrochemical/IR spectroscopic methods. In the latter case, the similarity of the pattern of CO stretching frequencies of the IR spectra of isolated, CO-bound FeMoco to that obtained from the intact, CO-bound enzyme [94] has been noted. The possible mechanisms of reactions of isolated FeMoco are discussed in more detail in Chapter 9.

8. METAL COMPLEXES WITH SULFUR LIGANDS THAT RELATE TO THE FUNCTION OF FeMoco

In the preceding section it was noted that although a number of structural characteristics of FeMoco are modelled by Fe-S-based cluster complexes, the crucial feature required to help in understanding the precise mode of action of FeMoco, namely binding of N_2 to any such cluster, has not yet been achieved. Moreover, the binding of N_2 to sulfur-ligated metal centres in general has proved to be a difficult synthetic problem, and although such compounds have been obtained, their number is limited [13,18]. The sulfur-donor co-ligands used to obtain well-defined N_2 complexes include thiolates, thioethers and dithiocarbamates, but not, as yet, sulfide.

The mononuclear complexes obtained, which have N_2 bound in the end-on mode (Figure 1a), have thiolate or macrocyclic polythioethers as coligands. They are $[OsCl(SC_6F_5)(N_2)(PMe_2Ph)_3]$ [95], $[Re(SC_6H_2Pr^i_3\text{-}2,4,6)_3(PPh_3)]$ [96] and *trans*-$[Mo(N_2)_2(Me_8[16]S_4)]$ [97] (Figure 12). The last complex, which is an analogue of the diphosphine complexes discussed in Section 1.2, gives, like them, ammonia on treatment with mineral acid, but the reaction leads to degradation of the molybdenum complex and a catalytic or cyclic system based on this compound does not seem achievable. No iron complexes of N_2 with sulfur-donor complexes have been reported and the low affinity of such a site as $Fe(NS_3)$ [$NS_3 = N(CH_2CH_2S)_3^{3-}$] for N_2 has been demonstrated by DFT calculation [98].

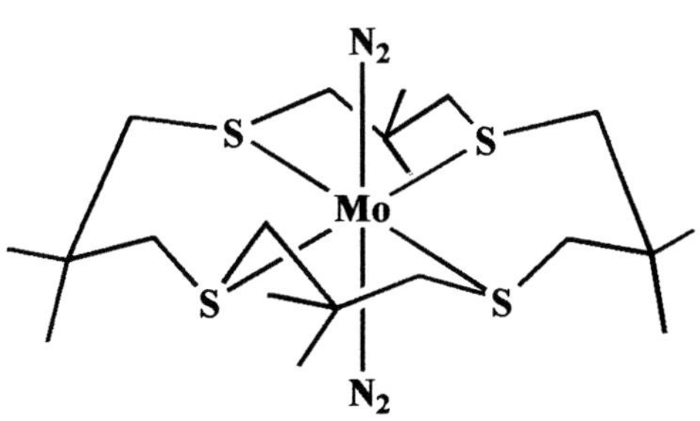

Figure 12: structure of *trans*-$[Mo(N_2)_2(Me_8[16]S_4)]$.

Dinuclear complexes with bridging N_2 have also been observed, but stable examples are few in number. None are reported for molybdenum, although the tungsten complex $[\{W(\eta^5\text{-}C_5Me_5)(Me)_2(SR)\}_2(\mu\text{-}N_2)]$ is known [99]. For vanadium a complex formulated as $[\{V(NS_3)\}_2(\mu\text{-}N_2)]$ has only transient existence [100] but stable compounds have been isolated for the heavier members of the Group 5 metals, namely $[\{M(S_2CNEt_2)_3\}_2(\mu\text{-}N_2)]$ (M = Nb or Ta) [22] (Figure 13) and $[\{Ta\{(SR)_3(THF)\}_2(\mu\text{-}N_2)]$ [25]. The dithiocarbamate compounds give a stoichiometric yield of hydrazine on protonation *via* intermediates such as $[\{Nb(S_2CNEt_2)_3\}_2(\mu\text{-}NHNH_n)]$ (n = 0, 1 or 2), which have been detected spectroscopically or isolated and characterised structurally [100, 101].

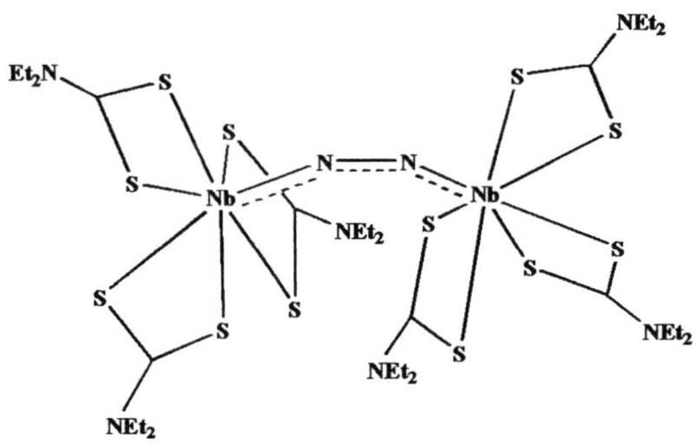

Figure 13. Structure of [{Nb(S_2CNEt_2)$_3$}$_2$(μ-N_2)].

For iron, no stable dinuclear complexes of N_2 have been reported, but μ-N_2H_2 complexes such as [{Fe(PPr_3)('S_4')}$_2$(μ-N_2H_2)] (see Figure 14) have been structurally characterised. This compound was oxidised in order to attempt to generate [{Fe(PPr_3)('S_4')}$_2$(μ-N_2)], but although this compound appeared to be generated, it was extremely labile above -78° C [102, 103]. These data were used as the basis for a proposal that FeMoco operates by opening one side of the Fe-S-Fe system to bind N_2 in the first step of its function (see also Chapter 9).

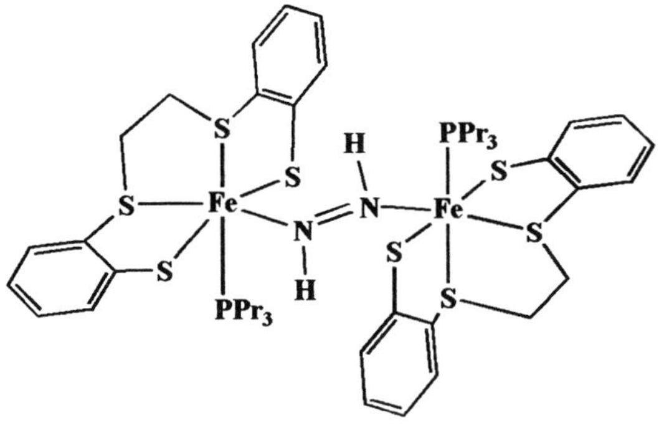

Figure 14. Structure of [{Fe(PPr_3)('S_4')}$_2$(μ-N_2H_2)].

9. CONCLUSIONS AND FUTURE PROSPECTS

Despite the range of detailed work discussed in this Chapter it is still not clear where N_2 interacts with FeMoco during nitrogenase action. Although Fe is favoured as the binding site, this is essentially on the basis of circumstantial evidence and Mo could be involved directly in the reduction of N_2, or in further reduction of partially reduced nitrogen species after initial reaction on Fe. To clarify this situation we still require chemical models for the unusual structural features of the FeMoco and P clusters of nitrogenase, as well as examples of N_2 bound to sulfide-ligated metal sites. Nevertheless, the problem of understanding nitrogenase has stimulated the development of a wide range of hitherto unknown chemistry of dinitrogen [13-20] and of Fe/S-type clusters [27,48,63-73]. Clearly there remains much essential work to be done to extend these areas so that detailed knowledge of nitrogenase action can be attained, and new catalysts for a variety of processes can be synthesised.

10. REFERENCES

1. W. A. Bulen and J. R. LeComte, Proc. Natl. Acad. Sci. USA 56 (1966) 979.
2. R. R. Eady, Chem. Rev 96 (1996) 3013.
3. K. Schneider, U. Gollan, S. Selsemeier, W. Plass and A. Müller, Naturwiss. 81, (1994) 405.
4. M. Ribe, D. Gadkari and O. Meyer, J. Biol. Chem. 272 (1997) 26627.
5. M. C. Durrant, Inorg. Chem. Comm. 4 (2001) 60 and references therein.
6 R. C. Burns and R. W. F. Hardy, Nitrogen Fixation in Bacteria and Higher Plants, Springer, New York, 1975.
7. J. R. Postgate, The Fundamentals of Nitrogen Fixation, Cambridge University Press, Cambridge, 1982, p. 20.
8 B. K. Burgess, Molybdenum Enzymes, T. G. Spiro ed., John Wiley, New York, NY, 1985, p. 161.
9. G. N. Schrauzer, Angew. Chem., Int. Ed. Engl. 14 (1975) 514.
10. G. N. Schrauzer in J. Chatt, L. M. da Camera Pina and R. L. Richards (eds.), New Trends in the Chemistry of Nitrogen Fixation, Academic Press, London, 1980, p. 103.
11. M. J. Dilworth, Biochem. Biophys. Acta 127 (1966) 288.
12. A E. Shilov in reference 10, p. 121.
13. T. A. Bazhenova and A .E. Shilov, Coord. Chem. Rev. 144 (1995) 69.
14. N. P. Luneva, S. A. Mironova, A .E. Shilov, M. Y. Antipin and Y. T. Struchkov, Angew. Chem., Int. Ed. Engl. 32 (1993) 1178.
15. J. Chatt, J. R. Dilworth and R. L. Richards, Chem. Rev. 78 (1978) 589.
16. R. A. Henderson, G. J. Leigh and C. J. Pickett, Adv. Inorg. Chem. Radiochem. 27 (1983) 197.
17. G. J. Leigh, Acc. Chem. Res. 25 (1992) 177.
18. M. Hidai and Y. Mizobe, Chem. Rev. 95 (1995) 1115.
19. M. Hidai, Coord. Chem. Rev. 185-186 (1999) 99.
20. R. L. Richards, Coord. Chem. Rev. 154 (1996) 83.
21. D. Rehder, Coord, Chem. Rev., 182 (1998) 297.
22. J. R. Dilworth, R. A. Henderson, A. Hills, D. L. Hughes, C. McDonald, A. N. Stephens and D. R. Walton, J. Chem. Soc., Dalton Trans. (1990) 1077.

23. T. Yoshida, T. Adachi, M. Kaminaka and T. Ueda, J. Am. Chem. Soc. 110 (1988) 4872.
24. D. Cruz-Garritz,, H. Torrens, J. Leal and R. L. Richards, Trans. Met. Chem. 8 (1983) 127.
25. R. R. Schrock, M. Wesolek, A. H. Liu, K. C. Wallace and J. C. Dewan, Inorg. Chem. 27 (1988) 2050.
26 J. R. Dilworth, J. Hu, R. M. Thompson and D. L. Hughes, J. Chem. Soc. Chem. Commun. (1992) 551.
27 D. J. Evans, R. A. Henderson and B. E. Smith, Bioinorganic Catalysis, 2nd ed., J. Reedijk and J. Bouwmann (eds.), Marcel Dekker Inc. New York, NY, 1999, p. 153 and references therein.
28. J. B. Howard and D. C. Rees, Chem. Rev. 96 (1996) 2965.
29. M. M. Giordiadis, H. Komiya, P. Chakabarti, D. Woo, J. J. Kornuc and D. C. Rees, Science 257 (1992) 1653.
30. J. Kim and D. C. Rees, Science 257 (1992) 1677.
31. D. C. Rees, M. K. Chan and J. Kim, Adv. Inorg. Chem. 40 (1993) 89.
32. S. M. Mayer, D. M. Lawson, C. A. Gormal, S. M. Rowe and B. E. Smith, J. Mol. Biol. 292 (1999) 871.
33. J. Chen, J. Christiansen, N. Campobasso, J.T. Bolin, R.C. Tittsworth, B.J. Hales, J.J. Rehr, and S.P. Cramer, Angew. Chem., Int. Ed Engl., 32 (1993) 1592.
34. J. T. Bolin, N. Campobasso, S. W. Muchmore, T. V. Morgan and L. E. Mortenson, E.I. Stiefel, D. Coucouvanis and W.E. Newton (eds.), American Chemical Society: Washington, DC, 1993, p. 186.
35. B. E. Smith, Adv. Inorg. Chem. 47 (1999) 159.
36. J. W.Peters, M. H. B. Stowell, S. M. Soltis, M. G. Finnegan, M. K. Johnson and D. C. Rees, Biochemistry 36 (1997) 1181.
37. K. K. Stavrev and M. C. Zerner, Chem. Eur. J. 2 (1996) 83.
38. R. L. Richards, Biological Nitrogen for the 21st Century, C. Elmerich, A. Kondorosi and W. E. Newton (eds.), Kluwer Academic, Dordrecht, 1997, p. 17.
39. J. Christianson, R. C. Tittsworth, B. J. Hales and S. P. Cramer, J. Am. Chem. Soc. 117 (1995) 10017.
40. H. B Deng and R. Hoffmann, Angew. Chem., Int. Ed. Engl. 32 (1993) 1062.
41. R. J. Deeth and C. N. Field, J. Chem. Soc., Dalton Trans. (1994) 1943.
42. I. Dance, J. Chem. Soc., Chem. Commun., (1997) 165.
43. P. E. M. Siegbahn, J. Westerberg, M. Svensson and R. H. Crabtree, J. Phys. Chem. B 102 (1998) 1615.
44. T. H. Rod, B. Hammer and J. K. Nørskov, Phys. Rev. Lett. 82 (1999) 4054.
45. C. J. Pickett, JBIC, 1 (1996) 601.
46. R. K. Szilagyi, D. G. Musaev and K. Morokuma, Inorg. Chem. 40 (2001) 766.
47. M. C. Durrant, Biochem. J., 355 (2001) 569.
48. R. H. Holm, Adv. Inorg. Chem., 38 (1992) 1.

49. *A. Manodori, G. Cecchini, I. Schröder, R. P. Gunsalus, M. T. Werth and M. K. Johnson, Biochemistry 31 (1992) 2703.*
50. S. D'Addario, F. Demartin, L. Grossi, M. C. Lapalucci, F. Laschi, G. Longoni and P. Zanello Inorg. Chem., 32 (1993) 1153.
51. M. J. Scott and R. H. Holm, Angew. Chem., Int. Ed. Engl. 32 (1993) 564.
52. E. Norlander, S. C. Lee, W. Chen, Z. Y. Wu, C. R. Natolu, A. D. Cicco, A. Filipponi, B. Hedman, K. O. Hodgson and R. H. Holm, J. Am. Chem. Soc. 115 (1993) 5549.
53. W. Chen, F. M. MacDonnell, M. J. Scott and R. H. Holm, Inorg. Chem. 33 (1994) 5809.
54. C. Gosh, B. M. Segal, J. Huang, J. R. Long and R. H. Holm, J. Am. Chem. Soc. 118 (1996) 11844.
55. P. P. Power and S .C. Shoner, Angew. Chem., Int. Ed. Engl. 30 (1991) 330.
56. F. M. MacDonnell, K. Ruhlandt-Senge, J. J. Ellison, R. H. Holm and P. P. Power, Inorg. Chem. 34 (1995) 1815.
57. D J. Evans, D. L. Hughes and J. Silver, Inorg. Chem. 36 (1997) 747.
58. B. S. Snyder, M. S. Reynolds, I. Noda and R. H. Holm, Inorg. Chem. 27 (1988) 595.
59. B. S. Snyder and R. H. Holm, Inorg. Chem. 27 (1988) 2339.
60. B. S. Snyder and R. H. Holm, Inorg. Chem. 29 (1990) 274.
61. I. Noda, B. S. Snyder and R.H. Holm, Inorg. Chem. 225 (1986) 3851.
62. D. Coucouvanis, Acc.Chem. Res. 24 (1991) 1.
63. R.H. Holm and E.D. Simhon, in T.G. Spiro ed., Molybdenum Enzymes, John Wiley, New York, NY, 1985, p. 1.
64. M. J. Carney, J. A. Kovacs, Y.-P. Zhang, G C. Papaefthymiou, K. Spartalian, R. B. Frankel and R. H. Holm, Inorg. Chem. 26 (1987) 719.
65. R. E. Palermo, R. Singh, J. K. Bashkin and R. H. Holm, J. Am. Chem. Soc. 106 (1984) 2600.
66. J. E. Barclay, D. J. Evans, G. Garcia, M. D. Santana, M. C. Torralba and J. M. Yago, J. Chem. Soc., Dalton Trans. (1995) 1965.
67. K. D. Demadis and D. Coucouvanis, Inorg. Chem. 34 (1995) 436.
68. J. Liang, M. M. Madden, V. K. Shah, and R. H. Burris, Biochemistry 29 (1990) 8577.
69. K. D. Demadis, C. F. Campana and D. Coucouvanis, J. Am. Chem. Soc. 117 (1995) 7832.
70. J. Han, K. Beck, N. Ockwig and D. Coucouvanis, J. Am. Chem. Soc. 121 (1999) 10448.
71. F. Osterloh, B. M. Segal, C. Achim and R. H. Holm, Inorg. Chem. 39 (2000) 980.
72. F. Osterloh, Y. Sanakis, R. J. Staples and R .H. Holm, Angew. Chem., Int. Ed. Engl. 38 (1999) 2066.
73. F. Osterloh, C. Achim and R. H. Holm, Inorg. Chem. 40 (2001) 224.
74. K. Tananka, Y. Hozumi and T. Tanaka, Chem. Lett. (1982) 1203.

75. Y. Hozumi, Y. Imasaka, K. Tananka, and T. Tanaka, Chem. Lett. (1983) 897.
76. R .N. F. Thorneley, R. R. Eady and D .J. Lowe, Nature (London) 272 (1978) 557.
77. M. J. Dilworth and R. R. Eady, Biochem. J, 277 (1991) 46.
78. S. M. Malinak, K. D. Demadis and D. Coucouvanis, J. Am. Chem. Soc. 117 (1995) 3126.
79. K. D. Demadis, S. M. Malinak and D. Coucouvanis, Inorg. Chem. 35 (1996) 4038.
80. D. Coucouvanis, JBIC, 1 (1996) 594.
81. R.S. McMillan, J. Renaud, J.G. Reynolds and R.H. Holm, J. Inorg. Biochem., 11 (1979) 213.
82. T. Itoh, T. Negano and M. Hirobe, Tet. Lett. 21 (1980) 1343.
83. N. Oguni, S. Schimazu and A. Nakamura, J. Mol. Catal. 22 (1983) 1.
84. L. J. Jaughlin and D. Coucouvanis, J. Am. Chem. Soc. 117 (1995) 3118.
85. K. Tanaka, M. Tanaka and T. Tanaka, Chem. Lett. (1981) 895.
86. K. L. C. Grönberg, R. A. Henderson and K. E. Oglieve, J. Chem. Soc., Dalton Trans. (1998) 3093.
87. V. K. Shah and W. J. Brill, Proc. Natl. Acad. Sci. USA 74 (1977) 3249.
88. A. J. M. Richards, D. J. Lowe, R. L. Richards, A. J. Thomson and B. E. Smith, Biochem. J. 297 (1994) 373.
89. H. I. Liu, A. Filipponi, N. Gavini, B. K. Burgess, B. Hedman, A. Di Cico, C. R. Natoli and K. O. Hodgson, J. Am. Chem. Soc. 116 (1994) 2418.
90. I. Harvey, R. W. Strange, R. Schneider, C. A. Gormal, C. D. Garner, S. S. Hasnain, R. L. Richards and B. E. Smith, Inorg. Chim. Acta 275-276 (1998) 150.
91. T. A. Bazhenova, M. A. Bazhenova, S. A. Miranova, G. N. Petrova, A. K. Shilova, N. I. Shuvalova and A. E. Shilov, Inorg. Chim. Acta, 270 (1998) 221.
92. T. Le Gall, S. K. Ibrahim, C. A. Gormal, B. E. Smith and C. J. Pickett, Chem. Commun. (1999) 773.
93. S. K. Ibrahim, K.Vincent, C. A. Gormal, B. E. Smith, S. P. Best and C. J. Pickett, Chem. Commun, (1999) 1019.
94. S. J. George, G. A. Ashby, C. W. Wharton and R N. F. Thorneley, J. Am. Chem. Soc. 119 (1997) 6450.
94. D. Cruz-Garritz, S. Gelover, H. Torrens, J. Leal and R. L. Richards, J. Chem. Soc., Dalton Trans. (1988) 2393.
96. J. R. Dilworth, J. Hu, J. R. Miller, D. L. Hughes, J. A. Zubieta and Qin Chen, J. Chem. Soc., Dalton Trans. (1995) 3153.
97. T. Yoshida, T. Adachi, T. Ueda, M. Kaminaka, N. Sasaki, T. Higuchi, T. Aoshima, I. Mega, Y. Mizobe and M. Hidai, Angew. Chem., Int. Ed. Engl. 28 (1989) 1040.
98. S. C. Davies, M. C. Durrant, D. L. Hughes, R. L. Richards and J. R. Sanders, J. Chem. Soc., Dalton Trans. (2000) 4694.
99. M. B. O'Regan, A. H. Liu, W. C. Finch, R R. Schrock and W. M. Davies, J. Am. Chem. Soc. 112 (1990) 4331.

100. S. C. Davies, D. L. Hughes, Z. Janas, L. B. Jerzykiewicz, R. L. Richards, J. R. Sanders and P. Sobota, Inorg. Chem. 39 (2000) 3485.
101. R. A. Henderson, S. H. Morgan and A. N. Stephens, J. Chem. Soc., Dalton Trans. (1990) 1101.
102. D. Sellmann and J. Sutter, JBIC 1 (1996) 587.
103. D. Sellmann and J. Sutter, Acc. Chem. Res. 30 (1997) 460.

Nitrogen Fixation at the Millennium.
G. Jeffery Leigh, Editor.

CHAPTER 13

Quantification of Nitrogen Fixation

M.B. Peoples[a], R.M. Boddey[b] and D.F. Herridge[c]
[a] CSIRO Plant Industry, GPO Box 1600 Canberra, ACT 2601, Australia,
[b] Embrapa-Agrobiologia, Km 47, Seropédica, 23890-000, Rio de Janeiro, Brazil
[c] NSW Agriculture, RMB 944 Tamworth, NSW 2340, Australia

1. INTRODUCTION

The fixation of atmospheric N_2 in terrestrial ecosystems can occur in the free-living state with some diazotrophs or via associative relationships on the roots or within the tissues of plants, while other organisms require a symbiosis with a specific host plant. Biological N_2 fixation (BNF) by these diverse systems may contribute directly to agricultural production where the fixed nitrogen (N) is harvested in grain or other food for human or animal consumption, and indirectly by adding N to the soil for the benefit of companion plant species or following crops. Much has been written about the desirability of harnessing BNF to provide a renewable source of N for the development of sustainable farming systems. However, it is not possible to undertake experimentation to identify and understand the principal factors regulating the N_2 fixation process, nor can BNF be managed for the benefit of agriculture, unless it can be reliably measured.

The different approaches used to provide measures of N_2 fixation have been extensively reviewed. While some of these reviews have focussed specifically on methodologies suitable for non-legume associations [1,2], most have concentrated on procedures to measure N_2 fixation resulting from the symbiosis between legumes and rhizobia since of all the N_2-fixing associations, annual and perennial legumes are responsible for contributing the largest amounts fixed N to agroecosystems [3-21].

It is not our intention to reproduce exhaustive critiques of all the various techniques that have been applied in the past. Such details can be found in the many comprehensive reviews listed above. Rather, we will provide only a broad overview of the most commonly used methodologies, and then direct more attention to those techniques we believe are likely to be of most value for N_2 fixation research in the coming millennium.

2. OVERVIEW OF APPROACHES USED IN THE PAST

Many early determinations of BNF for legumes were based solely on their N yield. This approach overestimated the amounts of N_2 fixed as it erroneously assumed that all the N present originated from N_2 fixation. A number of alternative methodologies have subsequently been developed to redress this situation. However, despite extensive debate there still remains considerable disagreement between researchers as to the most appropriate method to use in field-based studies. Each method has its own unique advantages and limitations (Table 1), and there is no *single correct* way of quantifying BNF that can be expected to provide an accurate measure of N_2 fixation under all conditions [1,4,9-14].

2.1 Acetylene reduction assay

The acetylene reduction assay aimed to overcome deficiencies in earlier N yield-based estimates of BNF by relating indirect measures of the activity of nitrogenase (the enzyme responsible for N_2 fixation) for detached nodules or recovered root systems to rates of N_2 fixation (Table 1). The acetylene reduction assay proved to be a sensitive diagnostic tool for detecting nitrogenase activity, and a useful technique for selecting species or cultivars for use as non-N_2-fixing reference plants [1]. However, usually it is necessary, or at least common practice, to uproot plants to assay nodulated root systems in a closed flask. Such disturbance can alter the partial pressure of O_2 around the nodules and can drastically lower nitrogenase activity [21]. This and other errors inherent in the methodology in conjunction with marked diurnal and seasonal changes in enzyme activity, variable relationships between ethylene production and N_2 fixed, and difficulties in quantitatively recovering whole root systems have resulted in the acetylene reduction assay now being considered too impractical to provide field determinations of BNF (Table 1, [7,13-14,21]).

2.2 N-Difference

The N-difference procedure attempts to separate plant N into fractions originating from soil N or atmospheric N_2. This is achieved by comparing the amount of N accumulated by the N_2-fixing plant under investigation with a neighbouring non-N_2-fixing control. The N present in the control is taken to represent a measure of the amount of soil mineral N available for plant growth during the growing season. It is assumed that the N_2-fixing species takes up the same amount of soil N. The amount of N_2 fixed is then calculated by difference:

$$N_2 \text{ fixed} = (\text{N yield fixing species}) - (\text{N yield non-fixing control}) \quad (1)$$

The N-difference method is a simple approach that can be applied when facilities for only total N analyses are available (Table 1). Unfortunately, there can be differences between fixing and non-fixing plants in root morphology, and/or their relative capacities to exploit soil N. The net result can be differences in residual concentrations of inorganic N remaining after fixing and non-fixing species [22]. Consequently, the fundamental assumptions underlying the N-difference method may be invalid. Modifications to the N-difference method to include measurements of soil mineral N under the fixing and non-fixing control at the end of growth might improve estimates of BNF [13], but not if the inorganic N resulted from species differences in root turnover or rhizodeposition [18].

2.3 Techniques that estimate plant reliance upon N_2 for growth

The use of various ^{15}N-based procedures, or the xylem ureide method allows researchers to determine the proportion of the plant's N derived from atmospheric N_2 (%Ndfa, sometimes also described as Pfix, or P_{atm}). The principles behind these techniques and their use in the calculation of amounts of N_2 fixed will be summarised here and their application will described in more detail in section 3.

2.3.1 Using analyses of N-isotope composition

There are two stable isotopes of N, ^{14}N and ^{15}N. The heavy isotope of N, ^{15}N, occurs in atmospheric N_2 at a constant abundance of 0.3663 atoms% ^{15}N. One of the key requirements for the success of ^{15}N-based methodologies is that the ^{15}N concentration in plant-available soil N differs significantly from atmospheric N_2. The difference in ^{15}N composition between soil N and N_2 is commonly increased for N_2 fixation studies by applying ^{15}N-enriched (between 5 and 95 atoms% ^{15}N) inorganic or organic compounds to the soil [1,3,5-14,16,20]. However, with attention to analytical procedures and a suitably precise mass spectrometer, it should also be possible to utilise the slight enrichment in ^{15}N (commonly 0.001 to 0.007 atoms% ^{15}N above atmospheric N_2) that occurs naturally in many soils [4,9-19]. The small differences in abundance of ^{15}N between plant-available soil N and atmospheric N_2 result from the net effect of fractionation between the ^{14}N and ^{15}N, which occurs during almost all soil N transformations. Gaseous losses of N via ammonia volatilisation and denitrification for example tend to be dominated by ^{14}N (*i.e.* depleted in ^{15}N), and this favours the retention of ^{15}N in soil [4,17,23-24].

Regardless of whether ^{15}N enrichment procedures, or the natural abundance approach is utilised, there is a requirement to sample the isotopic composition of the plant-available soil N pool. Estimates of %Ndfa are generally calculated as:

$$\%Ndfa = 100.(x - y)/(x - B) \qquad (2)$$

where:

x = ^{15}N content of the plant-available soil N

y = ^{15}N content of the N_2-fixing species.

B = a measure of the ^{15}N content of the target species fully dependent on N_2 fixation for growth. This is generally assumed to be zero in enrichment studies, but it is a factor that needs to be considered when using ^{15}N natural abundance (see section 3.2).

A measure of the ^{15}N content of the plant-available soil N (x in Equation 2) can be determined from soil extracts and the use of models to calculate temporal changes in isotopic composition [6], but in practice it is usually achieved by sampling a non-N_2-fixing 'reference' plant totally reliant upon soil mineral N for growth that is also growing at the same location as the putative N_2-fixing species (Table 1).

The various ^{15}N approaches are commonly described, as 'isotope dilution techniques', since the unlabelled atmospheric N_2 that is biologically fixed 'dilutes' the ^{15}N labelled soil N taken up by the N_2-fixing species during growth. It is the 'dilution' of ^{15}N in the fixing species relative to the non-fixing reference that provides the basis of the calculations used to provide a time-integrated (*i.e.* 'average') estimate of %Ndfa over a period of growth. This allows seasonal N_2 fixation to be calculated for an annual from a single analysis of the ^{15}N content of reference and N_2-fixing species at around the time of peak biomass (Figure 1).

In the case of non-legumes, evidence of associative N_2 fixation using ^{15}N-based methodologies requires that a lower ^{15}N enrichment also be combined with a higher N yield in the test plant compared to the reference [1]. Ideally this should occur under conditions where the enrichment of the available soil N pool does not change in time or space (Table 1).

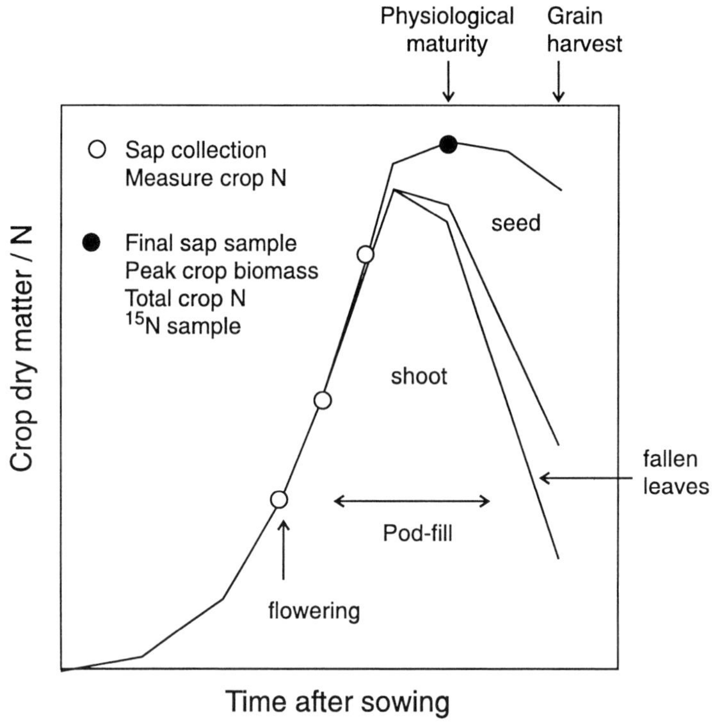

Figure 1. Idealized representation of the pattern of dry matter and N accumulation by a crop legume [25]. Typical timings of xylem sap collections during growth for ureide analyses are shown (O), as is the sampling at physiological maturity to measure peak crop biomass, total N, and ^{15}N (●).

Advantages	Limitations
Acetylene reduction assay	
Relatively inexpensive. Simple and rapid. Very sensitive.	Indirect - need conversion factor to relate ethylene production to N_2 fixation activity. Point-in-time estimate - need multiple assays to account for diurnal and seasonal changes. Acetylene may inhibit nitrogenase activity. Incomplete recovery or damage of root system can cause errors.
N-difference	
Adjusts for soil derived N Inexpensive and simple.	Requires non-N_2-fixing control. N_2-fixing plant and control do not always use the same amount of soil N.
Isotope dilution techniques	
Provides time-integrated estimate of % plant N derived from N_2.	Expensive equipment requirements. Need a non-N_2-fixing reference plant.
(a) ^{15}N enrichment	
Potentially very accurate	Requires addition of ^{15}N enriched material. Fixing and non-fixing plants must absorb the same relative amounts of N from soil and ^{15}N. Errors can arise if patterns of growth and/or rooting depth differ between the fixing and non-fixing species, and if ^{15}N enrichment of plant-available soil N changes with time and/or depth.
(b) ^{15}N natural abundance	
No addition of ^{15}N required. Can be used on-farm provided reference plants are present.	Requires a precise mass spectrometer. May be insensitive for low levels of %Ndfa if ^{15}N abundance of soil close to atmospheric N_2. Need to allow for isotopic fractionation. Field variability can sometimes be large.
Ureide xylem sap method	
Relatively inexpensive. Simple and rapid. Provides an estimate of % plant reliance upon N_2 for growth. No special experimental design required - can be used on-farm.	Indirect – must establish relationship between sap content and plant reliance upon N_2 fixation. Calibrations may change with development. Short-term estimate - need multiple sampling. Use restricted to warm-season legumes. Calibrations may differ with different species.

Table 1. Advantages and limitations of methods used for field estimates of N_2 fixation.

2.3.2 Using analyses of xylem sap composition

The xylem sap of field-grown legumes carries N-containing compounds originating from nodules as assimilation products of N_2 fixation, and soil mineral N taken up by the roots. In many tropical and sub-tropical legumes the main product of N_2 fixation exported from the nodules is in the form of allantoin and allantoic acid (ureides). In these same species, solutes derived from soil mineral N are transported in the xylem as nitrate or as organic products of nitrate reduction (amino acids). Because there are substantial differences in the principal forms of N transported in the xylem between highly symbiotic plants and unnodulated or poorly fixing legumes, it is possible to use the abundance of ureides relative to the other N components of xylem sap as an indirect measure of the plant's dependence upon N_2 fixation for growth [10,13].

Techniques have been developed to sample xylem sap from field-grown legumes, and the three N-solutes present (ureides, nitrate, and amino acids) can be relatively easily and rapidly analysed individually using colorimetric assays in a test-tube. However, interpretation of the results requires the xylem sap composition to be first calibrated with ^{15}N-based estimates of %Ndfa for the test legume species of interest in a glasshouse experiment (Table 1, [13,26-27]).

There are several perceived limitations in the sap-based method (Table 1). Firstly, its use is restricted to ureide-exporting legume species. Secondly, the relationships between sap composition and %Ndfa should ideally be calibrated for each legume species studied. Thirdly, the analyses of xylem sap represents only a 'point-in-time' estimate of the legume's symbiotic dependence at, or shortly before, the time of sap sampling, so unlike ^{15}N-based techniques repeated sampling of xylem composition and plant N may be required during a growing season (Figure 1). These issues will be discussed in greater detail in section 3.3.

2.3.3 Calculating amounts of N fixed from %Ndfa

Once the plant N can be partitioned into fixed and soil-derived fractions on the basis of %Ndfa, the amounts of N_2 fixed can be calculated as:

$$\text{Amount of } N_2 \text{ fixed} = (\text{plant N}) \times (\%\text{Ndfa}) \tag{3}$$

where the amounts of plant N are derived from measures of dry matter (DM) and tissue N content :

$$\text{Legume N} = (\text{plant DM}) \times (\%\text{N}) \tag{4}$$

Estimates of plant DM have almost always been determined from measures of shoot biomass. In the past it was generally assumed that N in the roots represented only a small fraction (10-15%) of the total plant N, and it was believed that shoot N provided a reasonable surrogate measure of whole plant N [28]. However, recent studies have suggested that, at least in legumes, such shoot-based estimates of plant N can greatly underestimate the total contributions fixed N to agricultural systems. Evidence is now emerging that suggests that N either associated with or derived from nodulated roots often represent 30-50% of the total plant N of pasture and crop legumes [18,29-31]. Therefore, total inputs of fixed N could be in the order of 50-100% greater than that determined from shoot-based measurements. Undoubtedly this would have major implications for the conclusions drawn about the relative size of potential rotational benefits derived from BNF that might otherwise be calculated from N-balance determinations of the net residual inputs of fixed N by crop legumes following grain harvest [31]. Consistent underestimation of plant N by ignoring the contribution of roots, and difficulties in accurately quantifying plant biomass in field experiments are likely to be the largest sources of error faced when quantifying BNF regardless of the technique used to estimate %Ndfa [25,28].

3. DETAILED EVALUATION OF METHODS FOR THE NEW MILLENIUM

It is our belief that the ^{15}N-enrichment approach, natural abundance technique, and the ureide assay represent the most useful techniques to assess BNF in the new millennium. The ^{15}N-enrichment procedure has proved to be a viable and valuable approach that is particularly well suited for annual legumes growing in soils low in N where %Ndfa can be expected to be high. The ^{15}N natural abundance technique and ureide analyses of legume xylem sap on the other hand can provide on-farm measures of N_2 fixation without perturbing the soil-crop system, or the need for special pre-crop treatments (Table 1). The ability to monitor N_2 fixation in farmers' crops represents an invaluable source of information which can be used to quantify the impact of BNF on nutrient cycling in farmers' fields, and to identify the major on-farm constraints or opportunities to improve inputs of fixed N [28,32].

3.1 ^{15}N-enriched methodologies

The ^{15}N enrichment technique is most suitable for use in pot or field experiments specifically deigned and planned for its application. It is much more difficult to apply for existing crops on farmers' properties or in natural ecosystems. This is mainly because it is necessary to artificially widen the difference in ^{15}N content between the soil and atmospheric N_2 by applying ^{15}N-enriched materials to the soil N in which both the N_2-fixing species and reference plants are growing to levels at least an order of magnitude higher than those that might be occurring naturally. The larger the ^{15}N enrichment of the plant-available soil N pool that can be achieved, the greater is the potential accuracy of subsequent calculations.

3.1.1 ^{15}N-enriching the soil N pool

Ideally the soil N should be labelled uniformly throughout the rooting zone, in order that the resultant ^{15}N enrichment of soil mineral N remains approximately stable with time over the period of the study (line 'A' in Figure 2). This proves to be extremely difficult to achieve in field experiments, particularly where long-lived perennials are

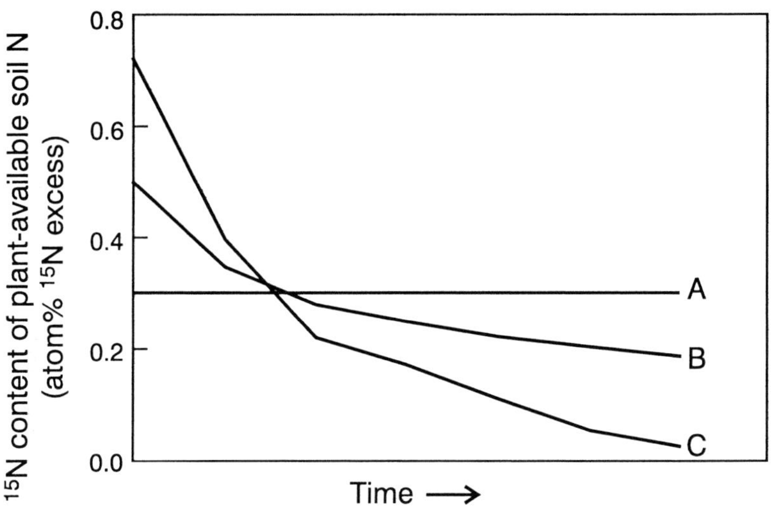

Figure 2. Hypothetical decline in soil atom% ^{15}N excess (*i.e.* ^{15}N enrichment above background levels) with time. Line 'A' indicates the theoretical ideal situation, 'B' represents changes in enrichment with a slow-release form of ^{15}N, and 'C' shows a typical decline following application of ^{15}N enriched fertiliser to the soil surface. Adapted from [8].

involved [14,16]. Until approximately 10 years ago, most studies enriched the soil N with a single dose of a soluble ^{15}N-labelled fertiliser, most commonly ammonium sulphate, at or near the time of sowing the experiment [5]. However, Witty [33] clearly demonstrated that although the addition of soluble labelled N (NH_4^+ or NO_3^-) results in an immediate enrichment of the soil mineral N pool close to the soil surface, the enrichment declines with time as N is assimilated by plant roots or N is mineralised from unenriched native soil organic matter (line 'C', Figure 2). To apply the ^{15}Ndilution technique it is necessary to estimate the ^{15}N enrichment of the N taken up from the labelled soil by the N_2-fixing species during its growth cycle. Most often this has been attempted by simultaneously planting a non-N_2-fixing reference crop in soil labelled with ^{15}N enriched fertiliser and harvesting the reference for N and ^{15}N analysis at the same time as the fixing species of interest. It is evident that if the reference crop removed different quantities of soil mineral N from the soil at times which differed from that of the target crop then the ^{15}N enrichment of the reference crop would not be equal to that N accumulated from the soil by the target crop. This difference in N uptake patterns between different fixing and non-fixing species and the magnitude of the error in the resulting estimates of the contribution of BNF, has been the subject of lengthy debate in the reviews of the application of this technique [1,5-11,13-14,20]. After evaluating various studies where the ^{15}N enrichment of several different reference crops growing in close proximity and labelled with ^{15}N in exactly the same manner, Boddey *et al.* [3] concluded that the strategy of using a single dose of ^{15}N-labelled fertiliser and a single reference crop yielded results of limited, sometimes useless, value, especially if BNF contributions were small.

Several different techniques have been used to slow the rate of decrease in ^{15}N enrichment with time of the soil mineral N pool (line 'B' in Figure 2). Pelleting of ^{15}N-labelled ammonium sulphate with gypsum was initially recommended [33], but later studies have tended to utilise techniques where sucrose, cellulose, straw or other C sources have been mixed with labelled fertiliser [20,34]. The C substrate is used to stimulate microbial growth within the soil and the ^{15}N label is incorporated into relatively stable pools of ^{15}N-labelled organic N. The expectation being that the label will be released slowly over time. Another possible means of achieving a 'slow-release' of ^{15}N label is to incorporate ^{15}N-enriched organic matter well in advance of experimentation or to take advantage of relatively homogeneously distributed 'residual ^{15}N' remaining in soil from a previous experiment [5]. This strategy has been used to quantify BNF contributions to grasses and cereals in concrete cylinders or tanks of soil [35]. The other main alternative approach used has been to add small additions of ^{15}N-labelled fertiliser at regular intervals through out plant growth [3,16,34,36]. This technique has mostly been applied to quantify BNF with forages or woody perennials as it can be applied to established experimental plots. All of these techniques should theoretically result in slower changes in ^{15}N enrichment of the mineral N pool during plant growth and hence species with different soil N uptake patterns should have smaller differences between them in ^{15}N enrichment of the N derived from soil. Unfortunately, only a few studies have compared different ^{15}N application techniques within the same experimental design, but the results generally confirm that more accurate estimates of BNF contributions to the legumes are obtained when such techniques are used [3,33-34].

3.1.2 Choice of reference species

Estimates of N_2 fixation could theoretically be calculated on the basis of direct measures of the ^{15}N content of soil mineral-N. It has been proposed that the ^{15}N enrichment of the N derived from the ^{15}N-labelled soil by a N_2-fixing species could be estimated from multiple harvests to assess total N accumulation and ^{15}N enrichment combined with soil sampling to assess simultaneously the same parameters of the soil mineral N [6]. Using a curve fitting approach for the total and labelled N accumulation and the decline in ^{15}N enrichment of the soil mineral N, it should be possible to compute the quantity and ^{15}N enrichment of the N derived from the soil throughout growth. The big practical disadvantage of the modelling technique is that there does not seem to be any technique available to predict what sampling depth would yield simulations of the actual soil uptake by any plant species since rooting depth is dynamic throughout growth. Furthermore, it is unlikely that they take up equal amounts of N from each layer down the profile of active roots, such that sampling to the maximum rooting depth is unlikely to resolve the problem. Furthermore, several repeated plant harvests (at least 5 or 6) are required during growth and many soil samples have to be taken. This increases plot size, is labour intensive and the analysis of the ^{15}N enrichment of soil mineral N is usually more problematic than of plant total N. Especially today when almost all mass spectrometers are coupled to total N analysers which require small samples of solid material for combustion.

Because of these complications it is more usual and convenient to use a non-N_2-fixing reference plant totally reliant upon soil N for growth to provide an integrated measure of the ^{15}N composition of plant-available soil N. It is assumed that the ratio of $^{15}N/^{14}N$ of the soil N assimilated by the reference plant is identical to that taken up by the fixing species. Apart from an absolute requirement for reference plants not to fix atmospheric N_2, potential reference species should ideally [16,20]:

(a) explore and exploit the same soil N pool as the fixing species because of similarities in root morphology and physiology;
(b) have the same duration of growth and pattern of N uptake as the fixing species;
(c) show similar patterns of root breakdown and turnover of below-ground biomass.

Given these requirements the selection of an appropriate reference can be problematic since these factors are difficult to predict or quantify in advance of experimentation, and in practice the choice of an appropriate reference plant is the single most important source of error affecting the accuracy of estimates of BNF. If, for example, the reference predominantly assimilates soil N from the highly enriched zone at the soil surface (Figure 3a), any unenriched soil N taken up by a deeper rooted legume further down the soil profile beneath the enriched zone will upon analysis be indistinguishable from fixed N, and the calculated BNF determination will be greatly overestimated.

An example of the impact of different rooting depths can be seen in Table 2 where the potential reference plants Guinea grass (*Panicum maximum*) or the non-nodulating tree legume *Senna spectabilis* were compared in a field study with the N_2-fixing tree legume gliricidia (*Gliricidia sepium*). The majority of the grass roots were located in

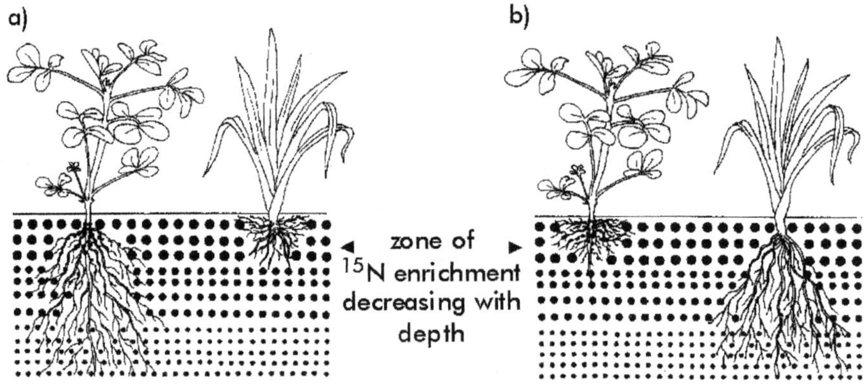

Figure 3. Diagrammatic representation of a nodulated legume and non-legume reference plant growing in a hypothetical ^{15}N enrichment study illustrating differences in comparative root growth. Application of ^{15}N enriched fertiliser to the soil surface has created a highly enriched zone. The enrichment decreases down the soil profile until finally natural abundance levels are reached at depth. Adapted from [14].

the top 20 cm, while both the fixing and non-fixing tree legumes had roots to at least 150-200 cm. The resultant enrichment in the Guinea grass 3 months after the initial application of ^{15}N was substantially higher than detected in the *Senna*, and consequently the estimate of %Ndfa for gliricidia was somewhat higher using the grass rather than the non-fixing tree as a reference source (Table 2).

Months since ^{15}N last applied	Reference plants		N-fixing species		
	Guinea grass	*Senna*	Gliricidia		
	^{15}N excess[a]		^{15}N excess[a]	% Ndfa	
				using grass	using *Senna*
3	0.299	0.203	0.103	65	49
6[b]	0.338	0.227	0.028	92	88

[a] ^{15}N enrichment above the ^{15}N content of atmospheric N_2 (0.3663 atoms% ^{15}N).
[b] 9 months after the initial application of ^{15}N-enriched fertiliser.

Table 2. Examples of levels of atom% ^{15}N excess achieved in regrowth of N_2-fixing gliricidia and non-N_2-fixing reference plants in a tree legume study using multiple applications of ^{15}N enriched fertiliser [16]. Estimates of %Ndfa are presented using ^{15}N data from both reference species, adapted from.

In the reverse situation where the roots of the reference might be recovering soil N for growth deeper in the soil profile than the legume (Figure 3b), the calculated determination of N_2 fixation will be underestimated.

One important strategy to improve the reliability of BNF estimates using the ^{15}N enrichment technique has been to use more than one reference plant [3, 35,37]. If the different reference species differ in their patterns of soil N uptake then they will attain different ^{15}N enrichments when grown in soils that are not uniform temporally or spatially in ^{15}N enrichment. This can result in different estimates of %Ndfa (see 3 month data in Table 2). The magnitude of the differences between these estimates is an approximate measure of the accuracy of the estimate. The potential impact of any variations between the fixing species and reference in root distribution and/or pattern of N uptake will be most important when attempting to quantify low levels of N_2 fixation. However, the absolute ^{15}N enrichment of the reference becomes less relevant when the legume is deriving a high proportion of its N requirements from atmospheric N_2 [16]. This can be seen in the reference comparison presented in Table 2 for the sampling 9 months after the initial ^{15}N application.

3.2 ^{15}N natural abundance approach

The natural abundance of ^{15}N is generally expressed as parts per thousand ($\delta^{15}N$ or ‰) deviation from atmospheric N_2 (by definition, the $\delta^{15}N$ of atmospheric N_2 is zero):

$$\delta^{15}N = 1000\ (\text{atoms\%}\ ^{15}N_{sample}) - (0.3663\ \text{atoms\%}^{15}N)\ /\ (0.3663\ \text{atoms\%}^{15}N) \qquad (5)$$

Ideally the natural abundance ($\delta^{15}N$) technique will work most efficiently if:

(i) there are just two pools of N for legume growth – a plant-available form of soil N and N_2 from the air;

(ii) the two pools are sufficiently different in ^{15}N abundance to allow accurate measurement with a mass spectrometer; and

(iii) the biological variability of these abundances is small compared to the difference between them.

Potentially the ^{15}N natural abundance approach has a number of distinct advantages over related ^{15}N enrichment procedures. In particular, the flexibility of the natural abundance method allows N_2 fixation to be monitored in almost any situation where both fixing and non-fixing reference plants are present since nothing needs to be added or disturbed before measurements can be undertaken. However, the technique also has a number of important limitations that should be recognised before it is applied. These are related to:

(a) the need to adjust for isotopic fractionation by the N_2-fixing species;

(b) levels of, and/or variability in, ^{15}N abundance;

(c) choice of appropriate reference species; and

(d) field sampling of plant material.

The following sections will focus primarily on the use of the natural abundance technique to provide measures of N_2 fixation for both legumes and non-legume species in agricultural systems. Some of the potential difficulties that can be encountered using the methodology to monitor BNF in natural communities have been specifically addressed elsewhere [4,12,24].

3.2.1 Isotopic fractionation

Isotopic discrimination and fractionation is generally ignored in enrichment studies (*i.e.* B in Equation 2 is assumed to be zero), but it can be an important factor to consider when using the natural abundance method to assess BNF since measurable fractionation of ^{14}N and ^{15}N does occur. For example, in many legume species the shoots of effectively fixing plants are often depleted in $\delta^{15}N$ relative to atmospheric N_2, while the nodules are enriched in $\delta^{15}N$ [4,16,19]. Consequently, when N_2 fixation is to be calculated on the basis of shoot-only analyses, a correction needs to be made for within-plant discrimination effects. Including an adjustment for isotopic fractionation or the use of different B values in the calculations is generally of little importance in determining %Ndfa or the amounts of N_2 fixed at low levels of fixation, but will be important when plant reliance upon N_2 fixation is high and the natural abundance of the N_2-fixing species approaches or falls below 0‰ (Table 3).

Sampled material			Estimate of N_2 fixation			
Reference	Chickpea		$B = 0$ ‰[a]		$B = -2.1$ ‰[b]	
$\delta^{15}N$ (‰)	$\delta^{15}N$ (‰)	shoot N (kgN/ha)	Ndfa (%)	Amount (kgN/ha)	Ndfa (%)	Amount (kgN/ha)
8.70	8.20	84	6	5	5	4
6.70	5.87	86	12	11	9	8
6.80	5.36	76	21	16	16	12
10.04	6.98	183	30	55	25	46
12.20	6.42	37	47	17	40	15
10.10	3.60	102	64	65	53	54
5.09	1.02	114	80	91	57	65
7.70	0.30	113	96	108	76	86
6.30	-0.14	96	100[c]	96	77	74
6.31	-0.83	114	100[c]	114	85	95
2.95	-1.45	59	100[c]	59	87	51

[a] Assuming no fractionation.
[b] Using the *B* value for chickpea determined by Doughton *et al.* [38].
[c] Designated 100%, although Equation 2 would calculate physiologically impossible estimates of %Ndfa >100%.

Table 3. The effect of adjusting for isotopic fractionation on calculations of the proportion (%Ndfa) and amounts of chickpea shoot N derived from N_2 fixation when applying the ^{15}N natural abundance technique to experimental and commercial crops.

For most nodulating symbiotic N_2-fixing systems, it is possible to grow effectively inoculated plants solely on N_2 fixation in N-free hydroponics or in an inert medium such as sand with N-free nutrients. Analyses of the $\delta^{15}N$ of shoots from plants grown in this way can then used to provide a measure of '*B*' to be inserted into Equation 2. Unfortunately it is not possible to grow non-nodulating species in the total absence of combined N, so for convenience a *B* value of 0‰ is generally used for associative N_2-fixing systems. While there may currently be little alternative to this approach, an assumption that no isotopic fractionation occurs as a result of N_2 fixation by non-symbiotic associations may not be totally appropriate since recent studies have indicated that the $\delta^{15}N$ of free-living N_2-fixing bacteria generally tends to be depleted in ^{15}N [2].

Ideally a *B* value should be prepared for each new species studied since different N_2-fixing associations can exhibit differences in the slight enrichment or depletion of $\delta^{15}N$ of various plant parts. For example, although there is generally little impact of host cultivar on the $\delta^{15}N$ of shoots, there can be large effects of rhizobial strains used to form the symbiosis in some legume species [4,13-19]. For example, in white clover, the shoot $\delta^{15}N$ were found to be very similar across several varieties inoculated with a single rhizobial strain (-1.58‰ to −1.48‰), but ranged from −3.93‰ to −1.48‰ (unrelated to the amount of N accumulated) across nine effective strains (median −2.08‰) which included one commercial inoculant strain and eight indigenous strains isolated from dairy pastures in southern Victoria, Australia [39]. Such results introduce considerable uncertainty about the choice of *B* value for use with legume material collected from field sites where many unknown rhizobial strains might be involved. A further dilemma can arise when the levels of $\delta^{15}N$ detected in legume samples fall below that of the presumed *B* value determined in the glasshouse from fully symbiotic plants. Under such circumstances %Ndfa could arbitrarily be designated as 100%. This may not greatly influence subsequent calculations of amounts of N_2 fixed, but clearly the use of an inappropriate estimate for *B* could lead to errors in data interpretation. Such findings indicate a need to reassess how *B* values might be determined for legume systems.

One alternative approach to determining a more 'realistic' *B* value may be to inoculate legumes with mixed rhizobial inoculants prepared from a diverse range of field sites [19]. However, nodule occupancy by rhizobial strains in the field can be variable and unpredictable and studies with mixed rhizobial strains in the glasshouse may not necessarily mimic plants in the field. Another strategy that has been proposed is to use an indirect means of determining isotopic fractionation under field conditions [38]. In this technique, both ^{15}N enrichment and natural abundance methods are used in parallel under the same conditions. Legume reliance upon N_2 fixation is calculated by both methods. The *B* value is adjusted in Equation 2 until the estimates of %Ndfa by natural abundance best match those based on the ^{15}N enrichment method. This procedure would be best suited where many N_2 fixation measurements are to be undertaken at a single location. The number of enrichment comparisons required need only be sufficient to establish a reliable *B* value, with the bulk of the remaining estimates of %Ndfa undertaken without the expense of applying ^{15}N-enriched fertiliser using the natural abundance method. The main disadvantage of this comparative approach is that it requires accurate data sets from both methods. These cannot always be guaranteed considering the multiplicity of potential errors. However, selective use of high %Ndfa data for estimating *B* could minimise such errors.

For logistical reasons neither of these options would be practical for on-farm surveys of N_2 fixation by commercial crops. Presumably some legumes exhibit $\delta^{15}N$ values below the glasshouse-derived *B* because differences in isotopic discrimination were induced by rhizobial strains and/or soil characteristics experienced by the farmers' crops. One possible solution may be to collate and examine all the $\delta^{15}N$ data collected in a district or region and designate a %Ndfa value of 100% to the legume crops exhibiting the lowest $\delta^{15}N$ value (*i.e.* assume this value approximate *B*, [15]). Where %Ndfa determinations have been calculated using such field 'adjusted' *B* values in Nepal and Pakistan, they have given similar measures of N_2 fixation to that derived using a N-difference technique. This provides circumstantial evidence to suggest that this arbitrary adjustment of *B* might be appropriate.

3.2.2 Levels and variability in ^{15}N natural abundance

It is difficult to predict what level of $\delta^{15}N$ the soil N assimilated by plants might be for any particular field site. The $\delta^{15}N$ of soil total N is dominated by the isotopic signature of stable forms of N that are not likely to change rapidly with time [24]. Consequently the measures of $\delta^{15}N$ in soil total N may not necessarily be reliable guide to the $\delta^{15}N$ of the N utilised by plants during a growing season. Evidence is emerging that suggests that some plants are capable of assimilating N from the soil or decaying organic matter as forms of N other than nitrate or ammonium via ericaceous-, ecto-, or endo-mycorrhizae [2]. These different forms of N in soil may also differ in $\delta^{15}N$ signature both in time and space. This can have important consequences for the validity in applying the natural abundance methodology to quantify BNF. Particularly if the major sources of soil N assimilated by the putative N_2-fixing species differ from the non-N_2-fixing reference. However, this is more likely to be problematic in natural ecosystems than farming systems [2,4,12].

Although the levels of ^{15}N detected in non-N_2-fixing reference species growing in different agro-ecosystems around the world cover a wide range, the $\delta^{15}N$ of plant-available soil N in cropping soils often tends to be higher than in soils under pasture, or agroforestry (Table 4). Presumably this is associated with the greater cycling, turnover and losses (or removal) of N in cropping systems. The levels observed are generally not related to concentrations of soil mineral N (4,16), and there seems to be no consistent effect of prior cropping history (Table 5). However, the $\delta^{15}N$ of plant-available soil N does seem to be influenced by factors such as soil age (older soils tending to have higher $\delta^{15}N$), annual rainfall, frequency of fire, waterlogging, denitrification, and applications of fertiliser-N or composted organic matter [23-24, 40-41]. The wide range of $\delta^{15}N$ measured in the cropping systems of south-east Asia for example (Table 4), reflects data collected from: (a) lowland rice-based farming systems where volatilisation and denitrification of N from flooded rice paddies favours losses of ^{14}N and elevation of ^{15}N in the soil, and (b) fields receiving applications of high rates of fertiliser N which is often depleted in ^{15}N.

The accuracy of the natural abundance technique ultimately depends upon the levels and uniformity of the $\delta^{15}N$ in the plant-available soil N pool. Since the relationship between legume $\delta^{15}N$ and %Ndfa is linear for a given reference $\delta^{15}N$, the sensitivity of the estimated %Ndfa will also be proportional to the magnitude of reference $\delta^{15}N$. Therefore, just as with the ^{15}N-enriched methodologies, the higher the reference ^{15}N content, the more precise the estimate at low levels of %Ndfa. Unkovich and co-workers [19] have suggested that given an analytical precision of ± 0.2‰, a soil enrichment of at least 2‰ (ie 10 times the precision of measurement) would be required to detect a theoretical change in Ndfa of around 10%.

Data collected by the authors and collated from the literature (Table 4) for more than 1,000 experimental trials and farmers' fields in Australia, south-east Asia (China, Indonesia, Malaysia, Philippines, Thailand, Vietnam), Africa (Nigeria, South Africa),

Geographical location	Land use	Number of observations	Level of $\delta^{15}N$ detected (‰)	
			Range	Mean
Australia	Cropping	285	+0.4 to +17.5	+7.5
	Pasture	388	-1.1 to +10.5	+3.5
	Agroforestry	27	+0.9 to +8.6	+4.3
SE Asia	Cropping	245	-3.8 to +21.6	+5.7
	Agroforestry	58	+1.0 to +9.5	+4.6
S Asia	Cropping	244	-0.7 to +14.1	+2.2
Middle East	Cropping	47	-0.2 to +9.9	+3.1
Africa	Cropping	38	+ 2.7 to +12.6	+6.8
S America	Cropping	15	+4.4 to +26.5	+8.9
	Agroforestry	25	+1.5 to +8.5	+4.5
Europe	Cropping	7	+2.2 to +9.4	+4.3
	Pasture	6	+0.65 to +5.8	+3.9

Table 4. Examples of levels of ^{15}N natural abundance ($\delta^{15}N$) detected in non-N_2-fixing reference plants growing under different patterns of land use, collated from [2, 15-16,19,39] and includes additional unpublished data.

Pre-crop treatment	Level of ^{15}N natural abundance detected (‰)					
	Site location *					
	(A)	(B)	(C)	(D)	(E)	(F)
Winter cereal	3.4	10.6	12.0	9.9		6.7
Summer cereal				9.7	6.3	6.1
Legume crop				7.3		6.5
Pasture	3.7					6.3
Fallow		7.9	5.3		11.9	

* Victoria: (A) Horsham [36°40'S, 142°17'E]; New South Wales: (B) Yanco [34°37'S, 146°25'E], (C) Trangie [32°2'S, 147°59'E], (D) Breeza [31°16'S,150°28'E]; Queensland: (E) Toowoomba [28°32'S, 151°58'E], (F) Kingaroy [26°33'S, 151°50'E]

Table 5. Examples of levels of ^{15}N natural abundance measured in non-N_2-fixing reference plants following different cropping sequences in Australian farming systems over a 1500km transect from temperate winter-dominant rainfall environments in the south to sub-tropical climates of the north.

and South America (Brazil), suggest that despite a wide range of $\delta^{15}N$ detected in reference plants, most values commonly fall between +3‰ to +9‰ [2,15,19]. Generally less than 10% of the sites examined in these particular regions would be considered unsuitable for application of the natural abundance technique using the criteria of 2‰ as a minimum threshold [15,19]. These conclusions are supported by more limited data collected in Europe (Denmark, Germany, Portugal). However, surveys in the rainfed cropping areas of the middle-east (Lebanon, Syria) and south Asia (Bangladesh, India, Nepal, Pakistan) revealed that >50% of on-farm sites had $\delta^{15}N$ values <2‰ (Table 4). These soils were often impoverished and deficient in most if not all macro- and micro-nutrients and presumably the very low plant-available soil $\delta^{15}N$ values reflected the low organic fertility, low intensity of N-cycling processes occurring in many farmers' fields, and/or dominant soil N transformation which do not favour the relative enrichment of soil mineral N with ^{15}N.

Initially it was thought that the low levels of $\delta^{15}N$ detected in the reference plants would greatly restrict the use of the ^{15}N natural abundance procedure for farmers' crops the middle-east and south Asia [15]. However, when the data were examined in more detail it was clear that in many cases the difference in the $\delta^{15}N$ between the reference and their matching legume samples were often in the order of 1 to 3‰ (Figure 4). This difference in natural abundance should be easily and accurately quantifiable on the mass spectrometer. The examples presented in Table 6 suggest that reference $\delta^{15}N$ values <2‰ can still give estimates of %Ndfa of similar precision to reference plants with higher levels of ^{15}N abundance. The standard errors (SE) for the %Ndfa estimates were reasonably constant over ranges of reference $\delta^{15}N$ < 0‰ to > 4‰, and the actual values of %Ndfa were unaffected by the magnitude of the reference $\delta^{15}N$. Therefore, it may not be necessary to always invoke a rule of rejecting estimates of %Ndfa based on reference $\delta^{15}N$ values of < 2‰. However, it is also apparent that there must be a lower acceptable limit for reference $\delta^{15}N$. Fortunately the legumes had a relatively high reliance upon fixation for growth and the difference between reference and legume $\delta^{15}N$ were generally been large enough such that data interpretation remained unambiguous for the majority of on-farm studies in south Asia (Figure 4) and other environments [39] where reference plants have exhibited very low $\delta^{15}N$. But clearly the lower the reference $\delta^{15}N$ the more difficult data interpretation is likely to be. Researchers should be wary of using data where differences between reference $\delta^{15}N$ and the presumed B value is <1‰. For example, the lowest derived B value for farmers' lentil crops examined in south Asia was -1.5‰, and data were rejected whenever the reference $\delta^{15}N$ fell below -0.5‰. This situation occurred in <5% of the 244 sites sampled.

Measures of the $\delta^{15}N$ of reference plants have demonstrated that levels of plant-available soil $\delta^{15}N$ can be relatively uniform across a field site (see Australian data in Table 7). Yet there is evidence that short distance site-to-site variability in soil $\delta^{15}N$ also exist (data from Pakistan in Table 7). This might occur in even a 'low-variability' field [44]. Reasons for this heterogeneity in the $\delta^{15}N$ signature of plant-available soil N are

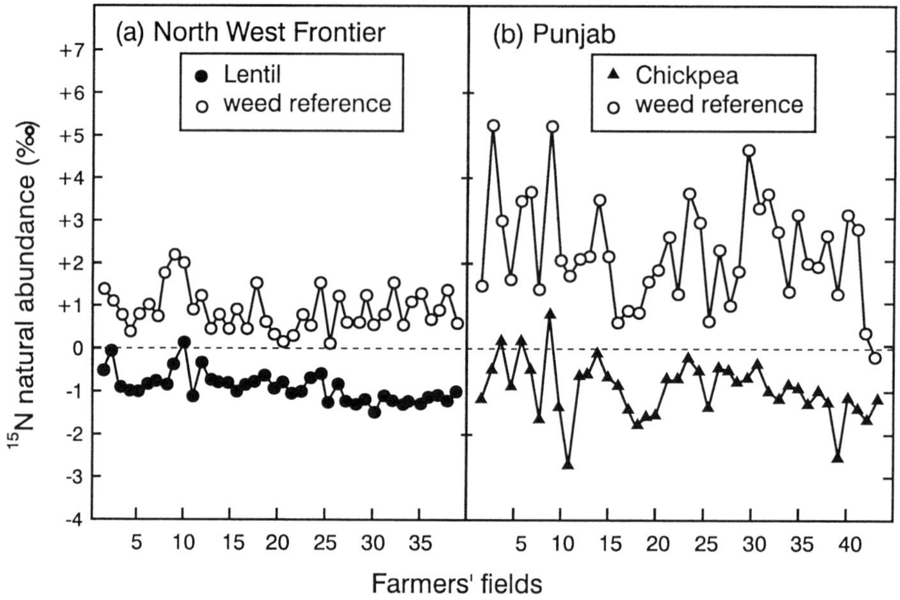

Figure 4. Levels of $\delta^{15}N$ detected in legume crops and weed references growing in farmers' fields in two difference regions of Pakistan, the (a) North West Frontier Province [42], and (b) Punjab [43]. All values are the means of 5 replicates. Each point data point represents a different farmers' field.

Ref $\delta^{15}N$ (‰)	Lentil Number of fields	Lentil Mean Ndfa (%)	Lentil SE	Mung bean and black gram Number of fields	Mung bean and black gram Mean Ndfa (%)	Mung bean and black gram SE
< 0	0			4	52	± 8.5
0 to +1	24	78	± 6.7	15	56	± 7.4
+1 to +2	13	89	± 5.7	13	45	± 9.5
+2 to +3	2	63	± 7.0	4	33	± 17.8
+3 to +4	0			0		
> +4	0			4	36	± 6.0

Table 6. Precision of estimates of the proportion of legume N derived from N_2 fixation (%Ndfa) for different values of the non-N_2-fixing reference species collected from farmers' fields in Pakistan [42].

Details	Reference $\delta^{15}N$ (‰)	Details	Reference $\delta^{15}N$ (‰)
Australia		Pakistan	
Farm location:	6.52	Farm location:	4.06
Croppa Creek	7.43	Swat Valley	2.45
Crop area = 50 ha	6.63	Crop area = 0.66 ha	4.13
	7.24		1.11
Distance between	6.34	Distance between	4.53
samples = 225m	7.50	samples = 30m	1.92
	6.37		1.13
	6.15		-0.01
	7.09		-0.11
	5.83		0.80
mean ± SE	6.71 ±0.18	mean ± SE	2.00 ±0.55

Table 7. Examples of the levels of $\delta^{15}N$ detected in reference plants systematically collected from within farmers' crops in northern New South Wales in Australia and in the North West Frontier Province in Pakistan. Sampling points were located at regular intervals along fixed transects crossing each field in a 'W' shape as described by Schwenke *et al.* [32].

not always obvious. In the case of some of the farming systems of south Asia, the application of farmyard manures containing complex inorganic and organic N sources to cropping soils and/or hand broadcasting urea fertiliser may contribute to some of the observed variability. Alternatively heterogeneity in $\delta^{15}N$ might be associated with variations in topography inducing localised differences in water content, soil mineral-N, waterlogging and denitrification [40,45-46]. Since considerable fractionation occurs during animal metabolism and N leaving grazing animals via urine is depleted in ^{15}N relative to the herbage consumed while faeces tends to be slightly enriched [23,47], some of the variability in reference $\delta^{15}N$ may be contributed from plants assimilating N derived from livestock excreta [48]. This can introduce complications in the application the natural abundance to intensive grazing systems with dairy cattle [39,47-48] where large concentrations of urinary N (equivalent to 600-1200 kg N/ha) and faecal N (up to 2000 kg N/ha) may be returned to 20-50% of the land area [29]. However, this potential source of heterogeneity appears to be less of a problem with more broad-acre pasture systems; particularly those involving sheep [18-19,29].

3.2.3 Choice of reference species

The ^{15}N content of the ideal reference plant is assumed to provide an integrated determination of the $\delta^{15}N$ of the soil N available for growth by the N_2-fixing species over the duration of the study. The choice of reference plant will be most important if %Ndfa is low and if the ^{15}N natural abundance of soil N changes with depth and/or time. There appears to be little consensus in the observed trends for the $\delta^{15}N$ of plant-available soil N across soil types or environments. The $\delta^{15}N$ of mineral N can be relatively uniform down a soil profile [4,16] and the $\delta^{15}N$ values for reference plants can remain relatively constant for prolonged periods, or change only gradually over time [4,13,16,36,49]. In this situation the impact of an inappropriate choice of reference is likely to be small. However, the $\delta^{15}N$ of available soil N has been observed change rapidly with depth and the $\delta^{15}N$ values detected in reference plants can fluctuate within and between growing seasons [4,16]. Under these latter conditions significant errors in determinations of N_2 fixation might be expected if the fixing and non-fixing reference species exhibit different rooting depths and/or N uptake characteristics. Certainly there are cases where measurably different $\delta^{15}N$ values have been detected between alternative reference plants [2,16,19], but it is reassuring to find that in both legume and non-legume systems different potential reference plants often have similar $\delta^{15}N$ values [2,11,13-16,50]. Collectively the data suggest that the choice of reference plant is less critical with the ^{15}N natural abundance procedure than with ^{15}N-enrichment techniques. However, it is important not to become complacent, and it is advisable whenever possible to always sample a range of reference species [15], particularly when assessing for potential associative N_2 fixation [2].

3.2.4 Sampling to quantify plant $\delta^{15}N$ and N_2 fixation

Levels of $\delta^{15}N$ can differ in annual cropping systems between various plant components of both field-grown N_2-fixing legumes and reference plants (Figure 5, also [15]), and it is generally recommended to analyse whole shoots when using ^{15}N natural abundance to assess BNF. However, in the case of N_2-fixing species that have the potential to accumulate large amounts of biomass it may not always be convenient or possible to quantitatively sample the entire plant. While some work with trees indicate that $\delta^{15}N$ can vary between different plant organs, a number of other studies with sugarcane and a range of woody perennials suggest that subsampling of individual plant parts often provides a viable indicator of whole plant ^{15}N abundance [2,4,16,36].

If determinations of total amounts of N_2 fixed are not crucial an alternative approach which might be useful to assess levels of %Ndfa achieved by commercial legume crops could be to analyse the $\delta^{15}N$ of seed collected at crop maturity. Grain N at final harvest often represents a major proportion of total above-ground crop N in legumes (Figure 1). Therefore, analyses of collected reference material in conjunction with grain $\delta^{15}N$ content could provide a guide to the symbiotic status of the whole crop (*e.g.* compare the $\delta^{15}N$ data for fruit with whole plant for faba bean and chickpea in Figure 5). One further advantage of analysing seed at crop harvest is that any late contribution of fixed N during seed maturation would also be included [51].

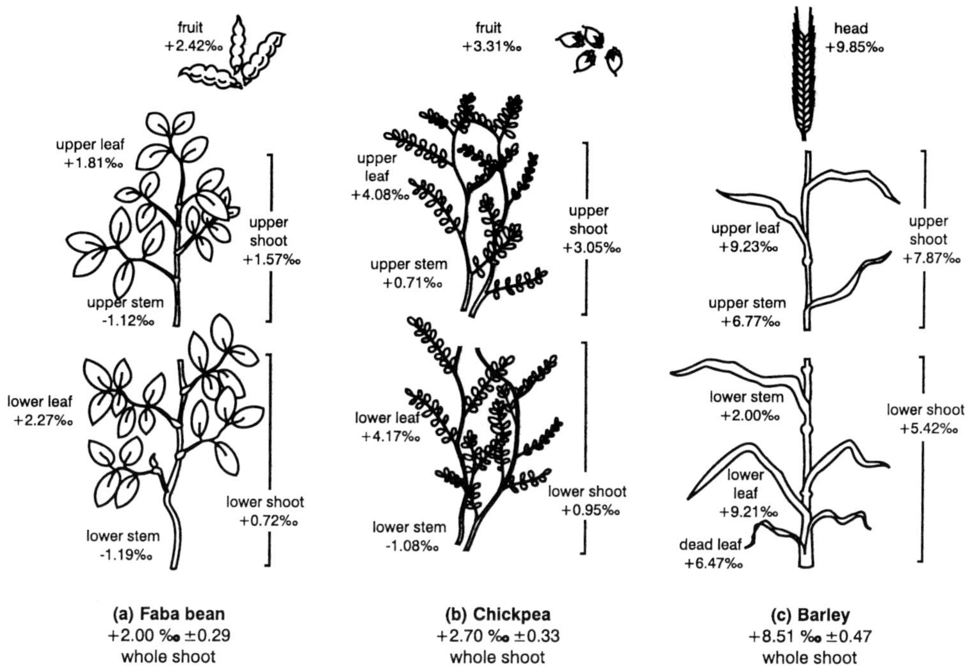

Figure 5. Diagrammatic representation comparing the mean levels of $\delta^{15}N$ detected in the different leaf and stem strata, shoot fractions and reproductive parts with measures of whole shoots of field-grown (a) faba bean, (b) chickpea, and (c) barley sampled during the early stages of seed-filling from adjacent dryland crops growing at Breeza, New South Wales, Australia.

When combining the collected reference and legume data to calculate %Ndfa it may not always be possible to avoid variability resulting from 'noise' in plant-available soil $\delta^{15}N$. Various protocols have been suggested involving increased replication, bulking replicates, or strategic sampling to improve precision of measurements [17,19]. However, high variability in plant-available soil $\delta^{15}N$ remains an insurmountable problem where reference and N_2-fixing plots are physically separated in experimentation. Ideally potential errors can be minimised if the non-N_2-fixing reference plants can be grown in close proximity to the fixing species [44]. This is usually possible in pasture systems where legumes are generally growing in association with a range of non-legume species, but this can also be achieved for crop legumes where they are intercropped with cereals in the farming systems of Asia or by using in-crop weeds (2,15-19,28-30,32,50). Collection of legume samples with 'paired' reference material from neighbouring non-legume intercrops or weeds may overcome problems of variability in soil $\delta^{15}N$ between replicates and result in acceptably accurate estimates of %Ndfa (Table 8).

Within field replicate sample	Reference $\delta^{15}N$ (‰)	Chickpea $\delta^{15}N$ (‰)	Estimate of Ndfa (%)
Lebanon			
1	0.85	-0.32	40
2	1.85	-0.05	48
3	3.06	0.86	43
4	3.52	0.35	56
mean ± SE			47 ± 3.5
Nepal			
1	2.75	-0.64	70
2	3.23	-0.93	78
3	4.71	-0.35	74
4	6.16	0.00	73
5	7.49	0.18	75
mean ± SE			74 ± 1.3

Table 8. Estimates of %Ndfa for chickpea where the $\delta^{15}N$ of weed reference plants collected within farmers' crops is variable. Replicate chickpea and paired weed samples were collected from farmers' crops growing at Nasrieh in Lebanon, and the Banke district of the Terai region of Nepal. The value of -2.10‰ was used as an estimate of *B* for Equation 2 [38].

3.3 Ureide sap-based method

Although the ^{15}N methods described above are arguably the standard methods for measuring N_2 fixation, the high cost of isotopes and mass spectrometry restrict their widespread use. For laboratories that don't have ready access to ^{15}N methodologies, the ureide technique may provide an alternative, at least for the measurement of N_2 fixation by a number of tropical and sub-tropical legumes species that transport ureides in their xylem sap. The great advantage of the method is that it is technically-simple and within reach of any laboratory with a supply of glassware, chemicals and a simple colorimeter.

3.3.1 Ureides in plants

The presence of the ureides, allantoin and allantoic acid, in plants was established almost 80 years ago, although another 30 years passed before their role in transport of N in the xylem stream was clearly defined [52,53]. Japanese studies in the mid-1970's showed a strong relationship between root nodulation and ureide contents of plant tissues [54]. Subsequent $^{15}N_2$-labelling experiments [55,56] confirmed that the ureides were synthesised principally in the nodules of N_2-fixing plants and were the major transport compounds involved in the export of fixed N from the nodules to shoot tissues. In the following decade, much of the ureide research focussed on the sites and mechanisms of allantoin and allantoic acid synthesis and utilisation in legumes [57].

Not all legumes export fixed N from their nodules as ureides (Table 9). The majority of nodulated legumes transport the amides, asparagine and glutamine, in xylem sap [53,57-58]. The ureide exporters are restricted to species in the Phaseoleae and Desmodieae tribes, within the Papilionoideae sub-family [58].

Amides (asparagine, glutamine)	Ureides (allanoin, allantoic acid)
Chickpea (*Cicer arietinum*)	Soybean (*Glycine max*)
Lentil (*Lens culinaris*)	Pigeon pea (*Cajanus cajan*)
Pea (*Pisum sativum*)	Mung bean (*Vigna radiata*)
Faba bean (*Vicia faba*)	Black gram (*Vigna mungo*)
Lupin (*Lupinus angustifolius*)	Cowpea (*Vigna unguiculata*)
Groundnut (*Arachis hypogaea*)	Rice bean (*Vigna umbellate*)
	Common bean (*Phaseolus vulgaris*)

Table 9. Examples of legumes that transport either amides or ureides in xylem sap as the dominant products of N_2 fixation exported from nodules. An expanded list of ureide exporting legumes can be found elsewhere [9,11,13].

3.3.2 Sap and tissue ureides as a measure of N_2 fixation

Independent research programs during the 1970s and early 1980s in the U.S.A., Australia, India, Africa and the U.K. capitalised on the earlier physiological and biochemical studies to develop quantitative ureide-based assays of legume N_2 fixation (*e.g.* 60-63). The principle of the assay is that the composition of N solutes in xylem sap (and aqueous stem extracts) changes from one dominated by ureides in N_2-dependent plants to one dominated by nitrate and amino-N in plants utilising soil N (Figure 6). Predictably, in calibration experiments, the relationship between the proportion of xylem sap N as ureides (termed relative ureide-N) and %Ndfa was shown to be close to unity [27].

3.3.3 Sampling of legumes for N solutes

Xylem sap can be collected as it exudes from the root stumps of decapitated plants (root-bleeding sap, RBS) or by vacuum extraction from cut stem segments (vacuum-extracted sap, VES). Collecting RBS from legumes growing in the humid tropics has not presented any problems; however, in other less humid environments the technique has been problematic. Vacuum extraction was developed as a consequence. Using this approach a mild vacuum is applied to the lower end of the detached shoot and small lengths of stem are cut successively from the top end to allow the xylem contents to be displaced by air and drawn through. The VES is trapped in a container connected in series to the source of the vacuum and the shoot. The trapped exudate proved to be very similar in composition to root-bleeding (xylem) sap (Figure 6).

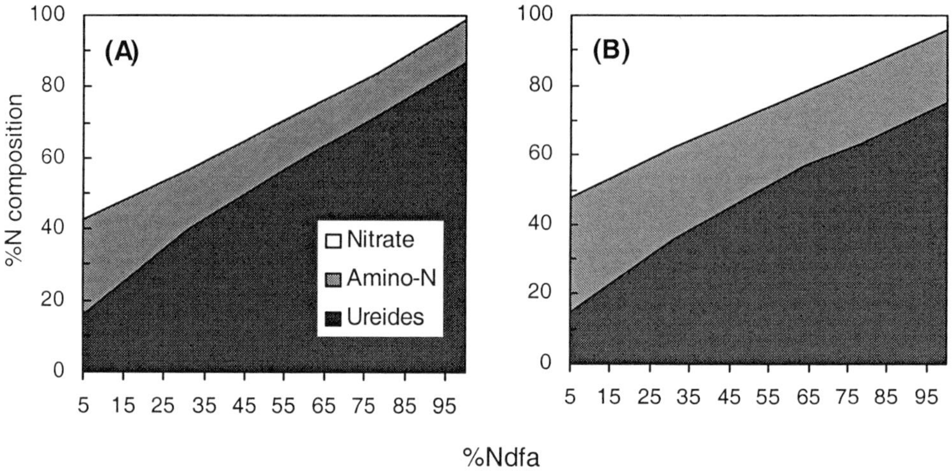

Figure 6. Changes in the composition of N solutes in (A) root-bleeding and (B) vacuum-extracted xylem sap of nodulated soybean, supplied with different concentrations of nitrate-^{15}N concentrations to generate different values for %Ndfa [27].

Aqueous extracts of the stem have also been used as a source of ureides in the measurement of N_2 fixation [26,64]. The stem extract assay appears to be as quantitative as assays based on xylem sap analysis [27], although the greater ease of sampling may be outweighed by the additional drying, grinding and extraction steps in the preparation of the samples for analysis.

3.3.4 Calibrating the ureide method for different legume species

Calibration experiments typically involve growing plants of the particular legume in large pots containing an N-free medium such as sand and supplied with one of a range of ^{15}N-labelled nitrate nutrient solutions (*e.g.* from zero to 8 or 10 mM nitrate [27,65]) to induce different levels of plant reliance upon BNF for growth. The plants are sampled at intervals throughout growth for measurement of %Ndfa using ^{15}N isotope dilution, plant N and ureides, amino-N and nitrate in xylem sap and plant tissues.

Relationships between relative ureide-N in VES of cowpea and mungbean and black gram are shown in Figure 7. Because low levels of ureides are detected in non N_2-fixing plants (typically about 10%), the calibration lines do not pass through the origin. The slopes of the lines, however, can vary considerably between species (Figure 7). Calibrations have now been published for a number of species of crop and forage and shrub legumes (Table 10).

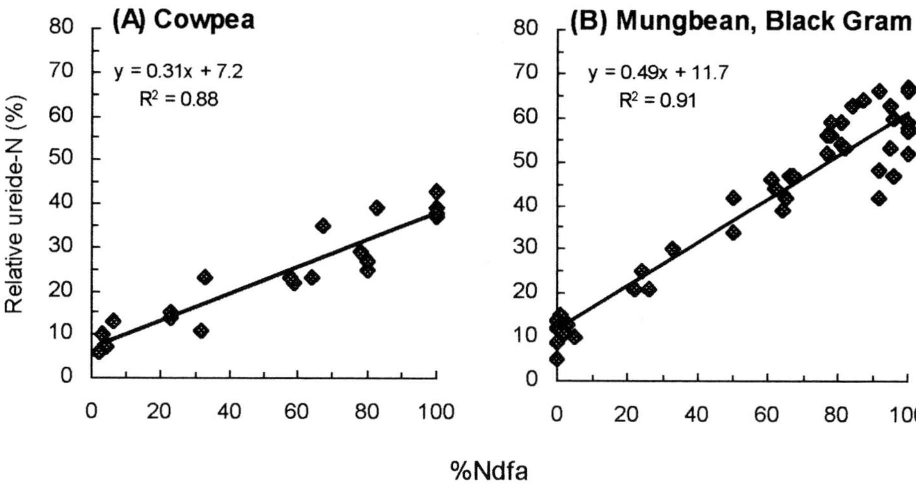

Figure 7. Calibrated relationships between %Ndfa and relative abundance of ureide-N in vacuum-extracted sap of (A) cowpea and (B) mungbean and black gram [66].

3.3.5 Sampling protocols and sources of error

The ureide assay of N_2 fixation based on VES or aqueous extracts of plant stems would be used most in field situations with RBS relevant in glasshouse studies or to plants in the humid tropics. With vacuum extraction, the vacuum can be generated using a hand, battery- or generator-powered vacuum pump. With the larger volume pumps, a number of operators doing the extraction can be serviced through a manifold system. With such a system, sap can be extracted from as many as 1500 plants a day (*i.e.* 10-15 plants/plot and 100-150 plots). Normally, the sap collected from the 10-15 plants in a plot would be bulked to a single plot sample.

Sampling of plants should be restricted to the 6-hour period between 10 am and 4 pm, to avoid diurnal variations in relative ureide-N [65,67]. The time between plant sampling (*i.e.* cutting the shoot) and xylem sap extraction should be restricted to <10 minutes, because of the differential effect of wilting on ureide and other N solutes. Samples, once collected, can be stabilised in ice or diluted 1:1 with ethanol [67]. There appears to be negligible effect of plant cultivar or rhizobial strain on the calibrated relationship between relative ureide-N and %Ndfa [13,27,65]. Sap collection can be non-destructive of the plant, if only the top of the stem or large lateral branch is used for extraction and the intact bottom section left to continue growing. Non-destructive sampling proved valuable in assessing N_2 fixation of F_2 soybean lines in a breeding program [68].

Collection of VES is particularly convenient for twining, forage legumes and woody legumes. With the latter, substantial volumes of sap (0.2—1.4 ml), were collected from 1-m lengths of branches of *Desmodium rensonii* and *Gliricidia sepium*, with no evidence of branch-to-branch variation in relative ureide-N [69].

Species	Calibration equation (y = RU-N; x = %Ndfa)	Sap/ tissue	Ref.
	Crop legumes		
Soybean	$y = 0.64x + 7.7$ (veg., flower.)	VES	27
	$y = 0.64x + 15.9$ (pod-fill)	VES	
	$y = 0.83x + 4.8$ (veg., flower.)	RBS	
	$y = 0.67x + 21.3$ (pod-fill)	RBS	
Pigeonpea	$y = 0.51x + 5.8$ (veg., flower.)	VES	65
	$y = 0.51x + 13.5$ (pod-fill)	VES	
Rice bean	$y = 0.0066x^2 - 0.06x + 7.6$	VES	50
	$y = 0.0086x^2 - 0.12x + 13.9$	RBS	
Cowpea	$y = 0.31x + 7.2$	VES	66
Mungbean, Black gram	$y = 0.49x + 11.7$	VES	
Cowpea, mungbean, black gram	$y = 0.75x + 8.6$	RBS	
Common bean	$y = 0.005x^2 + 0.06x + 8.5$	VES	70
	$y = 0.007x^2 + 0.02x + 7.6$	RBS	
	Forage and tree legumes		
Desmodium ovalifolium	$y = 0.89x - 2.1$	Stem	36
Centrosema spp.	$y = 0.85x - 2.6$	Stem	
Desmodium rensonii	$y = 0.45x - 18.3$	VES	69
Codariocalyx gyroides	$y = 0.28x + 8.5$	VES	

Table 10. List of species for which the ureide method has been calibrated. In each case, %Ndfa was calculated using ^{15}N isotope dilution.

3.3.6 Determining %Ndfa and cross-checking with ^{15}N

In a comprehensive review of legume N_2 fixation, Unkovich and Pate [18] stated that the main drawback of the ureide method was that each measurement reflected N_2 fixation at, or shortly before, the time of assay. The most useful information about N_2 fixation, however, is likely to be that describing the total proportion and amount of N fixed for a designated growth period. In the case of the tree and perennial fodder legumes, this may mean %Ndfa and total N fixed values for time-intervals between coppicing, cutting or grazing. Peoples *et al.* [36], in a study of N_2 fixation by tree legumes in northern Australia, reported very similar %Ndfa values for codariocalyx (*Codariocalyx gyroides*) using xylem ureide and ^{15}N methods (Figure 8A). This was despite the fact that the point-of-time ureide assay provided a %Ndfa value at the specific time of sampling, in contrast to ^{15}N-derived value that was an average for the time interval between the current and previous samplings.

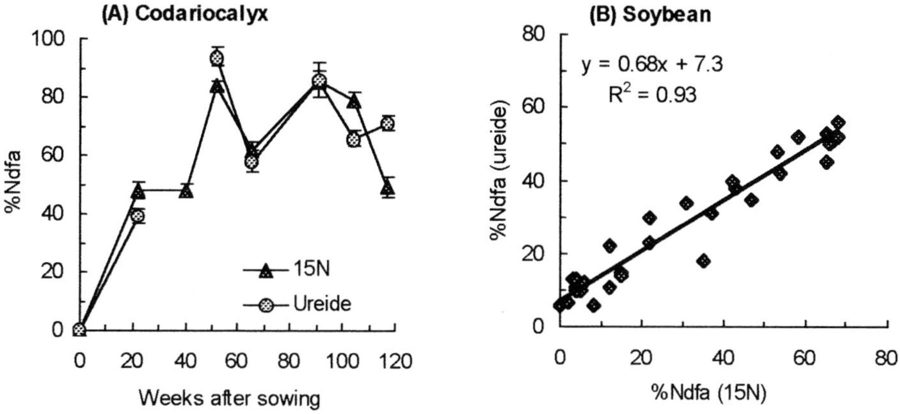

Figure 8. Cross checking of estimates of %Ndfa using ureide and ^{15}N methods for (A) *Codariocalyx gyroides* grown in the field in northern Australia [36], and (B) soybean genotypes at five field sites in the northern grains belt of Australia [49].

Similarly, Alves *et al.* [26] reported almost identical estimates of %Ndfa for the perennial forage legume, *Desmodium ovalifolium*, from the ureide assay based on N-solute analysis of aqueous stem extracts and ^{15}N isotope dilution. Values for %Ndfa, depending on reference species and inoculation, were 32-33% and 52-56% (ureide) and 30-40% and 53% (^{15}N). From these examples we could surmise that the ureide method provides similar estimates of %Ndfa to the ^{15}N methods for the perennial legumes when the objective is to assess N_2 fixation during specific growth periods.

For the annual crop legumes, the situation is a little different. The most valuable N_2 fixation values are the amounts of N fixed during the whole growth cycle. Recommended protocols to obtain such information involve multiple (up to 6-8) samplings at regular intervals from mid vegetative to late reproductive stages (*e.g.* Figure 1). Using those protocols, Herridge *et al.* [49] reported reasonably good agreement between ureide- and ^{15}N-determined %Ndfa estimates of soybean genotypes grown at five field sites in the northern grains belt of Australia (Figure 8B). Estimates of %Ndfa were in the ranges 6-56% (ureide) and 0-68% (^{15}N). At the higher levels of N_2 fixation, ureide-determined %Ndfa tended to underestimate ^{15}N-determined %Ndfa.

Sampat and colleagues quantified %Ndfa for different genotypes of the major crop legumes of northern Thailand from multiple (5—7) samplings of root-bleeding xylem sap and from natural ^{15}N abundance analysis of shoots at maximum biomass (Sampat, Peoples and Herridge unpublished data). Values for mungbean were 83% (^{15}N) and 76-87% (ureides), for black gram were 75-77% (^{15}N) and 80-85% (ureides), and for soybean were 70-81% (^{15}N) and 72-75% (ureides). Again, there was close agreement between the two methods.

3.3.7 Simplifying sampling protocols: the ureide assay for the future?

Clearly, the ureide method would be more convenient if the number of samplings was reduced to only one or two. The ^{15}N methods integrate all N_2 fixation activity from the onset of growth until the time of sampling, so estimates of total N fixed can be achieved by sampling the plants only once at the time of maximum biomass N (Figure 1). A single sampling of xylem sap for N-solute analysis would not necessarily coincide with the sampling for maximum biomass N and ^{15}N, but might be undertaken when the point-of-time %Ndfa had a similar value to the integrative, ^{15}N-determined %Ndfa.

Evidence that a single ureide sampling may provide a reliable estimate of overall %Ndfa was found in the published data of Herridge and Rose [71] for 13 soybean genotypes at three sites and Song *et al.* [72] involving nine genotypes of soybean at one site. In both cases, relative ureide-N values from early pod-fill samplings (physiological stage R2-R4, [73]) were highly correlated with $\delta^{15}N$ values of shoots and grain. Data from the Herridge and Rose study indicated a regression coefficient of 0.93 (Figure 9A). Agreement between ureide- and ^{15}N-derived estimates of %Ndfa was good.

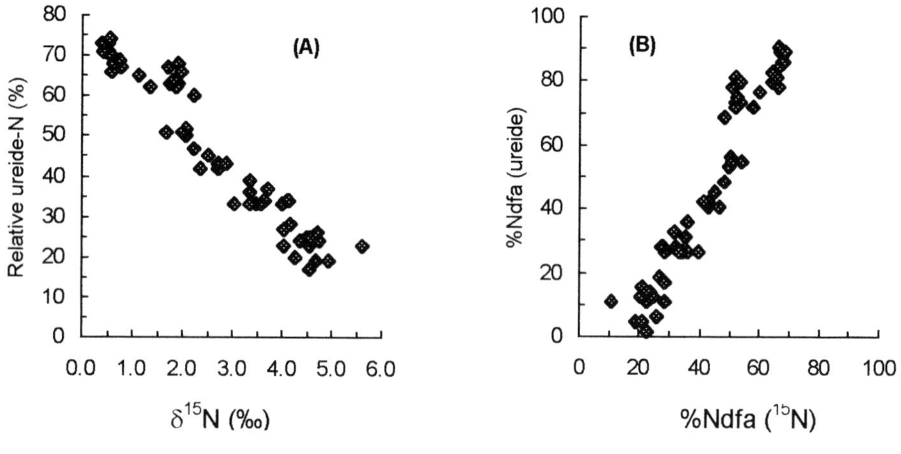

Figure 9. Cross-checking the natural ^{15}N abundance and ureide methods for assessing N_2 fixation of genotypes of soybean, grown in the field at Breeza, Australia. (A) Natural ^{15}N abundance values of harvested grain and relative ureide-N values of xylem sap collected during early pod-fill (physiological stage R3, [73]), and (B) %Ndfa estimates from grain ^{15}N and xylem sap ureides [71].

More recently, Herridge and Peoples [74] reported two studies involving soybean, cowpea, mungbean and black gram that clearly showed %Ndfa could be accurately estimated from a single sampling of xylem ureides during early pod-fill (R3.5). Estimates were compared to ^{15}N-determined %Ndfa, assessed at the time of maximum biomass N (R6-7). It is interesting that LaRue and colleagues [63,75] concluded almost

20 years ago that the ureide concentrations and relative ureide contents in petioles and young stems at R3 were best correlated with seasonal N_2 fixation.

Furthermore, the data from all species indicated that the temporal changes in relative ureide-N and the %Ndfa values calculated from them were relatively uniform, although ranging from gradual to sharp. At least under the experimental conditions of the reported study, relative ureide-N and %Ndfa (ureide) did not fluctuate to any degree in the short term. Such observations were supported by published profiles of relative ureide-N of xylem sap and stem segments and %Ndfa (ureide) of field-grown legumes that were similarly smooth [49,63-64,76]. The sharp changes in %Ndfa (ureide) of the short-duration cowpea, mungbean and black gram during pod-fill highlighted the need for accurate assessment of phenology and timely sampling. Substantial error could result from sampling just a week too early or too late. There was a marginal benefit in combining two sap samplings (R3.5 + R4, [74]) in terms of both accuracy and precision. The value of those benefits would need to be balanced against the additional cost of the extra sampling and analysis. Finally, a single sampling, or possibly two samplings, of xylem sap to determine seasonal %Ndfa simplifies the protocol of the technique and may encourage expanded use.

4. CONCLUSIONS

It is evident from the above review that considerable progress has been made over the last 10-20 years with techniques to provide field estimates of BNF for legumes and non-legumes. The possibility of using differences in ^{15}N natural abundance between fixing and non-fixing species to determine %Ndfa was first seriously investigated in the 1980's [11,13-14,17,51]. At around the same time, viable protocols were being developed for the application of the ureide sap method to field-grown legumes [64,67]. The great advantage of both these techniques is that they can be utilised to monitor inputs of fixed N by farmers' crops and pastures, and complements a recent shift of emphasis in BNF research from experimental trials to on-farm studies [*e.g.* 30,32,39,42-43,77-78].

The ^{15}N enrichment technique has its origins in the late 1950's [79], but with the increasing availability of ^{15}N-labelled fertilisers and isotope-ratio mass spectrometers, ^{15}N enrichment methodologies became widely used in the 1970's and 1980's [1,5,11,20,33]. In recent years a more critical attitude has been taken towards the application of this technique, using either multiple reference plants [3,34,37], or the simulation of soil N uptake from measurements of plant N and ^{15}N accumulation and ^{15}N enrichment of soil mineral N [6]. The use of the ^{15}N labelling of leaves or other aerial tissues to estimate the contribution of below-ground N to crop rotations is also another recent advance which has shown that these quantities of N can be highly significant [28-31].

The choice of which technique, or techniques, to apply to field studies of BNF depends firstly on the objectives of the study, and secondly on the facilities available to the investigators. It is generally considered advisable, wherever possible, to utilise more than one methodology (*e.g.* [36,49,80]). With the ^{15}N natural abundance approach the accuracy of N_2 fixation measurements tend to be most uncertain when %Ndfa is high because of difficulties in estimating a value to describe the extent of isotopic fractionation that occurs when the target species is totally dependent upon fixation for growth (*i.e.* problems in determining an appropriate B value for use in the calculations [2,4,15-16,19,38]). Errors with the ^{15}N enrichment method on the other hand, will be greatest when %Ndfa is low [16,80]. So if estimates of BNF obtained with both natural abundance and enrichment agree closely, then there can be some confidence in the result. In the same way similar estimates of N_2 fixation derived from both ureide and ^{15}N methodologies suggest some degree of reliability since the techniques are based on very different principles. The dilemma arises when estimates of %Ndfa differ between different approaches to quantify BNF. It is exceedingly difficult under such circumstances to distinguish which value might be closest to reality without a good understanding of the relative strengths and weaknesses of the various methodologies.

In the near future it is likely that the methods most utilised for measurement of the contributions of fixed N to legumes will be the ^{15}N natural abundance and ureide techniques, and for field studies on non-legumes the ^{15}N natural abundance approach. It is hoped that this review has pointed out the key difficulties and limitations of each of these methods so that more reliable estimates of BNF inputs into farming systems can be achieved.

5. REFERENCES

1. R. M. Boddey, CRC Crit. Rev. Plant Sci. 6 (1987) 209.
2. R. M. Boddey, J. C. Polidoro, A. S. Resende, B. J. R. Alves and S. Urquiaga, Aust. J. Plant Physiol. 28 (2001) 889.
3. R. M. Boddey, O. C. Oliveira, B. J. R. Alves, and S. Urquiaga, Fert. Res. 42 (1995) 77.
4. R. M. Boddey, M. B. Peoples, B. Palmer, and P .J. Dart, Nutr. Cycl. Agroecosys. 57 (2000) 235.
5. P. M. Chalk, Soil Biol. Biochem.17 (1985) 389.
6. P. M. Chalk and J. K. Ladha, Soil Biol. Biochem. 31 (1999) 1901.
7. S. K. A. Danso, Fert. Res. 42 (1995) 33.
8. S. K. A. Danso, G. Hardarson, and F. Zapata, Plant Soil 152 (1993) 25.
9. K. E. Giller and K. J. Wilson, Nitrogen Fixation in Tropical Cropping Systems, C.A.B. International, Wallingford, U.K., 1991.
10. D. F. Herridge and S. K. A. Danso, Plant Soil 174, (1995) 51.
11. S. F. Ledgard and M. B. Peoples, Advances in Nitrogen Cycling in Agricultural Ecosystems, J. R. Wilson (Ed.), Wallingford, UK: C.A.B. International. 1988, p.351.
12. J. S. Pate and M. J. Unkovich, Physiological Plant Ecology, M. C. Press, J. D. Scholes and M. G. Baker (eds.), Blackwell Sci. Ltd., Oxford, U.K. 1999, p.153.
13. M. B. Peoples and D. F. Herridge, Adv. Agron. 44, (1990) 155.

14. M. B. Peoples, D. F. Herridge and F. J. Bergersen, Sustainable Agriculture: Green Manure in Rice Farming, IRRI, Philippines, 1988, p.223.
15. M. B. Peoples, G. L. Turner, Z. Shah, S. H. Shah, M. Aslam, S. Ali, S.L. Maskey, S. Bhattarai, F. Afandi, G. D. Schwenke and D. F. Herridge, Extending Nitrogen Fixation Research to Farmers' Fields, O. P. Rupela, C. Johansen and D. F. Herridge, (eds.) ICRISAT, Patancheru, India, 1997, p.57.
16. M. B. Peoples, B. Palmer and R. M. Boddey, Stable Isotope Techniques in the Study of Biological Processes and Functioning of Ecosystems, M. Unkovich, J. Pate, A. McNeill D. J. Gibbs (eds.), Kluwer Academic Publ., Nordrecht, The Netherlands, 2001, p.119.
17. G. Shearer and D. H. Kohl, Aust. J. Plant Physiol. 13 (1986) 699.
18. M. J. Unkovich and J. S. Pate, Field Crops Res. 65 (2000) 211.
19. M. J. Unkovich and J. S. Pate, Stable Isotope Techniques in the Study of Biological Processes and Functioning of Ecosystems, M, Unkovich, J. Pate, A. McNeill D. J. Gibbs (Eds.), Kluwer Academic Publ., Nordrecht, The Netherlands, 2001, p.103.
20. J. F. Witty, R. J. Rennie and C. A. Atkins, World Crops: Cool Season Food Legumes, R. J. Summerfield (ed.), Kluwer Academic Press, London, UK, 1988, p.715.
21. J. F. Witty and F. R. Minchin, Nitrogen Fixation by Legumes in Mediterranean Agriculture, D. P. Beck and L. A. Materon (Eds), ICARDA, Aleppo, Syria, 1988, p.331.
22. M. B. Peoples, D. F. Herridge and J. K. Ladha, Plant Soil 174 (1995) 3.
23. L. L. Handley, A. T. Austin, D. Robinson, C. M. Scrimgeour, J. A. Raven, T. H. E. Heaton, S. Schmidt and G. R. Stewart, Aust. J. Plant Physiol. 26 (1999) 185.
24. P. Högberg, New Phytol. 137 (1997) 179.
25. M. B. Peoples, G. D. Schwenke and D. F. Herridge, Extending Nitrogen Fixation Research to Farmers' Fields, O. P. Rupela, C. Johansen and D. F. Herridge (eds.), ICRISAT, Patancheru, India, 1997, p.141.
26. B. J. R. Alves, A. S. Resende, S. Urquiaga and R. M. Boddey, Nutr. Cycl. Agroecosys. 56 (2000) 165.
27. D. F. Herridge and M. B. Peoples, Plant Physiol. 93 (1990) 495.
28. M. B. Peoples and D. F. Herridge, Nitrogen Fixation: From Molecules to Crop Productivity, F. O. Pedrosa, M. Hungria, M. G. Yates and W. E. Newton (eds.), Kluwer Academic Publishers, Netherlands, 2000, p.519.
29. M. B. Peoples and J. A. Baldock, Aust. J. Exp. Agric. 41 (2001) 327.
30. I. J. Rochester, M. B. Peoples, G. A. Constable and R. R. Gault, Aust. J. Exp. Agric. 38 (1998) 253.
31. C. A. Russell and I. R. P. Fillery, Aust. J. Agric. Res. 47 (1996) 1047.
32. G. D. Schwenke, M. B. Peoples, G. L. Turner and D. F. Herridge, Aust. J. Exp. Agric. 38 (1998) 61.
33. J. F. Witty, Soil Biol. Biochem. 15 (1983) 631.
34. M.S. Viera-Vargas, O.C. de Oliveira, S.M. Souto, G. Cadisch, S. Urquiaga and R.M. Boddey, Soil Biol. Biochem. 27 (1995) 1185.
35. S. Urquiaga, K. H. S. Cruz and R. M. Boddey, Soil Sci. Soc. Am. J. 56 (1992) 105.

36. M. B. Peoples, B. Palmer, D. M. Lilley, M. D. Lam and D. F. Herridge, Plant Soil 182 (1996) 125.
37. R. M. Boddey, S. Urquiaga, A. R. Suhet, J. R. Peres and M. C. P. Neves, Soil Biol. Biochem. 22 (1990) 649.
38. J. A. Doughton, I. Vallis and P. G. Saffigna, Plant Soil 144 (1992) 23.
39. P. A. Riffkin, P. E. Quigley, G. A. Kearney, F. J. Cameron, R. R. Gault, M. B. Peoples and J. E. Thies, Aust. J. Agric. Res. 50 (1999) 261.
40. A. S. Blackmer and J. M. Bremner, Soil Biol. Biochem. 9 (1977) 73.
41. D. F. Herridge, H. Marcellos, W. L. Felton, G. L. Turner and M. B. Peoples, Soil Biol. Biochem. 27 (1995) 545.
42. Z. Shah, S. Shah, D. F. Herridge and M. B. Peoples, Extending Nitrogen Fixation Research to Farmers' Fields, O. P. Rupela, C. Johansen and D. F. Herridge (eds.), ICRISAT, Patancheru, India, 1997, p.333.
43. M. Aslam, I. A. Mahmood, S. Ahmad, M. B. Peoples and D. F. Herridge, Extending Nitrogen Fixation Research to Farmers' Fields, O. P. Rupela, C. Johansen and D. F. Herridge (eds.), ICRISAT, Patancheru, India, 1997, p.353.
44. R. A. Sutherland, C. van Kessel and D. J. Pennock, Soil Sci. Soc. Am. J. 55 (1991) 1339.
45. G. L. Androsoff, C. van Kessel and D. J. Pennock, Biol. Fertil. Soils 20 (1995) 33.
46. F. C. Stevenson, J. D. Knight and C. van Kessel, Soil Sci. Soc. Am. J. 59 (1995) 1603.
47. K. W. Steele and R. M. Daniel, J. Agric. Sci. (Camb.) 90 (1978) 7.
48. J. Eriksen and H. Hogh-Jensen, Plant Soil 205 (1998) 67.
49. D. F. Herridge, F. J. Bergersen and M. B. Peoples, Plant Physiol. 93 (1990) 708.
50. B. Rerkasem, K. Rerkasem, M. B. Peoples, D. F. Herridge and F. J. Bergersen, Plant Soil 108 (1988) 125.
51. F. J. Bergersen, J. Brockwell, R. R. Gault, L. Morthorpe, M. B. Peoples and G. L. Turner, Aust. J. Agric. Res. 40 (1989) 763.
52. E. G. Bollard, Aust. J. Biol. Sci. 10 (1957) 292.
53. J. S. Pate, Ann. Rev. Plant Physiol. 31 (1980) 313.
54. S. Tajima and Y. Yamamoto, Plant Cell Physiol. 16 (1975) 271.
55. T. Matsumoto, M. Yatazawa and Y. Yamamoto, Plant Cell Physiol. 18 (1977) 459.
56. D. F. Herridge, C. A. Atkins, J. S. Pate and R. M. Rainbird, Plant Physiol. 62 (1978) 495.
57. K. R. Schubert, Ann. Rev. Plant Physiol. 37 (1986) 539.
58. D. F. Herridge, M. N. Sudin, J. S. Pate and M. B. Peoples, World Crops: Cool Season Food Legumes, R. J. Summerfield (ed.), Kluwer Academic Publishers, The Netherlands, 1988, p. 793.
59. J. I. Sprent, Nodulation in Legumes. The Cromwell Press Ltd., U.K., 2001.
60. D. F. Herridge, Current Perspectives in Nitrogen Fixation, A. H. Gibson and W. E. Newton (eds.), Aust. Acad. Sci., Canberra, 1981, p. 484.
61. P. R. McClure, D. W. Israel and R. J. Volk, Plant Physiol. 66 (1980) 720.
62. J. S. Pate, C. A. Atkins, S. T. White, R. M. Rainbird and K. C. Woo, Plant Physiol. 65 (1980) 961.

63. T. G. Patterson and T. A. LaRue, Crop Sci. 23 (1983) 825.
64. D. F. Herridge, Plant Physiol. 70 (1982) 7.
65. M. B. Peoples, D. M. Hebb, A. H. Gibson and D. F. Herridge, J. Exp. Bot. 40 (1989) 535.
66. D. F. Herridge and M. B. Peoples, Comm. Soil Sci. Plant Anal. 33 (2002) 425.
67. D. F. Herridge, P. O'Connell and K. Donnelly, J. Exp. Bot. 39 (1988) 12.
68. D. F. Herridge and I. A. Rose, Crop Sci. 24 (1994) 360.
69. D. F. Herridge, B. Palmer, D. P. Nurhayati and M. B. Peoples, Soil Biol. Biochem. 28 (1996) 281.
70. D. F. Herridge, J. F. Holland, I. A. Rose and R. J. Redden, Improving Yield and Nitrogen Fixation of Grain Legumes in Tropics and Sub-Tropics of Asia, IAEA, Austria, 1988, p. 9.
71. D. F. Herridge and I. A. Rose, Field Crops Res. 65 (2000) 229.
72. L. Song, B. J. Carroll, P. M. Gresshoff and D. F. Herridge, Soil Biol. Biochem. 27 (1995) 563.
73. W. R. Fehr, C. E. Caviness, D. T. Burmood and J. S. Pennington, Crop Sci. 11 (1971) 929.
74. D. F. Herridge and M. B. Peoples, Plant Soil 238 (2002) 57.
75. R. A. Glenister and T. A. LaRue, Exp. Agric. 22 (1986) 405.
76. J. Ying, D. F. Herridge, M. B. Peoples and B. Rerkasem, Plant Soil 147 (1992) 235.
77. S. L. Maskey, S. Bhattarai, M. B. Peoples and D. F. Herridge, Field Crops Res. 70 (2001) 209.
78. M. B. Peoples, A. M. Bowman, R. R. Gault, D. F. Herridge, M. H. McCallum, K. M. McCormick, R. M. Norton, I. J. Rochester, G. J. Scammell and G. D. Schwenke, Plant Soil 228 (2001) 29.
79. C. McAuliffe, D. S. Chamblee, H. Uribe-Arango and W. W. Woodhouse Jr, Agron. J. 50 (1958) 334.
80. J. A. Doughton, P. G. Saffigna, I. Vallis and R. J. Mayer, Aust. J. Agric. Res. 45 (1995) 225.

Nitrogen Fixation at the Millennium.
G. Jeffery Leigh, Editor.

CHAPTER 14

Nitrogen Fixation and Agricultural Practice

G.W. O'Hara[1], J.G.Howieson[1] and P.H.Graham[2]
[1]Centre for Rhizobium Studies, Murdoch University, Murdoch, 6150, WA, Australia.
[2] Department of Soil, Water and Climate, University of Minnesota, St Paul, MN, USA.

1. INTRODUCTION

Production of high-quality, protein-rich foods is necessarily dependent upon the availability of adequate supplies of nitrogen in soils. At present, the world's population of six billion people consumes an average of nearly 11 g of nitrogen per person per day, giving an annual consumption of about 24 Mt of nitrogen [1].

After carbon and water, nitrogen is the major limiting nutrient for plant growth [2] and crop productivity [3]. Biological fixation of atmospheric nitrogen and the application of chemical fertilizers together contribute most of the nitrogen available in soils for agricultural crops [4]. Overwhelmingly, most nitrogen fixation in agricultural practice is from symbiotic legumes, estimated to amount to at least 70 million metric tons of nitrogen per year globally [5]. In addition, quantifiable nitrogen fixation is reported from actinorhizal non-legumes such as *Alnus* and *Casuarina* in association with the actinomycete *Frankia* [6], and with sugar-cane in association with a range of endosymbionts and rhizospheric bacteria [7]. Furthermore, azorhizobia isolated from *Sesbania aculeata* and *Aeschynomene fluminensis* colonize rice endophytically [8] although their contribution to rice growth and capacity for nitrogen fixation in this setting remain to be established. Set against this, in 1990, world consumption of fertilizer nitrogen was approximately 80 million tons whose production utilised 1.4% of the fossil fuels consumed that year [9].

Nitrogen-fixing crop, forage and pasture legumes are essential components for sustaining productivity in many agricultural systems [10]. A clear benefit from using these symbiotic legumes in many of these agricultural systems is the fixation of atmospheric nitrogen by the root-nodule bacteria located in the nodules formed on legume roots [11]. During the 20th century inoculation of legumes with root-nodule bacteria developed in many parts of the world to become one of the most cost-effective of all agricultural practises [12]. This chapter will focus on the key roles for nitrogen-fixing crop and pasture legumes in agriculture. These roles will be discussed in terms of the legume hosts, their associated root-nodule bacteria and the environmental constraints that limit legume productivity by affecting symbiotic nitrogen fixation. For a broader account, see Sessitsch *et al.* [13], who review aspects of non-legume and non-rhizobial nitrogen fixation. The purpose of this present chapter is to outline some important areas of our current understanding of the factors influencing the use and productivity of symbiotic legumes, and also to highlight some important developments for improved legume productivity.

2. THE MACROSYMBIONT

2.1 Historical perspective

Legumes are conveniently classified taxonomically into one family (Leguminosae), with three sub-families, the Caesalpinioideae, Mimosoideae and Papilionoideae [14]. Legumes are believed to have evolved in the late Cretaceous period and subsequently to have spread to all continents except Antarctica [14].

Since antiquity legume species have been used as a primary protein source, forage or herb, on all continents, and across a diversity of agricultural and social settings. Pliny the Elder in the first century AD [15] referred to the forage *Melilotus* as a valuable horse fodder, and in the same era Virgil recorded the improved quality in wheat fields planted after 'lupine' [16]. The cool-season pulses *Cicer arietinum* L. (chickpea), *Lens culinaris* Medik. (lentil), *Pisum sativum* L. (field pea) and *Vicia faba* L. (faba bean) have, for millenia, provided the major protein component in human diets in West Asia, North Africa and the Indian sub-continent. *Phaseolus vulgaris* L. (common bean) appears to have been domesticated in Andean South America about 4400 years ago and has been cultivated alongside maize in Central America for 2500 years [17]. In the Mediterranean basin self-regenerating herbaceous legumes (*e.g*,, *Trifolium* spp., *Medicago* spp.) have been primary components of permanent non-arable grasslands for thousands of years and this continues almost unchanged in parts of the Mediterranean basin [18]. Similarly, *Trifolium alexandrinum* L. (berseem or Egyptian clover) was first cultivated as an annually re-sown forage legume in the Nile delta more than 5,000 years ago, as it is today [19]. So legume usage in agriculture is an ancient and well-proven practise that has relevance in the modern world as we search for sustainable ways to feed a voracious population that is expected to reach 10 billion by 2040 [2].

2.2 The value of legumes in agriculture

Although, legumes are utilised in agriculture because they enhance the productivity and potential sustainability of farming systems, they seldom dominate farming systems. This is practical recognition that, considered in isolation, legumes might not always be (transparently) the most profitable land-use option. Often legume yields or price for their grains are lower than for some non-legume commodities such as cereals or oilseeds. The decision to grow legumes in integrated farming is a complex one and can be aided by bio-economic modelling [20]. A series of flow diagrams connecting bio-economic modelling with likely legume responses to inoculation has also been developed [13].

Cereals, oilseeds, grasses and herbs produce higher protein grains and higher yields when grown after, or in conjunction with, symbiotic legumes in depauperate soils [21]. The primary agronomic reason for these responses is nitrogen fixation by the legumes and the transfer of some of the fixed nitrogen to the non-legumes. The transfer of symbiotically fixed nitrogen from legumes through soil microbial processes to non-legumes can readily be quantified using isotopic techniques, such as the natural abundance of the ^{15}N isotope [22].

The second major agronomic benefit resulting from utilising legumes in rotational farming systems is the control of cereal or oilseed diseases and pests [23]. Many pests and diseases that become dominant in monoculture agriculture are host-specific. However, the absence of a preferred host for a whole season substantially impedes their persistence and prevalence [24]. Thus, in legume-based rotations, cereal or oilseed disease and insect damage are reduced because of a lack of build-up of inoculum or insect numbers during the legume phase. This reduces the reliance on expensive chemical fungicides and pesticides to maximise yields.

The major and secondary benefits surrounding the use of legumes in agriculture have been reviewed by Howieson *et al.* [25] and are expanded upon in Table 1.

2.3 Environmental factors limiting use of legumes in agriculture

Several key abiotic factors limit legume productivity and nitrogen fixation in world agriculture. These may be summarised as extremes of temperature, water availability and nutrient availability, and toxic soil factors such as sodicity and pH.

2.3.1 Temperature

In the tropics, soil temperatures may exceed 50 °C at 1 cm depth and effects of this stress upon nitrogen fixation directly, as well as upon nodulation and plant growth, are well documented [26]. In Mediterranean-type climates, high temperature (above 30 °C) associated with water deficit during spring and early summer is the primary determinant of the termination of the growing season and hence nitrogen fixation in annual legumes. The rapidity with which this combination of temperature and water stress causes plant senescence depends upon rooting depth of the legume, access to subsoil moisture and the genetic capacity of the plant to tolerate transient drought stress.

In contrast, in Mediterranean and temperate agriculture, low temperatures during autumn and winter are a substantial impediment to legume growth and nodulation, with subsequent detrimental effects on seed yield. Whilst *Lupinus* spp. roots may tolerate minus 3.5 °C at 10 cm depth [27], temperatures as low as this in the root-

Specific legume value	Foragee	Pulse
N_2-fixation benefiting companion or rotational non-legumes	***	**
high protein human feed in green parts, seeds or tubers	*	***
high protein animal feed	***	*
increased soil fertility and structure, decreased erosion risk	***	*
in permanent settings, provide biodiversity, shelter (tree legumes) and natural beauty	***	*
high protein, high value cash crop	*	***
reduced growth of weed species	**	***
break disease and pest life cycles	**	***
capacity for high water use⁺	*	**
capacity for nutrient recycling⁺	*	**
deliver operational flexibility in farming systems	**	**

*, **, *** increasing applicable value
⁺greater for trees, and for perennial rather than for annual forage species

Table 1. Benefits from incorporation of legumes into farming systems.

zone can impede nodulation by reducing the production of biochemical factors required for nodule initiation by both the legume and the root-nodule bacteria [28]. In addition, low temperatures later during the growing season often cause substantial frost damage during anthesis and as a consequence reduce grain yields. The ability to survive beneath layers of snow is an essential requirement in spring-growing ley legumes in harsh temperate environments. This requirement has substantially affected the eco-geographic distribution of a wide range of forage legumes found on the rangeland steppes of north Africa and West Asia [29].

2.3.2 Nutrition

Mineral nutrients may limit nitrogen fixation in legumes through direct effects upon host-plant growth and rhizobia [30], or through indirect effects upon the symbiosis. Robson [31] summarised the nature of the interaction between nutrient supply and combined nitrogen on legume growth as a means for estimating symbiotic sensitivity to their supply or concentration. He concluded that Co and Mo are required in greater amounts for symbiotic nitrogen fixation than for host-plant growth, as were Cu, Ca and P under some circumstances. Although there is currently experimental evidence for specific requirements for 11 nutrients (B, Ca, Co, Cu, Fe, K, Mo, Ni, P, Se and Zn) for symbiotic development in some species of legume, only four of these elements (Ca, P, Fe and Mo) appear to cause significant limitations on the productivity of symbiotic legumes in some agricultural soils [30].

2.3.3 pH

One of the most widely researched facets of the legume symbiosis is the impact of pH upon it, in particular of acidity. This is because many landforms (be they ancient or modern) are naturally acidic. Parenthetically, acidification is a by-product of legume-based agriculture that is seen as a threat to its sustainability [32]. However, soil acidification through agriculture may be managed by the routine application of lime, or the selection of legumes which minimise acid production and nitrate leaching, and therefore minimise sub-surface acidification [33]. A large proportion of tropical soils are geologically old and have been exposed to high rainfall events that have left a cation exchange complex dominated by the strongly acidic cations of H, Al and Mn, rather than by the more benign ions of Ca and Mg [34]. Currently arid landscapes that are also geologically ancient, such as those in south-west Australia, resemble these tropical soils in their chemistry as it impacts legume growth and symbiotic function [31].

Legumes and rhizobia, as well as their mutual interaction, are variably susceptible to soil pH outside the 'normal' ranges of pH 5 to pH 8 [35]. Investigations into effects of acidity upon legume production must first consider which steps in the symbiosis, from rhizobial survival through nodulation to host plant growth, are most sensitive to this stress [36]. For example, the biochemical signalling between symbionts which precedes nodulation appears to be perturbed by pH stress in the *Medicago* [37] and *Trifolium* [38] symbioses at pH levels above which the plant can grow with combined nitrogen [39]. Whether this perturbation impacts mostly on the production of root exudates essential for nod-gene induction in rhizobia, by assimilation or metabolism of these exudates within the rhizobial cell, or by the subsequent recognition by the root of rhizobial nod-factor remains unclear. Nevertheless, we see substantial genetic variation in the capacity of *Medicago* species to nodulate under acid stress that might in future be harnessed to develop plant-based tolerance to acidity in the nodulation phase in commercial forage *Medicago* spp. [40].

At the other end of the scale, soil pH above 6.0 may reduce nodulation and limit nitrogen fixation in some alkaline sensitive legume species. For example, high pH caused impaired root-hair formation and disintegration of the root surface in *Lupinus angustifolius,* which reduced nodule initiation and decreased nitrogen fixation [41]. *Arachis hypogaea* is similarly sensitive to high pH or high bicarbonate levels in soil, whereas *L. pilosus* is not.

2.4 Legumes in agriculture for the 21st century

There is little doubt that the requirement to feed the world's expanding population will demand improved agricultural productivity. Legumes and their microsymbionts must necessarily play a significant role in increasing this productivity. Herridge and Rose [42] consider a mere 10 % increase in nitrogen fixation in grain legumes alone would lead to bio-economic benefits of nearly US$ 1 billion ($ 1x10^9). One approach to increase inputs of legume-fixed nitrogen into agricultural soils might be achieved by breeding legumes with increased nitrogen fixation supportive traits. An alternate strategy could be to expand the usage of legumes in world agriculture. We will examine these two possible routes for obtaining increased nitrogen fixation in agriculture.

2.4.1 Increased legume usage

Howieson *et al.* [25] report on the changing roles for legumes in Mediterranean agriculture. In selecting a "second generation" of legumes for ley-farming in Australia, scientists there have responded to changing economic, biological and environmental constraints upon legume productivity. Similar constraints to pasture production from existing legume-based pastures operate in the dryland 'secano interior' of Chile [43]. These constraints, in particular secondary salinity caused by rising water tables, the declining terms of trade for rural producers in non-subsidised economies and the development of herbicide resistant weeds, have together precipitated the development of alternative ley-legumes with substantially different agronomic traits. The discovery of new legumes for agriculture with traits such as deeper root systems, very high fecundity and levels of hard seed which survive ingestion, acid and grazing tolerance combined with efficient phosphate utilisation and insect tolerance has increased productivity from forage legumes used in crop rotations in ley farming in Australia [44]. Some measured productivity increases are of the order of 100% relative to the previously recommended legumes and much of this productivity has been from soils that are predominantly impoverished and respond greatly to legume cultivation [45]. Not coincidentally, the development of the new ley legumes was supported by parallel selection of well-adapted inoculum strains of root-nodule bacteria sourced from the Mediterranean basin [46].

The scientists involved in this paradigm shift in ley-legume usage expect the further development of legumes with traits that provide secondary processing options to potentiate diversification of rural enterprises. Examples of these include forage legumes with appropriate levels and types of condensed tannins that reduce the anthelmintic burden in grazing animals [47]. Forage legumes are also being sought from which dietary isoflavones can be extracted to deliver health benefits relating to hormone-dependent cancers and other degenerative diseases [48]. In addition, in anticipation of climate change and in recognition of the requirement to increase water transpiration in regions with rising water tables, sub-tropical legumes such as *Lotononis, Swainsona, Crotalaria* and the African species of *Trifolium* are being evaluated for their adaptation to agriculture in a Mediterranean climate. Thus, there appears to be a considerable dynamism in contemporary research to extend the usage of forage legumes and to increase their productivity, at least in Mediterranean-type agriculture where production is not subsidised and therefore must be efficient.

Improvement of forage legumes in the tropics has been less successful and there appears little optimism that this situation will rapidly change. The reasons for comparative lack of success are complex, yet Miles [49] lists inadequate assessment of the target environment in socio-economic terms, inadequate institutional base and true biological obstacles as the primary reasons. The biological constraints included the recognition that all legumes are C_3-type plants with restrictive photosynthetic and nitrogen-fixation properties not well suited to the heavy grazing and rapid regrowth requirements in the tropics. In contrast to the many annual species of legumes used for forage and pasture in temperate and Mediterranean-type agriculture, most tropical legumes are perennial species. This perennial habit limits their flexibility in intensive farming systems. Breeding C_4 tropical grasses has been highly successful in comparison to breeding tropical forage legumes and the introduction of temperate legumes and rhizobia into the permanent sub-tropical grasslands of South America also holds substantial promise [50].

There is also little likelihood of developing rapidly an expanded suite of grain legumes with which to increase productivity in world agriculture in the same way that the second-generation ley legumes have in Australia. The domestication of *Lupinus angustifolius* since 1960 has been notable [51] and there remains considerable potential in the New-World species of lupin such as *L. mutabilis*. Bennett *et al.* [52] review the wide range of minor and under-exploited pulses in the Mediterranean basin, yet the domestication process for these may take upwards of 40 years. Sprent [14] highlights a range of legume genera with agricultural potential that produce root-tubers, such as

Pachyhrizus and *Vigna.* However, for immediate gain, scientists may need to concentrate upon improving productivity and nitrogen fixation from the current suite of grain legumes of world importance including *Glycine max, Cicer arietinum, Phaseolus vulgaris, Pisum sativum, Vicia faba, Vignaspp, Cajanus spp* and *Lupinus spp*. There has already been considerable progress in improving productivity from some of these pulses through incorporation of traits such as disease resistance, winter hardiness, insect tolerance and improved harvest index [53].

2.4.2 Breeding legumes for increased nitrogen fixation

Most of the major pulse or grain legume species exhibit variability in traits relating directly to nitrogen fixation that have genetic heritability, sometimes at a single locus [10]. Traits such as time to formation of the first nodule, nodule number, nodule efficiency for nitrogen fixation in the presence of background soil nitrogen, interaction with rhizobial strain *etc*., are frequently reported, and may be actively selected to improve nitrogen fixation [54]. Even for established commercial pulses, yield improvements of up to 100% can be achieved through selection of elite inoculant strains of root-nodule bacteria well adapted to stressful soils to give early and sustained nodulation [55].

Herridge and Rose [42] have summarised the outcomes of various breeding efforts for increased nitrogen fixation in some grain crops in the three decades preceding the current millenium. Parameters important in breeding for this trait were considered to be yield, nitrate tolerance, hypernodulation, selective nodulation and promiscuous nodulation. A common problem in optimising the symbioses in grain legumes often was the presence in the soil of competitive, yet poorly effective, populations of microsymbiont. There has been progress in overcoming this constraint in Africa by breeding soybeans capable of selectively nodulating with indigenous and effective populations of bradyrhizobia [56]. This approach is also being explored in Australia to increase nitrogen fixation in *Medicago* and *Trifolium* pasture legumes severely challenged for fixation by ineffective background populations of rhizobia [57]. Likewise, in the USA where soybean nitrogen fixation is compromised by the ubiquitous but poorly effective USDA123 type strains, attempts are being made to develop cultivars capable of restricting entry of this class of bradyrhizobia into nodules in favour of more effective inocula [58]. Van Kessel and Hartley [59] advocate an agronomic approach to improving nitrogen fixation from grain legumes rather than a focus upon the nitrogen fixation process in isolation.

A common theme in successful programs to select or breed for improved legume symbioses seems to be to undertake the research in soils that represent the target farming soils for the legume. After background research to ensure legumes and/or inoculants are required in the farming systems of focus and are socio-economically relevant, Sessitsch *et al.* [13] advocated that a phase in the selection of elite rhizobial strains be undertaken *in situ* in representative soils. Similarly, Herridge and Rose [42] concluded that successful programs to breed or select for improved nitrogen fixation in grain legumes were those that were undertaken in soils that represented the farming soils of the particular country. The development of second-generation ley legumes in Western Australia, as discussed above, was the outcome of field-based research with optimised symbioses that strongly focussed upon farming systems inadequacies in the contemporary legume species [25].

These various examples of success in disparate programs provide the salient lessons that if we are to continue to improve inputs from symbiotic nitrogen fixation in agricultural practice in the current millennium, significant components of the research must be carried out *in situ* in low nitrogen soils, and that legume breeders must work in close association with rhizobiologists. Further impetus in the adoption of legumes into farming systems will be gained when major world economies abandon their subsidisation of cereal grain and oilseed production and thereby force producers to focus upon cost-effective inputs, such as legumes, for agricultural production.

3. THE MICROSYMBIONT

From their initial isolation by Beijerinck in 1888, until well into the 1950s, the main interest for study of root-nodule bacteria was with the more practical aspects of nodule formation and nitrogen (N_2) fixation in symbiosis. Recent improvements in molecular methodologies, coupled with a greater interest in basic research in countries where fertilizer nitrogen remains cheap and manure in excess, has dramatically changed that picture. Current emphases for research on these bacteria include the molecular signalling between them and their legume host plant, biodiversity and taxonomy, tolerance of stress, and genomics. In several of these areas the rhizobia serve as model organisms with implications for plant and animal disease.

In the following Sections we will attempt to present a cohesive picture of these nitrogen-fixing microsymbionts associated with nodule formation in legumes, and of their importance in agriculture.

3.1 Cross-inoculation grouping and speciation among root-nodule bacteria

Each rhizobial strain or nodule isolate has a finite host range, nodulating certain legumes but not others. In early studies this led to the concept of cross-inoculation, with legumes grouped according to the different rhizobia with which each formed nodules [60]. More than 20 cross-inoculation groups were identified, with the bacteria from the clover, medic, bean, lupin, pea and soybean groups named as separate species within the single genus *Rhizobium*. Host specificity is still important in taxonomy of root-nodule bacteria, but other traits are now given greater weight. This was partly because of the blurring of cross-inoculation boundaries that followed the study of a more diverse assortment of plant species, but also because of developments in computer-assisted and phylogenetic approaches to taxonomy. The latter often gave results at variance with those obtained on the basis of host range [61]. For example, *P. vulgaris* is now known to be nodulated by at least six different species of rhizobium [62]. Subdivisions among the root-nodule bacteria are still an issue for debate. Changes suggested during the past 40 years have divided the original genus *Rhizobium* into 6 genera and 30 species as shown below. A number of these divisions are still questioned [63].

Genus	**Species included**
Allorhizobium	*undicola*
Azorhizobium	*caulinodans*
Bradyrhizobium	*elkanii, japonicum, liaoningense*
Mesorhizobium	*amorphae, chacoense, ciceri, huakuii, loti, mediterraneum, plurifarium and tianshanense*
Rhizobium	*etli, galegae, gallicum, giardinii, hainanense, huautlense, leguminosarum, mongolense,* and *tropici*
Sinorhizobium	*arboris, fredii, kostiense, medicae, meliloti, saheli, teranga and xinjiangense.*

The rhizobia from *Lupinus, Arachis* and *Vigna* and many other tropical legumes are slow-growing, and belong in the genus *Bradyrhizobium*. However, they have not been awarded species names and are generally identified as *Bradyrhizobium spp* with the legume host identified in parentheses. Greater detail of each species including the hosts from which isolated, and original references, is provided by Martinez-Romero *et al.* [64] and Zakhia and de Lajudie [65].

Significant differences between fast- and slow-growing rhizobia have been recognized for some time [66]. A high percentage of strains in the latter group have more recently shown the ability to grow as hydrogen autotrophs [67], and to recycle hydrogen (H_2) lost during nitrogen fixation. Soybeans inoculated with strains having Hup activity yielded 9-11% more, and contained more N than plants inoculated with isogenic strains differing only in the absence of the *Hup* gene [68]. Some slow-growing rhizobia have also been shown to produce bacteriochlorophyll [69], and to bear a close relationship to *Rhodopseudomonas palustris* [61]. In contrast, *Rhizobium tropici* possesses unusual pH and temperature tolerances [70], with close relationships claimed to both *Brucella* [64] and *Agrobacterium* [63]. The confusing taxonomic and infective situation has not been improved by recent reports [71] in which species of methyl-oxidizing bacteria have been shown to nodulate legumes such as *Sesbania, Crotalaria* and *Lotononis.*

3.2. Nodule formation and development

Rhizobia can infect their legume hosts and induce root- or stem-nodule formation through three mechanisms:

root-hair penetration and infection thread formation, as occurs in clovers, beans and soybeans [72];

entry via wounds or sites of lateral root emergence as found in peanut, the pasture legume *Stylosanthes* and the aquatic species *Neptunia natans* [73]; or penetration of root primordia found on the stem of plants such as *Sesbania* [74].

The same strain may infect one host by root-hair penetration and another by wound entry; while predominantly stem-nodulating *Azorhizobium* and root-nodulating *Sinorhizobium* can each infect the other site [75]. Stem nodules formed by *S.teranga* and *S. saheli* are ineffective; root nodulation by *S.teranga* does not limit stem nodulation by *A.caulinodans* [76].

While cytological changes associated with root-hair infection and nodule development have been appreciated for some time [77], more recent progress has been in understanding the molecular signalling between host and rhizobia. Flavonoids secreted by the host root [78] and complex lipo-chitooligosaccharides or "nod-factors" produced by the rhizobia [79] play major roles in this signalling [80]. Several regulator molecules are also involved [81]. Agricultural implications exist at several levels.

Lipo-chitooligosaccharides act as powerful morphogens that at low concentrations can themselves effect many of the changes typical of early nodule development [82; reviewed in 72]. Nod factors have also been shown to enhance germination and early plant growth in a number of plants including non-legumes [83]. This is perhaps a link to the reputed role of methylobacter in maintaining seed viability.

In the northern USA and Canada, soybeans and peas can experience delays in nodulation and the onset of N_2 fixation. This has been attributed to the higher flavonoid levels needed for rhizobial *nod*-gene induction at low temperature [84], with the problem worsened under minimum tillage. Application of genistein to inoculant rhizobia or to soil has enhanced nodule number and mass per plant and improved early nitrogen fixation in soybean [85]. Flavonoid amendments have also increased soybean grain yield and protein content in Canadian studies [86] with shoot weight and total protein increased in the second year of evaluation by 28% and 70%, respectively. Products containing flavonoids are now commercially available.

Although the overall mass of nodules is generally a better indicator of potential for N_2 fixation per plant than is nodule number, interest in the latter continues because of its regulation by host, strain and environmental factors. Most regulatory studies have emphasized the host [87] and have implicated ethylene and available nitrogen as important factors [88]. How these factors might relate to strain differences in nodule number per plant is not known, though nodule number is also enhanced by inoculation delay [89]. The practical value of hypernodulation mutants is yet to be determined, but reference [90] notes the release of "Nitrobean 60" in Australia, and cites yield increases of 15%, combined with higher rates of N_2 fixation.

From the microbial perspective there appears to be greater potential in the identification of inoculant rhizobia that are "preferred" by their hosts. Strain preference in nodulation has been shown for clover [91], soybean [92] and bean [93] but not advanced to a meaningful level in the field.

Nucleotide sequencing, particularly of the symbiotic plasmid, has identified type III secretion proteins in *Rhizobium spp* NGR234, *M.loti* and *B. japonicum* [94] but not *S. meliloti* [95]. Such proteins are essential for chronic infection in a number of animal and plant pathogens. A high degree of conservation between *Brucella* and *Rhizobium* in genes coding for putative cytoplasmic membrane transport proteins [96] is suggestive of commonalities between pathogenic and symbiotic organisms in infection or the avoidance of detection.

3.3 Inoculation of legumes

Legume inoculation with root-nodule bacteria is an established and successful practice, especially with particular legumes in the more technically- developed countries [97]. When a new legume is introduced into a region, few of the soils will contain appropriate rhizobia, and inoculation will usually be needed. Under soil conditions of low nitrogen, yield increases following inoculation can then exceed 50%, with clear differences evident between inoculated and uninoculated plants. With rare exceptions [98], reinoculation in subsequent years will not be needed, and over time even uninoculated soils will tend to accumulate rhizobia, limiting inoculation response. A consequence, as shown by Hall and Clark [99] for soybean in Thailand, is that there will be greater interest in inoculation and inoculant technologies in the newer areas of production than in regions where the crop has been grown for some time.

Where inoculation is needed, the inoculant must both supply adequate numbers of rhizobia and use inoculant-quality strains having the following characteristics [100]: the ability to form highly effective nodules with all commonly used cultivars and species for which it is recommended; be competitive in nodule formation and persistent in the soil; the ability to tolerate soil environmental stresses such as acid soil pH and temperature; display good growth in simple, inexpensive culture media; be genetically stable and not be subject to mutation; and the ability to survive well on the seed prior to seed germination

Less than one strain in 100 is likely to meet these criteria, so rhizobia considered for use in inoculants must undergo exhaustive testing [101]. This needs to include growth-chamber or greenhouse and field evaluation of symbiotic performance with a range of cultivars and an evaluation of the other points listed above. Designation of an organism as being of inoculant quality should not be done lightly. Once added to the soil such organisms can be difficult to displace, as is evident from the dominance of soybean strains belonging to serogroup 123 in the soils of the American midwest [102].

Selection of a carrier that will maintain high cell count until the inoculant is used by the farmer is also critical. Milled peat is the preferred inoculant carrier [103], and is often sterilized by γ-irradiation or electron acceleration, before inoculation with rhizobia. Initially sterile carriers will generally support a higher population of rhizobia, contain fewer contaminants, and have longer shelf lives than those that are non sterile [104]. Since peat is not universally available, especially in the tropics, compost, bagasse derived from the milling of sugar cane, coal, polyacrylamide, vegetable oils, clays and clay granules, liquid preparations and soil have all been used in inoculant preparation.

The number of rhizobia inoculated per seed varies with seed size. In countries where inoculant quality is regulated by law (Canada, France, Uruguay) or by voluntary procedures taken by inoculant manufacturers (Australia, New Zealand, South Africa, Thailand) [97] the usual seed standard is for 10,000 rhizobia per seed in small-seeded legumes such as clover, but 1 million rhizobia per seed for larger-seeded legumes such as bean or soybean [103]. This number may need to be greater, and the method of inoculation changed, when soil or temperature conditions are adverse. Inoculants from many countries do not meet that standard [105]. Factors contributing to poor inoculant quality include inadequate testing of the rhizobia used, strain mutation or sym plasmid loss, use of unsuitable culture medium, inappropriate carrier materials, and poor storage and transport conditions. Catroux *et al.* [106] have noted that "we enter the era of biotechnology knowing more and more about the mechanisms of N_2 fixation at the gene level, but except for some manufacturers in developed countries... still missing good quality and reliable inoculants."

3.4 Strain persistence and competitiveness of soil rhizobia

At rates of 10^3 to 10^6 rhizobia applied per seed, inoculation should supply up to 8×10^{10} rhizobia per ha [107]. Where there are no soil problems and few indigenous rhizobia, this should be more than sufficient to ensure abundant nodulation, with most of the nodules due to the inoculant strains formed in the first year [108]. As these nodules senesce toward the end of the first growing season, further large numbers of the inoculant strain(s) will be released into the soil. Bever and Simms [109] suggest that bacteroids are incapable of reproduction, and thus that the kin of the inoculant strain(s) derive no significant benefit from these initially high nodule-occupancy rates. This ignores the high % viability of the rhizobia recovered from determinate nodules, and the survival of bacteria from the infected but non-nitrogen-fixing regions of the indeterminate nodule [110]. Recently, Muller *et al.* [111] have also reported a potential for redifferentiation in nodule bacteroids protected from osmotic shock by the high trehalose levels found in older nodules. Therefore, it is not surprising that a significant number of studies have documented inoculant-strain dominance in nodules 5-15 years after initial inoculation [112].

Inoculant strains do not always persist. Typical is a study by Brockwell *et al.* [see reference 108] in which 19 strains of *R. leguminosarum* bv *trifolii* were used separately to inoculate subterranean clover in three soils having different numbers of indigenous clover rhizobia. At Murrumbateman where the soil was essentially free of indigenous rhizobia, most inoculant strains produced essentially 100% of the nodules formed in the first year. In the second year however the fraction of nodules produced by the inoculant strains ranged from 19 to 94%. At the Deniliquin B site, where the population of indigenous rhizobia was higher, only 4 of the strains produced more than 50% of the nodules formed in the second year, and two strains produced none.

Differences in inoculant-strain contribution to nodule formation over time are usually attributed to strain variation in bacteroid viability, or to differences in strain persistence and competitiveness in soil. Rhizobia are generally considered excellent soil saprophytes and able to persist in soil for quite long periods in the absence of a suitable host [113]. This ability has been described as saprophytic competence [114], but what determines strain persistence in soil is still poorly defined. Marked strain-to-strain and species variation is clear. Second-year nodulation differences at Murrumbatemen were referred to above. Chatel *et al.* [114] found low persistence of *R. leguminosarum bv trifolii* in hot/dry sandy soils from Western Australia, but found populations of *Bradyrhizobium spp* dramatically superior in persistence under serradella. Maurice *et al.* [115] have noted change in the physiological state and infective ability of *B. japonicum* inoculants stored for long periods in liquid culture. Can similar changes occur in soil after long separation from their host?

An attractive hypothesis is that rhizobial persistence is affected by the ability to utilize common soil aromatic and hydroaromatic compounds. Parke and Ornstom [116] found significant differences among the rhizobia in the utilization of these compounds, and also noted that the enzymes for protocatechuate metabolism were constituitive in *Bradyrhizobium spp* but inducible in fast-growing rhizobia [117]. Species of rhizobia also differed in chemotaxis toward selected aromatic and hydroaromatic compounds [118]. Unfortunately, Rynne *et al.* [119] could demonstrate no correlation between catabolic ability and strain persistence. The possibility that unusual compounds exuded by the legume could create rhizosphere conditions favoring particular rhizobia, first suggested by van Egeraat in 1975, has found favor again in recent years. Opines [120] and mimosine [121] have been suggested as candidate molecules, but many other substances could have similar effects. These could include simple inorganic molecules such as phosphate [122]. Better methodologies and more detailed studies of rhizosphere community genetics are needed.

Howieson and colleagues developed a cross-row technique used in the selection of strains able to persist and colonize soils under unfavorable environmental conditions [123]. In this procedure, inoculated seeds are planted in rows 2 m apart, and allowed to grow for varying periods of time. Uninoculated and surface-sterilized seed of the same species is then planted at 90° to the original row and at distances up to 1 m from it. After 10-12 weeks plants from the cross rows are harvested, and their nodulation patterns with distance from the initial line of inoculation determined.

When rhizobia are uniformly distributed throughout the soil, competition for nodule occupancy depends on the numbers of each rhizobium present and their competitive ability with the host cultivar used. This has been modeled both by Ireland and Vincent [124] and by Amarger and Lobreau [125]. Many competition studies have been undertaken, the methods used to evaluate strain differences including serology [92,126], intrinsic or high-level antibiotic resistance [127], ineffective rhizobial marker strains [128], macroscopically distinguishable nodules [129] and *lac*- or *gus*- marked strains [130]. Where antibiotic-resistant mutants are utilized, adequate evaluation of parallel changes in symbiotic performance are essential [131]. Speed of nodulation, commonly assayed using a root-tip marking procedure modified from that of Bhuvaneswari *et al.* [132], has also been effective in identifying competitive strains [133]. Lupwayi *et al.* [134] compared 33 pairs of strains in competition experiments, noting that strains that were faster to nodulate were also more competitive in 10 of 15 comparisons. However, when there was no difference between strains in speed of nodulation, only two of 18 pairs showed similar competitive abilities. From the various speed-of-infection studies undertaken, it appears that rhizobia are not competing directly with each other, but rather attempting to complete a series of interactions with their host during a relatively narrow window of opportunity [135]. Strains able to complete these interactions will appear competitive; those that do not will appear to be weak competitors.

As evident from the preference in nodulation data already cited, host cultivar can have a marked influence on nodule occupancy. Early studies showed specificity in nodule formation for peas and soybean in the different areas of production of these crops [136]. Host x strain interaction is likely to become more evident with the greater use of molecular methodologies for isolate differentiation. Thus Bernal and Graham [137] noted major differences in the rhizobia recovered when local or introduced cultivars of *Phaseolus vulgaris* were used as trap hosts, while Mpepereki *et al.* [56] and Abaidoo *et al.* [138] found variation in the rhizobia associated with promiscuously nodulating soybean cultivars in Africa. Other traits likely to be important in strain competitiveness include age of the rhizobia, lysogenic or antibiotic interaction between the strains present [139] and environmental conditions (see below).

When seed-applied inoculant rhizobia are introduced into a soil that contains indigenous organisms, the seed-applied organisms form a significant proportion of nodules in the crown region of their host, near where they were placed, but few nodules outside that region [140]. This is not a competition problem, but rather one of limited strain mobility and inoculant dilution throughout the root system. It has been suggested that where the indigenous soil rhizobia exceed 10-100 cells g^{-1} soil, inoculation will not benefit the host [106,141]. While this is commonly true, few of the studies on which the assumption is based used high-potency inoculants or repeated inoculation over years. In contrast, Dunigan *et al.* [142] found average recovery of the inoculant strain to increase from less than 10 to almost 50% following three years of inoculation. Similar data were reported by Brutti *et al.* [143].

3.5 Rhizobial biodiversity and ecology

Improved fingerprinting techniques, coupled with increased interest in microorganisms as indicators of soil health [144], have resulted in an increased focus on rhizobial biodiversity and the ecology of these organisms in the soil. Approaches used have included restriction fragment length polymorphism (RFLP) [145], polymerase chain reaction analysis using random primers [146] or repetitive elements [147], multilocus enzyme electrophoresis [148], and SDS-PAGE analysis of total cell proteins [149]. A number of these procedures are considered by Thies *et al.* [150].

Methodologies have also been developed for the direct isolation of DNA or rRNA from soil, and for subsequent microbial community analysis without the need for isolation of individual organisms [151]. When this approach is used, the diversity of the soil microbial population is some two-hundred-fold greater than that of the bacteria cultured from the same soil. Unfortunately, no current method permits selective extraction of rhizobial DNA or rRNA from soil for community analysis, and all rhizobial diversity analysis done to date uses either cells trapped using specific legume hosts or selective culture medium. These approaches can influence the apparent diversity of rhizobia present in the soil. Abaidoo *et al.* [138] inoculated soil from 65 locations in Africa onto symbiotically promiscuous soybean cultivars and identified 11 phylogenetic clusters on the basis of RFLP of the 16S sRNA gene. Some isolates related to *B.japonicum* and *B.elkanii*, while a number grouped to miscellaneous *Bradyrhizobium* reference strains obtained from a variety of tropical legumes. In contrast, Yang *et al.* [152] estimated indigenous soybean rhizobia populations in Xinjiang soil, and found MPN counts obtained with cultivar Jing Dou 19 much higher than with Heinong 33. Strains isolated from Jing Dou 19 were essentially all fast growing, and did not nodulate either Heinong 33 or Williams.

Rhizobial biodiversity studies need to pay close attention to inoculation patterns, crop and varietal history in the region, agricultural practices and soil conditions. Rhizobia can be introduced naturally as seed-borne contaminants [153], and this must have been a factor in the spread of *R.etli,* the specific host of *P. vulgaris*, to Europe and Africa following the colonization of Latin America, where the crop originated. However, it is our experience that numbers of rhizobia carried even on freshly harvested seed are not high, so many seeds would need to be introduced to provide a significant diversity of rhizobia. Founder effects have only rarely been reported in rhizobia, but are likely to be quite common [154]. Similarly Hirsch *et al.* [155] found that application of heavy metal-containing manure had restricted the biodiversity of *R.leguminosarum* bv *trifolii* to a single biotype, while Bernal [156] noted changes in the populations of bean rhizobia recovered from soils in areas where fungicide and antibiotic seed dressings were common. Caballero-Mellado and Martinez-Romero also noted fertilizer effects on strain biodiversity in soil. Finally, and although rhizobia can differ markedly in acid pH tolerance [70], many of the rhizobia recovered from acid soils are not themselves acid tolerant [157]. Presumably these organisms can persist in niches of less acid pH, but are likely to be constrained in their activities in the soil. Trapping of rhizobia at other than the natural pH of the soil is likely to overestimate the importance of such organisms.

Given the restrictions indicated above, population diversity among the rhizobia for a particular legume species is likely to be greatest near the center of origin of that legume [158]. Pinero *et al.* [see reference 147] recorded a mean genetic distance per enzyme locus of 0.691 for 51 isolates of *R.etli* in the Mesoamerican center of origin for *Phaseolus vulgaris*; Souza *et al.* [see reference 147] grouped 372 bean rhizobia from this center into 7 clusters with 95 electrophoretic types. Even here, differences in diversity have been noted between the Mesoamerican and Andean centers of origin of the crop [see reference 146].

Environmental stresses can have major effects on the biodiversity of soil rhizobia, and on nodulation and N_2 fixation. Thus Brockwell *et al.* [see reference 108] recovered an average of 89,000 *S. meliloti* g^{-1} soil from soils of pH > 7.0, but only 37 cells g^{-1} soil from soils of pH < 6.0. Similarly in China, the proportion of nodules formed by fast-growing soybean rhizobia (presumably *S. fredii*) increased from only 10.1% at pH 6.0 to 65.9% at pH 7.0 [152]. Even more striking is the replacement of the normal symbiont for beans, *R.etli,* by the acid tolerant *R. tropici* under acid soil conditions in Africa and Brazil [159]. At neutral pH *R.etli* is much more competitive in the nodulation of *P .vulgaris* than is *R. tropici* [160]. However at acid pH more than 97% of the nodules produced on beans may be due to *R. tropici* [159]. This species is uncommon in the soils of Mesoamerica and Andean South America, and its origins remain obscure. However, both inoculant strains currently recommended for acid-soil conditions in Brazil (PRF81 and CIAT899) belong to this species.

Studies on acid pH tolerance in CIAT899 have shown that the organism will produce individual colonies on Keyser Munns medium, pH 4.25 [see reference 70] and induce nodulation of beans at pH 4.5 [161]. A shift to acid pH results in the production of a number of acid-shock proteins [162]. Acid pH tolerance in this organism needs further study, but appears to involve a significant change in membrane structure, a proton-translocating ATP-ase, and intracellular pH regulation [163]. Glutathione-deficient mutants of this strain are also pH sensitive [164]. Variation in acid pH tolerance has also been shown in *R. leguminosarum* bv. *trifolii* [165] and in rhizobia nodulating species of *Medicago,* including both strains of *S. meliloti* from acid soils in the Mediterranean [123] and *Rhizobium* spp from acidic soils in Uruguay and Argentina [166]. Dilworth *et al.* [167] have reviewed recent studies on acid tolerance in *S. meliloti*, and noted a range of responses to acid pH stress. These include a sensor-regulator gene pair, a copper-transporting ATP-ase, a gene involved in lipid metabolism, and a possible antiporter related to *nha*B. Heavy metals seem a recurrent theme in several of the pH responses studied. Strain TG2-6, a pH mutant of *S. meliloti,* is also copper sensitive [168], while Priefer *et al.* [169] have identified pH mutants of *R. leguminosarum bv phaseoli* affected in cobalamine biosynthesis.

Both high and low temperatures affect the survival and persistence of rhizobia, nodulation and N_2 fixation, with significant strain variation evident. Maximum soil temperatures in the tropics regularly exceed 40 °C to 5 cm depth [170]. Thus it is significant that Eaglesham *et al.* [171] found that more than 90% of cowpea rhizobia from the hot dry Sahel to grow well at 40 °C, whereas isolates from the more humid Onne region generally failed to grow at this temperature. High surface temperatures reduce strain survival, but might also contribute to the number of ineffective plasmid-cured rhizobia found in some soils [172]. Unsuitably high temperature is also a concern in the storage and shipment of inoculants.

Physiological adaptation to cold temperature has also been demonstrated [173], the last report describing differences in the number of cold-shock proteins produced by cold-tolerant and sensitive strains, and significant changes in membrane structure. High-molecular-weight membrane polypeptides have been implicated in the cold tolerance of *Rhizobium* [174].

4. CHALLENGES FOR RESEARCH AND DEVELOPMENT

Legumes provide the largest inputs of biologically fixed nitrogen for agriculture and it has been proposed that this input should routinely be >100 kg N/ha/year in farming systems incorporating legumes [175]. However, the amounts of nitrogen fixed by legumes are considerably less, as a consequence of environmental and biological constraints and agricultural practices that limit the presence of effective bacteria or affect levels of soil nitrogen [175]. There are many current challenges for improving inputs of legume-fixed nitrogen into farming systems. As an initial step, Sessitsch *et al.* (13) provide a matrix for sensible decision making in rationalising the adoption of legumes in farming systems and then in evaluating the need to inoculate them. Significant increases in nitrogen inputs may then be achieved in the short-term by the wider availability of high quality inoculants, improvements in crop agronomy, the introduction of legumes to new areas, and changes in residue management. In addition, rhizobial strain selection and advances in technologies to improve survival of inoculum have considerable potential in many situations. The problems associated with these issues on a global scale, and the deteriorating position for applied research related to biological nitrogen fixation, have been reviewed and discussed by Giller and colleagues [175]. They propose the need for strong policy intervention at an international level to increase awareness of the issues and maintain the benefits from using nitrogen-fixing legumes.

At a practical level, applied research is continuing to address some of the important challenges faced by farmers growing legumes. In many farming systems a diverse range of legumes is grown that can be nodulated by the same species of root-nodule bacteria. In these situations, where inoculation is beneficial and quality inoculants are available, there are clear benefits from having a single-inoculant strain that effectively nodulates the broad range of legume hosts. The availability of an effective broad-host strain benefits inoculum manufacturers and farmers, and avoids the potentially disastrous consequences of introducing an ineffective strain into the farming system. One example of a successful solution to this "host range problem" has been with the new Australian inoculant strain of *R. leguminosarum* bv. *viciae* (WSM1455) that nodulates effectively many pulse genera, including species of *Lens*, *Lathyrus*, *Vicia* and *Pisum* [11, 97].

Few of the herbaceous legume genera grown in agriculture for pasture and forage contain both annual and perennial species. In the Mediterranean-type regions of the world these genera include *Lotus, Medicago* and *Trifolium*. Many *Trifolium* spp. have significant roles in agriculture, and this genus is a focus for research programs in Australia, Uruguay and countries of the Mediterranean basin to develop new annual and perennial pasture species for sustainable rotation-based systems. A challenge to these programs, and one for which there is no immediate solution, is that the current inoculum-quality strains of *R. leguminosarum* bv. *trifolii* effective on annual *Trifolium* spp. are generally ineffective on perennial *Trifolium* spp., and vice versa [101]. One research approach is to obtain a strain of *R. leguminosarum* bv. *trifolii* that is effective on both the currently grown annual species of clover and also on the new annual and perennial *Trifolium* spp. being evaluated for introduction [176]. Comprehensive ecological studies of the interactions between legumes and their microsymbionts in both natural and managed ecosystems are required to address this issue. The objective of these research programs is to increase our understanding of symbiotic interactions of this genus as a model to develop strategies to improve the management of symbiotic interactions between legumes and their root-nodule bacteria to extend, rather than restrict, their use in agriculture.

Farmers require inoculant strains to survive in stressful soils in sufficient numbers to provide a population able to nodulate under environmental constraints such as pH, temperature and competition from less effective indigenous and naturalised strains. This last problem of competition is significant in many areas, not the least in soils of the tropics and sub-tropics. Several research programs around the world are addressing the problems of stress tolerance in root-nodule bacteria. Improvements in our understanding of the molecular and physiological mechanisms of stress-sensitivity in both symbionts will be important if we are to increase legume nitrogen fixation and productivity. In addition, there will be clear benefits to legume production from increasing the survival of the inoculant root-nodule bacteria on seed, or when delivered directly into soil. Enhanced formulations, granular inoculants, and seed coating techniques that protect the bacteria from environmental stress or physically separate them from toxic chemicals, such as fungicides applied to seed, offer new research directions [177].

Our detailed understanding of the biology of the root-nodule bacteria make it an excellent model for researching more general rhizosphere bacteria. The long term benefits of such research might be manifested in our ability to deliver novel products to the roots of legumes and other crops, or to understand how roots and rhizosphere organisms communicate with each other. To make progress in this field will require appropriate use of molecular techniques, for example, to transfer genes that encode products that will benefit legumes into root-nodule bacteria. The insertion and regulation of Bt-toxin genes for control of insect pests that attack legume roots and nodules, and the transfer to rhizobia of genes encoding plant growth hormones are just two possibilities that capitalise on our knowledge of rhizobia as successful colonizers of the soil and rhizosphere. An aligned area of research that applies our knowledge and techniques developed with the root-nodule bacteria might be in the selection and development of other soil microbes such as plant growth-promoting bacteria and microbial agents as inocula for biological control of pests, diseases and weeds. Research into these areas is currently underway in several groups throughout the world and it is expected that this research may provide beneficial outcomes resulting in improved sustainability and productivity in agricultural systems.

5. CONCLUSIONS

Pulse, pasture and forage legumes have significant roles in agriculture because they maintain productivity in farming systems as a result of their nitrogen-fixing symbioses with soil inhabiting root-nodule bacteria. They play an additional role in sustainability because their use in rotation systems can help control pests, diseases and weeds. As the demands for increased and more efficient agricultural productivity continue, new roles are emerging for nitrogen-fixing legumes in both established and new farming systems. These new roles include the continued expansion of pulse crops onto less fertile soils, and the exploration for new species of pasture and forage legumes with deep-rooting habits for control of soil moisture to combat salinity.

The successful use of nitrogen-fixing legumes in these current and new roles in agriculture will be dependent upon appropriate attention to the formation of effective symbioses with root-nodule bacteria. The integration of plant breeding and cultivar development with the selection of elite strains of root-nodule bacteria is a model that has returned substantial benefits to N-fixation in agricultural practice during the last decades of the previous millennium. Further benefits to farmers will be manifested through advances in inoculation technologies that deliver elite inoculum strains of root-nodule bacteria to the soil with enhanced survival. Finally, the use of N-fixing legumes in agriculture will be greatly enhanced when major world economies reduce their direct subsidisation of inefficient and unsustainable agricultural practices.

6. REFERENCES

1. C. R. Frink, P. E. Waggoner and J. H. Ausubel, Proc. Nat. Acad. Sci. USA 96 (1999) 1175.

2. A. van Kammen in Biological Fixation of Nitrogen for Ecology and Sustainable Agriculture, A. Legocki, H. Bothe and A. Puhler (eds.), Springer-Verlag, Berlin, 1997; C. P. Vance in Biological Fixation of Nitrogen for Ecology and Sustainable Agriculture, A. Legocki, H. Bothe and A. Puhler (eds.), Springer-Verlag, Berlin, 1997; C. P. Vance in Biotechnology and the Improvement of Forage Legumes, B.D. McKersie and D.C.W. Brown (eds.), CAB International, Wallingford, 1997; C. P. Vance, P. H. Graham and D. L. Allan in Nitrogen Fixation: from Molecules to Crop Productivity, F. O. Pedrosa, M Hungria, M. G. Yates and W. E. Newton (eds.), Kluwer, Dordrecht, 2000.
3. B. B. Bohlool, J. K Ladha, D. P. Garrity and T. George, Plant Soil 141 (1992) 1; M. B. Peoples, D. F. Herridge and J. K. Ladha, Plant Soil 174 (1995) 3.
4. J. I. Sprent and P. Sprent, Nitrogen Fixing Organisms, Chapman and Hall, London, 1990; K. E. Giller, Nitrogen Fixation in Tropical Cropping Systems, CAB International, Wallingford, 2001.
5. J. Brockwell, P. J. Bottomley and J. E. Thies, Plant Soil 174 (1995) 143.
6. C. S. Schwintzer and J. D. Tjepkema in The Biology and Biochemistry of Nitrogen Fixation, M. J. Dilworth and A. R. Glenn (eds.), Elsevier, Amsterdam, 1991.
7. J. I. Baldani, A. L. M. Oliveira, S. L. Guimaraes, V. L. D. Baldani, F. B. Reis Jr., L. G. Silva, V. M. Reis, K. R. S. Teixeira, and J. Döbereiner in Nitrogen Fixation: from Molecules to Crop Productivity, F. O. Pedrosa, M Hungria, M. G. Yates and W. E. Newton (eds.), Kluwer, Dordrecht, 2000.
8. J. K. Ladha, M. Garcia, S. Miyan, A. Padre, and I. Watanabe, Appl. Environ. Microbiol. 55 (1989) 454 ; Z. Y. Tan, F. L. Kan, G. X. Peng, E. T. Wang, B. Reinhold-Hurek, and W. X. Chen, Int. J. Syst. Evol. Microbiol. 51 (2001) 909.
9. I. R. Kennedy and E. C. Cocking (eds.), Biological nitrogen fixation; the global challenge and future needs. Proceedings, Rockefeller Foundation Bellagio Conference, 1997.
10. P. H. Graham and C. P. Vance, Field Crop Res. 65 (2000) 93.
11. J. G. Howieson, J. Malden, R.J. Yates and G.W. O'Hara, Symbiosis 28 (2000) 33; G. W. O'Hara, J. Crop Prod. 1 (1998) 1.
12. D. F. Herridge in Inoculants and Nitrogen Fixation of Legumes in Vietnam, D. F. Herridge (ed.), ACIAR, Canberra, 2002.
13. A. Sessitsch, J. G. Howieson, X. Perret, H. Antoun H.and E. Martínez-Romero, Crit. Rev. in Plant Sci. (2002).
14. J. I. Sprent, Nodulation in Legumes. Royal Botanical Gardens. Kew, 2001.
15. J. H. Kirkbridge in Trefoil: The Science and Technology of *Lotus,* P.R. Beuselinck (ed.), CSSA Special Publication 28, American Society of Agronomy, Madison, 1999.
16. J. Hamblin in Lupins as Crop Plants: Biology, Production and Utilization, J.S. Gladstones, C. Atkins and J. Hamblin (eds.), CAB International, Cambridge, 1998.
17. L. Kaplan and T. F. Lynch, Econ. Bot. 53 91999) 261.
18. C. Porqueddu, A. Loi and P. S. Cocks, J. Ag. Sci. 126 (1996) 161.

19. K. F. M. Reed, M. J. Mathison and E. J. Crawford in Persistence of Forage Legumes, G. C. Marten, A. G. Matches, R. F. Barnes, R. W. Brougham, R. J. Clements and G. W. Sheath (eds.), Persistence of Forage Legumes, American Society of Agronomy, Madison, 1989.
20. R. S. Kingwell and D. J. Pannell (eds.), MIDAS, a Bioeconomic Model of a Dryland Farm System. Pudoc, Wageningen, 1987.
21. E. L. Armstrong, D. P. Heenan, J. S. Pate and M. J. Unkovich, Aust. J. Ag. Res. 48 (1997) 39; F. D. Dakora and S. O. Keya, Soil Biol. Biochem. 29 (1997) 809; J. C. Rowland, M. G. Mason and J.Hamblin, Aust. J. Exp. Agric. 28 (1988) 91.
22. M. J. Unkovich and J. S. Pate, Field Crop. Res. 65 (2000) 211; C. van Kessel and C. Hartley, Field Crop. Res. 65 (2000) 165.
23. A, D. Robson in The Role of Legumes in the Farming Systems of the Mediterranean Areas, A. E. Osman, M. M. Ibrahim and M. A. Jones (eds.), ICARDA, The Netherlands, 1990.
24. M.W. Sweetingham, R. A. C. Jones and A. G. P. Brown, in Lupins as Crop Plants: Biology, Production and Utilization, J.S. Gladstones, C. Atkins and J. Hamblin, (eds.), CAB International, Cambridge, 1998.
25. J. G. Howieson, G. W. O'Hara and S. J. Carr, Field Crop. Res. 65 (2000)107.
26. M. Hungria and M. A. T. Vargas, Field Crop. Res. 65 (2000)151.
27. M. Dracup, N. C. Turner, C. Tang, M. Reader and J. Palta, in Lupins as Crop Plants: Biology, Production and Utilization, J.S. Gladstones, C. Atkins and J. Hamblin, (eds.), CAB International, Cambridge, 1998.
28. S. C. Peltzer, L. K. Abbott and C. A. Atkins, Aust. J. Ag. Res. 53 (2002) 355.
29. T. Ehrman and P. S. Cocks, J. App. Ecol. 27 (1990) 578.
30. G. W. O'Hara, Aust. J.Exp. Ag. 41 (2001) 417.
31. A. D. Robson in Mineral Nutrition of Legumes in Tropical and Sub-tropical Soils, C.S. Andrew and E.J. Kamprath (eds.), CSIRO, Melbourne,1978.
32. A. D. Robson and P. J. Bottomley in The Biology and Biochemistry of Nitrogen Fixation, M. J. Dilworth and A. R. Glenn (eds.), Elsevier, Amsterdam, 1991.
33. C. Tang, M. J. Unkovich and J. W. Bowden, New Phytol. 143 (1993) 513.
34. G. Uehara in Mineral Nutrition of Legumes in Tropical and Subtropical Soils, C. S. Andrew and E. J. Kamprath (eds.), CSIRO, Melbourne, 1978.
35. A. E. Richardson, M. A. Djordjevic, B. G. Rolfe and R. J. Simpson, Plant Soil 109 (1988) 37; M. J. Sadowsky and P. H. Graham in 'The *Rhizobiaceae*', H. P. Spaink, A. Kondorosi and P. J. Hooykaas (eds.), Kluwer, Dordrecht, 1998.
36. A. D. Robson, J. Aust. Inst. Ag. Sci. 35 (1969) 154.
37. J. G. Howieson, L. K. Abbott and A. D. Robson, Aust. J. Ag. Res. 43 (1992) 565.
38. I. A. McKay and M. A. Djordjevic, Appl. Environ. Microbiol. 59 (1993) 3385.
39. D. N. Munns, Plant Soil 32 (1970) 90.
40. J. G. Howieson and M. A. Ewing, Aust. J. Ag. Res. 40 (1989) 843; Y. Cheng, E. L. J. Watkin, G. W. O'Hara and J. G. Howieson, Plant Soil 238 (2002) 31.
41. C. Tang and A. D. Robson, Plant Soil 152 (1993) 269.

42. D. F. Herridge and I. Rose, Field Crop. Res. 65 (2000) 229.
43. C. Ovalle in Legumes for Mediterranean Forage Crops, Pastures and Alternative Uses, L. Sulas (ed.)., Cahiers Options Mediterranéennes, University of Sassari, 2000.
44. A. Loi, J. G. Howieson and S. J. Carr, Aust. J. Exp. Ag. 41 (2001) 1.
45. S. J. Carr, A. Loi, J. G. Howieson and C. Porqueddu in Proc. IX Meeting FAO-IHEAM on Mediterranean Pastures and Fodder Crops. FAO-CIHEAM, Institut Agronomique Mediterraneen de Zaragosa, Badajoz,1997.
46. J. G. Howieson in Plant Genetic Resources of the Mediterranean Basin, N. Maxted and S. J. Bennett (eds.), Kluwer, Dordrecht, 2000.
47. T. N. Barry, and W. C. McNabb in Tannins in Livestock and Human Nutrition, J. D. Brooker (ed.), ACIAR, Canberra, 1999.
48. F. O. Stephens, Med. J. Aust. 167 (1997) 138.
49. J. W. Miles in Proc. XIX International Grasslands Congress, J.A. Gomide, W.R.S. Mattos and S.C. da Silva (eds.), Brazilian Society of Animal Husbandry, Brazil, 2001.
50. D. Real, J. Franco and J. Crossa in Proc. XIX International Grasslands Congress, J. A. Gomide, W. R. S. Mattos and S. C. da Silva (eds.), Brazilian Society of Animal Husbandry, Brazil, 2001.
51. J. S. Gladstones in Lupins as Crop Plants: Biology, Production and Utilization, J. S. Gladstones, C. Atkins and J. Hamblin, (eds.), CAB International, Cambridge, 1998.
52. S. Bennett, C. Francis and B. Reid, in Plant Genetic Resources of the Mediterranean Basin, N. Maxted and S. J. Bennett (eds.), Kluwer, Dordrecht, 2000.
53. A. Abd-El-Moneim, B. Van Dorrestein, M. Baum and W. Mulugeta, Food Nut. 21 (2000) 493; K. M. M. Siddique, S. P. Loss, K. L. Regan and R. L. Jettner, Aust. J. Agric. Res. 50 (1999) 375; M. W. Perry, M. Dracup, P. Nelson, R. Jarvis, I. Rowland and R. J. French in Lupins as Crop Plants: Biology, Production and Utilization, J. S. Gladstones, C. Atkins and J. Hamblin, (eds.), CAB International, Cambridge, 1998.
54. J. E. Thies, P. W. Singleton and B. B. Bohlool, Soil Biol. Biochem. 27 (1995) 575; O. P. Rupela and C. Johansen, Soil Biol. Biochem. 27 (1995) 539.
55. J. G. Howieson, Soil Biol. Biochem. 27 (1995) 603.
56. S. Mpepereki, F. Javaheri, P. Davis and K. E. Giller, Field Crop. Res. 65 2000) 137; S. Mpepereki, F. Makonese and K. E. Giller in Nitrogen Fixation Global Perspectives, T. M, Finan, M. R. O'Brian, D. B. Layzell, J. K. Vessey and W. Newton (eds.), CAB International, Wallingford, 2002.
57. R. A. Ballard and N. Charman, Aust. J. Exp. Ag. 40 (2000) 939.
58. S. M. Lohrke, J. H. Orf and M. J. Sadowsky, Crop Sci. 36 (1996) 1217.
59. C. van Kessel and C. Hartley, Field Crop Res. 65 (2000) 165.
60. E. B. Fred, I. L. Baldwin and E. McCoy, Root-nodule Bacteria and Leguminous plants, University of Wisconsin Press, Madison, 1932.

61. P. H. Graham, M. J. Sadowsky, H. H. Keyser, Y. M. Barnett, R. S. Bradley, J. E. Cooper, D. J. De Ley, B. D. W. Jarvis, E. B. Roslycky, B. W. Strijdom and J. P. W. Young, Int. J. Syst. Bacteriol. 41 (1991) 582; J. P. W. Young and K. E. Haukka, New Phytol, 133 (1996) 87.
62. J. Michiels, B. Dombrecht, N. Vermeiren, C.W. Xi., E. Luyten and J. Vanderleyden, FEMS Microbial. Ecol. 26 (1998) 193; P.H. Graham, K. G. Ballen, C. Montealegre, R. K. Jones, B. Fischer and E. Luque inHighlights of Nitrogen Fixation Research, E. Martinez and G. Hernandez (eds.), Kluwer/Plenum Press, New York 1999.
63. J. M.Young, L. D. Kuykendall, E. Martinez-Romero, A. Kerr and H. Sawada, Int. J. Syst Evolut. Microbiol. 51 (2001) 89.
64. E. Martinez-Romero, J. Cabalero-Mellado, B. Gandara, M. A. Rogel, A. Lopez Merino, E. T. Wang, L. E. Fuentes-Ramirez, I. Toledo, L. Martinez, I. Hernandez-Lucas, J. Martinez-Romereo, in Nitrogen Fixation: from Molecules to Crop Productivity, F. O. Pedrosa, M Hungria, M. G. Yates and W. E. Newton (eds.), Kluwer, Dordrecht, 2000.
65. F. Zakhia and P de Lajudie, Agronomie 21 (2001) 569.
66. P. H. Graham, J. Gen. Microbiol. 35 (1964) 511.
67. R. J. Maier, N. E. R. Campbell, F. J. Hanus, F. B. Simpson, S. A. Russell and H. J. Evans, Proc. Nat. Acad. Sci (USA) 75 (1978) 3258; F. J. Hanus, R. J. Maier and H. J. Evans, Proc. Nat. Acad. Sci. USA 76 (1979) 1788.
68. H. J. Evans, F. J. Hanus in Proc. World Soybean Conf. III, R. Shibles (ed.), Westview Press, Colorado, 1985.
69. J. K. Ladha and R. B. So, Int. J. Syst. Bacteriol. 44 (1994) 62 ; R. B. So, J. K. Ladha and J. P. W. Young, Int. J. Syst. Bacteriol. 4 (1994) 392.
70. P. H. Graham, K. J. Draeger, M. L. Ferrey, M. J. Conroy, B. E. Hammer, E. Martinez, S. R. Aarons and C. Quinto, Can J. Microbiol. 40 (1994) 198; M. Hungria, D. S. Andrade, L. M. Chueire and M. Megias in Highlights of Nitrogen Fixation Research, E. Martinez and G. Hernandez (eds.), Kluwer/ Plenum Press, New York, 1999.
71. L. Moulin, A. Munives, B. Dreyfus and C. Boivin-Masson, Nature 411(2001) 948; L. Moulin, W. M. Chen, G. Bena, B. Dreyfus and C. Boivin-Masson in Nitrogen Fixation Global Perspectives, T. M, Finan, M. R. O'Brian, D. B. Layzell, J. K. Vessey and W. Newton (eds.), CAB International, Wallingford, 2002.
72. A. M. Hirsch, New Phytol. 122 (1992) 211; F. G. P. Lhuissier, N. C. A. de Ruijter, B. J. Sieberer, J. J. Esseling and A. M. C. Emons, Ann. Bot. 87 (2001) 289.
73. N. S. Subba Rao, P. F. Mateos, D. Baker, H. S. Pankratz, J. Palmer, F. B. Dazzo and J. I. Sprent, Planta 196 (1995) 311; W. M. Grant and A. T. Trese, Symbiosis 20 (1996) 247; F. C. Boogerd and D. VanRossum, FEMS Microbiol Rev. 21 (1997) 5.
74. C. Boivin, I. Ndouye, F. Molouba, P. Lajudie, N. Dupuy and B. Dreyfus, Crit. Rev. Plant Sci. 16 (1997) 1.

75. B. K. Robertson, B. Dreyfus and M. Alexander, Microbial Ecol. 29 (1995) 71; C. Boivin, I. Ndouye, F. Molouba, P. Lajudie, N. Dupuy and B. Dreyfus, Appl. Environ. Microbiol. 63 (1997) 1040.
76. K. Tomepke, B. Dreyfus and M. Holsters, Can. J. Microbiol. 42 (1996) 187.
77. G. Fahraeus, J. Gen. Microbiol. 16 (1957) 374; K.Sahlman and G. Farhaeus, J. Gen. Microbiol. 33 (1963) 425.
78. D. A. Phillips, C. A. Maxwell, C. M. Joseph and U. A. Hartwig in Nitrogen Fixation: Hundred Years After, H. Bothe F. J. de Bruin and W. E. Newton (eds.), Fischer Verlag, Stuttgart, 1988; U. A. Hartwig, C. A. Maxwell, C. M. Joseph and D. A. Phillips, Plant Physiol. 91 (1989) 1138; C. A. Maxwell and D. A. Phillips, Plant Physiol. 93 (1990) 1552; M. Hungria, C. M. Joseph and D. A. Phillips, Plant Physiol. 97 (1991) 759; see review by H.A. Stafford Bot. Review 63 (1997) 27.
79. P. Roche, P. Lerouge, C. Ponthus and J. C. Prome, J. Biol. Chem. 17 (1991) 10933; P. Roche, F. Debelle, Biochem. Soc. Trans. 20 (1992) 288 ; N. P. J. Price, B. Relic, Mol. Microbiol. 6 (1992) 2575; see review by S.R. Long, Plant Cell 8 (1996) 1895.
80. X. Perret, C. Staehelin and W. J. Broughton, Microbiol. Mol. Biol. Rev. 64 (2000) 180.
81. H. P. Spaink, Trend. in Plant Sci. 2 (1997) 203; F. Martinez Abarca, J.A. Herrera Cervera, P. Bueno, J. Sanjuan, T. Bisseling and J. Olivares, Mol. Plant-Micro. Int. 11 (1998) 153; R. H. Gulden and J. K. Vessey, Plant Soil 198 (1998) 127; F. Ligero, J.L. Poveda, P. M. Gresshoff and J. M. Caba, J. Plant Physiol. 154 (1999) 482.
82. H. H. Felle, E. Kondorosi and M.Schultze, Planta 209 (1999) 207; R. Heidstra and T. Bisseling, New Phytol. 133 (1996) 25; N. C. A. de Ruijter, T. Bisseling and A. M. C. Emons, Mol. Plant-Micro. Int. 12 (1999) 829; see review by F. G. P. Lhuissier, N. C. A. de Ruijter, B. J. Sieberer, J. J. Esseling and A. M. C. Emons, Ann. Bot. 87 (2001) 289.
83. B. Prithiviraj, X. Zhou, A. Souliemanov and D. L. Smith in 17th North American Conference on Symbiotic Nitrogen Fixation. Université Laval, Quebec 2000; D. L. Smith, B. Prithiviraj and F. Zhang in Nitrogen Fixation Global Perspectives, T. M, Finan, M. R. O'Brian, D. B. Layzell, J. K. Vessey and W. Newton (eds.), CAB International, Wallingford, 2002.
84. F. Zhang, T. C. Charles, B. Pan and D. L. Smith, Soil Biol. Biochem. 28 (1996)1579; A. A. Begum, S. Leibovitch, P. Migner and F. Zhang, J. Exp. Bot. 52 (2001) 1537.
85. F. Zhang and D. L. Smith, Plant Soil 192 (1997) 141.
86. A. M. Belkhier, X. M. Zhou and D. L. Smith, Plant Soil 229 (2001)41.
87. P. S. Nutman, Aust. J. Agric. Res. 18 (1967) 381; J. E. Harper, K. A. Corrigan, A. C. Barbera and M. H. Abdalla Crop Sci. 37 (1997) 1242; G. E. D. Olroyd, E. M. Engstrom and S. R. Long, Plant Cell 13(2001) 1835.
88. H. P. Spaink, Trend. Plant Sci. 2 (1997) 203; F. Ligero, J. L. Poveda, P. M. Gresshoff and J. M. Caba, J. Plant Physiol. 154 (1999) 482.
89. P. J. Dart and J. S. Pate, Aust J. Biol Sci. 12 (1959) 427.

90. C. R. Bhatia, K. Nichterlein and M. Maluszynski, Euphytica 120 (2001) 415.
91. G. Hardarson and D. G. Jones, Ann. Appl. Biol. 92 (1979) 329; D.G. Jones and G. Hardarson, Ann. Appl. Biol. 92 (1979) 211.
92. P. B. Cregan and H. H. Keyser, Crop Sci. 26 (1986) 911.
93. C. Montealegre, P. H. Graham and J. Kipe Nolt, Can. J. Microbiol. 41 (1995) 912; C. Montealegre and P. H. Graham, Can. J. Microbiol. 42 (1986) 844; J. C. Rosas, J. A. Castro, E. A. Robleto and J. Handelsman, Plant Soil 203 (1998) 71.
94. C. Freiberg, R. Fellay, A. Bairoch, W. J. Broughton, A. Rosenthal and X. Perret, Nature 387 (1997) 394 ; V. Viprey, A. Delgreco, W. Golinowski, W. J. Broughton and X. Perret, Mol. Microbiol. 28 (1998) 1381; T. Kaneko, Y. Nakamura, S. Sato, E. Asamizu, T. Kato, S. Sasamoto, A. Watanabe, K. Idesawa, A. Ishikawa, K. Kawashima, T. Kimura, Y. Kishida, C. Kiyokawa, M. Kohara, M. Matsumoto, A. Matsuno, Y. Mochizuki, S. Nakayama, N. Nakazaki, S. Shimpo, M. Sugimoto, C. Takeuchi, M. Yamada, and S. Tabata, DNA Res. 7 (2000) 331.
95. F. Galibert, T. M. Finan, Science 293 (2001) 668.
96. K. LeVier, R. W. Phillips, V. K. Grippe, R. M. Roop and G. C. Walker, Science 287 (2000) 2492; D. O. Sanchez, R. O. Sandomeni,, Infect. Immun. 69 (2001) 865.
97. D. F. Herridge, G. Gemell and E. Hartley in Inoculants and Nitrogen Fixation of Legumes in Vietnam, D. F. Herridge (ed.), ACIAR, Canberra, 2002.
98. M.A.T. Vargas and M. Hungria in Biologia dos solos de Cerrados, M. A. T. Vargas and M. Hungria eds., EMBRAPA CPAC, Planaltina, Brazil, 1997.
99. A. Hall and N. Clark, World Dev. 23 (1995) 1601.
100. J. A. Thompson in Report on the expert consultation on legume inoculant production and quality control, FAO, Rome, 1991.
101. G. W. O'Hara, R. Yates and J. G. Howieson in Inoculants and Nitrogen Fixation of Legumes in Vietnam, D. F. Herridge (ed.), ACIAR, Canberra, 2002; J. G. Howieson, J. Malden, R.J. Yates and G. W. O'Hara, Symbiosis 28 (2000) 33; J. G. Howieson, G. W. O'Hara and S. J. Carr, Field Crop. Res. 65 (2000)107.
102. W. R. Ellis, G. E. Ham and E. L. Schmidt, Agron. J. 76 (1984) 573; H. A. Mowad, W. R. Ellis and E. L. Schmidt, Appl. Environ. Microbiol. 47 (1984) 607.
103. J. H. G. Stephens and H. M. Rusk, Field Crop. Res. 65 (2000) 249.
104. R .J. Roughley and J. M. Vincent, J.Appl. Bacteriol. 30 (1967) 362.
105. M. Gomez, N. Silva, A. Hartmann, M. Sagardoy and G. Catroux, World J. Microbiol. Biotech. 13 (1997) 167.
106. G. Catroux, A. Hartman and C. Revellin, Plant Soil 230 (2001) 21.
107. J. Brockwell and P. J. Bottomley, Soil Biol. Biochem. 27 (1995) 683.
108. A. H. Gibson, R. A. Date, J. A. Ireland and J. Brockwell, Soil Biol. Biochem. 8 (1976) 395; J. Brockwell, R. R. Gault, M. Zorin and M. J. Roberts, Aust. J. Agric. Res. 33 (1982) 803.
109. J. D. Bever and E. L. Simms, Heredity 85 (2000) 366.

110. W. D. Sutton in Nitrogen Fixation, W. J. Broughton (ed.), Oxford University Press, Oxford, 1983.
111. J. Muller, T. Boller and A. Wiemken, J. Exp. Bot. 52 (2001) 943; J. Muller, A. Wiemken and T. Boller, J. Exp. Bot. 52 (2001) 2181.
112. A. Diatloff, Soil Biol. Biochem. 2 (1977) 85; B. Brunel, J. C. Cleyet-Marel, P. Normand and R. Bardin, Appl. Environ. Microbiol. 54 (1988) 2636 ; K. Lindstrom, P. Lipsanen and S. Kailalainen, Appl. Environ. Microbiol. 56 (1990) 444.
113. R. M. N. Kucey and M. F. Hynes, Can. J. Microbiol. 38 (1989) 661; P. J. Bottomley, Appl. Environ. Microbiol.60 (1992) 1754.
114. D. L. Chatel, R. M. Greenwood and C. A. Parker, Trans. Intern. Cong. Soil Sci. 2 (1968) 65.
115. S. Maurice, P. Beauclair, J. J. Giraud, G. Sommer, A. Hartmann and G. Catroux, World J. Microbiol. Biotech. 17 (2001) 635.
116. D. A. Parke and L. N. Ornstom, J. Gen. Microbiol. 130 (1984) 1743.
117. D. A. Parke and L. N. Ornstom, J. Bacteriol. 165 (1996) 288.
118. D. A. Parke, M. Rivelli and L. N. Orston, J. Bacteriol. 163 (1985) 417.
119. F. G. Rynne, A. R.Glenn and M. J. Dilworth, Soil Biol. Biochem. 26 (1994) 703.
120. S. Rossbach, D. A. Kulpa, U. Rossbach and F. J. de Bruijn, Mol. Gen. Genet. 245 (1994) 11.
121. M. Soedarjo, T. K. Hemschiedt and D. Borthakur, Appl. Environ. Microbiol. 60 (1994) 4268.
122 I. Christiansen, personal communication (2001).
123. J. G Howieson and M. A. Ewing, Aust. J. Agric. Res. 37 (1986) 55.
124. J. A. Ireland and J. M. Vincent in Trans. 9th Intern. Congr. Soil Sci. Adelaide Vol. 2, 1968.
125. N. Amarger and J. P. Lobreau, Appl. Environ. Microbiol. 44 (1982) 583.
126. B. B. Bohlool and E. L. Schmidt, Soil Sci. Soc. Amer. Proc. 37 (1973) 651.
127. M. Obaton, Bul. Ecol.Res. Comm. 17 (1973) 170; J. L. Beynon and D. P. Jocey, J. Gen. Microbiol. 118 (1980) 437.
128. L. A. de Oliveira and P. H. Graham, Arch. Microbiol. 153 (1990) 306; N. Amarger, Soil Biol. Biochem. 13 (1981) 475.
129. A. Ayanaba, R. A. Haugland, M. J. Sadowsky, R. G. Upchurch, K .D. Weiland and R. M. Zablotowicz, App. Environ. Microbiol. 52 (1986) 847.
130. W. Streit, K. Kosch and D. Werner, Biol. Fert. Soils. 14 (1992) 140; K. J. Wilson, A. Parra and L. Botero, Can. J. Microbiol. 45 (1999) 678.
131. E. A. Schwinghamer, Can. J. Microbiol. 14 (1968) 355.
132. T. V. Bhuvaneswari, A. A. Bhagwat and W. D. Bauer, Plant Physiol 68 (1981) 1144.
133. P. M. Stephens and J. E. Cooper, Soil Biol. Biochem. 20 (1988) 465; L. A. de Oliveira and P. H. Graham, Arch. Microbiol. 153 (1990) 311. T. R. McDermott and P.H. Graham, App. Environ. Microbiol. 56 (1990) 3035.
134. N. Z. Lupwayi, P. M. Stephens and M. J. Noonan, Symbiosis 21 (1996) 233.

135. P. H. Graham and T. R. McDermott (1989) in The Physiology, Biochemistry, Nutrition and Bioengineering of Soybeans: Implications for Future Management, R.D. Munson (ed.), FAR/PPI, 1989.
136. V. Ranga Rao, A. Ayanaba, A. R. J. Eaglesham and G Thottappilly, Trop. Agric. Trinidad 62 (1985) 125.
137. G. R. Bernal and P. H. Graham, Can. J. Microbiol. 47 (2001) 526.
138. R. C. Abaidoo, H. H. Keyser, P. W. Singleton and D. Borthakur, Int. J. Syst Evol. Microbiol. 50 (2000) 225.
139. E. A. Schwinghamer and D. J. Reinhardt, Aust. J. Biol. Sci. 16 (1963) 597; E. W. Triplett, Mol. Plant-Microbe Int. 3 (1990) 199.
140. T. R. McDermott and P. H. Graham, Appl. Environ. Microbiol. 55 (1989) 2493.
141. J. E. Thies, P. W. Singleton and B. B. Bohlool, Appl. Environ. Microbiol. 57 (1991) 19.
142. E. P. Dunigan, P. K.Bollich, Agron. J. 76 (1984) 463.
143. L. Brutti, E. Rivero, J. C. P. Basurco, M. Nicolàs,L. Iriarte, N. Abbiati, H. Ljunggren and A. Mårtensson Appl. Soil Ecol. 10 (1998) 87.
144. G. T. Hill, N. A. Mitkowski, et al., Appl. Soil Ecology 15 (2000) 25.
145. G. Laguerre, M.R. Allard, F. Revoy and N. Amarger, Appl. Environ. Microbiol. 60 (1994) 56; H .B. Krishnan and S. G. Pueppke, Plant Soil 161 (1994) 21.
146. S. Selenska Pobell, L. Gigova and N. Petrova, J. Appl. Bacteriol. 79 (1995) 425; M. Dye, L. Skot, L. R. Mytton, S. P. Harrison, J. J. Dooley and A. Cresswell, Can. J. Microbiol. 41 (1995) 336.
147. M. Schneider and F. J. De Bruijn, World J. Micro. Biotech. 12 (1996) 163; G. R. Bernal and P. H. Graham, Can. J. Microbiol. 47 (2001) 526.
148. D. Pinero, E. Martinez and R. K. Selander, Appl. Environ. Microbiol. 54 (1988) 2825; B. D. Eardly, Appl. Environ. Microbiol. 56 (1990) 187; V. L. Souza, L Eguiarte, G Avila, R Cappello, C Gallardo, J Montoya, and D PineroAppl. Environ. Microbiol. 60 (1994) 1260.
149. N. Dupuy, A. Willems, B. Pot, D. Dewettinck, I. Vandenbruaene, G. Maestrojuan, B. Dreyfus, K. Kersters, M. D. Collins, and M. Gillis, Int. J. Syst. Bacteriol. 44 (1994) 461.
150. J. E. Thies, E.M. Holmes and A. Vachot, Aust. J. Exp. Agric. 41 (2001) 299.
151. V. Torsvik, R. Sorheim and J. Goksoyr, J. Ind. Microbiol. 17 (1996) 170.
152. S. S. Yang, R. A. Bellogin, A. Buendía, M. Camacho, M. Chen, T. Cubo, A. Daza, C. L. Díaz, M. R. Espuny, R. Gutiérrez, M. Harteveld, X. H. Li, M. C. C. P. Lyra, N. Madinabeitia, C. Medina, L. Miao, F.J. Ollero, M. M. A. Olsthoorn, D. N. Rodríguez, C. Santamaría, H. R. M. Schlaman, H. P. Spaink, F. Temprano, J. E. Thomas-Oates, A. A. N. Van Brussel, J. M. Vinardell, F. Xie, J. Yang, H. Y. Zhang, J. Zhen, J. Zhou, J. E. Ruiz-Sainz, J. Biotech. 91 (2001) 243.
153. N. O. Perez Ramirez, M. A. Rogel, E. Wang, J. Z. Castellanos, and E. Martinez Romero, FEMS Microbial Ecol. 26 (1998) 289.
154. M. J. Hagen and J. L. Hamrick, Molec. Ecol. 5 (1996) 707.

155. P. R. Hirsch, New Phytol. 133 (1996) 159.
156. G. Bernal, personal communication (2001)
157. A. E. Richardson and R. J. Simpson, Soil Biol. Biochem. 21 (1989) 87.
158. T. A.Lie, D. Goktan, J. Pijnenborg and E. Anlarsal, Plant Soil 100 (1987) 171.
159. B. Anyango, K. J.Wilson, J. L. Beynon and K. E.Giller, Appl. Environ. Microbiol. 61 (1995) 4016; M. Hungria, D.S. Andrade, A. Colozzi and E. L. Balota, Pesq. Agropec. Brasil. 32 (1997) 807.
160. E. Martinez Romero and M. Rosenblueth, Appl. Environ. Microbiol. 56 (1990) 2384; M. H. Chaverra and P. H. Graham, Crop Sci. 32 (1992) 1432.
161. A. A. T. Vargas and P. H. Graham, Field Crops Res. 19 (1988) 91.
162. S. R. Aarons and P. H. Graham, Plant Soil 134 (1991) 145; B. Peick, P. Graumann, R. Schmid, M. Marahiel and D.Werner, Soil Biol. Biochem. 31 (1999) 189.
163. K. G. Ballen. P. H. Graham, R. K. Jones and J. H. Bowers, Can. J. Microbiol. 44 (1998) 582.
164. P. M. Riccillo, C .I. Muglia, F. J. de Bruijn, A. J. Roe, I. R. Booth and O. M. Aguilar, J. Bacteriol. 182 (2000) 1748.
165. E. L. J. Watkin, G. W. O'Hara and A. R. Glenn, Soil Biol. Biochem. 29 (1997) 1427; E. L. J. Watkin, G. W. O'Hara, J. G. Howieson and A. R. Glenn, Soil Biol. Biochem. 32 (2000) 1393.
166. M. F. Del Papa, L. J.Balague, S. Castro Sowinski, C. Wegener, E. Segundo, F. Martínez Abarca, N. Toro, K. Niehaus, A. Pühler, O. M. Aguilar, G. Martínez-Drets, and A. Lagares, Appl. Environ. Microbiol. 65 (1999) 1420; C. Wegener, S. Schroder, D. Kapp, A. Pühler, E. Segundo Lopez, F. Martinez-Abarca, N. Toro, M. F. Del Papa, L. J. Balagué, A. Lagares, G. Martinez-Drets, and K. Niehaus, Symbiosis 30 (2001) 141.
167. M. J. Dilworth, J. G. Howieson, W. G. Reeve, R. P. Tiwari, and A. R. Glenn, Aust. J. Exper. Agric. 41 (2001) 435.
168. R. P. Tiwari, W. G. Reeve, M. J. Dilworth and A. R. Glenn, Microbiol. 142 (1996) 601.
169. U. B.Priefer, J. Aurag, B. Boesten, I. Bouhmouch, R. Defez, A. Filali-Maltouf, M. Miklis, H. Moawad, B. Mouhsine, J. Prell, A. Schlüter, B. Senatore J. Biotechnol. 91 (2001) 223.
170. A. R. J. Eaglesham and A. Ayanaba in Selected topics in Biological Nitrogen Fixation, N. S. Subba Rao (ed.), Oxford/IBH Publishing, New Delhi, 1984.
171. A. R .J. Eaglesham, B. Seaman, H. Ahmad, S. Hassouna, A. Ayanaba and K.Mulongoy in Current Perspectives in Nitrogen Fixation, A. H. Gibson and W. E. Newton (eds.), Australian Academy of Science, Canberra, 1981.
172. L. Segovia, D. Pinero, R. Palacios and E. Martinez Romero, Appl. Environ. Microbiol. 57 (1991) 426.
173. J. Ek-Jander and G. Fahraeus in Biological Nitrogen Fixation in Natural and Agricultural Habitats, T. A. Lie and E. G. Mulder (eds.), Martinus Nijhoff, The Netherlands, 1970; P. Drouin, D. Prevost and H. Antoun, FEMS Microbial Ecol. 32 (2000) 111.
174. N. Sardesai and C. R. Babu, Microbiol. Res. 156 (2001) 279.

175. K. E. Giller and G. Cadisch, Plant Soil 174 (1995) 255; M. B. Peoples, K. E. Giller, D. F. Herridge and J. K. Vessey in Nitrogen Fixation Global Perspectives, T. M, Finan, M. R. O'Brian, D. B. Layzell, J. K. Vessey and W. Newton (eds.), CAB International, Wallingford, 2002.
176. R. Yates, E. di Mattia, D. Real, G. W. O'Hara and J. G. Howieson in Survival of Perennial Legumes in Dry Mediterranean Areas. S. J. Bennett (ed.), UWA Press, Perth, 2002.
177. P. Singleton, H. Keyser and E. Sande In Inoculants and Nitrogen Fixation of Legumes in Vietnam, D. F. Herridge (ed.), ACIAR, Canberra, 2002, and references therein.

Nitrogen Fixation at the Millennium.
G. Jeffery Leigh, Editor.

CHAPTER 15

Nitrogen Fixation in Rice

Pallavolu M. Reddy[a], Euan K. James[b] and Jagdish K. Ladha[a]

[a]International Rice Research Institute, DAPO Box 7777, Metro Manila, Philippines
[b]Centre for High Resolution Imaging & Processing, School of Life Sciences, University of Dundee, Dundee DD1 5EH, UK

1. INTRODUCTION

Rice is the basic food for nearly half the people on earth. The productivity of a rice crop is based on numerous variables including weather, soil type, moisture and nutrients. The most common nutrient that affects rice production is N. In the tropics, lowland rice (*Oryza sativa*) yields 2–3.5 t ha^{-1}, utilizing naturally available N derived from biological nitrogen fixation (BNF) by free-living and plant-associated diazotrophs, and from the mineralization of soil N. This system of rice cultivation over thousand years has sustained population, particularly in Asia, which is currently growing at 1.8% a year. For higher yields to support a rapidly increasing population, however, additional N must be applied. During the Green Revolution, and since the 1960s, the application of chemical N fertilizers boosted rice yields by 100–200% to match the demands of the increasing human population. Although rice production has so far kept up with population growth, new studies suggest that an additional 30% on top of the current rice supply will be needed during the next 30 years [1]. There is increasing pressure on rice-growing resources, primarily because agricultural lands are shrinking in area owing to increased urbanization, soil erosion etc, and present trends suggest that tomorrow's rice land will be under even greater pressure [2]. This is further exacerbated by the deleterious effects that modern agriculture itself has on environmental degradation, particularly those caused by excessive chemical fertilization. Rice production in the coming years is expected to lean toward more intensification in terms of increased cropping per year, and this will be facilitated by the greater use of high-input technologies. Further expansion and intensification of the irrigated rice area could greatly increase the application of

BNF system	Estimated range and Advantage	Constraint
Autochthonous (indigenous) systems	Easily manageable without incurring additional cost/effort	Low BNF potential; large G x E
Cyanobacteria/Phototrophic bacteria	10-80 kg N ha^{-1} $crop^{-1}$	P deficiency; grazer pressure
Heterotrophic bacteria	1-39 kg N ha^{-1} $crop^{-1}$	
Associative bacteria	1-70 kg N ha^{-1} $crop^{-1}$	
Allochthonous (exogenous) systems	High BNF potential	
Azolla	20-150 kg N ha^{-1} $crop^{-1}$	Development of inoculum through sporulation; insect damage; requires continuous water; P deficiency; labor intensive; poor adoptability
Legumes	10-450 kg N ha^{-1} $crop^{-1}$; well adapted to diverse situations	Seed production; labor intensive; poor adoptability; uneconomical in many situations
***In planta* (endogenous) systems**		
Rice-endophyte associations	Diazotrophs are protected from external competition	Nonspecific; low population of endophytes; low BNF potential; no active transfer of fixed N to plant
Rice-rhizobial symbiosis	Diazotrophs are protected from external competition; highly specific. High BNF potential	Complex genetic engineering; long-term and high-risk project
nif gene transfer	High BNF potential	Complex genetic engineering; long-term and high-risk project

Table 1. Existing and futuristic biological nitrogen fixation systems in rice production: advantages, constraints, and potential.

nitrogenous fertilizers, which in turn may have further negative environmental consequences.

Increased future demand for rice will, therefore, clearly entail an increased application of fertilizer N. It is estimated that twice as much fixed N will be required to raise rice production by 2020 to supplement the food requirements of the increasing human population [3]. The increase in requirement of fixed nitrogen for enhancing crop production will have to be met by industrial nitrogen fixation (through the Haber-Bosch process) in complementation with biological nitrogen fixation (BNF). Even though both sources of biologically and industrially fixed nitrogen are needed to meet the demand for food production, BNF has advantages of lower costs and reduced environmental hazards, and is more consistent with the development of sustainable agriculture. Therefore, in view of the importance of BNF in sustainable agriculture, it is imperative to improve existing biological nitrogen-fixing systems and to develop nitrogen-fixing non-legume crops, particularly important cereal crops such as rice. It is in this context that BNF-derived N assumes importance.

The submerged soils within which most rice is grown provide two of the essential prerequisites for optimum BNF, *viz.*, a low oxygen tension, and a consistent supply of carbon substrate *e.g.*, from rice root exudates. In rice ecosystems, depending on their origin diazotrophs can be broadly categorized into two existing BNF systems, with the possibility of a potential third system (Table 1). The existing systems consist of (1) autochthonous (indigenous) BNF systems comprising of heterotrophic and phototrophic bacteria and cyanobacteria native to soil-plant-floodwater and (2) allochthonous (exogenous) BNF system comprising of organisms, such as *Azolla*, which harbor symbiotic N_2-fixing cyanobacteria, and aquatic legumes, such as *Sesbania* and *Aeschynomene* spp. that form symbioses with heterotrophic and phototrophic rhizobia. The exogenous BNF systems are not ubiquitous, and hence need to be applied/inoculated on to rice fields to supplement nitrogen to the crop. The third system, an *in planta* (endogenous) BNF system, would come about from the transformation of rice to an autonomous N_2-fixing plant. The present review assesses the prospects and potentials of these various BNF systems for sustainable rice production.

2. AUTOCHTHONOUS (INDIGENOUS) BNF SYSTEMS

The indigenous free-living and associative BNF systems offer an alternative to the input-based exogenous BNF technologies [4].

2.1. Free-living phototrophs and heterotrophs

Diverse free-living N_2-fixing microorganisms (aerobes, facultative anaerobes, heterotrophs, phototrophs) grow in wetland rice fields and contribute to soil N. These BNF systems include cyanobacteria and photosynthetic bacteria inhabiting floodwater and the soil surface, and heterotrophic bacteria in the root zone and in the soil. Semi-quantitative data suggest that free-living diazotrophs have a low-to-moderate potential to supply N to rice (Table 1). Estimates using data obtained from acetylene reduction assays (ARA) have suggested that indigenous cyanobacteria may contribute as much as 80 kg N ha^{-1} $crop^{-1}$, and averaging 30 kg N ha^{-1} $crop^{-1}$ [5].

Heterotrophic rhizosphere bacteria appear to contribute less BNF than the phototrophs (Table 1). Various N-balance studies in which growth and N_2 fixation by cyanobacteria and other phototrophs were prevented by shading the soil/floodwater [6,7] suggest that heterotrophic BNF typically averages only 7 kg N ha^{-1} [8]. Using the available N of a stabilized ^{15}N-labeled soil as a control, Zhu [9] estimated that with no fertilizer and with photo-dependent BNF minimized, heterotrophic BNF could contribute about 16-21% of rice N, or 11-16 kg N ha^{-1} $crop^{-1}$.

2.2. Associative heterotrophs

Even though several types of diazotrophs inhabit the rhizosphere of rice [10], some bacteria, such as *Azospirillum* spp., *Azorhizobium caulinodans*, *Klebsiella* spp., *Pseudomonas diazotrophicus* and *Enterobacter* spp. form loose associations with the plant surface [10,11]. Nitrogen fixation by free-living diazotrophs occurs outside the plant, and hence it is not instantly available to the plant, and may be subjected to losses. On the other hand, and unlike with free-living diazotrophs, some of the N fixed by associative bacteria may still be immediately available to the rice plant. Soil and plant N balance studies (the sum of all gains and losses) in the greenhouse revealed significant differences in the ability of rice cultivars to support a positive N balance [8]. Nitrogen gains of up to 70 kg N ha^{-1} $crop^{-1}$ were recorded (Table 1); of this, 55 kg N ha^{-1} could be attributed to a cultivar effect on stimulation of BNF. Although the N balance study of App *et al.* [8] included the N contribution from the soil, it indicated for the first time that rice influences BNF, and that the ability differs widely between lines. Using ^{15}N-stabilized soil in pots, Shrestha and Ladha [12] and Malarvizhi and Ladha [13] quantified N in the plant derived from the atmosphere (Ndfa) for 69 rice lines of diverse background (indica, japonica, and javonica; traditional and improved) and duration (early, medium, and long). Ndfa was estimated based on ^{15}N enrichment of a reference cultivar of representative growth duration and that of available soil N. Percent Ndfa for various lines differed significantly, ranging from 1.3% to 26.6% [12,13]. Those with high Ndfa g^{-1} dry weight were mostly traditional (*viz.*, Hsiang Ai-Tsao-7, Pokkali, Biron; [12]), but some improved lines also had high associative N_2 fixation (*viz.*, IR42; [13]).

The possible mechanisms by which rice genotypic diversity associated with ^{15}N dilution may influence N_2-fixing bacteria in the rhizosphere are not yet elucidated. It has been hypothesized that diazotrophs associate with roots in response to a specific chemical compound/attractant excreted by the roots. Malic acid in the root exudates has been identified as the most important attractant to diazotrophs associated with C_4 crops [14]. In rice, a C_3 crop grown under flooded conditions, the main organic acids in root exudates are citric and oxalic acids [15,16], although the nature of the main attractant in the chemotaxis of diazotrophs to rice is not yet identified. Furthermore, although it has long since been reported that most (culturable) bacteria (80%) found in rice roots are N_2-fixing [17], the physiological and biochemical mechanisms that operate in this association have not yet been deciphered [11]. Hence, questions that still must be addressed are: (1) what component(s) of rice root exudates are the key chemotaxic attractants to diazotrophs, and what is the extent of genotypic variation in the excretion of these attractants, (2) what types of diazotroph proliferate in the rhizosphere and within the roots, and (3) what is the genetic background of the mechanism(s) underlying their attraction towards roots? The difficulties encountered in investigating these questions reside in the complexity of analyzing root exudates of the plant under soil conditions. Molecular approaches may help in identifying the genetic factors underlying these mechanisms and may help us to understand and manipulate associative N_2 fixation. Recently, Wu *et al.* [18] mapped the genes underlying rice genotypic variability in their ability to stimulate N_2 fixation. Seven restriction fragment length polymorphism marker loci on four chromosomes were identified to be significantly associated with the segregation of atom % ^{15}N excess in an F_2 population derived from Palawan x IR42 rice lines. Results indicate that the ability of rice to enhance N_2 fixation in the rhizosphere is controlled by multiple genes. The identification of quantitative-trait loci underlying this trait provides the first real evidence of the presence of genetic factors that interact with N_2-fixing bacteria in the rice rhizosphere.

3. ALLOCHTHONOUS (EXOGENOUS) BNF SYSTEMS

The indigenous BNF system can support only a low rice production system of less than 4 t ha^{-1} $crop^{-1}$. However, today's agriculture cannot depend only on the indigenous BNF system, as higher rice productivity is essential to meet growing demand and therefore higher inputs of N are required. To achieve this, exogenous BNF systems such as symbiotic *Azolla* and nodulated legumes, which are not ubiquitous, are generally applied/inoculated on to rice fields to supplement nitrogen to the crop. BNF can be a major source of N in rice production when symbiotic N_2-fixing systems are used, e.g. the amount of N input from a nodulated legume was reported to be as high as 458 kg N ha^{-1} [19].

3.1. ***Azolla***

Azolla is generally inoculated and grown as a cover crop for incorporating into the soil as a top-dressing in rice cultivation. *Azolla* in symbiosis with the cyanobacterium *Anabaena azollae*, under optimum conditions, can fix substantial amounts of N_2 through BNF, and release as much as 70% of the fixed N to the rice crop upon incorporation [5]. Using the δ ^{15}N method, Yoneyama *et al.* [20] estimated that 59-99% of *Azolla* N is derived from the atmosphere. In a separate study, based on the ^{15}N dilution method, Watanabe and Talukdar, and Kumarasinghe [cited in 21] found that as much as 80-85% of *Azolla* N is Ndfa.

The reported N content in maximum standing crops of *Azolla* ranged from 20 to 146 kg ha^{-1} (Table 1) and averaged 70 kg ha^{-1}, and N_2-fixing rate ranged from 0.4 to 3.6 kg N ha^{-1} d^{-1} and averaged 2 kg N ha^{-1} d^{-1} in a growing cycle of approximately 40 days. The N balance in the 0-50-cm soil profile after 27 cropping cycles confirmed an average annual gain of 76 kg N ha^{-1} under *Azolla* treatment [22].

Maintaining the *Azolla* inocula between cropping seasons is a major constraint to its wider adoption by rice farmers. The amount of inoculum required varies from 300-500 kg to 2-5 t fresh biomass ha^{-1}. The problem of maintaining *Azolla* vegetatively would be eliminated if mass quantities of spores could be obtained. Unfortunately, technologies to induce mass sporulation have not been developed, although some research has been carried out to address this problem [23]. The selection of superior germplasm, development of improved *Azolla* hybrids, and an understanding of the mechanism for inducing sporulation may help in developing *Azolla* as an adoptable technology.

3.2. Legumes

Sesbania rostrata, S. aculeata (*S. cannabina*) and *Crotalaria juncea* are common green-manure legumes grown in rice cropping systems of the tropics. *Sesbania* species adapt better than *Crotalaria* as green manures, due to their ability to withstand soil saturation and waterlogging, and to their tolerance to soil salinity. Currently, green-manure legumes in rice cropping systems are primarily grown in irrigated environments [24]. Rain-fed rice-environments prone to soil waterlogging appear to have the greatest potential for growing green-manure legumes in the future [25]. Flooding-tolerant, stem-nodulating legumes, such as *S. rostrata* and *Aeschynomene afraspera* have been found particularly promising for such environments [26]. The most promising forage legumes that can be inter-cropped with rice include milk vetch (*Astragalus sinicus*) [27], clitoria (*Clitoria ternatea*), desmanthus (*Desmanthus virgatus*), lablab bean (*Lablab purpureus*), siratro (*Macroptilium atropurpureum*) and sunhemp (*Crotalaria juncea*) [28]. Among the major food legumes grown on rice lands, mungbean (*Vigna radiata*), cowpea (*V. unguiculata*), chickpea (*Cicer arietinum*), groundnut (*Arachis hypogaea*), pigeonpea (*Cajanus cajan*), and soybean (*Glycine max*) are common in tropical countries [28].

Quantification studies on N_2 fixation by legumes in the tropics and subtropics have revealed a large range in estimates of the % Ndfa, e.g. 0-95% for soybean, 8-89% for cowpea, 22-92% for groundnut, and 10-88% for pigeonpea [19]. The amounts of N fixed by these legumes ranged from 10 to 450 kg N ha^{-1} (Table 1). Estimates of Ndfa tend to be higher for green-manure and forage legumes than for food legumes, e.g. the Ndfa for green-manure legumes normally exceeds 80% [29]. Nitrogen fixation by common green-manure legumes ranges from 70 kg ha^{-1} to 458 kg ha^{-1}, but more typically from 70 kg ha^{-1} to 188 kg ha^{-1} [30]. Forage legumes fix more N_2 (from 91 to 240 kg ha^{-1}, but more typically from 196 to 240 kg ha^{-1}) than grain legumes (from 37 to 167 kg ha^{-1}).

In most situations, the N_2 fixed by food or forage legumes probably does not increase the soil N pool and may not have considerable residual effect. This is because the N_2 fixed by these legumes is often less than the N removed with the harvest of grain or fodder. On the other hand, legumes grown solely for incorporation as green manures can increase the soil N pool, but only provided they effectively fix N_2, and also that losses of legume N are minimized. However, rice farmers are often reluctant to devote land and resources to grow legumes solely for green manure or for animal feeds, because they provide no immediate income or food, and yet they require labor. On the other hand, the intercropping of forage or grain legumes with rice during the dry season to provide food, fodder, and residual N for the next wet-season rice crop is found to be more attractive to farmers [31].

4. *IN PLANTA* (ENDOGENOUS) BNF SYSTEMS

Among the conventional BNF systems of rice, the free-living/associative diazotrophs have only low-to-moderate potential to supply N to rice, as N_2 fixed outside the plant is subject to greater losses. On the other hand, although green-manure crops, such as *Azolla* and *Sesbania* have high N supply potential to support higher grain yields of rice, farmers are reluctant to use them as there is no economic advantage over N fertilizer [32]. In addition, rice suffers from a mismatch of its N demand and N supplied as fertilizer, resulting in a 50-70% loss of applied N fertilizer. A regulated supply of N is likely to maximize the yield potential of rice, and so clearly rice needs a self-sustaining N-fixing system in which the demand controls the supply and very little or no loss of N occurs. If such a BNF system could be assembled within the rice plant it could enhance the potential for N supply, as the fixed N would be available directly to the plant [32-34]. In addition, there would be no additional economic burden on farmers. Several approaches toward developing rice capable of fixing N_2 are now being considered, such as the establishment of effective endophytic symbioses, the development of legume-like nodulation, and the introduction and expression of nitrogen fixation (*nif*) genes in rice plants [35]. In this section, we will consider the potential and current status of all three approaches that are aimed at developing N_2-fixing rice.

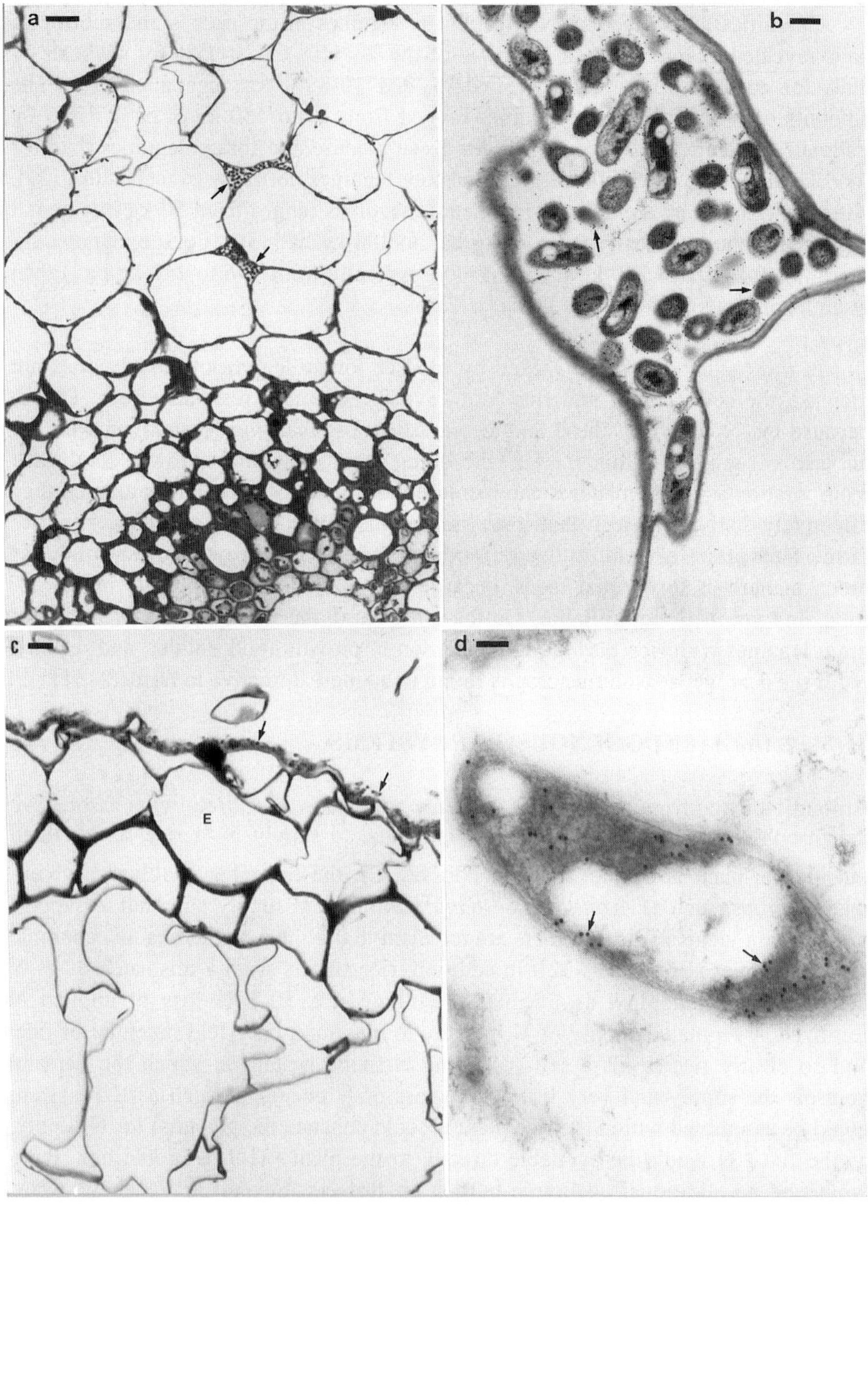
a
b
c
E
d

4.1. Development of rice-endophytic diazotroph associations

Some diazotrophs, such as (*Glucon*)*acetobacter diazotrophicus* and *Herbaspirillum* spp. can live endophytically within roots, stems and leaves of sugar cane. These bacteria, whose population ranges from 10^3 to 10^7 cells g^{-1} fresh weight of sugar cane, may contribute to the BNF measured in this crop in Brazil, which, according to N-balance studies, was found to range from 0 to 150 kg N ha^{-1} $year^{-1}$, depending on variety [36]. Another N_2-fixing endophyte that is of considerable interest is *Azoarcus*. This diazotroph was originally isolated from the roots of Kallar grass (*Leptochloa fusca*) [37], a forage crop which yields 20-40 t of hay ha^{-1} $year^{-1}$ without the addition of any N fertilizer in saline-sodic, alkaline soils with low fertility [38]. So far, it is not clear to what extent (*Glucon*)*acetobacter diazotrophicus*, *Herbaspirillum* spp. and *Azoarcus* are responsible for N_2 fixation in these plants, or if they fix N_2 in conjunction with other diazotrophs [39-42]. Nonetheless, by inhabiting the interior of the plants the endophytes are thought (1) to avoid competition with rhizosphere bacteria and (2) to derive nutrients directly from host plants [7,40,43-45]. In return, as the plant interior may provide an environment conducive to N_2 fixation by being low in O_2 and relatively high in carbon, the bacteria can fix N_2 more efficiently than rhizosphere diazotrophs and can also transfer fixed N products more efficiently to the host [40, 46].

Dobereiner *et al.* [47] and Boddey *et al.* [7] hypothesized that endophytic diazotrophs in certain rice genotypes may, in fact, be responsible for the substantial contributions of BNF to rice reported earlier [8,48]. If endophytic diazotrophs are partly or wholly responsible for BNF in sugarcane and Kallar grass, it is possible that rice varieties that show significant heterotrophic BNF [8,12,13,18] may also

Figure 1. Colonization of rice by *Herbaspirillum seropedicae*. a. Transverse section (TS) through a root of rice cv. IR72 at 2 d after inoculation with *H. seropedicae*. Numerous bacteria are present within cortical intercellular spaces (arrows). Bar = 10 µm; b. Transmission electron micrograph (TEM) of intercellular bacteria in a leaf of rice cv. IR72 at 7 d after inoculation with *H. seropedicae*. The bacteria have been immunogold labeled with an antibody raised against *H. seropedicae* (arrows). Bar = 500 nm; c. TS of a root of rice cv. Moroberekan at 10 d after inoculation with *H. seropedicae*. There are numerous bacteria on the surface of the root (arrows). E = epidermis. Ex = Exodermis. Bar = 5 µm; d. TEM of a bacterium on the surface of a root of rice cv. Moroberekan at 10 d after inoculation with *H. seropedicae*. The bacterium has been immunogold-labeled with an antibody raised against the Fe-protein of the nitrogenase enzyme complex (arrows). Bar = 100 nm.

obtain fixed N from bacteria living within their tissues. Numerous recent studies have shown that rice indeed harbors a wide range of endophytic diazotrophs [7,10,49-56] (Figs 1a, b). All or some of these may be responsible for supplying plants with fixed N. Inoculation experiments with the endophytes *Herbaspirillum seropedicae* Z67, *Rhizobium leguminosarum* bv. *trifolii* and *Serratia marcescens* IRBG500 in non-sterile soil under greenhouse conditions have shown that these diazotrophs can be readily introduced into the rice plant by applying bacterial cultures on seeds or roots prior to planting. Infection occurs through roots and the diazotrophs are translocated to all parts of the plant via the xylem, the aerenchyma and through interconnecting intercellular spaces [10,55,57,58] (Fig. 1a and b).

Although it appears that diazotrophs associated with rice, endophytic or otherwise, may also stimulate plant growth through mechanisms other than BNF [51,55,58-61], *e.g.*, *via* phytohormone production or by increasing root growth [62], although there is some direct evidence that they can fix N_2 in association with their hosts. *Azoarcus* cells harboring the *gusA* gene under the control of the *nifH* promoter, upon inoculation, were able to invade and colonize intercellular spaces in the cortical cells of rice roots and were found to express *gusA*, suggesting that the proper physiological environment for the derepression of N_2 fixation genes exists in such tissues [63]. When various *Herbaspirillum* and diazotrophic *Burkholderia* strains were inoculated onto seedlings of a Brazilian upland rice variety (Guarani) growing in N-free Hoagland solution containing ^{15}N-labeled N, the plants showed ^{15}N dilution values representing over 50% Ndfa, depending on the strain [64]. In a separate study, inoculation of the aluminum-tolerant (carbon-excreting) wetland rice variety Moroberekan with *H. seropedicae* Z67 resulted in significant incorporation of $^{15}N_2$ [58]. The plants were found to derive 4.7 % of their N from air over a 3 d period, although transmission electron microscopy and immunogold labeling using an anti-nitrogenase Fe (NifH) antibody showed that the bacteria on the surface of the roots of cv. Moroberekan were most likely responsible for the observed BNF (Figs 1c, d).

Endophytic bacterial associations with rice are generally non-specific, and the size of the bacterial population density in rice tissues is too low to support adequate N_2 fixation. Hence, it is important to develop strategies that enable enhanced diazotrophic bacterial colonization for significant endophytic BNF in rice. It is also important to have an efficient method of inoculation to enhance the chances of the inoculated diazotrophic endophyte to colonize selectively the rice plant, because inoculated microbes are usually out-competed by indigenous microflora in the rhizosphere of crop plants. This problem could be overcome by the creation of a "biased rhizosphere" [65], a novel procedure that can be used to encourage selectively the growth of introduced microorganisms in the rice rhizosphere [32]. This system involves the expression of specific genes in the transgenic rice plants that would enable the roots to produce the specific nutritional compound, which can be used only by the bacteria possessing genes for catabolism of this compound. If such catabolism genes are incorporated into efficient endophytic N_2 fixers and inoculated into the rhizospheric zone of the genetically modified rice plants that are able to produce unique nutritional compound, the prospects for the inoculated bacteria to survive and colonize the plant roots in the face of the competition offered by indigenous microflora will be considerably improved [32].

4.2. Development of endophytic/symbiotic associations via coculture

Experiments with establishing associations of cultured cells (tissues) of leguminous and nonleguminous plants with symbiotic bacteria could form a basis for obtaining diazotrophic bacterial associations in rice. The problems of inducing artificial endophytic associations were addressed through the regeneration of plants from cell lines possessing amenable properties to coexist with the diazotroph [66]. This was achieved by co-culturing the plant cells (in tissue culture) with a suitable diazotroph to encourage the production and proliferation of populations of plant cells (somaclonal variants) having the ability to grow in association with the selected diazotroph. The resultant populations of plant cell lines showed ability to coexist with the diazotroph, and the regenerated plants from such adapted plant cell lines were found to possess the capabilities to form endophytic associations with the diazotroph without eliciting a host defense response. Indeed, such an approach was successfully employed in producing endophytic/symbiotic associations between cyanobacteria and alfalfa and tobacco [67,68], even though we are not aware whether such associations could be stably maintained in field conditions. Nevertheless, it may be worthwhile to try a similar approach to generate rice plants with amenable properties to enter into endophytic associations with suitable diazotrophs [69]. However, it is essential to assess whether the 'acquired symbiotic trait' can be maintained stably in the plants derived through such a process.

4.3. Development of rice-rhizobia symbioses

Legume-rhizobial interactions culminate in the formation of structures known as nodules, where rhizobia, insulated from microbial competition, fix N_2. Currently there is much interest in determining whether rhizobia would be able to nodulate monocots such as rice, and carry out N_2 fixation [32,70,71]. In fact, the possibility of extending the host range of rhizobia to non-legumes was encouraged by the discoveries that *Parasponia*, a member of the Ulmaceae, forms nodules with *Rhizobium* [72].

Various investigations aimed at understanding legume-nodule development [73] form the basis for devising rational strategies to extend the host range of N_2-fixing symbioses to non-legume crops such as rice. A complex interplay between the legume host and its rhizobial partner leads to the formation of a morphologically defined specialized plant organ, the nodule, in which rhizobia carry out N_2 fixation. Rhizobia interact with leguminous plants to form N_2-fixing nodules through a process that begins with the secretion of flavonoids from roots and the consequent flavonoid-triggered *nod* gene expression in the microsymbiont leading to the production of Nod factors [74,75]. Rhizobial Nod factors, which play a crucial role in determining the fate of the symbiotic interaction, activate early nodulin (*ENOD*) genes and alter the growth pattern of two cell types in legume roots: in epidermal cells/root hairs new tip growth/deformation is initiated, and cortical cells are mitotically reactivated [74,76]. The latter results in the formation of a nodule primordium, whereas the induction of tip growth/deformation is instrumental in the infection process.

For rhizobia to infect the legume root, bacterial *nod* genes must be induced by plant-produced flavonoids. Our studies have demonstrated that roots of specific rice cultivars exude compounds that are able to induce, albeit to a low degree, the transcription of *nod* genes of rhizobia (Table 2) [77]. Results available so far indicate that rice root exudate contains only extremely low concentrations of the appropriate flavonoids that are able to activate *nod* gene expression in rhizobia. This problem could be overcome relatively easily through the genetic transformation of rice with appropriate genes in the polypropanoid pathway for enhancing the production of *nod* gene-inducing flavonoids in roots.

Rhizobia are able to invade rice roots through cracks in the epidermis and fissures created during emergence of lateral roots and colonize intercellular spaces in the epidermis, exodermis and cortex (Table 2) [51,78,79]; however, the invasion by rhizobial cells was found to be a *nod* gene-independent and non-specific process in rice [78]. In addition, no evidence of infection threads was observed [78]. This contrasts with the mode of rhizobial infection in legumes where *bona fide* infection threads ultimately develop as a route for rhizobial intercellular dissemination.

Nod factors excreted by rhizobia play a pivotal role in the early development of nodules on legumes, particularly on root-hair curling and infection-thread formation [74]. So, any demonstration that they can be recognized by rice would be a significant step toward devising strategies to make rice nodulate. Hence, root-hair curling or induction of cortical cell division in rice in response to Nod factors would imply that rice indeed has Nod factor receptors that could facilitate a symbiosis with rhizobia. However, the application of neither rhizobia nor of purified Nod factors has so far elicited root-hair deformation or cortical cell divisions leading to true nodule development in rice [78]. Similarly, there is no formation of infection threads through root hairs, nor the development of an intricate network of infection threads in the root cortex. Therefore, as the Nod factors do not appear to elicit any response in rice roots, it is essential to determine whether this is due to the absence of

Table 2. Comparison of the symbiotic sub-programs of legumes and rice.

Trait	Legume	Rice	Specific comments on the response of rice, and corresponding references
Mycorrhizal symbiosis			
Mycorrhizal infection	+	+	Through intercellular spaces in root epidermis [V. Gadkar, P.M. Reddy and J.K. Ladha, unpublished]
Mycorrhizal colonization	+	+	Mostly in lateral root cortical cells surrounding vascular bundles [80, V. Gadkar, P.M. Reddy and J.K. Ladha, unpublished]
Perifungal membrane or interface material in mycorrhizae	+	+	The interface layer around the infecting mycorrhizal fungus is synthesized *de novo* [80]
Rhizobial symbiosis			
Ability of root exudates to induce *nod* gene expression in rhizobia	+	+	Root exudates of several rice varieties have a limited ability to induce *nod* gene expression [77]
Nod factor perception	+	+	Rice roots perceive Nod factors as evidenced by Pr*ENOD12*-GUS expression [81,82]
Rhizobial attachment to root hairs/epidermal cells	+	+	As with legumes, rhizobial cells attach themselves to root hairs/epidermal cells in a polar orientation [78,83], but in rice it is a *nod* gene-independent process [78]
Root hair deformation	+	-	Neither rhizobia nor purified Nod factors can induce root hair deformation in rice [78]
Entry of rhizobia into roots	Infection	Invasion	Invasion is through lateral root racks/ epidermis, and it is a *nod* gene-independent and non-specific process in rice [78,79]
Infection thread formation	+	-	Rhizobia-mediated infection thread formation has never been observed in rice [57,78,83]

Table 2 (continued)

Trait	**Legume**	**Rice**	**Specific comments on the response of rice, and corresponding references**
Cortical cell divisions	+	-	Neither rhizobia [78,83], nor purified Nod factors [78], can induce cortical cell divisions in rice roots
Site of colonization	Nodules	Roots	Rhizobial colonization in rice is restricted to inter-cellular spaces, aerenchyma and xylem vessels [78,84]. In the wild rice *Oryza breviligulata*, bradyrhizobia colonize inter- and intra-cellularly [83]
Bacteroid formation	+	-	
Peribacteroid membrane	+	-	
Defense reaction in root tissues due to infection/invasion	Transient in initial stages of infection	Slight localized host reaction upon colonization	Occasionally, rhizobia solubilize intercellular fibrillar matrix and cell wall material, but any cell invasion leads to host cell death [78]. In *O. breviligulata*, however, bradyrhizobia colonize root cells apparently without causing defense reaction or cell death [83]
Presence of *ENOD* genes	+	+	Homologues of legume *ENOD2, ENOD5, ENOD12, ENOD14, ENOD40, ENOD55, ENOD70* and *ENOD93* have so far been detected in rice [87-89]
Pr*GmENOD40-GUS* expression	+	+	Spatial expression of Pr*GmENOD40* in rice is similar to that in legumes [77]
Changes induced due the over expression of legume *ENOD40*	Amyloplast deposition and extensive cortical cell divisions in roots in response to rhizobia/Nod factors	Amyloplast deposition in cortical and xylem/ phloem parenchyma cells in roots; formation of stunted laterals, and occasionally, multiple vascular bundles	[90]

+ Present; - Absent

receptors for recognition of Nod factors or to the degradation of Nod factors by root chitinases. To determine the former, Pr*MtENOD12*-GUS (as a molecular marker for Nod signal perception) was introduced in rice and its responsiveness to Nod factors was analyzed [81]. This showed that under N limiting conditions, Nod factors were able to activate the *MtENOD12* promoter in rice root cortical cells, as well as in endodermal, pericycle, and parenchymatous cells in the stele, but not in epidermal cells (Table 2). Werthmuller *et al.* [82] showed that the Nod factors in the presence of auxin. Both these findings demonstrated that rice root tissues have a mechanism to perceive Nod factors, and also that they possess a signal transduction system to enable the subsequent activation of the legume *ENOD12* gene promoter.

Specific plant genes, known nodulin genes, are induced during legume nodulation [73]. These nodulin genes are divided into early and late nodulin genes, according to the timing of their expression [91]. Early nodulin (*ENOD*) genes are associated with infection and nodule organogenesis, whereas late nodulin gene products are thought to be involved in nodule function and include oxygen transporter leghemoglobin, enzymes of carbon and nitrogen metabolism, and proteins located in the bacteroid interface [73]. Hence an essential requirement for rice to be engineered to interact with rhizobia will be the presence of many of the proteins known to be required for legume nodule development. It has recently been shown that cDNAs of several legume *ENOD* genes hybridized to DNA from a wide variety of rice genotypes, indicating that homologues of *ENOD* genes are indeed widespread in rice genotypes (Table 2) [88]. Moreover, the characterization of two of these *ENOD* homologues from rice, *OsENOD40* and *OsENOD93a*, revealed open reading frames (ORFs) encoding peptides with significant homologies to legume *ENOD40* and *ENOD93*, respectively [87,89,92]. With regard to late nodulin genes, homologues of several genes responsible for the synthesis of the enzymes of carbon and nitrogen metabolism in legume nodules are also found in rice genome [93]. In addition, globin genes are widespread in non-leguminous plants including rice, and they encode plant hemoglobins of cryptic function [94,95]. A challenging task, however, will be to direct the expression of all these genes in a coordinated manner at the site of “symbiosis” in rice.

Among *ENOD* genes that are activated rapidly upon rhizobial inoculation, *ENOD40* is expressed very early during nodule initiation [96,97], and it was found to play a crucial role in nodule organogenesis [98]. Recent studies have demonstrated that the homologues of legume *ENOD40* gene are widely distributed in plants including monocots such as rice [92]. To ascertain if regulation mechanisms related to the activation of *ENOD40* expression are shared between rice and legumes, we developed transgenic rice carrying the soybean *ENOD40* (*GmENOD40*) promoter fused to GUS, and analyzed the activity of the promoter in rice (Table 2). Our studies showed that in rice, as in legumes, even without the addition of Nod factors, the *GmENOD40* promoter was expressed at a basal level in root and stem vascular traces [90,99]. In a separate study, we examined the induction of the rice *ENOD40* promoter in soybean roots and found that its expression was also confined to the vascular tissue of roots as well as nodules [89]. This indicates that the promoter activity of rice *ENOD40* gene is essentially same as that of the soybean *ENOD40* gene and that *OsENOD40* and legume *ENOD40*s probably share common functions in the differentiation and/or functioning of vascular bundles [89].

The activation of cortical cells in legumes appears to involve the induction of at least one early nodulin gene, *ENOD40*, in the pericycle of the vascular bundle [96,97]. Transient expression of *ENOD40* driven by the 35S CaMV promoter was found to trigger localized division of cortical cells in alfalfa roots, with net amyloplast accumulation, suggesting a critical role for *ENOD40* in the development of the nodule primordium [98]. In rice, however, over-expression of 35S-*MtENOD40* resulted in major alterations in growth and development in root architecture, such as inhibition of the formation of the cortical schlerenchymatous layer, and occasional differentiation of multiple vascular bundles in roots (Table 2) [90,99]. Furthermore, the transgenic plants also exhibited sporadic development of globular hypertrophies from lateral root primordia. It is also intriguing that in rice roots, over-expression of *MtENOD40* led to extensive deposition of amyloplasts in cortical cells particularly around the stele. These findings suggest that at least some of the cellular manifestations resulting from the activity of *MtENOD40* appear to be similar in both alfalfa and rice.

In conclusion, certain critical differences exist between the association of rice and rhizobia relative to the root-nodule symbiosis with legumes (Table 2). In the legume-*Rhizobium* symbiosis, the host plant provides the genetic information for the development of nodules. The role of the rhizobia is to trigger host-plant genes to facilitate infection and help create a nodular niche for N_2 fixation [74,75]. A monocot such as rice would unlikely to possess the complete complement of genes or genetic/cellular programs involved in nodule organogenesis that could be triggered by rhizobia. Indeed, studies with rice have shown that in spite of some similarities there are many differences between legume-rhizobial symbioses and the rice-rhizobial association [77,78]. On the other hand, although rice does not yet form a symbiosis with rhizobia, it is able to do so with mycorrhizal fungi (Table 2) [80,100,101]. This suggests that rice possess the genetic information required for the successful symbiotic infection by the mycorrhizal fungi, involving the reactivation of cortical cells in the presence of a symbiont, the accepted penetration of the symbiont into host cells, and synthesis by the host cells of a membranous interface to allow for exchange of metabolites [102]. In legumes, many parallels can be drawn between root-nodule formation by rhizobia and colonization by mycorrhizal fungi. Recent studies on legumes have provided evidence for genetic links between the processes involved in nodulation and endomycorrhizae formation [103-109].

Investigations with the nodulation mutants of pea and alfalfa have demonstrated that the early nodulin genes *ENOD2*, *ENOD5*, *ENOD11*, *ENOD12*, and *ENOD40*, which control the initial steps of nodulation are also expressed during the early stages of mycorrhizal development [110,111]. Thus, because rice is able to form symbioses with mycorrhizal fungi, and because the formation of such a symbiosis of mycorrhizal fungi with legumes is mediated by *ENOD* genes, it is inferred to be likely that at least some of the genetic machinery required to promote endosymbiosis with rhizobia exists and functions in rice [77,78]. Our studies described above indeed provide strong evidence that rice possesses at least a part of the genetic program necessary for entering into symbiotic associations with rhizobia. A comparison of genetic programs for root endosymbioses in legumes and rice should allow a clear delineation of sets of genes involved in endomycorrhization, nodulation, and for both types of symbioses [102]. This information is required for exploiting and improving the existing symbiotic genetic program of rice in such a way as to make endosymbiotic associations with rhizobia a realistic goal for rice.

4.4. Need for bacterial compartmentalization for a successful root endosymbiosis

In genuine symbiotic plants such as legumes, actinorrhizal plants and *Gunnera*, microsymbiont infection and colonization in compatible host plants do not seem to elicit any significant defense response in the host plant. This may be largely because of the compartmentalization or sequestering of the infecting microsymbiont from host cells by a plant-derived membrane interface [73]. In rice, however, although rhizobia and other endophytes have the ability to invade and colonize the roots [51,55,78,79,83], they appear to evoke a localized plant-defense response culminating in the dissolution of the lamellar matrix during intercellular colonization, or the host cell death upon intracellular infection by the bacterial cells [78]. Unlike in legumes and other symbiotic plant systems, the bacterial cells residing in rice root tissues are never found to be encapsulated in a host-derived membrane, and the lack of membrane interface between the plant and bacterial cells may be responsible for the observed defense responses in the host tissues [77]. Such a defense response could probably be curtailed if rice were programmed to synthesize a membrane interface around the bacterial cells, upon their invasion, akin to the membrane sequestering the microsymbionts in legumes, etc.

In legumes, the host-derived peribacteroid membrane in nodules and perifungal membrane in mycorrhizae seem to share similarities in their chemical nature (glycoconjugates) [105] indicating commonalities, at least to some extent, in host cellular mechanisms that are involved in the *de novo* synthesis of the interface material in both these symbioses. A reason for optimism is that rice is able to synthesize an interface layer around the infecting fungus in mycorrhizal symbiosis [80] suggesting that rice has the genetic machinery for promoting the formation of the interface material around the infecting mycosymbiont. So the challenge is to identify why such a response (*i.e.*, the synthesis of peribacterial interface) is not triggered in rice upon rhizobial/endophyte invasion in order to contemplate necessary genetic amendments to enable the plant to enter into N_2-fixing root endosymbiosis.

4.5. *nif* gene transfer to rice

An alternative approach to achieve N_2 fixation in rice involves the transfer of N_2 fixation (nif) genes into the genome [112]. The genetic transformation of rice with nif genes should not only ensure its expression, but it also must provide for the protection of nitrogenase from inactivation by oxygen and for the supply of energy for its functioning. Potential locations for introducing foreign genes into eukaryotic cells are the nucleus, mitochondria and chloroplasts. Of all these three locations, the chloroplast appears to provide the most suitable environment for *nif*-gene expression in a plant cell [112,113], as DNA-sequence studies have demonstrated that signals for transcription and translation in chloroplasts most closely resemble those of prokaryotes [114]. On the other hand, for effective transcription of nif genes in chloroplasts, it may be necessary to substitute each nif gene promoter with an appropriate chloroplast promoter, or some other suitable analogous promoter recognized by chloroplast RNA polymerase. The localization of N_2 fixation within chloroplasts may allow some of the energetic costs of N assimilation to be met through the use of photosynthetically produced reductant. However, in order to avoid deactivation of nitrogenase by oxygen in chloroplasts, it may be necessary to establish temporal separation of N_2 fixation and photosynthesis by limiting nif gene expression to dark periods, or spatial separation of these processes could be achieved by expressing nitrogenase in non-photosynthetic root amyloplasts. Alternatively, the recent discovery of the novel superoxide-tolerant nitrogenase isolated from Streptomyces thermoautotrophicus [115] may allow the expression of active nitrogenase under oxic conditions within photosynthetic tissues [112].

Recently Dixon *et al.* [112], in their studies aimed at assessing the feasibility of developing autonomous N_2-fixing eukaryotes demonstrated the expression of the *nifH* gene within plastids of the model photosynthetic eukaryote *Chlamydomonas reinhardtii*. Developments in plant transformation technology may help to accelerate the introduction of a cluster of *nif* genes into plants, as the simultaneous transformation of multiple genes in cereals is now possible [116]. With current technology, it should also be possible to introduce *nif* genes directly into rice chloroplasts. Indeed, a method for stable chloroplast transformation in higher plants is already in place [117,118]. Recent advances in the field of plant molecular biology allow us to be optimistic about the prospects of *nif* gene transcription in plants, although the synthesis of an active nitrogenase enzyme is likely to be a far more complex task [119].

5. CONCLUDING REMARKS

Rice is the world's most important food crop. The rising demand for increased rice production by resource-poor farmers and concern for environmental degradation caused by heavy N fertilization in intensive agriculture has increased our interest in BNF. Some of the existing BNF systems, although they possess a high potential to replenish soil fertility, are not yet widely adopted due to various technological and socio-economical limiting factors.

A single technology - improved crop genotypes - has contributed most to productivity increases in modern agriculture. The adoption of a new genotype by farmers is usually very high because no additional cost is involved, and existing cropping systems and soil and water management practices are not affected. Hence, a rice genotype having autonomous N_2-fixing capability will be inherently advantageous and more suitable for wider adoption.

To achieve N_2 fixation in rice requires upstream research. Recent advances in our understanding of DNA sequences and gene function, and the ability to introduce new genes into rice by transformation have created an excellent opportunity to investigate the possibilities of N_2 fixation in rice. In the past six years, worldwide efforts have led to remarkable progress in the areas of rice-diazotroph associations and determination of the genetic predisposition of rice for symbiosis with rhizobia. In addition, intensive efforts are also under way to elucidate the genetic, biochemical, and physiological requirements for the assembly and function of the nitrogenase enzyme complex in plant cells. The strategies enabling rice to fix its own N_2 are extremely complex and long-term, but are likely to have an enormous impact if it is successful. Despite severe challenges, the development of this technology in one form or other may eventually be realized.

6. REFERENCES

1. M. V. Rosegrant, M. A. Sombilla and N. Perez, Global Food Projections to 2020: Implications for Investment, Food, Agriculture and the Environment Discussion Paper No.5, IFPRI, Washingon D.C., 1995.
2. D. J. Greenland, Sustainability of rice farming. CAB International, Wallingford, UK, 1997.
3. IRRI, Rice Research in a Time of Change: International Rice Research Institute's medium-term plan for 1994-1998, International Rice Research Institute, Los Banos, Philippines, 1993.
4. P .A. Roger and I. Watanabe, Fert. Res. 9 (1986) 39.
5. P. A. Roger and J. K. Ladha, Plant Soil 141 (1992) 41.
6. P. M. Chalk, Plant Soil 132 (1991) 29.

7. R. M. Boddey, O. C. de Oliveira, S. Urquiaga, V. M. Reis, F. L. Olivares, V. L. D. Baldani and J. Döbereiner, Plant Soil 174 (1995) 195.
8. A. App, I. Watanabe, T. S. Ventura, M. Bravo and C. D. Jurey, Soil Sci. 141 (1986) 448.
9. Z. Zhu, in: Progress in Irrigated Rice Research, International Rice Research Institute, Manila, Philippines, 1989, pp. 151-164.
10. E. K. James, P. Gyaneshwar, W. L. Barraquio, N. Mathan and J. K. Ladha, in: J. K. Ladha and P. M. Reddy (eds.), The Quest for Nitrogen Fxation in Rice, International Rice Research Institute, Makati City, Philippines, 2000, pp 119-140.
11. J. K. Ladha, Bionature 6 (1986) 47.
12. R. K. Shrestha and J. K. Ladha, Soil Sci. Soc. Am. J. 60 (1996) 1815.
13. P. Malarvizhi and J. K. Ladha, Soil Sci. Soc. Am. J. 63 (1999) 93.
14]. C. A. Neyra and R. H. Hagerman, Plant Physiol. 58 (1978) 726.
15. V. Vancura, Plant Soil 21 (1964) 231.
16 M. Lin and C .B. You, Sci. Agric. Sinica 22 (1989) 6 (english abstract).
17. I. Watanabe, W. L. Barraquio, M. de Guzman and D. A. Cabrera, Appl. Environ. Microbiol. 37 (1979) 813.
18 P. Wu, G. Zhang, J. K. Ladha, S. McCouch and N. Huang, Theor. Appl. Genet. 91 (1995) 117.
19. M. B. Peoples and D. F. Herridge, Adv. Agron. 44 (1990) 155.
20. T. Yoneyama, J. K. Ladha and I. Watanabe, J. Plant Physiol. 127 (1987) 251.
21. D. L. Eskew, in: Azolla Utilization, The International Rice Research Institute, Los Banos, Philippines, 1987, pp. 233-239.
22. J. K. Ladha, D. Dawe, T. S. Ventura, U. Singh, W. Ventura and I. Watanabe, Soil Sci. Soc. Am. J. 64 (2000) 1993.
23. R. E. Toia Jr., K. M. Buzby and G. A. Peters, New Phytol. 106 (1987) 271.
24. M. Becker, J. K. Ladha and M. Ali, Plant Soil 174 (1995) 181.
25. J. K. Ladha and D. P. Garrit, (eds.), Green-manure Production Systems for Asian Ricelands, International Rice Research Institute, Los Banos, Philippines, 1994.
26. J. K. Ladha, R. P. Pareek and M. Becker, Adv. Soil Sci. 2 (1992) 147.
27. R. S. Gu and Q. X. Wen, in: Proceedings of the Symposium on Paddy Soil (19-24 October 1980), Science Press, Beijing and Springer-Verlag, Berlin-Heidelberg-New York, 1981, pp 207-219.
28. V. R. Carangal, E. T. Rebancos Jr., E. C. Armada and P. L. Tengco, in: J. K. Ladha and D. P. Garrity (eds.), Green-manure Production Systems for Asian Ricelands, International Rice Research Institute, Los Banos, Philippines, 1994, pp 51-66.
29. R. K. Pareek, J. K. Ladha and I. Watanabe, Biol. Fertil. Soils 10 (1990) 77.
30. J. K. Ladha and D. K. Kundu, in: O. P. Rupela, C. Johansen and D. F. Herridge (eds.), Extending Nitrogen Fixation Research to Farmers' Fields, International Crops Research Institute for the Semi-Arid Tropics, Patancheru, Hyderabad, India, 1997, pp 76-102.

31. J. K. Ladha, D. K. Kundu, M. G. A. Coppenolle, M. B. Peoples, V. R. Carangal and P. J. Dart, Soil Sci. Soc. Am. J. 6 (1996) 183.
32. J. K. Ladha and P. M. Reddy, Geojournal 35 (1995) 363.
33. J. K. Ladha and P. M. Reddy, in: J.K. Ladha and P.M. Reddy (eds.), The Quest for Nitrogen Fxation in Rice, International Rice Research Institute, Makati City, Philippines, 2000, pp 33-46.
34. J. K. Ladha, G. J. D. Kirk, J. Bennett, S. Peng, C.K. Reddy, P. M. Reddy and U. Singh, Field Crops Res. 56 (1998) 41.
35. J. K. Ladha and P. M. Reddy (eds.), The Quest for Nitrogen Fxation in Rice, International Rice Research Institute, Makati City, Philippines, 2000.
36. S. Urquiaga, K. H. S. Cruz and R. M. Boddey, Soil Sci. Soc. Am. J. 56 (1992) 105.
37. B. Reinhold-Hurek and T. Hurek, Crit. Rev. Plant Sci. 17 (1998) 29.
38. G. R. Sandhu, Z. Aslam, M. Salim, A. Sattar, R. H. Qureshi, N. Ahmed and R. G. W. Jones, Plant Cell Environ. 4 (1981) 177.
39]. K. A. Malik, B. Rakhshanda, S. Mehnaz, G. Rasul, M. S. Mirza and S. Ali, Plant Soil 194 (1997) 37.
40. E. K. James and F. L. Olivares, Crit. Rev. Plant Sci. 17 (1998) 77.
41. L. E. Fuentes-Ramirez, J. Caballero-Mellado, J. Sepulveda and E. Martinez-Romero, FEMS Microbiol. Ecol. 29 (1999) 117.
42. E. K. James, Field Crops Res. 65 (2000) 197.
43. J. Döbereiner, V. L. D. Baldani and V.M. Reis, in: I. Fendrik, M. del Gallo, J. Vanderleyden and M. de Zamaroczy (eds.), *Azospirillum* VI and Related Microorganisms, Springer-Verlag, Berlin-Heidelberg, Germany, 1995, pp 3-14.
44. J. I. Baldani, L. Caruso, V. L. D. Baldani, S. R. Goi and J. Döbereiner, Soil Biol. Biochem. 29 (1997) 911.
45. B. Reinhold-Hurek and T. Hurek, Trends Microbiol. 6 (1998) 139.
46. J. I. Sprent and E.K. James, in: I. Fendrik, M. del Gallo, J. Vanderleyden and M. de Zamaroczy (eds.), *Azospirillum* VI and Related Microorganisms, Springer Verlag, Berlin-Heidelberg, Germany, 1995, pp 15-30.
47. J. Dobereiner, V. L. D. Baldani, F. Olivares and V. M. Reis, in: Sixth International Congress on BNF with Non-legumes, Egypt, 1993.
48. A. App, T. Santiago, C. Daez, C. Menguito, W. Ventura, A. Tirol, J. Po, I. Watanabe, S. K. De Datta and P. A. Roger, Field Crops Res. 9 (1984) 17.
49. W. L. Barraquio, L. Revilla and J. K. Ladha, Plant Soil 194 (1997) 15.
50. J. R. Stoltzfus, R So, P. P. Malarvizhi, J. K. Ladha and F. J. de Bruijn, Plant Soil 194 (1997) 25.
51. Y. G. Yanni, R. Y. Rizk, V. Corich, A. Squartini, K. Ninke, S. Philip-Hollingsworth, G. Orgambide, F. J. de Bruijn, J. R. Stoltzfus, D. Buckley, T. M. Schmidt, P. F. Mateos, J. K. Ladha and F. B. Dazzo, Plant Soil 194 (1997) 99.
52. J. R. Stoltzfus and F. J. de Bruijn, in: J. K. Ladha and P.M. Reddy (eds.), The Quest for Nitrogen Fxation in Rice, International Rice Research Institute, Makati City, Philippines, 2000, pp 63-91.

53. M. Engelhard, T. Hurek, and B. Reinhold-Hurek, Environ. Microbiol. 2 (2000) 131.
54. P. Gyaneshwar, P. M. Reddy and J. K. Ladha, J. Microbial. Biotechnol. 10 (2000) 694.
55. P. Gyaneshwar, E .K. James, N. Mathan, P. M. Reddy, B. Reinhold-Hurek and J. K. Ladha, J. Bacteriol. 183 (2001) 2634.
56. S. Mehnaz, M.S. Mirza, J. Haurat, R. Bally, P. Normand, A. Bano and K. A. Malik, Can. J. Microbiol. 47 (2001) 110.
57. J. Prayitno, J. Stefaniak, J. McIver, J. J. Weinman, F. B. Dazzo, J. K. Ladha, W. Barraquio, Y. G. Yanni and B. G. Rolfe, Aust. J. Plant Physiol. 26 (1999) 521.
58. P. Gyaneshwar, E. K. James, P.M. Reddy and J. K. Ladha, New Phytol. (2001) (submitted).
59. T. Hurek, B. Reinhold-Hurek, M. Van Montagu and E. Kellenberger, J. Bacteriol. 176 (1994) 1913.
60. V. Tran Van, O. Berge, S. Ngo Ke, J. Balandreau and T. Heulin, Plant Soil 218 (2000) 273.
61. C. Van Nieuwenhove, R. Merckx, L. Van Holm and K. Vlassak, Biol. Fertil. Soils 33 (2001) 25.
62. Y. Okon and C. A. Labandera-Gonzalez, Soil Biol. Biochem. 26 (1994) 1591.
63. T. Egener, T. Hurek and B. Reinhold-Hurek, Mol. Plant-Microb. Interact. 12 (1999) 813.
64. V. L. D. Baldani, J. I. Baldani and J. Döbereiner, Biol. Fertil. Soils 30 (2000) 485.
65. S. Rossbach, D. Kulpa, U. Rossbach and F. J. de Bruijn, Mol. Gen. Genet. 245 (1994) 11.
66. M. V. Gusev and T. G. Korzhenevskaya, in: A.N. Rai (ed.), Handbook of Symbiotic Cyanobacteria, CRC Press, Boca Raton, Florida, 1990, pp 173-230.
67. T. G. Korzhenevskaya, A. Yu. Skripnikov, R. G. Butenko and M. V. Gusev, Dokl. Akad. Nauk SSSR 284 (1985) 765.
68. M. V. Gusev, T. G. Korzhenevskaya, L. V. Pyvovarova, O. I. Baulina and R. G. Butenko, Planta 167 (1986) 1.
69. P. M. Reddy and J. K. Ladha, in: I. A. Tichonovich, N. A. Provorov, V. I. Romanov and W. E. Newton (eds.), Nitrogen Fixation: Fundamentals and Applications, Kluwer Academic Publishers, Dordrecht, The Netherlands, 1995, pp 629-633.
70. G S. Khush and J. Bennett (eds.), Nodulation and Nitrogen FIxation in Rice: Potential and Prospects. International Rice Research Institute, Manila, Philippines, 1992.
71. F.J. de Bruijn, Y. Ying and F.B. Dazzo, Plant Soil 174 (1995) 225.
72. M. J. Trinick, Nature 244 (1973) 459.
73. P. Mylona, K. Pawlowski and T. Bisseling, Plant Cell 7 (1995) 869.
74. J. Dénarié, F. Debelle and J-C. Prome, Annu. Rev. Biochem. 65 (1996) 503.
75. S. R. Long, Plant Cell 8 (1996) 1885.

76. G. Stacey, R. B. Day, P. M. Reddy, J. Cohn, S. Koh, M. Okada, Y. Ito, N. Shibuya and J. K. Ladha, in: J. K. Ladha and P. M. Reddy (eds.), The Quest for Nitrogen Fxation in Rice The Quest for Nitrogen Fixation in Rice The Quest for Nitrogen Fxation in Rice, International Rice Research Institute, Makati City, Philippines, 2000, pp 273-289.
77. P. M. Reddy, J. K. Ladha, H. Kouchi, G. Stacey, R. J. Hernandez-Oane, M. C. Ramos, R. B. So, O. R. Angeles, V. S. Sreevidya, R. B. Day, J. Cohn and S. Koh, in: J.K. Ladha and P.M. Reddy (eds.), The Quest for Nitrogen Fixation in Rice, International Rice Research Institute, Makati City, Philippines, 2000, pp 241-261.
78. P. M. Reddy, J. K. Ladha, R. B. So, R.J. Hernandez, M. C. Ramos, O. R. Angeles, F. B. Dazzo and F. J. de Bruijn, Plant Soil 194 (1997) 81.
79. G. Webster, C. Gough, J. Vasse, C. A. Bathchelor, K. J. O'Callaghan, S. L. Kothari, M. R. Davey, J. Dénarié and E. C. Cocking, Plant Soil 194 (1997) 115.
80. S. Gangopadhyay and K. M. Das, Indian Phytopath. 37 (1984) 34.
81. P. M. Reddy, J. K. Ladha, M. C. Ramos, F. Maillet, R. J. Hernandez, L. B. Torrizo, N. P. Oliva, S. K. Datta and K. Datta, Plant J. 14 (1998) 693.
82. D. Werthmuller, R. Terada, P. Bauer, M. Schultze, E. Kondorosi, A. Kondorosi, I. Potrykus and M. Sautter, in: C. Wijffelman, A. Wijfjes and B. Lugtenberg (eds.), Third European Nitrogen Fixation Conference: Abstracts (20-24 September 1998), Lunteren, The Netherlands, 1998, p 123.
83. C. Chaintreuil, E. Giraud, Y. Prin, J. Lorquin, A. Ba, M. Gillis, P. de Lajudie and B. Dreyfus, Appl. Environ. Microbiol. 66 (2000) 5437.
84. G. Gopalaswami, S. Kannaiyan, K. J. O'Callaghan, M. R. Davey and E. C. Cocking, Proc. Roy. Soc. Lond. B 257 (2000) 103.
85. F. B. Dazzo, Y. G. Yanni, R. Rizk, F. J. de Bruijn, J. Rademaker, A. Squartini, V. Corich, P. Mateos, E. Martínez-Molina, E. Velázquez, J. C. Biswas, R. J. Hernandez, J. K. Ladha, J. Hill, J. Weinman, B. G. Rolfe, M. Vega-Hernández, J. J. Bradford, R. I. Hollingsworth, P. Ostrom, E. Marshall, T. Jain, G. Orgambide, S. Philip-Hollingsworth, E. Triplett, K. A. Malik, J. Maya-Flores, A. Hartmann, M. Umali-Garcia and M. L. Izaguirre-Mayoral, in: J. K. Ladha and P.M. Reddy (eds.), The Quest for Nitrogen Fixation in Rice, International Rice Research Institute, Makati City, Philippines, 2000, pp 167-189.
86. S. Saxena, J.K. Ladha, P. Gyaneshwar, B. Reinhold-Hurek, R. J. Hernandez and J. C. Biswas, Ind J. Microbiol. 40 (2000) 15.
87. P. M. Reddy, H. Kouchi and J. K. Ladha, Biochim. Biophys. Acta 1443 (1998) 386.
88. P. M. Reddy, R. K. Aggarwal, M. C. Ramos, J. K. Ladha, D. S. Brar and H. Kouchi, Biochem. Biophys. Res. Commun. 258 (1999) 148.
89. H. Kouchi, K. Takane, R. B. So, J. K. Ladha and P. M. Reddy, Plant J. 18 (1999) 121.

90. P. M. Reddy, R. J. Hernandez-Oane, H. Kouchi, G. Stacey and J. K. Ladha, in: F. O. Pedrosa, M. Hungria, M. G. Yates and W. E. Newton (eds.), Nitrogen Fixation: From Molecules to Crop Productivity, Kluwer Academic Publishers, Dordrecht, The Netherlands, 2000, pp 331-332.
91. A. van Kammen, Plant Mol. Biol. Rep. 2 (1984) 43.
92. H. Kouchi, K. Takane, R. B. So, J. K. Ladha and P. M. Reddy, in: J.K. Ladha and P.M. Reddy (eds.), The Quest for Nitrogen Fixation in Rice, International Rice Research Institute, Makati City, Philippines, 2000, pp 263-272.
93. T. Sasaki, Proc. Natl. Acad. Sci. USA. 95 (1998) 2027.
94. J. Landsmann, E. S. Dennis, T. J. V. Higgins, C. A. Appleby, A. A. Kortt and W. J. Peacock, Nature 324 (1986) 166.
95. R. Arredondo-Peter, M. S. Hargrove, G. Sarath, J. F. Moran, J. Lohrman, J. S. Olson and R. V. Klucas, Plant Physiol. 115 (1997) 1259.
96. H. Kouchi and S. Hata, Mol. Gen. Genet. 238 (1993) 106.
97. W-C. Yang, P. Katinakis, P. Hendriks, A. Smolders, F. de Vries, J. Spee, A. van Kammen, T. Bisseling and H. Franssen, Plant J. 3 (1993) 573.
98. C. Charon, C. Johansson, E. Kondorosi, A. Kondorosi and M. Crespi, Proc. Natl. Acad. Sci. USA 94 (1997) 8901.
99. P. M. Reddy, J. K. Ladha, R. J. Hernandez-Oane and V. S. Sreevidya, in: G. S. Khush and D. S. Brar (eds.), Proceedings of the Fourth International Rice Genetics Symposium. International Rice Research Institute, Los Banos, Philippines, 2001 (in press).
100. M. B. Brown, T. H. Quimio and A. M. De Castro, Philippine Agriculturist 71 (1988) 317.
101. J. Secilia and D. J. Bagyaraj, Biol. Fertil. Soils 13 (1992) 108.
102. C. Gough and J. Dénarié, in: J. K. Ladha and P. M. Reddy (eds.), The Quest for Nitrogen Fiation in Rice, International Rice Research Institute, Makati City, Philippines, 2000, pp 311-326.
103. M. Sagan, D. Morandi, E. Tarenghi and G. Duc, Plant Sci. 111 (1995) 63.
104. M. Sagan, H. De Larembergue and D. Morandi, in: C. Elmerich, A. Kondorosi and W. E. Newton (eds.), Biological Nitrogen Fixation for the 21st Century: Proceedings of the 11th International Congress on Nitrogen Fixation (20-25 July 1997), Institut Pasteur, Paris, France, Kluwer Academic Publishers, Dordrecht, Netherlands, 1998, pp 317-318.
105. V. Gianinazzi-Pearson, Plant Cell 8 (1996) 1899.
106. S. J. Shirtliffe and J. K. Vessey, Plant Sci. 118 (1996) 209.
107. K. Szczyglowski, R. S. Shaw, J. Wopereis, S. Copeland, D. Hamburger, B. Kasiborski, F. B. Dazzo and F. J. de Bruijn, Mol. Plant-Microbe Interact. 11 (1998) 684.
108. E. Wegel, L. Schauser, N. Sandal, J. Stougaard and M. Parniske, Mol. Plant-Microbe Interact. 11 (1998) 933.
109. C. Albrecht, R. Geurts and T. Bisseling, EMBO J. 18 (1999) 281.

110. P. van Rhijn, Y. Fang, S. Galili, O. Shaul, N. Atzmon, S. Wininger, Y. Eshed, M. Lum, Y. Li, V. To, N. Fujishige, Y. Kapulnik and A. M. Hirsch, Proc. Natl. Acad. Sci. U.S.A. 94 (1997) 5467.
111. C. Albrecht, R. Geurts, F. Lapeyrie and T. Bisseling, Plant J. 15 (1998) 605.
112. R. Dixon, Q. Cheng and A. Day, in: J. K. Ladha and P. M. Reddy (eds.), The Quest for Nitrogen Fixation in Rice, International Rice Research Institute, Makati City, Philippines, 2000, pp 327-336.
113. M. Merrick and R. Dixon, Trends Biotech. 2 (1984) 162.
114].P. R. Whitfeld and W. Bottomley, Ann. Rev. Plant Physiol. 34 (1983) 297.
115. M. Ribbe, D. Gadkari and O. Meyer, J. Biol. Chem. 272 (1997) 26627.
116. A. Kohli, M. Leech, P. Vain, D. A. Laurie and P. Christou, Proc. Natl. Acad. Sci. USA 95 (1998) 7203.
117. J. M. Staub and P. Maliga, Plant Cell 4 (1992) 39.
118. S. Ruf, M. Hermann, I. J. Berger, H. Carrer and R. Bock, Nature Biotechnol. 19 (2001) 870.
119. R. Dixon, J. Vanderleyden, C. Elmerich, B. G. Rolfe and J. Sprent, in: R. Palacios, J. Mora and W. E. Newton (eds.), New Horizons in Nitrogen Fixation, Kluwer Academic Publishers, Dordrecht, The Netherlands, 1993, pp 765-768.

SUBJECT INDEX